D1547910

THE PETROCHEMICAL INDUSTRY

THE PETROCHEMICAL INDUSTRY
Market and Economics

Albert V. Hahn

Adelatec Technical and Management Services Company S.A.

In Collaboration with

ROGER WILLIAMS, Chairman
and
HERMAN ZABEL, Executive Vice-President

Roger Williams Technical & Economic Services, Inc.

McGRAW-HILL BOOK COMPANY   New York   St. Louis
San Francisco   London   Sydney   Toronto   Mexico   Panama

THE PETROCHEMICAL INDUSTRY

To Anne de Egry

PREFACE

During the last two decades, a radical change has taken place in the relative importance of the various sources of raw materials for the organic chemical industry. Coal and coal tar, and other materials of animal or vegetable origin, cultivated by man or occurring in nature, have been superseded as the primary sources of organics by petroleum and natural gas.

The main factor in bringing about this process of substitution has been the sustained worldwide rise in living standards and full employment. Coal mining, for example, or the extraction of natural products such as rubber, are activities in which the main cost component is cheap labor. The upward pressure on real wages all over the world have thus opened the door to highly capital-intensive techniques of producing synthetic replacements for the traditional raw materials from petroleum refinery fractions or natural gas. The trend towards petrochemistry as the basis for the heavy organic chemical industry seems irreversible, as long as reserves hold out and as long as living conditions in the developing countries are permitted to continue improving.

Petrochemical products were initially priced about the same as materials from other sources. But ever-larger production units, lower unit investment and production costs led to low price levels that would have seemed unbelievable a few years ago, but which nevertheless appear permanent. There are strong indications, however, that this process is reaching a plateau and that the time has come for taking stock. Now that the petrochemical industry has reached maturity, the major purpose of this book is to analyze its structure, its economics of production, and the effects of plant size and vertical integration.

The structure of the petrochemical industry can be studied by several lines of approach. In selecting a form of organization along "family tree" lines, my intention is to emphasize vertical integration as the dominant motivation in the chemical industry. There are certainly examples of companies that specialize in making products for a given industry, or concentrate on compounds with a common chemical function or made by a given unit process. Another consideration influencing the structure of the industry is end-product technology; companies may frequently decide to manufacture an entirely new product because they have specialized in a certain field of application, and must therefore protect themselves against competition by producing chemically unrelated raw materials but related as to their end uses. But the most important factor in determining who makes a product, and by which route, is the ability of companies to integrate vertically. Backwards integration allows them to face competition from larger and larger units, and forward integration lets them upgrade the products they already make. Moreover, there seems to be a steady increase in the number of products that can't be made competitively without captive sources of raw materials, or even of precursors made several stages upstream. Thus, even in the developed countries, it is becoming more difficult for producers to enter fields in which they do not have a raw-material or end-product position, not to mention the greater problems faced by nations just reaching the stage in their development where building up a chemical industry of their own seems worth considering.

Since no country has published so much on the economics of its chemical sector, I have based all my data concerning market structures on statistics released by the United States. This study should nevertheless be of interest to executives, planners, market researchers, and decision-making groups in other countries. Since only a small percentage of the U.S. national income is involved in international trade, the output of its chemical industry constitutes essentially the domestic requirements of the U.S. economy, and, as a consequence, useful parallels can frequently be drawn between actual demand levels or market distributions in the U.S. and potential conditions in the economies of other countries.

I wish to express my thanks to Paulo Vieira Belotti and Sebastiao Simoes Fo., for their constant moral support; to Miss G. Certier, for her help in the preparation of the manuscript; and to Dr. Thomas A. Unger, to whose many helpful comments this book owes a great deal.

Albert V. Hahn

CONTENTS

THE PETROCHEMICAL INDUSTRY

1

INTRODUCTION

1   GENERAL CONSIDERATIONS

The term *petrochemicals* has come to mean all chemical products derived from petroleum fractions and by-products or from natural gas constituents.

From its modest beginnings in the 1920's with the manufacture of isopropanol from refinery off-gas propylene, petrochemistry has by now not only made possible the almost total elimination of coal and coal-tar as sources of chemical raw materials, but has also gone a long way towards replacing such methods of obtaining organic chemicals as fermentation, extraction of compounds from materials occurring in nature, and chemical transformation of vegetable fats and oils.

Until the late 1930's petrochemistry was limited in its scope to the synthesis of oxygenated solvents, most of them previously obtained by fermentation. World War II ushered in the age of synthetic polymeric substitutes for natural and inorganic material — metals, leather, wood, glass, rubber, waxes, gums, fibers, glues, drying oils, etc. The production of these materials on a large scale sufficient to satisfy their enormous potential markets required raw materials far in excess of those available from refinery off-gas. Therefore, additional olefins began to be produced by cracking light saturated hydrocarbons present in the off-gas themselves, and later by resorting to similar materials recovered from natural gas.

A parallel phenomenon was the extremely rapid growth in the need for ammonia and nitrogen fertilizers all over the world. Whereas synthesis gas was originally obtained primarily from coal and by up-grading coke oven gases, the surge in ammonia requirements made it necessary to tap other raw material sources. In the regions of the world where natural gas was found, this alternate source of synthesis gas became the steam-reforming of methane.

So far, petrochemistry had become exclusively a source of aliphatic chemicals. The next step in an extraordinarily far-reaching series of consequences was the development of processes for extracting aromatic hydrocarbons from catalytic reformate. This was to be followed by methods for correcting the imbalance between toluene and benzene in reformed naphtha by dealkylating the former and producing additional benzene. With these developments, the elimination of coal as a necessary base for the synthetic organic chemical industry was practically completed.

The most economical techniques for producing olefins and synthesis gas are, respectively, cracking in a tubular furnace and steam reforming. For purely technical reasons these methods were restricted at first to materials no heavier than butane. A natural advantage was conferred on those regions of the world where natural gas was found, or those where liquid fuels had acquired such a large share of the total demand for energy that enough by-products were available for the chemical industry. The growth of petrochemistry in certain countries, especially Europe and Japan, was thus hindered by conditions that had to be satisfied both as to regional resource structure and the distribution of demand among the various forms of energy. In the early 1960's, one of the most important stages in the evolution of petrochemistry was reached. It became possible to apply the techniques of steam reforming and tubular furnace cracking to liquid feedstocks, thereby freeing the industry from the requirement of locating in the vicinity of petroleum refineries or in regions rich in natural gas. It could even be termed *chemical refinery*, a chemical complex feeding on liquid feedstocks that are totally converted to petrochemical raw materials. Figure 1-1 (p. 4) shows how the various units of such a chemical refinery are interrelated.

On the one hand, particularly for ammonia, the possibility of producing synthesis gas from naphtha at reasonable cost had immediate repercussions, especially in Europe. However, on the other hand, the possibility of making ammonia for $20/ton or less in regions where natural gas can be considered inexpensive may eventually produce a trend opposite to the prevailing one of regional autonomy. By manufacturing and shipping anhydrous ammonia on a very large scale, certain regions of the world, e.g., Mexico, Chile, Kuwait, may in effect become substantial exporters of natural gas. However, the strategic importance of ammonia is such that most countries have made a point of producing it domestically rather than relying on foreign, albeit cheaper, sources of supply. Therefore, the day when anhydrous ammonia becomes a widely traded commodity in the manner of petroleum itself may still be far off.

Despite the uncertainty in predictions of this type, it may be concluded that the identification of the petrochemical industry with the synthetic organic chemical industry as a whole is irreversible, and, furthermore, that new and so-far apparently improbable kinds of materials, not excluding foodstuffs, will eventually be derived from petroleum. Although coal is less than half the cost of most liquid fuels, the handling and processing of liquids is much cheaper. In addition, the extraction of coal is so much more labor-intensive in a world where the trend is towards capital-intensive techniques in all possible types of human endeavor, that the return to coal as the basis of the chemical industry seems unlikely at least as long as known petroleum reserves continue to increase.

A further trend within the chemical industry has been the extraordinary simplification of numerous organic syntheses made possible during the last ten years. This is due particularly to developments in catalysis and automatic control. Oxygenated, unsaturated and nitrogenated compounds, formerly obtained via routes involving several steps, are gradually being produced by direct oxidation, nitrilation, amination or dehydrogenation. Petrochemicals generally tend to be made from hydrocarbon raw materials having the same number of carbon atoms as the finished product. This, combined with the construction of ever larger production units, has been the cause of the drop in the price of organic chemicals to an extent that would have seemed unthinkable a few years ago.

2   PETROCHEMICAL FEEDSTOCKS

2.1 Petroleum Refining

A block-flow diagram of a typical petroleum refinery is shown in Fig. 1-1.

As crude oil enters the refinery, first it undergoes atmospheric distillation, which separates it into fuel gas, LPG, straight-run light and heavy naphtha, kerosene, gas oil and fuel oil. However, straight-run naphtha has a very low octane number, and in many

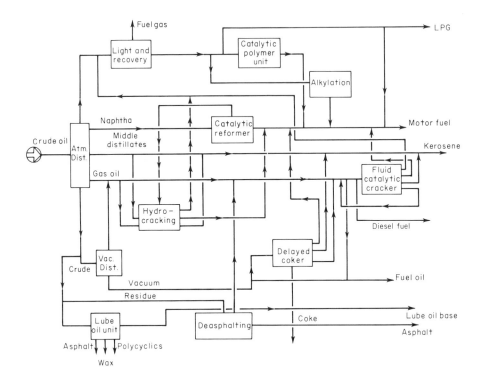

Fig. 1-1.

regions of the world, the market for gasoline is larger than can be recovered from crude oil simply by distillation, other operations perform the functions of improving the quality of the gasoline pool and of increasing the output of gasoline per barrel of feed.

Catalytic reforming of naphtha is a dehydrogenation, isomerization and cyclization process, in which cycloaliphatic (and to some extent even aliphatic) hydrocarbons are converted to aromatics having a much higher octane number than the saturated hydrocarbons from which they are derived. Therefore, this process is the main source of aromatic hydrocarbons from petroleum. In addition, reforming produces large amounts of propane, i-butane and especially n-butane. Propane can be incorporated into the LPG stream or fed to the cracking section of an olefin plant; isobutane is the primary raw material for making alkylation gasoline, originally used as aviation gas but nowadays in increasing demand by the automotive fuel pool; n-butane, the major constituent of LPG, can serve either as feedstock for butadiene manufacture or similarly cracked to olefins. Finally, catalytic reforming produces hydrogen, which several U.S. refineries transfer as feed to nearby ammonia units; however, the trend is towards the use of this hydrogen for treating other refinery streams, or for hydrocracking. In fact, some refineries already have a net hydrogen deficiency which must be compensated by additional hydrogen generation units.

Upgrading the gasoline-range fractions originally present in the crude petroleum does nothing to correct the imbalance in demand between the various fuel cuts. In the U.S. the demand for motor fuel has tended to grow faster than for fuel oil or diesel fuel. Accordingly, most refineries contain catalytic cracking units which convert heavy fractions to additional gasoline, which is high in olefin content, and makes for a high octane number. Part of the feed is converted to olefins below the boiling range of gasoline, and the catalytic cracker is thus a major producer of olefins in the typical refinery. Catalytic cycle oil, which would produce too much petroleum coke to be handled by a catalytic cracker, is often converted to additional gasoline by thermal cracking, which con-stitutes a further source of olefins. Recently, there has been a trend towards hydrocracking, a process in which vacuum gas oils or middle distillates are broken down in the presence of hydrogen in order to convert them to kerosene, diesel fuel or gasoline, de-pending on market requirements. The presence of hydrogen pro-motes the formation of saturated hydrocarbons instead of olefins. Therefore, the growth in hydrocracking capacity is somewhat of a threat to the supply of refinery propylene and butylenes.

Two processes convert C_3 and C_4 olefins to gasoline. Catalytic polymerization produces olefinic gasoline, high in octane number but unstable. Alkylation consists of reacting isobutane and C_3 or C_4 olefins. The alkylate is the best possible motor fuel, having both excellent stability and high octane number. It was originally developed for aviation gasoline, but since being replaced by jet fuel, the share of alkylate in the U.S. gasoline pool has been rising.

Catalytic polymerization is also the source of propylene trimer and tetramer and of "oxo" process feedstocks such as heptene. On the other hand, alkylate gasoline, formerly made almost ex-clusively from C_4 paraffins and olefins, is now also being made from propylene due to the constant demand for better motor fuel. This has already resulted in occasional local shortages of propylene for the chemical industry.

Kerosene is a source of detergent-range n-paraffins, which can be recovered by molecular-sieve adsorption. Gas oil is either treated to give diesel fuel, or fed to the cat-cracking unit. Finally, the atmospheric distillation residue is either consumed as fuel oil or distilled again under vacuum. Vacuum tower overhead is the main source of cat-cracker feedstock; the bottoms can be used either as fuel oil, or fed to a delayed coker, a method of obtaining additional cat-cracking feedstock from residual fuel oil. If fed directly to a cracking unit, vacuum residue would produce more coke than could be burned off. Delayed coking consists of heating the charge rapidly in a furnace, and then letting it soak in a pair of coke drums where carbon settles out. Apart from the coke itself, the products of this operation are gasoline, cat-cracker charge stock and gases containing some olefins. As alternatives, atmospheric bottoms can be charged to the vacuum tower of a lube-oil plant, and vacuum residue deasphalted with propane to give asphalt and gas-oil. About 40% of all U.S. refineries produce asphalt.

Finally, some petrochemical feedstocks originate in the chemical treating processes employed to upgrade various types of fuel other than gasoline. Thus, caustic treatment of cracked gasoline gives raw materials for cresylics recovery plants; the same process when applied to diesel fuel yields sodium naphthenates. Sulfuric acid treatment produces various types of petroleum sulfonates, used as surface-active agents in a variety of industries. Some sweetening processes recover mercaptans.

The most important links between the refinery and the chemical industry are thus related to the production of motor fuel. Olefins, however, are produced principally when it is desired to increase the gasoline yield from crude oil by converting heavier fractions to lighter ones. On the other hand, aromatic hydrocarbons are produced in connection with the upgrading of virgin naphtha. Accordingly, in the U.S. where the demand for light fuels is proportionately largest, refineries are relatively more important sources of olefins than those in Europe or Japan, where heavy fuels enjoy a greater share of the market. Converting them to lighter fractions is not required to the same extent as in the U.S. Outside the U.S. the most important source of raw materials for the organic chemical industry is the cracking of liquid feedstocks in the naphtha range or heavier. Naphtha is in excess outside the U.S. as a feedstock. It is relatively inexpensive and therefore convenient not only for olefins, but also as a raw material for synthesis gas.

2.2 Natural Gas

Natural gas is a mixture of hydrocarbons ranging from methane to C_7 or higher. The lighter hydrocarbons are paraffinic; those corresponding to a naphtha cut also consist mainly of paraffins but include a small percentage of naphthenes and occasionally aromatics. In addition to hydrocarbons, natural gas may contain as much as 45% acid gases, that is, H_2S and CO_2. Exploitation of natural gas reserves increases in difficulty with temperature, pressure, depth and acid gas concentration.

The purpose of processing natural gas is to produce a stream containing only ethane and methane, which in the cold regions of the world is used mainly for space heating in large urban centers. In countries where natural gas is plentiful, methane is also the major raw material for producing synthesis gas. Purified methane containing no heavier hydrocarbons is the raw material for making hydrogen cyanide, carbon disulfide and chlorinated methanes.

Natural gas constituents heavier than methane are also excellent petrochemical feedstocks. Ethane is the most desirable starting point for making ethylene when a minimum amount of byproducts is desired. Propane and butane, while valuable as LPG, can also be cracked to olefins, and butane can be dehydrogenated to butadiene. The fraction corresponding to naphtha (known as natural gasoline) has a low octane number and therefore is employed for vapor-pressure blending in gasoline or else mixed with the crude charge to a refinery. Due to its paraffinic character and low sulfur content it is an excellent feedstock for cracking to olefins and for steam-reforming. At least one company extracts cyclohexane from natural gasoline.

The abundance of natural gas in the U.S. was responsible for a long time for the extraordinarily low cost of petrochemical feed-stocks on the Gulf Coast. However, with the construction of pipe-lines to the consuming centers of the East and Midwest, natural gas acquired an opportunity cost and its price rose to the point where it has already begun to be displaced by liquid feedstocks for many petrochemical purposes, despite the fact that, under conditions of comparable raw material costs, a given product can be made more cheaply from methane than from any other feed. It was the switch from fuel oil and coal to natural gas, as the preferred source of domestic and industrial heat, that brought about the abundance of fuel oil in the U.S. fuel balance.

2.3 Feedstock Valuation

At the bottom of any discussion of petrochemical economics lies the problem of feedstock valuation.

Petroleum and natural gas derivatives can be attributed values that may differ widely depending on the distribution of the local market among the various fractions, transportation costs, availability of competing sources of energy or chemical feedstocks and the effect of taxation and import duties.

With these limitations in mind, the table below serves as a guide to preliminary economic evaluations of proposed petrochemical complexes in which both chemical feedstocks and fuel streams are to be produced.

	U.S. cost, ¢/lb
Fuel-gas (off-gas)	0.6 − 0.9
Ethane	0.8 − 1.2
Natural gas	0.45 − 0.9
Liquefied petroleum gas	1.4 − 1.7
Light naphtha	0.8 − 1.4
Regular gasoline	1.4
Premium gasoline	1.6
Kerosene	1.25
Gas oil	0.8 − 1.1
Fuel oil	0.5 − 0.7
Crude oil	0.7

Transportation costs may also vary as a result of numerous factors: flag; affiliation (conference vs. tramp ships); seasonality; degree of labor organization; contract vs. spot rates; quantities involved; size of vessel; subsidies; war; other government policies; etc. A detailed discussion of all these factors would be beyond the scope of this book. The following table is intended as an indication of the orders of magnitude involved.

3   IMPORTANCE OF VERTICAL INTEGRATION

The subject of economics in the organic chemical industry can be approached in several possible ways. The purpose of selecting a form of organization along family-tree lines is to emphasize the influence of vertical integration on the evolution of the chemical sector and on the decisions taken by the firms active in the organic field.

Other methods also may deserve consideration. Traditionally the subject of organic chemistry is approached according to chemical function. There are cases in industry of firms specializing in the production of compounds having a common chemical function or that are made by a given unit process. For example, some companies produce several nitration derivatives, or fatty amines, or isocyanates. In general, these cases fit into the scope of this

Transportation cost, ¢/lb

Means	Pipeline	Piggyback		Truck		Ocean liner			Railroad		Barge	
Distance, miles	1500	1000	2000	250	500	750	1500	5000	500	1500	250	1000
Products												
Methane	0.45	—	—	—	—	—	1.0					
Bulk solids	—	0.7	1.1	0.5	—	0.25	0.35	0.60	0.9	1.3	0.8	1.8
Liquids, nonrefrigerated	—	—	—	0.7	1.4	0.10	0.15	0.45	—	—	0.4	1.45
Liquids, refrigerated	—	—	—	—	—	0.35	0.55	1.15	—	—	0.9	2.8

book since they are often the result of a particular raw material position.

More important from the standpoint of the actual behavior of chemical companies is horizontal integration.   This can be a powerful motivation due either to a desire to provide hedges against changes in market structures (as is the case of firms that produce various types of polymers or synthetic fibers) or to complement a line of products (e.g., when a company making polyols decides also to produce isocyanates).

Despite the significance of these types of motivation in the chemical industry, however, the main influence in recent years has been the need to integrate vertically.   Firms that until recently were content to produce intermediates or end-products, have been under constant pressure either to integrate backwards by acquiring their own sources of raw materials, or to become the objects of the desire on the part of their suppliers to integrate forward by gaining control of their clients.   The percentage of captive utilization of most major chemical intermediates is growing steadily. This is mainly because, unit profits being as a rule higher at the finished-product end of the chain, large oil and chemical companies have rapidly enlarged the scope of their activities both by acquisition and by downward trend in chemical prices and a gradual increase in the relative power of companies that have consolidated their positions in this manner, compared with those content to maintain their original structure.   In the U.S., there are limitations to forward vertical integration by acquisition imposed by antitrust legislation, but these are beyond the scope of this book.

The topics have been grouped under eight headings, seven of which correspond to basic organic building blocks.   The eighth includes those petrochemicals which are impossible to classify elsewhere.   The seven building blocks are as follows:

> synthesis gas
> methane
> ethylene
> propylene
> C_4 hydrocarbons
> acetylene
> aromatic hydrocarbons

Many chemicals can be made from two or more building blocks. These have been classified according to such criteria as, which is the prevailing process, or, which process is being used in the most recent projects.   Some examples of products being made commercially from more than one basic petrochemical are shown below;

the building block underlined is the one under which the product was included.

Product	Possible headings
Acrylonitrile	Ethylene, <u>propylene</u>, acetylene
Vinyl chloride	<u>Ethylene</u>, acetylene
Adiponitrile	C_4 hydrocarbons, propylene, <u>aromatic hydrocarbons</u>
Chloroprene	C_4 hydrocarbons, <u>acetylene</u>
Acetic acid	Synthesis gas, acetylene, <u>ethylene</u>
Vinyl acetate	Ethylene, <u>acetylene</u>

A single product may derive from more than one basic raw material. These products have been classified according to such criteria as, which of the raw materials producers tend most obviously to make themselves, or, which raw material or intermediate contributes the most to production costs. Some examples are given in the table below.

Product	Possible headings
Styrene	Ethylene, <u>aromatic hydrocarbons</u>
SBR	C_4 hydrocarbons, aromatic hydrocarbons
Ethanolamines	<u>Ethylene</u>, Synthesis gas
"Oxo" chemicals	<u>Synthesis gas</u>, ethylene, propylene
DDT	<u>Ethylene</u>, aromatic hydrocarbons
2,4-D	Ethylene, <u>aromatic hydrocarbons</u>

Admittedly, some of these classifications are open to dispute, especially those included under "Acetylene." Perhaps a separate chapter should not have been devoted to acetylene since it has been replaced in almost all of its applications — to the point where a developing country intent on building up a balanced chemical industry may be able to dispense with acetylene altogether. The possibility of making acetylene for slightly more than the cost of ethylene has been widely discussed.

The total amounts of basic building blocks consumed by the U.S. chemical industry in 1965 are as follows:

Raw material	Consumption, 1965
Synthesis gas.....................	1800 MM* scfd**
Methane........................	400.0 MM lbs
Ethylene........................	8210.0 MM lbs
Propylene.......................	3350.0 MM lbs
C_4 hydrocarbons	4600.0 MM lbs
Acetylene.......................	975.0 MM lbs
Aromatic hydrocarbons	6750.0 MM lbs

*Throughout this book, the abbreviation "MM lbs" denotes millions of pounds.
**Standard cubic feet per day.

Assuming that all these products were to be produced from liquid feedstocks, this would correspond to around 550,000 bbl/day of naphtha, roughly 5% of all the petroleum refined in the U.S. Since crude oil contains between 20 and 40% of naphtha, 12.5 to 25% of the total naphtha content would have to be allocated to petrochemical production if no natural gas were available. Countries with poor oil resources, and which are just beginning to build up their chemical industry, may therefore have to take into account future feedstock requirements in planning their fuels and transportation policies.

4   ECONOMICS OF THE ORGANIC CHEMICAL INDUSTRY

This book discusses primarily the economics of both production and utilization of the products or groups of products. The following types of information have been given wherever possible:

1. total U.S. demand and market distribution, as of 1965;
2. U.S. manufacturers, their capacities, locations, routes, raw material positions and captive outlets, as of 1965;
3. relative merits and economics of various processes.

The discussions of market distribution and the competitive position of each product in its main outlets are intended as a guide to predictions regarding the future of the product in question. While these data regard only the U.S., their value to those concerned with chemical economics in other countries should not be underestimated since they are representative of the kind of "mass economy" most governments are pledged to promote and most peoples of the world have learned to demand of them.

The figures regarding U.S. producers, locations, capacities and so forth, are presented in support of the central theme of this book, namely, the overriding influence of the desire to integrate. Whereas the data regarding demand and market breakdowns should serve as a basis of quantitative forecasts, those regarding producers are most likely to interest such professional groups as security analysts and engineering company executives.

Finally, the economics of producing most of the heavy organic chemicals have been analyzed in order to substantiate qualitative statements concerning them, and to provide the background for predicting how prices can be expected to evolve as plants get larger and producers consolidate their raw materials positions. In this respect the cost-vs.-capacity graphs should be particularly helpful.

Calculations regarding process economics were based on a given plant capacity, and from them graphs were constructed showing how production costs are likely to vary with plant size. The necessary calculations have had to be based on certain assumptions regarding the cost of feedstocks, already discussed intermediates, utilities, labor and capital-related charges.

The assumptions concerning the value of intermediates are intended to reflect the conditions actually prevailing in industry in 1965. Thus, the producer of vinyl chloride is assumed to have a captive source of chlorine, the producer of acetone is considered as starting from captive isopropanol, and so forth. The situation assumed is stated in each case and can easily be adapted to suit other alternatives.

Utility costs were taken at the following values

Steam....................	$1.00/1000 lbs
Electricity	1.0 ¢/kWh
Cooling water	2.0 ¢/1000 gal
Treated water............	50 ¢/1000 gal
Refrigeration (32°F)	$3.00/million Btu
Plant fuel	35 ¢/million Btu

These figures are somewhat higher than those usually found in the literature. Since investment costs for process units are available mostly on a battery-limits basis, utility costs have been taken to reflect the capital-related charges corresponding to their part of total plant investment.

Total labor and overhead charges have been taken at $10/operating man-hour. This includes all plant overhead and central administration charges to the plant exclusive of sales and research costs. One operator per shift is thus the equivalent of 8000 man-hours, or $80,000 per annum.

Where there is no vertical integration sales and research costs can vary from as low as 3.58% of sales to as high as 28.5%. This variation has been discussed elsewhere.*

Capital costs given in this book reflect conditions in the U.S. in 1965. During that year the overall Marshall & Stevens equipment cost index (1926 = 100) was 245. This index is regularly reported in the magazine *Chemical Engineering*. The *Engineering News-Record* overall construction cost index (1913 = 100) was 970. Hence changes can be made to the capital figures used in this book by these indices.

The percentage attributable to capital-related charges was calculated on the basis of a situation considered typical with respect

*R. Williams, Jr., Why Cost Estimates Go Astray, *Chemical Engineering Progress,* April,1964.

to taxation, allowed depreciation, debt-equity ratio, interest rate and expected return on equity.
Let:

a = debt as fraction of total investment
i = interest on debt (declining balance)
θ_1 = expected payout time for equity years
θ_2 = period over which loan is repayable
d = allowable depreciation
t = income tax rate
m = maintenance
c = taxes and insurance
I = investment

The expected net cash-flow (NCF) will be

$$NCF = I\left(\frac{a}{\theta_2} + \frac{1-a}{\theta_1}\right)$$

The gross cash-flow (GCF) will be

$$GCF = I\left[\frac{\left(\dfrac{a}{\theta_2} + \dfrac{1-a}{\theta_1}\right)}{1-t} - d\right]$$

and total capital-related charges (C), including those items that can be expensed:

$$C = I\left(GCF + \frac{i}{2} + m + c\right)$$

Assuming the following values:

a = 0.50
i = 6%/yr
θ_1 = 5 years
θ_2 = 8 years
d = 10%/yr
t = 50%/yr
m = 5%/yr
c = 2.5%/yr

and substituting, one obtains:

GC F = 22.5%/yr

C = (22.5 + 3 + 5 + 2.5)% = 33%/yr

Therefore, capital-related charges have been taken at 33% of capital investment.   Production costs containing these capital charges correspond to the minimum sales or internal transfer price of the product in question.

Assuming comparable plant designs, the factors by which the cost of the various inputs other than raw materials must be multiplied in order to compare production costs in the U.S. with those in other parts of the world, have been estimated as follows:

	Capital charges	Utilities	Labor + overhead
U.S.	1.00	1.00	1.00
European Common Market....................	0.95	1.25	0.75
Underdeveloped countries (without financing)................................	1.25	1.40	0.50
Underdeveloped countries (including typical financing schemes)	1.45	1.40	0.50

Such conversion factors must be used with caution, since the values on which the original estimate is based already reflect a compromise between the cost of capital, utilities and labor. Also, if this relationship were to vary the compromise reached in the new situation might well differ. For example, from the table above, in Europe it would seem profitable up to a certain point to achieve heat economy at the expense of higher capital charges, etc.

A number of simplifying assumptions were made in constructing the cost vs. capacity graphs.

Unit raw material and utility costs were assumed not to vary with plant size, except in cases where two or more process units are necessarily integrated; for example, the cost of ethylbenzene fed to a styrene monomer plant was made to vary with the size of the latter. When several units typically receive their raw material from the same captive source, raw material costs were assumed independent of plant size. For example, in calculating the cost of bisphenol A, the value of phenol was considered unaffected by capacity.

Labor and overhead were considered constant and therefore independent of capacity. Due to the high degree of instrumentation necessary to the proper functioning of petrochemical plants, this is a reasonable assumption for continuous processes above a certain minimum capacity even in less developed economies.

Investment costs were assumed to vary with capacity according to a 0.6 exponent. This method is generally considered fairly reliable, at least for scaling up data concerning a given capacity.

Let it be supposed that the investment for a chemical process plant can be broken down into three classes of costs with respect to their dependence on capacity. The reactor section will generally vary in cost almost proportionately to plant size, although with discontinuities in the regions where each succeeding reactor must be added. Purification, intermediate storage, piping, and so forth, vary according to a 0.6 exponent. Finally, such items as instrumentation and engineering charges hardly vary at all. Assuming that costs of the first type vary with a 0.85 power, and those of the third type with a 0.15 power, and, furthermore, that the plant cost on which the original estimate was based can be divided into 25%, 50% and 25%, respectively, of the three types of outlay, a comparison between the straight application of a 0.6 exponent and the separate scaling up or down of the various components of the base-case investment gives the result shown in the following table:

Capacity (with respect to base case)	0.6 Exponent	0.85–0.60–0.15 Exponent method	% Error when applying 0.6 exponent
0.125			
0.25	0.290	0.370	-0.22
0.50	0.435	0.500	-13
0.75	0.66	0.695	-5
1.00	1.00	1.00	0.00
2.00	1.52	1.49	+2
4.00	2.29	2.27	+1
8.00	3.46	3.54	-2

Thus the 0.6 factor can be useful in determining how the production cost of a chemical product will be affected by the entry into the field of larger and larger units. Unfortunately, however, this is not a very safe guide to the planner in a less developed country, since working backwards from data concerning the large units being reacted in Europe and the U.S. to the capacities that a much smaller market would justify, is almost certain to lead to overoptimistic conclusions.

In some cases the graphs were extended downwards because accurate data at more than one point were available; otherwise, they were extended upwards as far as eight times and downwards to 0.5 times the capacity given in the text.

5   REFERENCES

Material regarding product uses, market distributions, process economics, plant capacities and nature of the processes employed by the various manufacturers was obtained by compilation of information published regularly in the following publications:

Chemical Engineering
Chemical Week
Industrial and Engineering Chemistry
Chemical and Engineering News
Chemical and Process Engineering
Chemical Engineering Progress
Hydrocarbon Processing and Petroleum Refiner
Oil and Gas Journal
European Chemical News
L'Industrie Chimique
Chemie-Ing. Technik
Modern Plastics
Adhesive Age
Soap and Sanitary Chemicals
American Dyestuff Reporter
Revue de l'Institut Français du Pétrole
Química e Derivados
Revue des Produits Chimiques
Oil, Paint & Drug Reporter
Nitrogen
Erdoel und Kohle
British Chemical Engineering
Textile Industries
L'Industrie Textile
Skinner's Record
Agricultural Chemicals

The following books and reference works were also consulted:

General Information

"Encyclopedia of Chemical Technology," R. E. Kirk and D. F. Othmer (eds.), Interscience Publishers, New York.
"Ullmanns Encyklopaedie der Technischen Chemie," W. Foerst (ed.), 3rd ed., Urban & Schwarzenberg.

Pesticides

"Pesticides, Plant Food Regulators and Food Additives," vols. II–IV, Academic Press, New York, 1964.

Rubber and Rubber Chemicals

Morton, M.: "Introduction to Rubber Technology," Reinhold Publishers, New York, 1959.

Surface-active Agents

Schwartz, A.M., J.P. Perry and J. Berck: "Surface-active Agents and Detergents," vol. II, Interscience Publishers, New York, 1958.

Drugs

Evers, N. and D. Caldwell: "The Chemistry of Drugs," E. Benn, 1959.

Surface Coatings

Parker, D.H.: "Principles of Surface Coating Technology," John Wiley & Sons, New York, 1965.

Polymers

"Polymer Processes," C.E. Schildknecht (ed.), Reinhold Publishers, New York, 1956.

Petroleum Refinery Economics

Nelson, W.L.: "Petroleum Refinery Engineering," 4th ed., McGraw-Hill Book Company, New York, 1958.

Petrochemical Processes

"Advances in Petroleum Chemistry and Refining," K.A. Kobe and J.J. McKetta, Jr. (eds.), vols I–IX, John Wiley & Sons, New York, 1958–64.

Substituted Aromatics

Groggins, P.H.: "Unit Processes in Organic Synthesis," 5th ed., McGraw-Hill Book Company, New York, 1958.

2
SYNTHESIS GAS

1 INTRODUCTION

The use of petroleum hydrocarbons to make synthesis gas (mixtures of $CO + H_2$ in any proportion) has made petroleum and natural gas the world's main source of ammonia, the source of almost all nitrogenous fertilizers. Considerable attention has been given to technology and improvements in catalysts and techniques for converting anything between methane and residual fuels to synthesis gas.

Although synthesis gas can be obtained from a number of sources — coke oven gas, electrolysis of water, catalytic reformer off-gas, water-gas — the two main methods for producing synthesis gas, steam reforming, or partial oxidation, start from gaseous or liquid hydrocarbons. Steam reforming, whereby steam reacts with the hydrocarbon feed in a fixed bed tubular furnace type reactor, is the most important.

$$CH_x + H_2O \longrightarrow CO + \left(1 + \frac{x}{2}\right)H_2$$

The reaction is endothermic, and heat of reaction is supplied by burning fuel in the reforming furnace.

Early catalysts were unable to promote this reaction when the feed was heavier than butane without excessive coke laydown in the tubes. Since paraffins lighter than butane are abundant only in regions where access to natural gas is easy, in the absence of gas reserves, there is likely to be a shortage of hydrocarbons in the LPG range as well as methane. Thus, at first this process was limited in practice to the natural-gas-producing regions of the world. With the development of catalysts (by H. Topsøe and ICI)

19

offering satisfactory operation on naphthas of up to 400°F end-point, steam reforming acquired an entirely new dimension in regions such as Europe and Japan, where natural gas is scarce and straight-run naphtha is the choice for petrochemical feedstock. In the U.S., the overwhelming majority of steam-reforming units are fed on methane. Regardless of the feed, the sulfur content at the inlet must be below 2 ppm in order to keep down catalyst poisoning. Due to the reformer catalyst costs, catalytic desulfurization is limited to feeds containing less than 120 ppm sulfur. Above this level one must resort to chemical treatment, which is even more costly.

There has been considerable progress in raising the pressure at which the steam-reformer reaction could be carried out economically. Over 50 percent of all synthesis gas is used to make ammonia and methanol, both of which are synthesized at pressures far in excess of the 600 psig that in 1965 was considered the practical limit for carrying out the steam-reforming reaction. This has provided a constant incentive to raise this pressure and thus cut down charge-compression costs. Higher pressures, however, drive the equilibrium of the reforming reaction to the left, and also favor the formation of coke and methane. Nevertheless, there has always been considerable speculation about the possibility of an ammonia plant with no feed compressor at all. In such a plant, feed gas would be delivered to the synthesis loop at the reaction pressure, limiting compression costs to those for nitrogen and for recirculation. It must be kept in mind that recirculation and refrigeration costs rise swiftly with an increase in the methane content of the feed to the synthesis section.

The second method for obtaining synthesis gas is partial oxidation of the feed in a deficiency of pure oxygen:

$$CH_x + \frac{1}{2}O_2 \longrightarrow CO + \frac{x}{2}H_2$$

In the Texaco process, the feed is preheated and fed into the reactor with oxygen from air plant. The C:O ratio is kept close to stoichiometric, that is, 1:1. Too little oxygen increases coke formation, and excess would mean higher raw material consumption per unit of synthesis gas. The reactor effluent is quenched in a recirculating stream of water to avoid excessive coke formation. A slip-stream of this quench-water is treated for carbon removal, either by filtration or by extraction. This operation is partly responsible for the difference in capital requirements between

partial oxidation and steam reforming. However, this does not
hold if water is so cheap that a once-through quench system is
feasible. A flow-sheet of a partial oxidation unit is shown in Fig.
2-1.

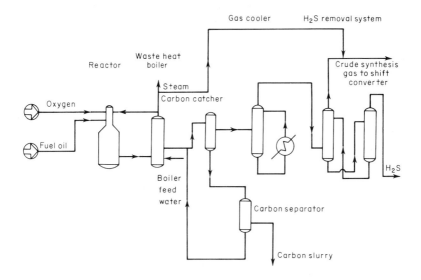

Fig. 2-1 Synthesis Gas (Partial Oxidation).

This method has the serious disadvantage of requiring an
air-separation plant, hence in general involves higher production costs than the steam reforming method. Its main attraction
is that there are no raw material limitations. Any material, from
methane to fuel oil, can be fed to such a process. Except in
comparatively rare cases where fuel oil is the only available
petrochemical feedstock, the world-wide trend has been towards
steam reforming especially since use of naphtha has become common.

Naphtha reforming requires a steam-to-carbon ratio of 7 or 8:1.
Thus, it is inherently more expensive than methane reforming,
where this ratio is around 3:1. However, naphtha reforming usually
must be compared with fuel oil partial oxidation, and not with
methane reforming. The latter option is usually ruled out by the
prevailing raw material structure. Improvements in catalyst
performance are gradually reducing dilution steam requirements.

Whereas the ICI and Topsøe steam-reforming processes are in essence similar to conventional methane reforming, the Topsøe-SEA process, used in several European countries, is a combination of partial oxidation and steam reforming. It has the advantage of being employed when pure hydrogen is desired.

After preheating, the desulfurized reformer feed consisting of hydrocarbon raw material, air or oxygen, and steam (generated by cooling the reaction products) enters the top of the reaction vessel. Part of the hydrocarbon reacts with oxygen and generates the heat necessary for steam reforming the rest of the feed further down the reactor. The gaseous stream is quenched and sent on to a waste-heat recovery section.

Other sources of synthesis gas and hydrogen should be mentioned. The BASF acetylene process yields a $2H_2:1CO$ off-gas which several producers use to make methanol, or subject to low-temperature fractionation for separate utilization of H_2 and CO. Ethanol dehydrogenation to acetaldehyde produces hydrogen, which after purification is often used in the same plant to produce butanol or 2-ethylhexanol. Acetone from isopropanol also gives hydrogen, which can be used to make MIBK or hexylene glycol. Caustic-chlorine plants produce hydrogen, about 5 ft^3/pound chlorine, which several U. S. plants use as ammonia-plant feedstock. Although coke-oven gas hydrogen is rarely used in the U. S., except for fuel, it is an important source of synthesis gas in Japan and Europe. The same can be said of water-gas, produced by the following reaction:

$$C + H_2O \longrightarrow CO + H_2$$

After the generation section proper, crude synthesis gas must be processed for delivery in conditions of purity and H_2-CO mol-ratio suitable for further processing. The five types of synthesis gas are given below with respect to composition.

Gas mixtures produced from synthesis gas	Main uses
$3H_2 : 1N_2$	Ammonia
$2H_2 : 1CO$	Methanol
$1H_2 : 1CO$	Oxo reaction
Pure CO, or CO + H_2O	Phosgene; formic, oxalic, acetic, glycolic, propionic and neo-acids; acrylates
Pure H_2	Hydrogen peroxide, various reduction and hydrogenation processes

Pure hydrogen can be obtained from crude synthesis gas in three additional steps, each of which may be accomplished in one or two operations, depending on the degree of purity required. However, the trend is towards better catalysts and absorbents able to do a satisfactory job for most purposes in a single operation. The first of these steps is shift conversion:

$$CO + H_2O \longrightarrow CO_2 + H_2$$

Steam is introduced into the gas stream, which then enters a catalytic reactor. The classical shift-conversion catalysts were able to convert around 90 percent of the CO in the crude synthesis gas to hydrogen, thus reducing the CO content to 3 percent. This high CO content cannot be tolerated in the synthesis loop. It was removed by nitrogen washing, in the case of partial oxidation units, or by an intermediate CO_2 removal step followed by another shift conversion-CO_2 sequence. Recently, however, more active catalysts have been developed that can reduce the CO content to around 0.3 percent without intermediate CO_2 removal. This is accomplished in two stages: the first at 750°F and the second at around 460°F where equilibrium favors the formation of CO_2 and hydrogen. These more active catalysts impose strict limitations on the sulfur content of the gas. Even if the feed has been purified to a 1 ppm level or less, the sulfur picked up from the furnace refractories can suffice to poison the conversion catalyst. Next, CO_2 is eliminated by absorption followed by regeneration of the absorbent. Since one of the main cost items in the production of synthesis gas is the utilities charge for stripping CO_2 from the liquid in which it has been absorbed, there is a constant search for compounds combining high capacity for absorbing CO_2 with ease of regeneration at low solvent losses. The classical method for removing CO_2 is absorption in monoethanolamine (MEA), or in diethanolamine when H_2S is present (MEA is subject to excessive degradation in this case), and regeneration by heating. Two pounds steam/pound CO_2 removed are needed for regeneration, which is more than any other method. However, the trend to higher reforming pressures has made waste heat available at higher levels. Next in importance is a family of processes based on absorption of CO_2 in a hot solution of potassium carbonate. Since absorption can be carried out at high temperatures, heat requirements are about 40 percent lower than for MEA. The original technique has been improved in several ways. For example, the Giammarco-Vetrocoke process uses potassium carbonate promoted with arsenic oxide to increase the rate of reaction thereby reducing investment costs for a given throughput. Even more recent but already quite successful is the "Sulfinol" process, which uses tetrahydrothiopene oxide ("Sulfo-lane") as the solvent.

Shift conversion still leaves some CO in the gas. Means of removing this must be found, since CO is poison to most catalysts, and, furthermore, is reactive enough to produce unwanted side-reactions. This conversion can be carried out in several ways. If pure nitrogen is available, the CO -free gas steam can be washed in liquid nitrogen; if not, catalytic methanation converts the remaining CO to methane, which can be tolerated even though it lowers somewhat the hydrogen purity.

When ammonia synthesis gas ($N_2 + 3H_2$) is to be produced, the required nitrogen can be introduced in two ways. Since partial oxidation requires pure oxygen, and pure nitrogen is therefore usually available, the gases after CO_2 removal are washed in liquid nitrogen under conditions yielding an exit gas having the desired 3:1 H_2-to-N_2 mol ratio. When crude synthesis gas has been generated by steam reforming, nitrogen can be introduced by "secondary reforming." This involves admitting air into the reformer effluent stream and removing the oxygen by combustion of part of the product until the $(H_2 + CO):N_2$ mol ratio reaches 3:1. When the hydrocarbon raw material is relatively expensive, it may be preferable to add an air plant (even though there is no need for the oxygen), and use a nitrogen wash. However, in the U. S. where most plants are being built to reform cheap natural gas, secondary reforming is the usual procedure. A flowsheet of a methane steam-reforming unit is shown in Fig. 2-2.

The manner of obtaining methanol synthesis gas ($CO + 2H_2$) depends on how the crude gas was generated. If methane is the reforming feedstock, the reformer effluent contains a H_2:CO ratio of around 3:1. This ratio must be corrected by addition of CO_2 from an outside source, usually recovery from the reformer furnace combustion gases. If, on the other hand, partial oxidation was used, the mol ratio is below the required 2:1 and is corrected by shift-converting a slip-stream and removing the resulting CO_2 before recombining the two streams.

To produce 1:1 synthesis gas for making oxo chemicals, CO_2 recirculation may also be employed, but there are two preferable alternatives. The first is low-temperature fractionation of a gas obtained by steam-reforming into pure hydrogen (used later in the process) and a 1:1 H_2:CO mixture. The second is partial oxidation of a heavy liquid fraction which in itself is more costly but produces a 1:1 gas directly with no need for fractionation. Some

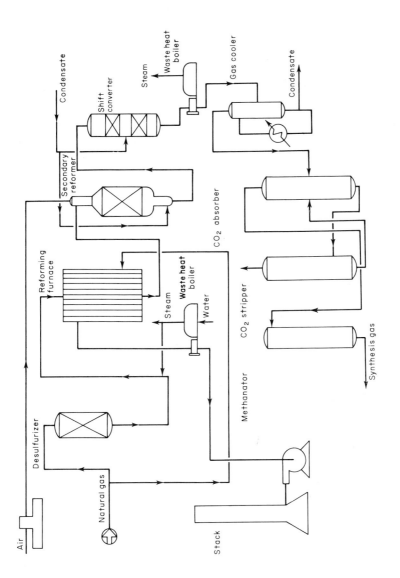

Fig. 2-2 Ammonia Synthesis Gas (Steam Reforming).

recent oxo processes produce alcohols in one step from an olefin and a 2:1 H_2:CO synthesis gas.

Finally, pure CO is obtained from crude synthesis gas by low-temperature fractionation. A considerable part of the pure CO consumed in the U. S. is derived from inorganic sources, e.g., phosphorus and calcium carbide furnaces, ferroalloy plants and BOF (Basic Oxygen Furnace) steel mills. Nevertheless, there is appreciable production of CO via methane steam-reforming as can be seen from the number of phosgene plants recently built on the Gulf Coast.

The economics of large-scale synthesis gas production are very dependent on fuel costs and investment requirements. Some examples are given in Table 2-1 as guides to other situations (five cases were considered).

a) *Methane reforming*
b) *Naphtha reforming*
c) *Fuel oil partial oxidation*
d) *Reformer off-gas purification by low-temperature fractionation*
e) *Topsøe-SEA autothermal process using naphtha*

In all cases, production of 25 million scfd was taken as a basis. In general, steam reforming is the natural choice both in oil-producing and oil-importing regions. Although the cheapest source of hydrogen is catalytic reformer off-gas, this source is slowly being exhausted as refineries tend to use their hydrogen captively for refining operations — hydrocracking and hydrotreating — or for such quasi-refining operations as cyclohexane production or dealkylation. It should be remembered that the lighter the refinery product the higher the H:C ratio; in the U. S., where the emphasis is on gasoline production, some refineries are no longer self-sufficient in hydrogen and therefore have been compelled to install independent generation facilities.

The demand for synthesis gas in the U. S., aside from about 290.0 million scfd of liquid H_2 for military uses, was about 2710 million scfd in 1964, distributed as follows:

End-uses	MM scfd, 1964
Ammonia	1400
Refinery consumption	930
Methanol	260
Other chemical uses	120
	2710

Table 2-1.

	Case				
	a	b	c	d	e
Investment, $ million	2.45	3.25	4.70	0.6	1.25
Requirements/1000 ft^3					
Hydrocarbon feed	440×10^6 Btu	26.2 lbs	23.0 lbs	1000 ft^3	17.5 lbs
Oxygen			22.0 lbs	−	18.5 lbs
Cost, ¢/1000 ft^3					
Hydrocarbon feed:					
naphtha—0.8 ¢/lb					
natural gas—40 ¢/10^6 Btu					
fuel oil—$2.30/bbl	18	21	16	9	14
Oxygen ($9.00/ton)	−	−	10	−	8
Utilities (net)...........	2	3	4	1	1
Labor and overhead.......	1	1	3	1	2
Capital charges	10	13	19	3	5
Total	31	38	52	14	30

The sources of synthesis gas for ammonia and methanol produced in 1965 can be broken down as follows:

Process	% Ammonia	% Methanol
Steam reforming	81	} 90
Partial oxidation	7	
Refinery off-gas	5	−
Coke-oven gas	1	−
BASF-process off-gas	−	10
Electrolysis (caustic-chlorine)	4	−
Others	2	−

These sources account for 100 percent of all ammonia produced, and 95 percent of methanol output, the remaining 5 percent coming from direct oxidation of butane and propane.

2   AMMONIA

Hydrogen and nitrogen react under pressure in the presence of an iron catalyst to form ammonia:

$$3H_2 + N_2 \longrightarrow 2NH_3$$

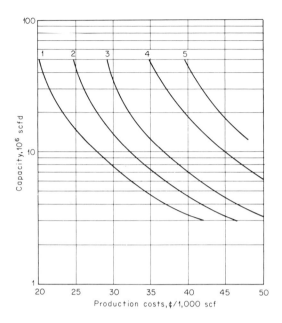

Fig. 2-3 Synthesis Gas (Steam Reforming).   1-Natural Gas 20¢/1000 scf; 2-Natural Gas 30¢/1000 scf; 3-Natural Gas 40¢/1000 scf; 4-Naphtha 0.8¢/lb; 5-Naphtha 1.0¢/lb.

The reaction is exothermic; temperature in the reactor is controlled by heating the cold feed against the reaction products before it enters the catalyst bed.

The numerous ammonia processes available at present differ from each other mainly by the type of compromise reached between high- and low-pressure operation. At high pressures, equilibrium is shifted to the right, increasing conversion per pass, reducing recycle costs but also increasing feed compression expenses. Equipment diameters are reduced, but this is compensated by greater wall thicknesses throughout, which affect cost more than proportionately to equipment weight. The trend is definitely towards lower pressures; the early Claude plants operated at around 1000 atmospheres, later plants employing pressures of 600 atmospheres, began to gain popularity, as did a number of processes in the 3500-5000 psi range.

Around 1964, plants began to be built with capacities unheard of only a few years earlier. This raised a great deal of interest in the use of turbine-driven centrifugals instead of motor-driven reciprocal feed compressors, and consequently in synthesis loops operating at even lower pressures. Centrifugal compressors require a minimum gas volume at the entrance of any given stage which can only be met at high capacities and low loop pressures. Therefore they have several advantages over reciprocal machines, in that: 1) they are cheaper; 2) they require much less costly and less frequent maintenance; 3) they require no lubrication and thus eliminate oil entrainment into the synthesis loop, which can foul the catalyst and contaminate the product; and 4) which is most important, they can be turbine-driven by the steam generated in the high-pressure steam-reforming unit that precedes the synthesis loop and still supply exhaust steam at a level sufficient to regenerate the CO_2-absorption solution. Thus, the use of centrifugal compressors has made possible lower investments, maintenance costs and energy requirements per ton of ammonia.

Since the synthesis reaction is exothermic, the reactor must be designed so as to prevent the temperature from rising above the level where the catalyst would suffer. This is achieved either by generating steam in the converter, or by designing it as a feed-effluent exchanger in which part of the cold feed is injected counter-current into the catalyst bed. A recent innovation is the radial-flow converter, in which the gas stream flows through the converter perpendicularly to its axis. This type of reactor has a lower pressure drop for a given space velocity, and permits the use of finer and therefore more active catalysts. This makes possible higher outputs for a given volume of catalyst, or, consequently, for a given reactor size. The progress made in the development of more active catalysts derives from the hope expressed by some ammonia plant designers of building 2000-3000 ton-per-day converters. The significance of these improvements in ammonia technology can be appreciated when one considers that, as late as 1960, a 200 ton-per-day ammonia converter was considered fairly large.

Another important development has been the increase in catalyst life; modern catalysts last five to seven years, which has been made possible by reducing to less than 10 ppm the oxygen and CO content in the synthesis gas. This represents savings of almost $2 per ton of ammonia compared to the catalyst cost around 1960.

A typical ammonia synthesis-loop flowsheet is shown in Fig. 2-4. Feed and recycle gas are cooled and sent to the secondary separator, where ammonia contained in the recycle is condensed out, as is any oil that may have been carried in by the fresh feed and which would otherwise foul the catalyst. The level to which this stream must be cooled is a function of the operating pressure; 32°F is common for the 4000-5000 psi processes, 0°F to -20°F is the range for the larger low-pressure plants. The gases are then introduced into the converter. Reactor effluent, containing 14-18 percent ammonia, is cooled against water (with generation of steam) and sent to the primary separator, where most of the ammonia condenses out. The off-gas from this separator is the recycle stream, and at this point a slip-stream is removed in order to purge from the system the inert gases introduced with the feed. Under any given set of conditions — price of raw material, fuel and power costs — there exists an optimum concentration of inert gases in the recycle stream. Above this optimum, recompression and refrigeration costs begin to rise, and below this concentration, purging the inerts involves too high a loss of synthesis gas. Usually the optimum lies in the 12-15 percent inerts range, 10 to 15 times their concentration in the fresh feed. This slip-stream is absorbed in water, giving aqua-ammonia, which can be stripped and the ammonia recovered as a liquid. Ammonia from the two separators goes to a letdown tank, where the liquid is released and also sent to an ammonia absorption system; if the synthesis gas is pure enough, and the purge stream consequently contains only a minor amount of product, it may be found that recovery is not economically justified.

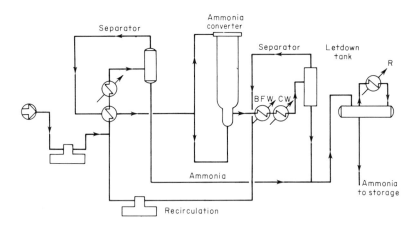

Fig. 2-4 Ammonia Synthesis.

There are two apparently contradictory trends at work in ammonia production. On the one hand, plants are getting larger, as is the case in the entire chemical industry; on the other hand, however, when natural gas is cheap and the demand for ammonia is geographically concentrated, the type of small, packaged ammonia plant offered by several U. S. engineering firms becomes viable. The number of very large plants is definitely on the increase, but small plants will nevertheless continue to be built. There will always be new sources of almost costless hydrogen, such as from chlorine plants; perhaps even more important are the savings it is possible to realize in the distribution costs. These amount to around $50/ton for a merchant producer selling to the market, and can be cut to about $25/ton for a farm-cooperative; this difference alone often outweighs the savings in production costs made possible by the 1000 tons/day and larger plants.

Ammonia is in the process of becoming a commodity (just as petroleum) with the oil-producing countries that have no captive requirements of any importance for natural gas. They convert it to ammonia and ship it, refrigerated and in anhydrous form, to the oil-poor regions of the world. The world price of nitrogen in anhydrous ammonia is around $80/ton, whereas solid fertilizers cost $200/ton of nitrogen; thus considerable savings can be realized by the oil-importing countries by purchasing ammonia and converting it to solid fertilizers themselves. Plants producing ammonia primarily for export operate in Trinidad, and are likely to be built in Mexico, Kuwait and Chile.

Despite the mounting emphasis on naphtha, natural gas and fuel oil as sources of synthesis gas, 40 percent of all the ammonia produced in the world is still based on coal. Steel plants recover ammonium sulfate from coke-oven gas by absorption in sulfuric acid; coke-oven gas itself contains about 50 percent hydrogen, which can be separated by low-temperature fractionation, thus simultaneously upgrading the remaining fuel gas, which has its heating value per unit volume raised. There is also still some ammonia being produced from water gas (coke plus steam), although no new projects of this type have been reported for several years. Another 40 percent of the world's ammonia output is derived from natural gas, and the remaining 20 percent comes from refinery off-gas (catalytic reforming hydrogen, and partial oxidation or steam-reforming of fuel-gas streams) from naphtha, fuel oil, and other heavy stream, and from electrolysis hydrogen.

The world-wide nitrogen demand is growing at 8 percent per year. It is being used increasingly in concentrated form, while the relative importance of materials such as ammonium sulfate is declining. In the U. S., for example, a rising percentage of all fertilizers is being consumed in the form of granulated complex fertilizers applied at planting time, often containing equal proportions of the three macronutrients — N, P, and K — followed by side-dressing with pure nitrogen, in the form of urea, N-solutions or directly applied anhydrous ammonia, later in the season. In general, granulated and concentrated forms of plant nutrients are more expensive to manufacture per pound of N, and require more capital-intensive production techniques.   The savings appear later — in handling, transportation, avoidance of losses due to wind, rain, etc.

Despite the much more favorable raw material position of the U. S., Europe is still the main exporting region of nitrogenous fertilizers.   The importing regions are chiefly Asia and Africa, which in 1965 produced, respectively, 80 percent and 40 percent of their requirements.   Thus, by 1965, there was no real sign of large-scale exports of anhydrous ammonia from developing areas such as the Middle East and South America that were predicted above.   The world's sources of nitrogenous fertilizers are broken down below; ammonium sulfate, saltpeter and calcium cyanamid are losing ground, while urea, ammonium phosphates and granulated fertilizers are gaining; other forms of nitrogen are maintaining their share.

Type of nitrogen fertilizer (World)	*% of total N*
Ammonium sulfate	24
Ammonium nitrate	28
Naturally-occurring nitrates	3
Nitrocalcium	4
Calcium cyanamid	2
Urea	9
Ammonium phosphates, ammonium chloride	3
Granulated fertilizers	9
Nitrogen solutions and direct applications	17
Manure	1

The consumption of ammonia in 1965 was 16.5 billion pounds in the U. S., and the demand has been growing at 10 percent annually.   Farm policy has been to encourage higher yields from lower acreage by means of more intensive cultivation methods. It has been estimated that since one dollar spent on fertilizers yields a profit of around $3, and the total amount spent on fertilizers was $2.0 billion in 1965, fertilizers were responsible for 50 percent of the $12.0 billion net farm income that year.   The pattern of fertilizer consumption in the U. S. is quite different from that of the rest of the world due to the obvious predominance of the more capital-intensive application techniques. Whereas application in liquid form accounts for only 17 percent of the world's total nitrogen utilization, in the U. S. it represents 62 percent. The total consumption of ammonia can be broken down as follows.

End-uses		Ammonia MM lbs/yr
Fertilizers:		12,800
solid fertilizers	4900	
solutions	4600	
direct application	3300	
Nonfertilizer uses:		3700
plastic, resins, fibers	870	
chemicals	660	
explosives	500	
others	1670	
		16,500

Other uses are mainly in refrigeration, and as a weak base, for example, in the textile industry.

The economics of ammonia production vary greatly with raw material costs, and with the complicated interrelationships between method of synthesis gas generation and heat recovery, synthesis-loop pressure, power costs, type of drives, etc. The higher reforming pressures made possible by today's improved catalyst allow a greatly improved degree of power and heat recovery. The economics of the 1000-ton-per-day plant reflect improvements in engineering over the older processes of about $4.20 per ton in combined feed and power savings. The estimates below were made for four cases, all of which include synthesis-gas generating facilities. The first example corresponds to a large European plant; the second to that of a Gulf Coast producer; the third represents a plant producing captive ammonia for a large chemical manufacturer; the fourth is that of a small plant having access to hydrogen practically at its fuel value.

Case	Feed	Capacity	Synthesis gas process
a-	Naphtha	600 T/day	High-pressure steam reforming
b-	Natural gas	1000 T/day (low-pressure loop)	High-pressure steam reforming
c-	Natural gas	200 T/day	High-pressure steam reforming
d-	Reformer off-gas	80 T/day	Low-temperature fractionation

	Cases			
	a	b	c	d
Investment requirements, million $	11.5	13.8	5.85	2.8
Hydrocarbon requirements per ton	1780 lbs	35 Mscf	30 Mscf	63 Mscf
Power	720 kWh/ton	50 kWh/ton	720 kWh/ton	720 kWh/ton
Hydrocarbon costs	1 ¢/lb	25 ¢/Mscf	25 ¢/Mscf	14 ¢/Mscf

	Cost $/ton NH$_3$			
	Cases			
	a	b	c	d
Feed	17.80	8.80	7.50	9.00
Power	5.80	0.40	5.40	
Catalyst and chemicals	1.10	1.10	1.10	7.50
Other utilities	2.90	2.30	2.30	
Labor and overhead	2.00	1.20	6.00	3.20
Capital charges	19.10	13.80	29.20	23.00
	48.70	27.60	51.50	42.70

For our purposes, we have taken captive ammonia at 2.5 cents per pound ($50.00 per ton) and merchant ammonia at 4.0 cents per pound.

A list of the principal U.S. producers of ammonia and announced plants as of mid-1965 follows:

United States Synthetic Ammonia Capacity[*]

	Estimated capacity (1000 tons/yr NH$_3$)			
	1964	1965	1966	1967
Air Products				
New Orleans, La.	—	—	210	210
Allied Chemical				
Geismar, La.	—	—	—	350
La Platte, Neb.	80	185	185	185
South Point, Ohio	320	320	320	320
Hopewell, Va.	400	400	350	350
American Cyanamid				
Fortier, La.	78	78	425	425

[*]*Chemical Week*, September 11, 1965, pages 53-4.

United States Synthetic Ammonia Capacity (Continued)

	Estimated capacity (1000 tons/yr NH_3)			
	1964	1965	1966	1967
American Oil (Tuloma Oil) Texas City, Texas	220	220	220	220
Ammonia, Inc. (International Minerals) Bonnie, Fla.	70	100	100	100
Apache Powder Benson, Ariz.	15	15	15	15
Arkla Chemical (Arkansas Louisiana Gas) Helena, Ark.	—	—	210	210
Armour Agricultural Cherokee, Ala. Crystal City, Mo.	134 86	134 86	134 86	134 86
Atlantic Refining Point Breeze, Pa.	60	60	60	60
Best Fertilizers (Occidental Petroleum) Plainview, Texas	70	70	70	70
Borden Chemical Geismar, La.	—	—	350	350
California Ammonia (Best Fertilizers operator) Lathrop, Calif.	50	100	100	100
California Chemical (Standard Oil of California) Richmond, Calif. Fort Madison, Ia. Pascagoula, Miss.	130 105 —	130 105 —	130 105 —	130 105 525
Calumet Nitrogen (Tuloma Gas and Sinclair) Hammond, Ind.	127	127	127	127
Central Nitrogen (50% owned by FS Services) Terre Haute, Ind.	125	125	125	125
Cities Service (Tennessee Corp.) Tampa, Fla.	125	125	125	125
Cities Service (Petroleum Chemicals) Lake Charles, La.	140	140	140	380
Coastal Chemical (Mississippi Chemical) Pascagoula, Miss.	180	180	180	180
Collier Carbon & Chemical (Union Oil) Brea, Calif.	120	120	265	265
Columbia Nitrogen (Pittsburgh Plate Glass and DSM Chemicals, Netherlands) Augusta, Ga.	126	126	126	126
Commercial Solvents Sterlington, La.	144	144	144	485

United States Synthetic Ammonia Capacity (Continued)

	Estimated capacity (1000 tons/yr NH$_3$)			
	1964	1965	1966	1967
Consumers Cooperative				
Fort Dodge, Ia.	—	—	210	210
Hastings, Neb.	146	146	146	146
Joplin, Mo.	—	—	35	35
Cooperative Farm Chemical				
Lawrence, Kan.	200	200	200	200
Continental Oil				
Blytheville, Ark.	—	—	347	347
Diamond Alkali				
Deer Park, Texas	35	35	35	35
Dow Chemical				
Freeport, Texas	110	110	220	220
Midland, Mich.	36	36	36	36
Pittsburg, Calif.	12	12	12	12
DuPont				
Beaumont, Texas	—	—	350	350
Belle, W. Va.	270	270	350	350
Gibbstown, N. J.	75	75	75	75
Victoria, Texas	80	80	80	80
Escambia Chemical				
Pensacola, Fla.	80	80	80	80
Farmers Chemical				
Tyner, Tenn.	70	140	140	140
Fel-Tex (Farmers Union State Exchange and Farmers Elevator Service)				
Fremont, Neb.	—	42	42	42
Felton Oil				
Olean, N.Y.	—	—	83	83
First Nitrogen (FS Services and First Mississippi)				
Donaldsonville, La.	—	—	350	350
FMC				
South Charleston, W. Va.	24	24	24	24
Frontier Chemical				
Wichita, Kan.	15	30	30	30
General Exploration				
Mercedes, Calif.	—	—	70	70
Grace Chemical				
Big Spring, Texas	75	94	94	94
Woodstock, Tenn.	180	255	255	255
Green Valley				
Creston, Ia.	—	35	35	35

United States Synthetic Ammonia Capacity (Continued)

	Estimated capacity (1000 tons/yr NH_3)			
	1964	1965	1966	1967
Hawkeye Chemical (Skelly Oil and Swift)				
Clinton, Ia	140	140	140	140
Hercules Powder				
Hercules, Calif.	55	55	115	115
Louisiana, Mo.	45	45	115	115
Hooker Chemical				
Tacoma, Wash.	22	22	22	22
Ketona Chemical (Hercules Powder and Alabama By-products)				
Tarrant, Ala.	47	47	47	47
Mississippi Chemical (jointly with Coastal Chemical)				
Yazoo City, Miss.	115	465	465	350
Monsanto				
El Dorado, Ark.	255	255	255	255
Luling, La.	220	440	440	440
Muscatine, Ia.	88	88	88	88
New Jersey Zinc				
Palmerton, Pa.	35	35	35	35
Nipak (Lone Star Producing)				
Kerens, Texas	–	119	119	119
Pryor, Okla.	105	105	165	165
Nitrin (International Minerals and Northern Natural Gas)				
Cordova, Ill.	145	145	145	145
Northern Chemical Industries				
Searsport, Me.	40	40	40	40
Northern Illinois Gas (Apple River Chemical)				
East Dubuque, Ill.	–	–	245	245
Odessa Natural Gas (El Paso Natural Gas)				
Odessa, Texas	22	130	130	130
Olin Mathieson				
Lake Charles, La.	113	613	500	500
Oregon Chemical Fertilizer				
Umatilla, Ore.	–	–	36	36
Pennsalt				
Wyandotte, Mich.	35	35	35	35
Portland, Ore.	15	15	15	15
Phillips Petroleum				
Beatrice, Neb.	–	–	220	220
Etter, Texas	208	208	208	208
Pasadena, Texas	219	255	255	255

United States Synthetic Ammonia Capacity (Continued)

	Estimated capacity (1000 tons/yr NH$_3$)			
	1964	1965	1966	1967
Phillips Pacific (15% owned by Phillips Petroleum) Kennewick, Wash.	86	139	139	139
Pittsburgh Plate Glass Natrium, W. Va.	55	55	55	55
Pure Oil Worland, Wyo.	11	11	11	11
Reserve Oil and Gas Hanford, Calif.	—	23	23	23
Rohm & Haas Deer Park, Texas	50	50	50	50
St. Paul Ammonia Pine Bend, Minn.	90	90	90	90
Shamrock Oil & Gas Dumas, Texas	70	70	140	140
Shell Chemical Pittsburg, Calif.	112	112	112	112
St. Helens, Ore.	—	—	75	75
Ventura, Calif.	105	160	160	160
J. R. Simplot Pocatello, Ida.	—	55	55	55
Smith-Douglass (Borden) San Jacinto, Texas	44	44	44	44
Socony Mobil (Mobil Chemical) Beaumont, Texas	—	—	250	250
Solar Nitrogen Joplin, Mo.	150	150	150	150
Lima, Ohio	136	136	136	136
Southern Farm Supply (Techne) Plainview, Texas	21	21	21	21
Southern Nitrogen Savannah, Ga.	150	150	150	150
Southwestern Nitrochemical (First Mississippi and Arizona Agrochemical) Chandler, Ariz.	30	43	43	43
Spencer Chemical (Gulf Oil) Henderson, Ky.	105	105	105	105
Pittsburgh, Kan.	190	190	190	190
Vicksburg, Miss.	77	77	77	77
Sun Oil Marcus Hook, Pa.	110	110	110	110
Tenneco Chemical Houston, Texas	125	125	125	125

United States Synthetic Ammonia Capacity (Continued)

	Estimated capacity (1000 tons/yr NH$_3$)			
	1964	1965	1966	1967
Tennessee Valley Authority				
Wilson Dam, Ala.	90	90	90	90
Terra Chemicals				
Port Neal, Ia.	—	—	210	210
Texaco				
Lockport, Ill.	77	77	77	77
Union Carbide				
Texas City, Texas	—	90	90	90
U. S. Industrial Chemical (National Distillers)				
Tuscola, Ill.	80	80	80	80
U. S. Steel				
Clairton, Pa.	—	—	490	490
Geneva, Utah	72	72	72	72
Valley Nitrogen Producers				
Fresno, Calif.	90	150	150	150
Imperial Valley, Calif.	—	—	210	210
Western Ammonia				
Dimmitt, Texas	27	27	27	27
Wycon Chemical (Colorado Interstate Oil & Gas)				
Cheyenne, Wyo.	—	20	20	20

The production costs for ammonia from various feedstocks and at various prices and capacities are shown in Fig. 2-5 (see p. 42).

2.1 Nitrogen Fertilizers

The main functions of phosphorus and potassium are to promote the synthesis of sugars and starches, the staples essential to man's bare survival. Nitrogen is needed for the growth and formation of leaves, fruits and proteins, often considered "luxury" foodstuffs. Accordingly, a rising standard of living in any economy invariably causes a shift in the basic ratio between the overall consumption of N, P$_2$O$_5$ and K$_2$O in favor of nitrogen. Thus, in the U. S., this basic ratio was around 1:2:1 in 1940, but by 1965 had moved closer to 5:4:3. In the early 1960's, when priorities in the Soviet Union were shifted towards consumer goods, even methanol plants were converted to ammonia production in an effort to increase the supply of fruit and vegetables. The world as a whole consumed roughly equal amounts of the three macronutrients, but in under-developed regions phosphorus is still the most important of the three. In the advanced regions, the demand for nitrogen is fast outgrowing that for phosphorus and potassium.

The demand in the U. S. for fertilizer nitrogen was 4.6 million tons for the year ending June 30, 1965; distributed as follows:

Form of nitrogen—U.S.	% of total
Ammonium sulfate	3.5
Ammonium nitrate	13.0
Nitrogen solutions	16.0
Urea	4.0
Anhydrous ammonia	27.5
Ammonium phosphates	2.5
Mixtures	33.5
	100.0

Source: U.S. Department of Agriculture.

In agriculturally advanced countries, most of the total phosphorus and potassium requirements are usually applied to a given crop in the form of mixed fertilizers in granulated or powdered form. The majority of the nitrogen is added later in the season in the form of side-dressing. The overall basic ratio for the mixed fertilizers used in the U. S. is around 1:2:2, and the proportion of all P_2O_5 and K_2O applied in this form is 80 and 87 percent, respectively. Only 30 percent of all nitrogen, on the other hand, is applied in the form of mixtures; the remaining 70 percent is used "directly," that is, alone.

The fixed cost of making solid nitrogen fertilizers is a decreasing function of the total nutrient content. However, this does not consider the fact that the higher-analysis forms require more expensive raw materials. A comparison of the capital-related charges involved in making various solid nitrogen fertilizers, based on a production of 7500 tons per year of nutrient, illustrates this point.

	Total % nutrient content	$/ton nutrient capital charges
Ammonium sulfate	20.5	47.5
Ammonium nitrate	33.5	35.0
Urea	44.5	17.0
Diammonium phosphate	64.0	14.0

The total cost of converting a given amount of hydrocarbon raw material into nitrogen depends somewhat on the type of conversion downstream from the ammonia plant. The estimate below was made for a complex based on a 650 tons per day NH_3 plant, of which one third is converted to urea and the rest to ammonium nitrate and N-solutions.   Assuming natural gas at 30 cents per 1000 ft^3 and a total investment of $45 million, conversion costs per ton of nutrient can be calculated as follows:

	$/ton N
Natural gas	11.90
Operating costs	42.60
Capital charges	79.00
Total	$133.50

2.2 Mixed Fertilizers

Between 60 and 65 percent of the total macronutrients consumed in the U. S. are applied as mixed fertilizers. Of this, a percentage that is expected to reach 80 percent by 1970 is in the form of granulated products, and the remaining 20 percent as simple mixtures of solid powdered materials.

Granulation presents several important advantages.   The product can be applied even in the presence of strong winds, without the losses experienced with powdered materials. Granule size can be regulated according to rainfall.   In very wet conditions, for example, granules can be made large enough to prevent the nutrients being leached out as rapidly as would be the case with powdered fertilizers.   Also, granulated fertilizers can be stored longer without the danger of caking, thereby permitting the producer to plan his output evenly throughout the year and enabling the farmer to purchase at times other than the planting season.

The simplest way to obtain granulated fertilizers is to mix the solid materials *intimately*, and humidify them slightly.   Then, the granules are formed, dried, cooled and classified, e.g., on a tilting-pan mixer.

The mere granulation of solid raw materials, however, does not enable the producer to avail himself of the large price differential between nitrogen in the form of anhydrous ammonia, and solid materials, the latter costing almost twice as much. Accordingly, most of the granulated fertilizer in the U. S. is made by one of the

several versions of the TVA ammoniation process. Normal superphosphate can be ammoniated up to a 5 percent N content and
additional sulfuric acid can be added and neutralized with more
ammonia in order to dry the material through the heat evolved by
the reaction. Other solids besides superphosphate can be added
to make up the desired formula. The most versatile flowsheets
include an ammoniator, a granulator, a drier and a cooler. However, simplified and therefore less flexible schemes exist in which
granulation takes place in the ammoniator, or from which the drier
is excluded.

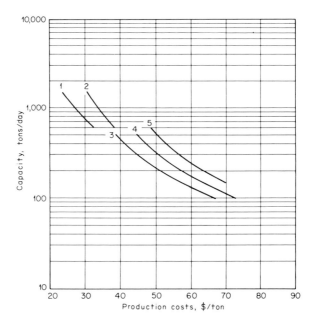

Fig. 2-5 Production Costs of Ammonia. 1-Natural Gas 25¢/1000 scf,
centrifugal compressors; 2-Natural Gas 45¢/1000 scf, centrifugal compressors; 3-Natural Gas 25¢/1000 scf, reciprocal compressors; 4-Natural
Gas 45¢/1000 scf, reciprocal compressors; 5-Naphtha 1.0¢/lb.

The most common formulas are 5-10-10, 4-12-12, 5-20-20
and 10-10-10, and the overall basic ratio is close to 1:2:2 since,
as was pointed out earlier, most of the nitrogen requirements are
applied later in the season.

A type of process that has acquired some importance in sulfur-
poor regions, and which periodically becomes of interest all over

the world during sulfur shortage or price increase, is based on the solubilization of phosphate rock with nitric acid instead of sulfuric or phosphoric. Plants of this type operate in the U. S. (e.g., Fort Madison, Iowa), Central America and India. Nitric acid reacts with phosphate rock:

$$2HNO_3 + Ca_3(PO_4)_2 \longrightarrow CaHPO_4 + Ca(NO_3)_2$$

Calcium nitrate, however, is highly hygroscopic and makes the product unsuitable as a commercial fertilizer. It is transformed into a less soluble calcium salt by the addition of ammonia and another acid:

$$Ca(NO_3)_2 + 2NH_3 + H_2SO_4 \longrightarrow CaSO_4 + 2NH_4NO_3$$

$$Ca(NO_3)_2 + 2NH_3 + 2H_3PO_4 \longrightarrow Ca(H_2PO_4)_3 + 2NH_4NO_3$$

$$Ca(NO_3)_2 + 2NH_3 + CO_2 + H_2O \longrightarrow CaCO_3 + 2NH_4NO_3$$

The advantage of CO_2 over sulfuric or phosphoric acid is that the direct or indirect use of sulfur is eliminated altogether; its source is the ammonia plant that is normally part of a fertilizer complex. This process can produce granulated complex fertilizers having a total nutrient content of around 36 percent, against 28 to 30 percent for the sulfuric acid process. The best-known version of this method is the *Potasse et Engrais Chimiques* process, in which nitric acid, phosphate rock, ammonia and the second acid are combined in a cascaded reactor system to form a slurry. After addition of potash, the slurry is granulated, dried and classified as in other processes.

Finally, granulated complex fertilizers can be made by ammoniating triple superphosphate to attain a total nutrient concentration of around 45 percent. In general, total integrated investment costs are higher for a given overall nutrient output the greater the maximum nutrient concentration obtainable by the process.

2.3 Urea

Urea is made invariably in the vicinity of ammonia plants since the synthesis gas section supplies the necessary carbon dioxide stream. The process takes place in two steps; first, ammonia and CO_2 react at high pressures to produce ammonium carbamate:

$$CO_2 + 2NH_3 \longrightarrow NH_2COONH_4$$

which, in the second step is thermally decomposed into urea and water:

$$NH_2COONH_4 \longrightarrow NH_2CONH_2 + H_2O$$

Urea contains a theoretical 45 percent of N, against 35 percent for the next most concentrated solid, ammonium nitrate. It was prevented from becoming a large-scale nitrogen fertilizer for many years due to the highly corrosive nature of both urea and hot ammonium carbamate, especially in the reactor and the pressure-reducing valves. However, since the 1950's, improvements in process design and especially in construction materials have enabled large plants to be built and urea to become the fastest-growing of all solid forms of nitrogen.

The earlier urea plants were mostly of the "once-through" variety, with unreacted ammonia being converted to other forms of solid fertilizer by absorption in sulfuric or nitric acid. Today, the majority of new urea plants are self-contained in that the entire ammonia and CO_2 feeds are recovered and eventually converted to urea. Some plants employ partial recycle, with NH_3 from the primary decomposer being liquefied and returned to the reactor, but that from the secondary decomposer sent on to other fertilizer units. Processes no longer differ in the basic manner of recovering the feed gases. The processes that come into consideration for new projects all involve absorption of unreacted ammonia and CO_2 as an aqueous carbamate solution for recycle to the reactor. The various processes differ from one another as to reaction conditions, i.e., feed ratio, temperature and pressure techniques for reducing corrosion.

The three most widely used contemporary urea processes are Chemico, Stamicarbon and Toyo Koatsu. The first is offered by the Chemical Construction Company, and the other two by a number of engineering companies in several countries. Since 1961, 60 percent of the plants built used the Stamicarbon process and 40 percent the other two. More recently, however, Toyo Koatsu has been increasing its share of the business, having been awarded a number of single-train 500-ton-per-day plants and a 1000-ton-per-day plant in the Soviet Union which is to be designed with some single-train sections, although reactors of this size are still considered unfeasible.

The various processes differ very little. Ammonia, CO_2 and recycle solution are fed to the reactor at pressures around 3500 psi and temperatures of about 400° F. A large ammonia excess is used to improve conversion and reduce corrosion. The heat of absorption of NH_3 and CO_2 in water is used at present to provide the heat required to decompose the carbamate. Water formed in the reaction is removed from the system either by crystallizing out the biuret (an undesirable condensation product formed from two mols of urea) or by evaporation. Evaporation can reduce the biuret content only to around 0.6 percent, since at the temperatures used to evaporate the water some of it necessarily forms. Falling-film evaporators have been developed to reduce the time necessary for concentrating urea solution of 80-90 percent to the 99.5 percent required for prilling. Often, evaporation is carried out in two stages — the first to 90 percent in a conventional (and cheaper) evaporator, the second in one of the falling-film type.

The Toyo process owes part of its success to the technique of eliminating biuret by crystallization instead of attempting to avoid its formation. Vacuum crystallization and centrifuging produce urea containing less than 0.3 percent biuret. The rest of the biuret remains in the mother-liquors, which are returned to the reactor where operating conditions favor its reconversion to urea. The additional equipment is said to increase the investment requirement by 6 percent compared with other solution-recycle processes. Thus, the cost of urea by the Toyo process would be $3.00 per ton higher compared to the other two. Used principally as a fertilizer, urea is finally prilled in a tower where drops of molten urea are cooled against a rising air current. Chemico has also introduced crystallization in its newest plants.

A recent development of the Stamicarbon process replaces the primary decomposer by a stripper in which the net CO_2 fed to the system decomposes ammonium carbamate in the reactor effluent to NH_3 and CO_2, the overhead being returned to the reactor. Both reactor and stripper operate at 2000 psi, with savings in equipment and utility costs reported to be 10 and 35 percent, respectively.

In 1965, SNAM Progetti improved the three processes by decomposing ammonium carbamate in the presence of ammonia. The decomposition pressure was raised appreciably, thereby permitting the absorption of unreacted feedstocks at higher temperatures. Steam is thus provided for concentrating the final urea solution. Decomposition in the presence of ammonia also lowers biuret formation and inhibits corrosion. Therefore commercial stainless steel can be recommended as a material of construction. These modifications improve both steam consumption and investment requirements. Also, centrifugal pumps may be used to replace the expensive reciprocating pumps used in the processes that decompose ammonium carbamate at low pressures. Power costs are also reduced, but not significantly since the major item, CO_2 compression, remains unchanged.

Figure 2-6 describes a possible flowsheet for producing urea prills.

The cheapest way to make urea is in a once-through plant. However, this is not necessarily the most economical means of transforming a given amount of ammonia into solid fertilizer. A once-through urea plant requires 40 percent less investment and produces urea at a cost 13 percent below that of a total-recycle plant of the same capacity. Nevertheless, the overwhelming majority of all plants built since 1960 have been total-recycle. A 300-ton-per-day Stamicarbon plant requires an investment of $3.5 million. With the SNAM plant, the cost has been estimated at $2.75 million. Production costs can be estimated as follows:

	Process	
	Stamicarbon	SNAM
	(Cost, $/ton urea)	
Ammonia ($50/ton)	29.0	29.0
Carbon dioxide ($10/ton)	7.5	7.5
Utilities	5.0	3.5
Labor and overhead	4.0	4.0
Capital charges	11.5	9.0
	57.0	53.0

Urea production in 1964 was 2420 million pounds and distributed as follows:

End-uses	MM pounds, 1964
Nitrogen solutions	870.0
Solid fertilizers	1090.0
Animal feeds	240.0
Industrial chemicals	220.0
	2420.0

Source: United States Tariff Commission.

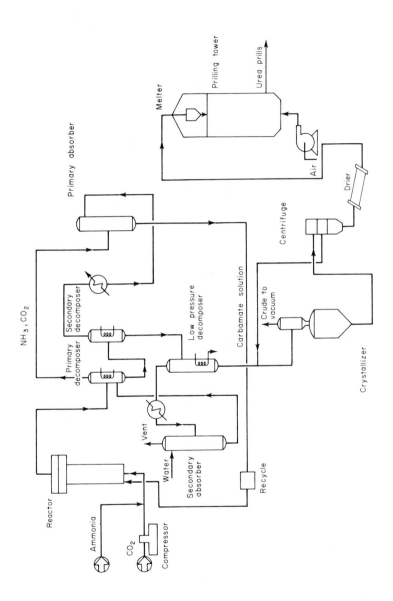

Fig. 2-6 Urea (Total Recycle).

List of Principal and Announced Urea Plants as of Mid-1964

Company	Location	Capacity in 1000 tons/yr	
		Jan. 1965	Jan. 1966
Allied Chemical	La Platte, Neb.	115	115
	South Point, Ohio	115	115
Armour	Cherokee, Ala.	18	18
Best Fertilizers	Lathrop, Calif.	*	*
Columbia Nitrogen	Augusta, Ga.	25	25
Cooperative Farm Chemicals	Lawrence, Kan.	33	33
John Deere Chemicals	Pryor, Okla.	165	165
DuPont	Belle, W. Va.	205	205
Escambia Chemical	Pensacola, Fla.	20	20
Farmers Chemical Assn.	Tyner, Tenn.	–	35
W. R. Grace	Woodstock, Tenn.	120	120
Gulf Oil	Henderson, Ky.	36	36
	Vicksburg, Miss.	15	15
Hawkeye Chemical	Clinton, Iowa	25	25
Hercules Powder	Hercules, Calif.	40	40
	Louisiana, Mo.	40	40
Ketona Chemical	Ketona, Ala.	7	7
Lone Star Producing Co.	Kerens, Texas	80	80
Mississippi Chemical	Yazoo City, Miss.	36	36
Monsanto Company	El Dorado, Ark.	36	36
Nitrin Inc.	Cordova, Ill.	24	24
Olin Mathieson	Lake Charles, La.	–	140
Phillips Pacific Chemical	Kennewick, Wash.	10	70*
Premier Petrochemical	Pasadena, Texas	70	70
Shell Chemical	Ventura, Calif.	88	88
Solar Nitrogen Chemical	Joplin, Mo.	52	52
	Lima, Ohio	60	60
Southern Nitrogen	Savannah, Ga.	22	50
Sun-Olin Chemical	Claymont, Del.	73	73

*Capacity uncertain.
Source: *Chemical Week*, April 25, 1964.

Urea is used as a feed supplement to the extent of around eight pounds per year per head of cattle. In liquid fertilizers, urea raises the upper limit for stable N solutions. In solids, it prevents caking due to the formation of ammonium chloride in mixtures of other nitrogen fertilizers and potassium chloride. "Ureaforms," condensates of urea and formaldehyde, are a special type of urea-derived fertilizer. They are employed when it is desired to release urea at a controlled uniform rate, for example, on golf-courses. The main use of solid, prilled urea is as side-dressing.

Urea-formaldehyde resins

Urea reacts with formaldehyde to form dimethylol-urea, which can undergo polymerization by loss of water:

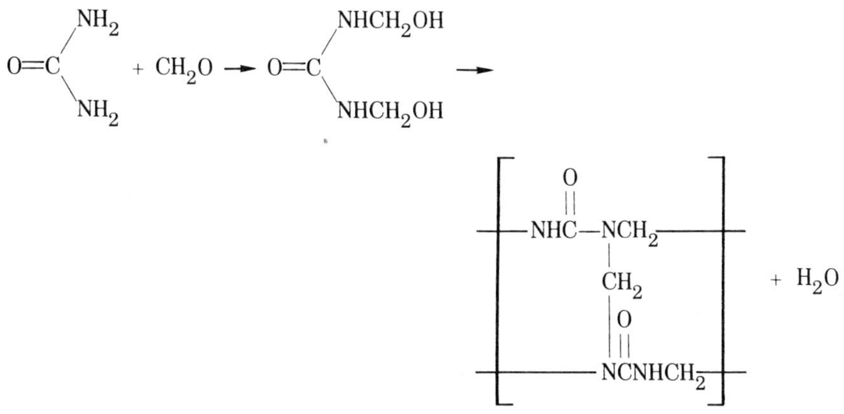

$$
O\!=\!C\Big\langle {{}^{NH_2}_{NH_2}} \quad + \ CH_2O \ \longrightarrow \ O\!=\!C\Big\langle {{}^{NHCH_2OH}_{NHCH_2OH}} \quad \longrightarrow
$$

$$
\left[\begin{array}{c} O \\ \| \\ -NHC\!-\!NCH_2\!- \\ | \\ CH_2 \\ | \\ O \\ \|\| \\ NCNHCH_2 \end{array} \right] \ + \ H_2O
$$

Urea-formaldehyde resins constitute about 62 percent of the total production of amino-resins in the U. S. The remainder is accounted for largely by melamine-formaldehyde resins, with minor amounts of dimethylhydantoin, aniline and other amines reacted with aldehydes in a similar manner.

By far the most important application of urea-formaldehyde resins is in the manufacture of plywood. As these resins are easily hydrolysed, their use is limited to interior-grade plywood, where they compete with protein glues. However, due to the drop in price of phenol, the tendency has been to use phenolic resins in interior as well as exterior grades of plywood, especially in products made from softwoods. Interior plywoods represent 30 percent of the 11.0 billion square feet of plywood made in 1964.

Particle or chip board is made from wood chips, bonded by urea or phenol-formaldehyde adhesives. It is 10 percent cheaper than plywood and is being employed increasingly in such indoor applications as furniture, counters and flooring support. In 1965, about 0.9 million square feet of particle board was made. But this product requires eight percent by weight of resin, against 2 percent for plywood. Furniture accounts for 50 percent of the market for particle board and indoor construction applications roughly around 30 percent.

Furfuryl alcohol-modified urea resins, advantageous mainly in that they permit rapid core production, are used to bind foundry cores formed in preheated pattern boxes as well as for self-curing binders using phosphoric acid as a curing catalyst. Their chief drawback is their tendency to form pin-holes in the casting due to excessive gas evolution.

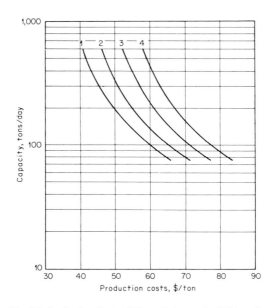

Fig. 2-7 Production Costs of Urea. 1-Ammonia=$30/ton; 2-ammonia=$40/ton; 3-ammonia=$50/ton; 4-ammonia=$60/ton.

Urea-formaldehyde molding powders are used primarily where particularly short molding cycles, and therefore low unit costs, are desired, e.g., wiring devices and plastic closures.

A total of 55.0 million pounds per year of all resins (primarily amino-resins) are used to impart wet-strength to paper. They require curing and thus make waste reclaiming difficult. These resins also improve paper sheet formation, and prevent accumulation of static electricity.

During 1965, a total of 72.0 million pounds per year of textile crease-proofing chemicals were used in the U. S. These resins have enabled cotton to maintain its status in the apparel market against competition from synthetic fibers. Approximately half of all cotton is treated with some kind of crease-proofing finish, which constitutes 4-5 percent by weight of the fiber. The use of urea- and melamine-formaldehyde resins in this application is limited to colored goods. This is due to the low-chlorine-resistance of the -NH groups, which eventually produces yellowing.

For white goods, e.g., men's shirts, more expensive resins are used. Also, there are indications that the overall share of urea-formaldehyde compounds is declining in favor of ethylene diamine derivatives of urea, e.g., ethylene urea. A breakdown of the demand for textile crease-proofing is as follows:

Type of finish	MM lbs/yr
Urea-formaldehyde	18.0
Melamine-formaldehyde	8.0
Triazone	6.0
Triazine	10.0
Carbamate	4.0
Permanent-crease	14.0
Dimethylol-ethylene-urea	12.0

In protective coatings, amino resins are used to modify alkyds by improving surface hardness and reducing baking times. The proportion of amino resins varies with the particular application, from around 10 percent in automobile finishes to 35 percent in washing-machine coatings. The most important feature of alkyds modified by amino resins is their resistance to yellowing. Therefore, they are employed primarily in white finishes, for example, for household utilities — refrigerators, washing-machines, kitchen ranges, etc. Melamine resins are much more resistant to hydrolysis and thus are preferred for such applications as washing machines. To make them compatible with alkyds as well as soluble in aromatic diluents (which are much cheaper than polar solvents), amino resins are etherified with butanol or isobutanol.

$$\left[\begin{array}{l} -NCH_2OR \\ \quad | \\ \quad C{=}O \\ \quad | \\ HN{-}CH_2{-}N{-} \\ \qquad\qquad | \\ \qquad\qquad CH_2OR \end{array} \right]$$

In 1964, consumption of urea-formaldehyde resins was 368.0 million pounds. The breakdown is given in the following.

End-uses	MM lbs, 1964
Plywood	110.0
Particle-board	95.0
Other bonding uses	5.0
Closures	19.0
Wiring devices	21.0
Other molding applications	15.0
Textile resine	18.0
Paper resins	39.0
Coatings	13.0
Urea-formaldehyde-furfuryl alcohol foundry resins	23.0
Miscellaneous	10.0
	368.0

The principal U. S. producers of amino resins follow:

Producer	Capacity MM lbs/yr
Allied	60.0
American Cyanamid	150.0
Borden	90.0
Brown	20.0
Catalin	20.0
Hercules	15.0
Monsanto	50.0
Perkins Glue	25.0
Reichhold	50.0
Rohm & Haas	20.0
Scott Paper	15.0

Sulfamic acid

Sulfamic acid is made by the action of oleum on urea:

$$NH_2CONH_2 + H_2SO_4 + SO_3 \longrightarrow 2NH_2SO_3H + CO_2$$

The ammonium or amine salts of sulfamic acid are of considerable importance as a fire-retardant for paper and synthetic fibers. These salts are also used to kill brush and in equipment and industrial cleaning formulations. A major paper producer in the Pacific Northwest has begun to add sulfamic acid to calcium hypochlorite during bleaching. In many reactions, sulfamic acid can be used in place of chlorosulfonic acid, which is cheaper but has the disadvantage of giving off HCl. Sulfamic acid gives off ammonia, thus reducing corrosion and therefore equipment costs.

DuPont is the only producer of sulfamic acid in the U.S.; consumption was estimated at 60.0 million pounds in 1964.

Melamine

Although melamine is still predominantly a derivative of cheap electric power, it is well on the way to becoming a petrochemical. In the traditional route, calcium carbide is converted to calcium cyanamide and then successively to cyanamide, dicyandiamide and melamine.

$$CaC_2 + N_2 \longrightarrow CaCN_2 \xrightarrow{H_2O} Ca(HCN_2)_2$$

$$Ca(HCN_2)_2 + H_2O + CO_2 \longrightarrow 2NCNH_2 + CaCO_3$$

$$2NCNH_2 \longrightarrow NH_2C{-}NH{-}CN$$
$$\underset{NH}{\overset{\|}{}}$$

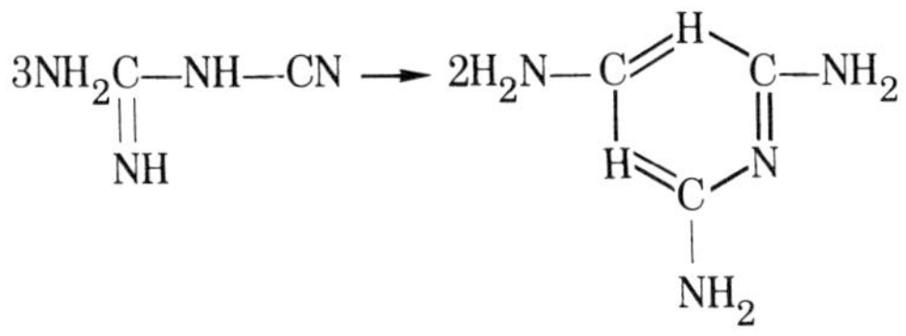

The final reaction is exothermic. In the original Cyanamid process, this heat was removed by carrying out the reaction in the presence of methanol. Conversion is carried out at high pressures to inhibit formation of deamination products. The French SPA process employs dry trimerization in the presence of ammonia.

The recent trend, both in the U.S. (Allied Chemical) and elsewhere, is to start from urea:

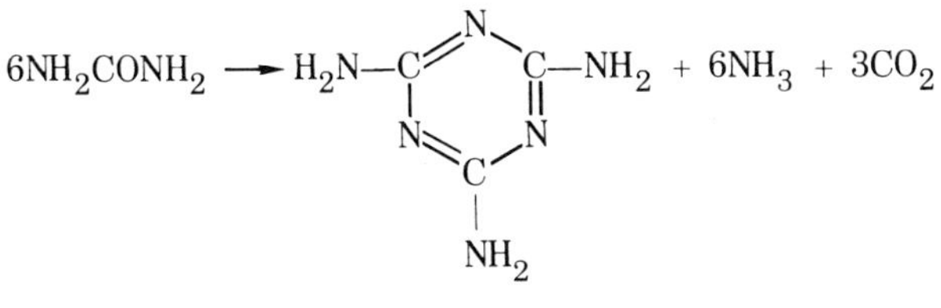

Due to equilibrium considerations, the reaction must be carried out at a high temperature. But high temperatures favor the decomposition of urea to ammonia and CO_2. Thus these processes also operate at high pressures and in the presence of excess ammonia. Both BASF and Oesterreichische Stickstoffwerke (OSW) have announced low-pressure routes. In the OSW process the reaction is carried out in two stages:

$$CO(NH_2)_2 \longrightarrow HNCO + NH_3$$

$$6HNCO \longrightarrow C_3H_6N_6 + 3CO_2$$

The second reaction is carried out catalytically at above the sublimation temperature of melamine. The product is recovered by quenching and centrifuging. Yields on urea are around 90 percent.

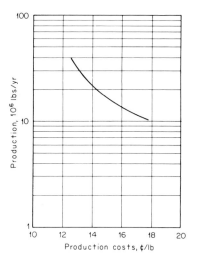

Fig. 2-8 Production of Melamine.

Since NH_3 and CO_2, as well as solutions containing ammonium carbonate and carbamate, appear as byproducts at various points, plants making melamine by one of these routes are located necessarily in the vicinity of a urea plant, to which these streams are recycled.

Finally, there has been considerable interest in processes starting from HCN, especially since the Sohio acrylonitrile route, which yields HCN as a byproduct, has become established all over the world.

The process used by Standard Oil of Ohio involves electrolytic conversion of HCN to cyanogen bromide; this is converted by ammonolysis to cyanamide, which trimerizes to melamine.

$$HCN + Br_2 \longrightarrow BrCN + HBr$$

$$BrCN + NH_3 \longrightarrow NH_2CN + HBr$$

$$3NH_2CN \longrightarrow C_3H_6N_6$$

By 1965, of the nonelectrochemical routes to melamine, only those starting from urea had found commercial application.

From the literature (which is scarce), the investment for a 20.0 million pounds-per-year melamine plant can be estimated at $2.8 million, with production costs as shown in the following.

	Cost, ¢/lb Melamine
Urea (at $60/ton)..............	10.0
Less: Ammonia ($50/ton).......	(-2.5)
CO$_2$ ($10/ton)...........	(-0.5)
Net raw materials	7.0
Labor and overhead...........	1.2
Utilities (estimate)...........	1.5
Capital charges..............	4.6
	14.3

Melamine is used primarily in the manufacture of amino resins. The reaction is analogous to that between urea and formaldehyde. Melamine resins steadily increased their share of the total amino-resins consumption from 18 percent in 1950 to 33 percent in 1964. They are more expensive than urea resins, but are generally more resistant to heat and hydrolysis.

Their principal use is as molding powders, which contain around 35 percent α-cellulose and 65 percent resin. Almost 90 percent of these molding powders are used to make dinnerware, mainly for households (85 percent), but lately also in institutional applications. Melamine dinnerware can be made sufficiently decorative for a large segment of the consumer market; the problem of its susceptibility to staining by coffee and tea has been overcome by substituting benzoguanamine for part of the melamine. Another important use is as a top layer in phenolic decorative laminates; melamine-formaldehyde has better physical properties than phenolics, and it can be colored, which phenolics cannot. In coatings, melamine resins are preferred to urea when color retention at high temperature is sought, for example, on stoves, or resistance to hydrolysis, or on washing machines. The demand for melamine resins in 1965 was 170.0 million pounds (equivalent to 95.0 million of melamine crystal), distributed as follows.

End use	MM *lbs 1964*
Molding powders	50.0
Textile resins	24.0
Laminates	70.0
Paper wet-strength resins	12.0
Adhesives	14.0
Coatings	10.0
	180.0

Chlorinated melamine is used in industrial cleaning formulations for dairy and brewery equipment, and as a crosslinking agent for olefin elastomers such as ethylene-propylene and ethylene-butylene rubbers.

Methylolmelamine is used as a synthetic tanning agent for white leather.

The U.S. manufacturers of melamine are listed below. All but Allied Chemical start from dicyandiamide. Much "dicy" is made in Canada and other cheap power regions, and transported to melamine plants located near the large urban concentrations.

Producers	Location	Capacity, MM lbs/yr
Allied Chemical	South Point, Ohio	20.0
American Cyanamid	Wallingford, Conn.	25.0
Fisher Chemicals	Willow Island, W. Va.	75.0
Monsanto	Everett, Mass.	10.0
Reichhold	Carteret, N. J.	20.0

2.4   Nitric Acid

Strictly speaking, nitric acid is not a petrochemical. No part of the hydrocarbon molecule remains in the nitric acid. Yet it is a large user of ammonia.

Nitric acid is made by burning ammonia in air over a platinum-rhodium gauze, oxidizing the resulting nitric oxide to nitrogen dioxide, and absorbing the dioxide in water:

$$4NH_3 + 5O_2 \rightarrow 4NO + 6H_2O$$

$$2NO + O_2 \rightarrow 2NO_2$$

$$3NO_2 + H_2O \rightarrow 2HNO_3 + NO$$

The nitric oxide formed in the last reaction is recycled, thus all the NO is finally converted to nitric acid.

Two process variables (oxidation and absorption pressures) account for the main differences between the various nitric acid processes.   High oxidation pressures drive the reaction equilibrium to the left, so five or six times as many catalyst layers must be used as at atmospheric pressure; thus the contact time is increased, which favors the decomposition of both ammonia and nitric oxide to nitrogen and lowers yields by as much as 6 percent.   Catalyst losses at high pressures are 5 to 10 times as heavy as at atmospheric oxidation, and gauze changes are much more frequent.   In addition, high pressures require higher oxidation temperatures.

Absorption of nitric oxide in water occurs very rapidly, but the rate of reoxidation of NO to NO_2 is proportional to the square of the absorption pressure.   Therefore it is advantageous to absorb at high pressure; for example, at 3 atmos., for a given capacity the total volume of the absorption section must be around 5 times larger than at 8 atmos.

Some processes, notably those offered by the European engineering companies, use atmospheric oxidation followed by compression up to the desired absorption pressure. The main inconvenience is that the stainless-steel compressor required for handling nitrogen oxides is very costly.   In general, the various processes represent compromises between the advantages of high- and low-pressure oxidation and absorption.

In Europe, low-pressure oxidation is preferred since it maximizes HNO_3 yields in a region where ammonia is relatively expensive.   In the U.S. where ammonia is cheap, the favorable effect of high pressures on capital investment seems to outweigh that of lower yield.   Thus, the main low-pressure-oxidation processes are European: Uhde, Kuhlmann, Stamicarbon, Pechiney-St. Gobain. The high-pressure processes originate in the U.S.: Chemico, DuPont, Girdler.

Figure 2-9 shows a flow-sheet of the Kuhlmann process, one of the most successful low-pressure oxidation—medium-pressure absorption systems. Ordinarily, this combination involves the highest investment costs, but considerable savings are achieved in the absorption section by means of a special tower design. Oxidation from NO to NO_2 can be carried out in the presence of nitric acid solution.   Cooling water is the major source of utility charges, and these are kept to a minimum by expanding the liquid ammonia feed in the absorption column. This technique also permits production of 75 percent nitric acid, higher than any other process.

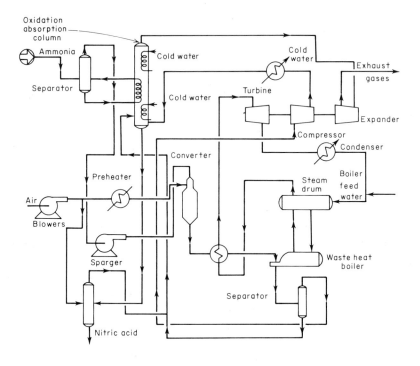

Fig. 2-9 Nitric Acid (Kuhlmann Process).

A summary of the relationships between the economics of the various schemes is given in the following table.

For the more concentrated nitric acid required for most nitration processes, the azeotrope can be overcome by means of desiccants such as sulfuric acid or magnesium nitrate. These remove enough water to cross the constant boiling-point line. Sulfuric acid is preferred for large plants, and for reconcentrating nitric acid contaminated with sulfate ions, which precludes the use of magnesium nitrate. A German process for making 99 percent HNO_3 crosses the azeotrope by a two-tower distillation system, the second tower under vacuum.

Since the current preference both here and abroad is for medium-pressure oxidation and absorption, economics for a plant using such a process to turn out 55-60 percent acid are evaluated below. Investment for a 300 tons-per-day plant (100 percent basis) is $3.1 million, and production costs are as follows.

Comparison between Different Nitric Acid Process Alternatives
Basis: 300 tons/day nitric acid

Case	Pressure, atmospheres		Investment index	Yields	Catalyst loss $/ton HNO$_3$	Capital charges $/ton HNO$_3$	Ammonia $/ton HNO$_3$	Ammonia = $50/ton	
	Oxidation	Absorption						Typical net utility costs $/ton HNO$_3$	Production costs $\triangle$ $/ton HNO$_3$
1	8	8	100	92%	2.20	9.50	14.60	1.10	+1.60
2	4	4	107	95%	0.80	10.20	14.20	0.60	0.00
3	0	0	112	97%	0.40	10.70	13.90	1.50	+0.70
4	0	4	125	97%	0.40	11.90	13.90	(−0.20)	+0.20

$/ton (100%)

Ammonia (at $50/ton)	14.20
Catalyst and chemicals	0.60
Net utility costs	0.60
Labor and overhead.	2.40
Capital charges	10.20
	28.00

Most nitric acid plants in the U.S. are located at the same site as the ammonia plant supplying the feed. Nevertheless several producers have found it profitable to produce ammonia and nitric acid at different locations thus taking advantage of the reduction in transportation costs; a ton of 57 percent nitric acid requires only 0.17 tons of ammonia.

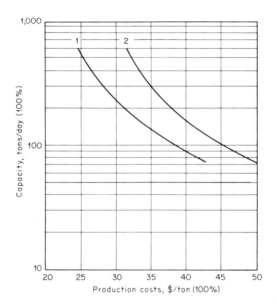

Fig. 2-10 Nitric Acid. 1-Captive ammonia, $50/ton; 2-Merchant ammonia, $75/ton.

Nitric acid is used principally in the manufacture of ammonium nitrate, which is used mainly as a fertilizer. However, ammonium nitrate is also the most important of all commercial explosives. In fertilizer formulations, ammonium nitrate provides a reasonably concentrated form of nitrogen (35 percent). Half of this nitrogen in nitrate form, which, although easier to leach from the soil than ammoniacal nitrogen, is preferred in acid soils. Nitric acid has also been used, in the U.S. but mainly in countries poor in sulfur, to solubilize phosphate rock. In the chemical industry, nitric acid performs two main functions: nitration, such as in manufacturing TNT, aniline, TDI, nitrochlorobenzenes, or nitroparaffins; and oxidation, in the manufacture of adipic acid, methyl methacrylate (Escambia process) or DMT (DuPont process). In 1965, approximately 5,150 million pounds (100 percent basis) of nitric acid was consumed in the U.S. The distribution is as follows.

End-use	MM lbs (100%) 1964
Ammonium nitrate	3300
Nitration	500
Oxidation	800
Other uses	550
	5150

Ammonium nitrate

Ammonium nitrate is used in various forms. As a solid it is usually sold in the form of prills coated with diatomaceous earth, or mixed with lime ("nitrolime") to reduce the explosion hazard. Together with ammonia and urea, it also goes into nitrogen solutions, which under pressure can contain up to 45 percent N.

In one common process, 57 percent HNO_3 is neutralized in one or two stages, heat of neutralization being sufficient to evaporate most of the water brought into the system by nitric acid. Low-pressure steam thus generated can be used to evaporate the liquid ammonia feed. The result of this neutralization is 83 percent ammonium nitrate, which can be used as such or sent on to evaporation and prilling. A flowsheet is shown in Fig. 2-11.

The demand for ammonium nitrate reached 4,200 million pounds in 1965, with the following distribution.

End-use	MM pounds 1965
Solid fertilizer	3120
Nitrogen solutions	550
Explosives	510
Others	20
	4200

Ammonium nitrate explosives, mainly in the form of a stable mixture with 5 percent fuel oil which can be detonated with TNT, has been steadily displacing dynamite as the most important commercial explosive. Although it accounts for about 82 percent by weight of all explosives consumed in the U.S., in terms of value ammonium nitrate represents only 55 percent of the explosives market. Its toxic products preclude its use in pit mining. Many dry-hole strip-mining operations run their own nitrate-fuel-oil-mixing plants to ensure a more homogeneous product and thus reduce the number of misfirings. Wet-hole strip-mining operations employ explosive slurries containing 15 percent water, 65 percent ammonium nitrate, and 20 percent TNT as a sensitizer.

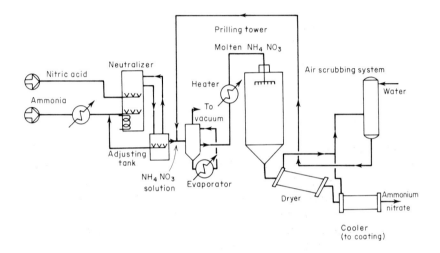

Fig. 2-11 Ammonium Nitrate.

The U.S. explosives market can be broken down as follows.

End-use	% of total (by wt)
Coal mining	28
Metal mining	16
Quarrying and nonmetals mining	27
Highway construction	23
Seismographic applications	6

The investment for a 225,000 tons-per-year ammonium nitrate plant is $3.8 million with production costs as follows (for a 33.5 percent N product).

	Cost, $/ton, ammonium nitrate
Nitric acid (at $28/ton)	21.10
Ammonia (at $50/ton)	10.20
Utilities	2.30
Labor and overhead	1.10
Capital charges	5.60
	40.30

This is equivalent to $120 per ton of nitrogen.

Ammonium nitrate is used also in the manufacture of nitrous oxide, well known as a general anesthetic popularly known as "laughing gas."

$$NH_4NO_3 \xrightarrow{\Delta} N_2O + 2H_2O$$

At present, most nitrous oxide is made by catalytic oxidation of ammonia:

$$2NH_3 + 2O_2 \longrightarrow N_2O + 3H_2O$$

In 1964, about 15.0 million pounds per year had been used, mainly as an aerosol propellent but also as an anesthetic.

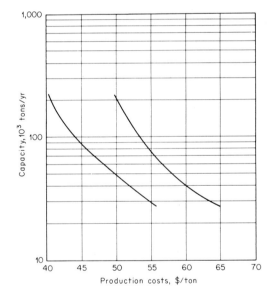

Fig. 2-12 Ammonium Nitrate. 1-Captive ammonia, $50/ton; 2-Merchant ammonia, $75/ton.

Potassium nitrate

Potassium nitrate is made from nitric acid and potassium chloride by the overall reaction:

$$4HNO_3 + 2KCl \longrightarrow Cl_2 + 2KNO_3 + 2NO_2 + 2H_2O$$

The process consists of reacting dilute nitric acid with potash, which produces nitrosyl chloride and chlorine.

$$\text{(dil.)}$$
$$3KCl + 4HNO_3 \longrightarrow NOCl + Cl_2 + 3KNO_3 + 2H_2O$$

This is then oxidized with concentrated nitric acid to chlorine and nitrogen dioxide, which is recycled to the nitric facilities:

$$\text{(concen.)}$$
$$4HNO_3 + 2NOCl \longrightarrow Cl_2 + 6NO_2 + 2H_2O$$

Concentrated nitric acid is produced by evaporating 80 percent acid from a solution of KNO_3 in dilute nitric acid. The presence of the potassium salts shifts the H_2O-HNO_3 azeotrope from 76 percent to 85 percent. The center of the process is a specially conceived column reactor, from which NO_2 and chlorine are removed as overhead, while KNO_3 dissolved in nitric acid is obtained as bottoms. This solution is evaporated to remove water formed in the process, and the product is finally crystallized out from the concentrated nitric acid, which is returned to the main reactor.

The only manufacturer of potassium nitrate, a binary fertilizer material, is Southwest Potash Corporation, with a 60,000 tons per year plant at Vicksburg, Mississippi.

Nitroparaffins

Propane reacts with nitric acid at high temperatures to give a mixture of nitromethane, nitroethane and nitropropanes in proportions that vary with reaction temperature. Higher temperature levels favor the formation of the lower homologs.

$$RH + HNO_3 \longrightarrow RNO_2 + H_2O$$

The reaction is carried out between an excess of propane and 75 percent nitric acid, the water contained in the nitric acid being evaporated by the heat of reaction. Yield on nitric acid is around 30 percent. Commercial Solvents operates two nitroparaffin plants (at Peoria, Ill., and Sterlington, La.), having a combined capacity of 20.0 million pounds per year.

Nitroparaffins have a great variety of applications, both as such and as intermediates. They are solvents for such specialized applications as Friedel Crafts reactions, epoxy resins, vinyl spray formulations, and certain types of cellulose acetobutyrates. Other uses are as lube-oil additives and as retarders in rubber processing. Nitromethane is an aerosol propellant.

As intermediates, nitroparaffins have numerous applications. They can be hydrolyzed to hydroxylamine sulfate and a carboxylic acid. This reaction is usually carried out with 1-nitropropane.

$$CH_3CH_2CH_2NO_2 + H_2SO_4 \xrightarrow{+H_2O} CH_3CH_2COOH + HONH_2 \cdot H_2SO_4$$

Hydroxylamine is used mainly to make oximes, such as methyl ethyl ketoxime.

Nitroparaffins react with formaldehyde to give the following type of products:

$$RCH_2NO_2 + CH_2O \rightarrow \begin{matrix} RCHNO_2 \\ | \\ CH_2OH \end{matrix}$$

$$\begin{matrix} RCHNO_2 \\ | \\ CH_2OH \end{matrix} \xrightarrow{H_2} \begin{matrix} R{-}CHNH_2 \\ | \\ CH_2OH \end{matrix}$$

These can be reduced to the respective alkanolamines, which find applications in the surfactant field similar to those of mono-ethanolamine. They are also intermediates for certain insecticides used mainly on vegetables and in household products, having the formula:

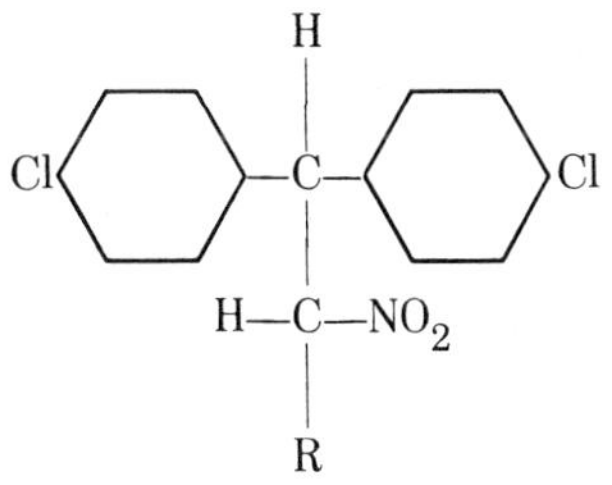

Aminomethylolpropane (AMP) is also a viscosity stabilizer for alkyd resins.

Nitroethane ,is the starting point for making a group of nitro-plasticizers for military applications.

2.5 Ammonium Phosphates

Since its introduction in the mid-1950's, the growth in the use of ammonium phosphates has been spectacular. They are made by reacting ammonia and phosphoric acid:

$$NH_3 + H_2PO_4 \rightarrow NH_4H_2PO_4$$

$$2NH_3 + H_3PO_4 \rightarrow (NH_4)_2HPO_4$$

At first, monoammonium phosphate (MAP) was preferred because of the ammonia losses experienced when making diammonium phosphate (DAP), which required an expensive recovery system. Also, the P_2O_5 source was limited to dry-process phosphoric acid, as the iron and aluminum present in wet-process acid tended to form gummy precipitates.   Later, DAP began to be made from wet-process acid previously subjected to purification, by precipitation and filtration of Fe and Al phosphates. The most recent method, developed by TVA, consists of carrying out the reaction in two steps.   Preneutralization gives a partially neutralized slurry filtration of Fe and Al phosphates. The most recent method, developed by TVA, consists of carrying out the reaction in two steps. Preneutralization gives a partially neutralized slurry for subsequent reaction in a TVA drum ammoniator.   The ammoniated product is dried, screened, milled and cooled in the same way as other granulated fertilizers. The Dorr process consists of three reactors in series, and gives a 20:80 mixture of MAP and DAP. Some steel plants have switched to phosphoric acid for absorbing the ammonia contained in their coke-oven gases, thereby becoming producers of ammonium phosphates. The usual analysis of DAP is 16-48-0 or 18-46-0 (theoretical would be 21-53-0) which amounts to 64 percent nutrients — more than any other solid fertilizer material.

The U.S. manufacturers of ammonium phosphates are listed below.

Producer	Location	Capacity, thousands of short tons	Remarks
AFC	Edison, Calif.	40	
American Cyanamid	Bradley, Fla.	120	Adding 200

Arkla Chemical Corp.	Helena, Ark.	—	100 in 1966
Armour	Cherokee, Ala.	180	
	Bartow, Fla.	30	
Borden	Texas City, Texas	52	
(Smith Douglass)	Streator, Ill.	12	
Central Farmers			
(El Paso Natural Gas)	Soda Springs, Ida.	80	
Cities Service			
(Tennessee Corp.)	Tampa, Fla.	225	
Coastal Chemical	Pascagoula, Miss.	190	Adding 265 in 1965
Colorado Fuel & Iron	Pueblo, Colo.	14	
Consumers Cooperative Assn.	Pierce, Fla.	100	
Continental Oil			
(American Agricultural Chemicals)	Pierce, Fla.	175	
Deere	Tulsa, Okla.	75	
Farmers Chemical	Joplin, Mo.	120	
Ford Motor Company	Dearborn, Mich.	20	
W. R. Grace	Bartow, Fla.	100	
(Davison Div.)	Ridgewood, Fla.	—	200 in 1965
Hooker Chemical	Marseilles, Ill.	160	
(National Phosphates)	Taft, La.	—	350 in 1965
International Minerals & Chemicals	Bonnie, Fla.	500	
Kaiser Steel	Fontana, Calif.	15	
Lone Star Producing	Kerens, Texas	100	
Mobil Oil			Adding 40 in 1965
(Virginia-Carolina)	Nichols, Fla.	60	
Monsanto	Luling, La.	—	240 in 1965
Northwest Cooperative Mills	St. Paul, Minn.	90	
Occidental Petroleum	Lathrop, Calif.	40	
(Best Fertilizers)	Plainview, Texas	20	
Olin-Mathieson	Pasadena, Texas	440	
	Joliet, Ill.	170	
Shell Chemical	Pittsburg, Calif.	15	
Simplot	Pocatello, Idaho	75	
Southwestern Agricultural	Chandler, Ariz.	40	
Stauffer Chemical			
(Western Phosphates)	Garfield, Utah	80	
(Bunker Hill)	Kellogg, Ida.	50	
(Victor)	Chicago Heights, Ill.	20	Chiefly nonfertilizer
(Domingo)	Long Beach, Calif.	30	
(with Phillips Petrol.)	Pasadena, Texas	100	
Swift	Harvey, La.	140	
	Agricola, Fla.	100	
Tennessee Valley Authority	Muscle Shoals, Ala.	40	
U. S. I. Chemicals	Danville, Ill.	30	
Valley Nitrogen Producers	Helm, Calif.	100	

Source: *Chemical & Engineering News*, February 1, 1965.

A 120,000 tons-per-year plant requires an investment of $3.2 million.

2.6 Ammonium Sulfate

A substantial portion of the ammonium sulfate is produced in the U. S. from coke-oven gas purification. This supply is lagging the rest of the economy, as has been the case with all coal-tar derivatives. It is also a byproduct from caprolactam manufacture, to the extent of 4.5 pounds ammonium sulfate per pound of product. Some synthetic $(NH_4)_2SO_4$ is still made by neutralizing sulfuric acid with ammonia, followed by continuous crystallization, centrifuging and drying.

Approximately 975,000 tons are produced per year, and the prospect of an increase is unlikely due to the low (20.5 percent) N content. It will always be required as a source of sulfur, which is essential to the formation of plant proteins and certain amino acids.

2.7 Ammonium Chloride

As a fertilizer, ammonium chloride is suited especially well for use in irrigation systems. In Japan its use is most widespread. Much of the production of NH_4Cl in Japan is tied to soda ash. Due to the shortage of salt there, the incentive is obvious for finding a way to use the Cl part of the salt molecule instead of wasting it as calcium chloride. This has been accomplished by coupling Japan's Solvay plants to ammonia units, thereby replacing limestone as the source of CO_2 for carbonation and, instead of distilling off the ammonia recovered as NH_4Cl by means of calcium hydroxide (a by-product of the CO_2—generating lime kiln), the ammonium chloride is removed from the process as such and used as fertilizer. Hence it is considered possible to convert anhydrous ammonia to solid fertilizer for $35 per ton N.

In the U. S., 55.0 million pounds per year of ammonium chloride are consumed, with more than half going into dry-cell batteries. Pennsalt and Allied Chemical are the two principal manufacturers.

2.8 Hydrazine

Although hydrazine can be made from urea

$$NH_2CONH_2 + NaOCl \longrightarrow H_2NNH_2 + NaCl + CO_2$$

the most widely employed route (the Raschig process) involves indirect oxidation of ammonia:

$$NH_3 + NaOCl \longrightarrow NH_2Cl + NaOH$$
$$\text{chloramine}$$

$$NH_2Cl + NH_3 \longrightarrow H_2N\text{-}NH_2 + HCl$$
$$\text{hydrazine}$$

A large excess of ammonium hydroxide and fresh sodium hypochlorite are first reacted to form chloramine, which is then converted to crude hydrazine by rapid heating. The reactor effluent, containing about 2.5 percent of product, is stripped of unreacted ammonia. The hydrazine solution, which has by then been concentrated to 3 to 4 percent, is fed to an evaporator where sodium chloride is removed and hydrazine boiled off. Fractionation in a series of three columns gives a final product consisting of pure hydrazine hydrate, which can be converted to anhydrous hydrazine by extractive distillation; the yield on ammonia is 60 percent.

Bergbauforschung GmbH, in Germany, has developed a different kind of indirect ammonia oxidation process. Ammonia, chlorine and a ketone react to form a diazacyclopropane, which is hydrolyzed by sulfuric acid to hydrozine sulfate, with regeneration of the ketone.

$$Cl_2 + 4NH_3 + RCOR' \longrightarrow 2NH_4Cl + H_2O + \begin{array}{c} R \\ \diagdown C \diagup \diagdown NH \\ R' \diagdown NH \end{array}$$

$$H_2O + \begin{array}{c} R' \diagdown \diagup NH \\ C \\ R \diagdown NH \end{array} \xrightarrow{H_2SO_4} H_2N\text{-}NH_2 \cdot H_2SO_4 + RCOR'$$

The process is used commercially by Fisons in Great Britain. In addition to much higher yields (90 percent), the Bergbau process involves lower investments and much lower steam requirements than the Raschig route. The hydrolysis reactor effluent contains 7 percent hydrazine against 2.5 percent for the Raschig process, thus steam consumption is reduced by a factor of about three. Also the ketone from which the product must be separated has a lower latent heat of vaporization than water. Ketone losses are 0.32 pounds per pound of product when using methyl ethyl ketone and 0.1 pounds per pound for a C_5 ketone. Ammonium chloride is the main byproduct.

The principal use for hydrazine is as a military missile fuel where anhydrous hydrazine is required. In 1964, this accounted for over 16.0 million pounds. Hydrazine has, however, a number of interesting nonmilitary applications, which consume another 4.5 million pounds per year. The most important of these is maleic hydrazide, a plant-growth regulator used for tobacco suckering and tree pruning. Hydrazine and urea react to form semicarbazide:

$$H_2N\text{-}NH_2 + NH_2CONH_2 \longrightarrow NH_2NHCONH_2 + NH_3$$

which finds some use in the pharmaceutical field, e.g., in the manufacture of nitrofuran drugs, but it is more important as an intermediate for azodicarbonamide, obtained by oxidation:

$$2NH_2NHCONH_2 \xrightarrow{(O)} NH_2CON\text{=}NCONH_2 + 2NH_3$$

This compound, of which some 2.0 million pounds per year are used in the U. S., is used mainly as a blowing agent for making vinyl foams, in particular, calendered vinyl leather. It is also used for polyethylene foams, which are expected to increase in demand as packaging materials. Azodicarbonamide acts by releasing nitrogen, and produces a foam that is said to be more agreeable to the touch than that obtained from other blowing agents. Semicarbazide also finds some use in making plastic foams.

Other uses include wash-and-wear finishes, pharmaceuticals, anti-oxidants used in foam and hydrazine monobromide, and soldering flux. A breakdown of these applications for hydrazine is as follows:

	MM lbs 1964
Maleic hydrazide	1.8
Blowing agents	1.2
Boiler feed water treatment	0.5
Pharmaceuticals	0.4
Others	0.6
	4.5

A 3.0 million pounds-per-year hydrazine plant using the Raschig process was reported to cost $4.0 million. U. S. producers of hydrazine are given in the following.

Producer	Location	Capacity MM lbs/yr
Naugatuck (U. S. Rubber)	Geismar, La.	3.0
Fairmount	Newark, N. J.	1.0
Olin	Lake Charles, La.	2.5
	Saltville, Va.	24.0*
National Polychemicals	Wilmington, Mass.	0.5

*Plant operated for government by Olin for anhydrous hydrazine.

2.9   Sodium Nitrite

Although it is not a petrochemical, sodium nitrite is made by burning ammonia and absorbing the nitric oxide in a soda ash solution, under oxidizing conditions:

$$2NO + Na_2CO_3 \xrightarrow{(O)} 2NaNO_2 + CO_2$$

It is used in the dye industry to form diazonium chlorides from aromatic amines:

$$Ar-NH_2 + NaNO_2 + 2HCl \longrightarrow Ar-N\equiv N^+Cl^- + 2H_2O$$

This reaction is the basis for the manufacture of azo dyes, of which some 42.0 million pounds per year are consumed in the U. S. requiring 15.0 million pounds of sodium nitrite per year. Its most important use, however, is as a corrosion inhibitor for iron and steel in neutral or alkaline media, such as in paper, paint and metal cutting fluids. It is also added to crude oil to protect refinery equipment by dissolving in the water contained in the oil, and to automobile antifreeze.

Sodium nitrite is also used in the manufacture of synthetic rubber; as a component of fused heat-transfer salt mixtures; and in the meat-packing industry. The total U. S. consumption is around 100.0 million pounds per year.

2.10   Carbon Dioxide

Some companies obtain carbon dioxide from large fermentation plants, e.g., breweries. Others make it by complete combustion of a suitable fuel, followed by removal and purification of the carbon dioxide. In the U. S. and in Europe, CO_2 for soda-ash manufacture is obtained by decomposition of limestone.

A substantial proportion of the solid and bottled CO_2 produced in the U. S. comes from ammonia-plant off-gas; also, this is the source of the CO_2 used to make urea.

In addition to the uses of CO_2 as an intermediate (urea, salicylic acid, various bicarbonates), 1.06 million tons of CO_2 were consumed in the U. S. during 1964 as a service chemical, of which 60 percent was in the form of dry ice. This demand can be broken down as follows.

End-use	Thousand tons 1964
Refrigeration, blow-molding, food processing (bakery goods, meats), rubber trim-flashing	430.0
Fire extinguishers, vacuum-packing (coffee, cheese), inert atmosphere for welding	110.0
Water treatment, neutralization in general	100.0
Pressure transferring, aerosol propellent, purging	60.0
Carbonated beverages	220.0
Miscellaneous	140.0
	1060.0

The investment for a 35,000 tons-per-year CO_2 plant from ammonia-plant off-gas is \$2.0 million. Production costs are around \$29 per ton for liquid CO_2 and \$41 per ton for dry ice.

Four companies, each of which operates several plants in different parts of the country, account for 83 percent of the CO_2 market.

Producer	% of CO_2 market
General Dynamics	32
Air Reduction	26
Chemetron	18
Olin	7
Others	17

3   METHANOL

Methanol is produced by the exothermic reaction between hydrogen and carbon monoxide or dioxide over a copper or Cr-Zn catalyst at pressures between 4000 and 6500 psig, and temperatures in the 750° F range.

$$2H_2 + CO \longrightarrow CH_3OH$$

$$3H_2 + CO_2 \longrightarrow CH_3OH + H_2O$$

A flow-sheet is shown in Fig. 2-13. Feed gases are compressed up to the synthesis loop pressure and enter the reactor. The effluent is condensed and sent to a separator vessel, where unreacted gases, except for a purge stream, are removed and recycled to the reactor. The liquid is depressured to remove and vent condensed gases, and the resulting crude methanol goes to a fractionation system where light materials (2-3 percent, mainly methane

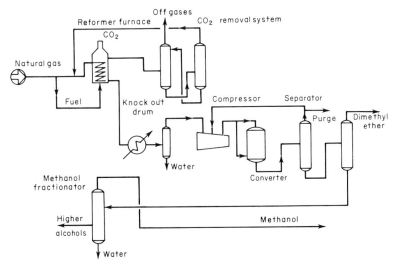

Fig. 2-13 Methanol.

and dimethyl ether) are first removed and sent to fuel. The stabilized stream is fractionated into pure methanol, dimethyl ether,
heavy alcohols (ethyl, isobutyl and higher) and waste water. The
process is sufficiently similar to that for making ammonia that
plants can be designed to make one or the other depending only on
the catalyst in the reactor. Materials of construction in the synthesis loop are subject to more restrictions than in the case of ammonia because the presence of oxygenated compounds increases
corrosion.

The investment for a methanol plant is about 20 percent higher
than for an ammonia plant of equal capacity. A 200-ton-per-day
methanol plant, including synthesis gas generation from natural
gas, requires an investment of $5.0 million. The production costs
are as follows.

Input	Cost, ¢/lb methanol
Raw materials, including fuel (30¢/10^6 scf)	0.6
Utilities (other than fuel)	0.7
Labor and overhead	0.2
Capital charges	1.2
	2.7

Since many plants operating in the U. S. are smaller than that in
the example given, captive methanol will be taken at 3.0 cents per

pound for the purposes of other economic evaluations. In the future, methanol costs may drop considerably as capacities grow, and as it becomes possible to design plants around turbine-driven centrifugal compressors just as in the case of ammonia. The first plant of this kind was announced in 1966 by Borden for construction at Geismar, La.; it was to have a capacity of 500.0 million pounds per year.

The demand for methanol in the U. S. was 2280 million pounds in 1965, and is growing at a rate of 4.5 percent per year. This demand can be broken down as follows.

End-use	MM lbs, 1965
Formaldehyde and formaldehyde stabilizer	1270.0
Methyl esters, amines, and other chemicals	1040.0
Solvent	200.0
Export and other	380.0
Total	2890.0

Source: *Oil, Paint & Drug Reporter,* November 29, 1965.

The amount converted to formaldehyde includes the raw material for making ethylene glycol by the DuPont process, as well as for poly-acetals. The largest consumer in the ester-amine category is poly-esters, most methanol being recovered during polyester production,

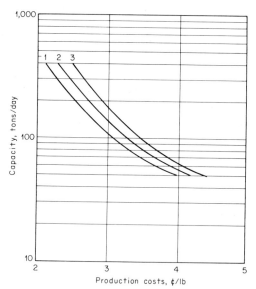

Fig. 2-14 Methanol. 1-Natural Gas, 20¢/1000 scf, steam re-forming; 2-Natural Gas, 30¢/1000 scf, steam reforming; 3-Natural Gas, 40¢/1000 scf, steam reforming.

purified and reused. The other major chemical use is methylamines. For nonchemical uses, the demand for nonpermanent antifreeze is down to only a fraction of what it was earlier, the market having been almost entirely taken over by ethylene glycol. The demand for methanol as an antidetonant for jet fuel is also declining. Miscellaneous uses include the 3 percent added to ethanol as a denaturant in some formulas, heater fluids for trucks and railroad cars, and others.

Manufacturers of methanol in the U.S. (1965) are given below.

Producer	Location	Capacity MM lbs/yr
Allied Chemical	South Point, Ohio	175.0
Borden	Geismar, La.	400.0
Celanese	Bishop, Texas	165.0
Commercial Solvents	Sterlington, La.	300.0
DuPont	Huron, Ohio	200.0
	Orange, Texas	1000.0
Escambia	Pensacola, Fla.	210.0
Gulf	Military, Kan.	60.0
Hercules	Hercules, Cal.	56.0
	Louisiana, Mo.	56.0
Monsanto	Texas City, Texas	165.0
Rohm & Haas	Houston, Texas	145.0
Tenneco	Houston, Texas	170.0
Union Carbide	S. Charleston, W. Va.	80.0
	Texas City, Texas	280.0

Source: *Oil, Paint & Drug Reporter,* November 29, 1965.

Almost all these producers are manufacturers and consumers of formaldehyde, which is transported mostly as a 37 percent solution. Although methanol plants tend to be located in regions where cheap natural gas is available, formaldehyde plants are found largely near well-populated centers.

3.1 Formaldehyde

Methanol can be oxidized to formaldehyde by air:

$$2CH_3OH + O_2 \rightarrow 2CH_2O + H_2O$$

This route accounts for 90 percent of U. S. formaldehyde capacity with the rest obtained by direct oxidation of hydrocarbons.

In the original technique for making formaldehyde, a silver-gauze catalyst was mounted in a converter somewhat resembling an ammonia burner that was used for making nitric acid. These converters have the disadvantage of giving yields around 80-85 percent, and that a considerable amount of unreacted methanol must be recovered by fractionation and recycled.   However, recently, three process licensors — Reichhold and Lummus in the U. S. and Montecatini in Italy — have developed fixed-bed catalytic processes using tubular reactors filled with an iron-molybdenum catalyst. These catalysts are said to give a 92 percent yield or better, with most of the losses being due to formation of CO and CO_2 and not of formic acid as in the silver-gauze processes. This type of catalyst has many advantages over the burner-type process.   Choosing a suitable coolant, the heat of reaction can be recovered completely instead of partly going to waste by quenching; secondary absorption under pressure becomes unnecessary, thus eliminating both equipment requirements and utility consumption for compression; fractionation of the methanol-formaldehyde mixture is eliminated; and the removal of formic acid by ion-exchange can be dispensed with, depending on the applications for which the product is intended. A flowsheet of such a process appears in Fig. 2-15.

Considerable effort has been directed towards developing a process for direct oxidation of methane to formaldehyde. Although by 1965 no commercial application of this route existed, Huettenwerk Oberhausen had reported obtaining 70 percent yields on a pilot scale using a Cu-Sn catalyst.

Using methanol, the investment for a 100.0 million pounds per year (100 percent) formaldehyde unit based on one of the iron-

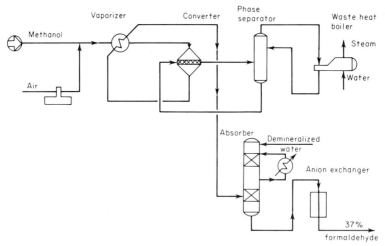

Fig. 2-15 Formaldehyde (Metal Oxide Catalyst Process).

molybdenum catalysts is around $0.45 million. The production costs are given in the following.

	Captive methanol (3.0 ¢/lb)	Merchant methanol (4.5 ¢/lb)
Methanol	3.5	5.2
Catalyst and chemicals	0.3	0.3
Utilities (including demineralized process water)	0.4	0.4
Labor and overhead	0.8	0.8
Capital charges	1.5	1.5
	6.5	8.2

Thus, even if based on noncaptive methanol, formaldehyde production is a profitable undertaking. Investment costs for smaller plants would not decrease much below $0.3 million. Also, captive production costs would equal the price of merchant formaldehyde at around 4.0 million pounds per year, 100 percent formaldehyde.

Despite these economics, by 1965, of 16 producers of formaldehyde in the U.S. only five did not manufacture methanol also. Their capacities were as follows (expressed as 100 percent formaldehyde):

Producer	Location	Capacity 1965 MM lbs/yr (100% formaldehyde)
Allied Chemical	South Point, Ohio	115.0
Borden	Bainbridge, N. Y.	15.0
	Demopolis, Ala.	40.0
	Fayetteville, N. C.	41.0
	Fremont, Calif.	22.0
	Kent, Wash.	20.0
	Sheboygan, Wisc.	16.0
	Springfield, Ore.	41.0
Celanese	Bishop, Texas	190.0
Commercial Solvents	Agnew, Calif.	11.0
	Sterlington, La.	15.0
DuPont	Belle, W. Va.	150.0
	Perth Amboy, N. J.	55.0
	Toledo, Ohio	45.0
Georgia Pacific	Coos Bay, Ore.	7.5
Gulf	Calumet City, Ill.	39.0
Hercules	Hercules, Calif.	28.0
	Louisiana, Mo.	37.0
Hooker	N. Tonawanda, N. Y.	37.0
Monsanto	Springfield, Mass.	90.0

Producer	Location	Capacity 1965 MM lbs/yr (100% formaldehyde)
Reichhold	Hampton, S. C.	15.0
	Kansas City, Kan.	11.0
	Tacoma, Wash.	11.0
	Tuscaloosa, Ala.	24.0
Rohm & Haas	Bristol, Pa.	9.0
	Philadelphia, Pa.	9.0
Tenneco	Fords, N. J.	27.0
	Garfield, N. J.	65.0
Trojan Powder	Seiple, Pa.	18.5
Union Carbide	Bound Brook, N. J.	55.0

Source: *Chemical Week*, December 5, 1964.
Note: A number of new plants were to come on stream in 1966.

Although only four of the 15 methanol plants in operation had capacities below 100.0 million pounds per year, only three of 32 aldehyde plants exceeded this figure. Furthermore, only five of the 15 methanol plants fed formaldehyde plants at the same location. Formaldehyde is made predominantly in the vicinity of large users, e.g., the plywood manufacturers of the Pacific Northwest.

Consumption of formaldehyde in the U. S. reached 980.0 million pounds (100 percent) in 1964, with the following distribution.

End-use	MM lbs, 1964
Phenol-formaldehyde resins	195.0
Urea-formaldehyde resins and condensates	280.0
Melamine formaldehyde	60.0
Pentaerythritol	80.0
Hexamethylene tetramine	55.0
Urea-formaldehyde fertilizers	27.0
Polyacetals	40.0
Ethylene glycol and glycolic acid	140.0
EDTA and other sequestrants	13.0
Other uses	90.0

Formaldehyde is sold mostly at 37 to 45 percent solution in water because it tends to polymerize at higher concentrations. However, this can be obviated by adding a stabilizer; at low concentrations this is mostly methanol. Concentration can be carried out up to a certain level by atmospheric distillation; to achieve higher concentrations, vacuum or azeotropic distillation, or solvent extraction, must be employed. However, it is difficult to keep the

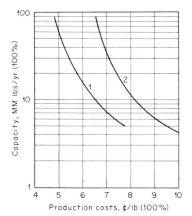

Fig. 2-16 Formaldehyde. 1-Captive Methanol, 3.0¢/lb; 2-Merchant Methanol, 4.5¢/lb.

temperature above 212° F to avoid polymerization. Dioxane has been suggested as an extractant for making either a concentrated formaldehyde solution or paraformaldehyde. The latter is a solid polymeric form containing less than 10 percent water. It is used as a formaldehyde source when making resins for electrical applications (where conductivity and therefore the presence of ions, must be kept to a minimum), as a starch preservative in drilling muds, and in other applications where water cannot be tolerated.

Hexamethylene tetramine

Formaldehyde reacts with ammonia to form hexamethylene tetramine (HMTA), also known as "hexamine" in the plastics industry and "urotropin" in the pharmaceutical field:

$$6CH_2O + 4NH_3 \longrightarrow \text{(HMTA structure)} + 6H_2O$$

The reaction is rapid and almost quantitative with yields above 96 percent. Formaldehyde in solution reacts with anhydrous ammonia in a cooled reactor. The product then is purified by evaporation, centrifuging, and finally dried. Except for a bleed-stream,

the centrifuge wash liquors are recycled. Producers of both hexamine and formaldehyde have the option of absorbing the methanol
oxidation reactor effluent in a reactor containing just enough water
to permit the heat of reaction and of absorption of the reactants
in water to drive off the water in the feed plus that formed in the
reaction. A slurry of hexamine in water is removed from the reaction vessel and purified by centrifuging. The reactor off-gases
still contain some methanol from the formaldehyde unit which can
be recovered, while the rest of the stream is burned. A flowsheet
of this process is shown in Fig. 2-17.

Up to 1965, hexamine consumption was 45.0 million pounds per
year. Its most important application is as a source of formaldehyde for crosslinking phenolic molding powders, shell molding
resins and two-step curing resins for chip board.

Quaternization of hexamine with an alkyl chloride gives a
family of bactericides (developed by Dow) for use in latex paints
and as a dermatitis preventive in water-soluble cutting oils.

Nitrosation yields N,N –dinitrosopentamethylene tetramine, sold
by DuPont as "Unicel," a blowing agent for making sponge rubber
and silicone elastomer foams. Consumption of this product is
steady at around 4.0 million pounds per year.

Hexamine is well known pharmaceutically for its use in formulations to combat urinary tract infections; it is also an intermediate in the production of chloramphenicol.

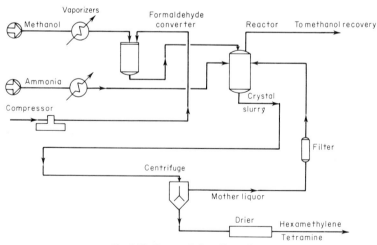

Fig. 2-17 Hexamethylene Tetramine.

Peacetime demand can be broken down as follows:

End-use	MM lbs, 1965
Crosslinking agent for phenolics	36.0
Blowing agents	4.0
Chloramphenicol	3.5
Other uses	2.5
	46.0

A large part of the capacity is idle in peacetime, as HMTA is the starting point for the high explosive known as R.D.X. Hexamine manufacturers are listed below; all but one are ammonia and formaldehyde producers.

Producer	Location	Capacity MM lbs/yr	Captive sources		
			Methanol	Formaldehyde	Ammonia
Borden	Demopolis, Ala.	18.0	x	x	x
DuPont	Perth Amboy, N. J.	8.0	x	x	x
Hooker	N. Tonawanda, N. Y.	25.0	—	x	x
Plastics					
Engineering	Sheboygan, Wisc.	8.0	—	—	—
Tenneco	Fords, N. J.	18.0	x	x	x
Union Carbide	Bound Brook, N. J.	6.0	x	x	x

Source: *Oil, Paint & Drug Reporter,* June 31, 1966.

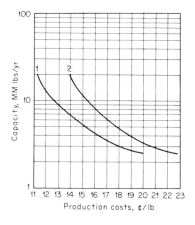

Fig. 2-18 Hexamethylene Tetramine.   1-Captive Methanol, 3.0¢/lb; 2-Merchant Methanol, 4.5¢/lb.

Investment for a 5.0 million pounds-per-year plant is \$0.4 million, and the production costs are as follows:

	Cost, ¢/lb HMTA
Methanol (captive, 3.0 ¢/lb)	5.3
Ammonia (captive, \$50/ton)	1.3
Utilities, catalyst, and chemicals	2.1
Labor and overhead	4.0
Capital charges	2.6
	15.3

Pentaerythritol

Formaldehyde condenses with acetaldehyde in the presence of a basic catalyst to form pentaerythritol as the principal product:

$$4CH_2O + CH_3CHO + NaOH \longrightarrow C(CH_2OH)_4 + HCOONa$$
$$\text{pentaerythritol}$$

Numerous side-reactions occur, mostly other condensations and autocondensations. Etherification of two mols of pentaerythritol gives dipentaerythritol, the main byproduct:

$$(CH_2OH)_3C-CH_2-O-CH_2C(CH_2OH)_3$$
$$\text{Dipentaerythritol}$$

A flow-diagram of the process is shown in Fig. 2-19. After condensation and precipitation of the CO_2 formed in the reaction with lime (some producers use soda-ash), the numerous byproducts are removed by a long series of purification steps. After the first crystallization-dissolution-filtration sequence, the product is dissolved in hot water and passed through an ion-exchange purification unit which removes the last traces of formic acid. The concentration of this solution is sufficient, thus another evaporation step is unnecessary. Deionization is followed by vacuum crystallization, redissolution, filtration and drying. The yield is a technical product containing about 11 percent dipentaerythritol. Some producers treat part of their output further to obtain both products in pure form, especially when the desired end-product is explosive grade pentaerythritol.

Since more than 0.3 pound of formic acid is produced per pound of pentaerythritol, large plants must recover this byproduct in order to operate profitably. Economics for a 10.0 million pounds-per-year plant are evaluated below. Investment requirements are

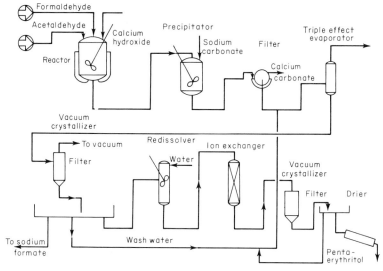

Fig. 2-19 Pentaerythritol.

$2.5 million; formaldehyde is assumed captive at 6.75 cents per pound (100 percent), acetaldehyde at 10.0 cents per pound and formic acid credited at 8.0 cents per pound. Production costs under these assumptions are as follows.

	Cost, ¢/lb pentaerythritol	
Formaldehyde	8.7	
Acetaldehyde	3.8	
Catalyst and chemicals	1.9	
Formic acid credit	(−2.4)	
Net raw materials		12.0
Utilities		1.6
Labor and overhead		1.6
Capital charges		8.3
		23.5

The most important end-use of pentaerythritol (in peacetime, since it is also a raw material for the explosive known as PETN) is in the manufacture of alkyd resins, in competition with glycerol. Although pentaerythritol gives alkyds having somewhat better properties, e.g., a higher melting point, it is virtually interchangeable with glycerol. The higher functionality of pentaerythritol permits alkyd resin formulation with a larger proportion of oil, which justifies a price-differential over glycerol. Most pentaerythritol alkyds are modified with amino-resins and go into baking enamels.

Next in importance are pentaerythritol rosin esters, which are used in floor-polish and in flexographic inks. Reconstituted drying oils are made by splitting an oil into fatty acids and glycerol, separating the acids in order to obtain a concentrate of those having the best drying properties, and reesterifying them with an alcohol other than glycerol — usually pentaerythritol. Another application of pentaerythritol esters is in the manufacture of fire-retardant paints which swell when heated, thereby protecting the substratum. Pentaerythritol esters are used as high-pressure lubricants, in producing PVC plasticizers of low volatility suitable for use in wire insulation. For this latter application the esters of di- and tripentaerythritol are preferred. Thus, another reason for further purification of technical pentaerythritol is to isolate the higher homologs. Ethylene and propylene oxide adducts of pentaerythritol are used as polyols for urethane foam manufacture. About 1.5 million pounds per year of pentaerythritol acetals are used to make chlorine-resistant textile finishes.

Chlorinated polyethers are made from pentaerythritol by the following series of reactions:

$$C(CH_2OH)_4 + 4CH_3COOH \longrightarrow C(CH_2OOCCH_3)_4 + 4H_2O$$

$$C(CH_2OOCCH_3)_4 + HCl \longrightarrow (ClCH_2)_3C-O-\overset{\overset{\displaystyle O}{\|}}{C}-CH_3 \xrightarrow{NaOH}$$

$$\begin{matrix} ClH_2C \\ \diagdown \\ \diagup \\ ClH_2C \end{matrix} \overset{CH_2}{\underset{CH_2}{\diagup}} C \diagdown O \diagup \longrightarrow \left[\begin{matrix} ClH_2C \\ \diagdown \\ -O-CH_2-C-CH_2- \\ \diagup \\ CH_2Cl \end{matrix} \right]_n$$

These resins are made by Hercules at Parlin, New Jersey, with a production capacity of 2.5 million pounds per year. They are used primarily in process equipment, as a compromise between PVC and the much more expensive fluorocarbons. Their main applications are in corrosion-resistant coatings (30 percent), pipe linings (40 percent) and injection-molded parts for pumps and valves; in most fields, these resins are losing ground to other superior materials.

In 1965, the demand for pentaerythritol was 82.0 million pounds, distributed as follows.

End-use	MM lbs, 1965
Alkyd resins	66
Synthetic lubes	4
Pentaerythritol tetranitrate	3
Chlorinated polyethers, esters and others	9
Total	82

Source: *Oil, Paint & Drug Reporter*, June 6, 1966.

Producers of pentaerythritol in the U.S. are listed below. Most of the output is sold, but Hercules and Reichhold manufacture a number of rosin esters and other derivatives that make up a captive market of around 15.0 million pounds per year.

Producer	Location	Capacity MM lbs/yr
Hercules*	Louisiana, Mo.	40.0
Celanese	Bishop, Texas	25.0
Tenneco	Fords, N. J.	25.0
Reichhold	Tuscaloosa, Ala.	12.0
Trojan Powder	Allentown, Pa.	12.0
Delaware Chemical**	New Castle, Del.	6.0
Commercial Solvents	Agnew, Calif.	0.5
		106.5

*Hercules is adding 5 million lbs/yr.
**To be completed in mid-1967.
Source: *Oil, Paint & Drug Reporter*, June 6, 1966.

Many producers began operations as a result of the demands for pentaerythritol tetranitrate during World War II.

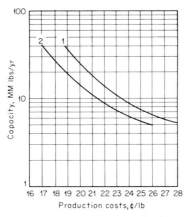

Fig. 2-20 Pentaerythritol. 1-Captive formaldehyde, 5.25¢/lb (100%); 2-Captive formaldehyde, 6.5¢/lb (100%).

Polyacetals

Compounds containing carbonyl groups polymerize by forming polyacetals:

$$RCHO \longrightarrow \left[\begin{array}{c} H \\ | \\ C \\ | \\ R \end{array} -O- \begin{array}{c} H \\ | \\ C \\ | \\ R \end{array} -O \right]$$

The stablest polymer of this type is polyformaldehyde or poly-oxymethylene, and it is thus the only polyacetal to have reached commercial status.   Although numerous companies have become interested in these polymers, by 1965 only three had entered the field on an industrial scale: DuPont, with a polyformaldehyde resin called "Delrin"; Celanese, with a copolymer of trioxane, the trimer of formaldehyde, and ethylene oxide (or 1,3-dioxolane, the reaction product of ethylene glycol and formaldehyde), called "Celcon."   For example:

The third producer is Tenneco, through its subsidiary Heyden-Newport, under the trade-name "Polyfyde."

The properties of these resins place them among the engineering plastics, that is, polymers that are able to compete with metals in a number of applications. While these resins are more expensive than nonferrous metals, lower density and cheaper finishing compensate for this to the point where polyacetals are replacing die-cast zinc and brass.   The melting point of these resins is approximately $350°F$, and their significant property is fatigue resistance. The outstanding advantage of the copolymer-type acetal resins is that their highest continuous-use temperature is $40°F$ above that of pure polyoxymethylene, which is $180°F$. Both types show excellent re-covery from tensile stress, dielectric properties and room-temperature solvent resistance. Copolymers are also more stable to hydrolysis than homopolymers. A significant feature is that they are potentially much cheaper than nylon, polycarbonates or ABS resins although, by 1965, the latter were being sold for half the price of polyacetals, and had about five times the market.

Domestic consumption of polyacetal resins reached 28.6 million pounds in 1965, and is growing at 15 percent per year. The main applications were as follows.

End-uses	MM lbs, 1965
Automotive parts (directional signal, switches, fender extensions, etc.)	6.3
Industrial components (gears, conveyor slats, etc.)	6.3
Hardware and plumbing (sink strainers, ball cock valves, faucet valves)	4.0

Communication equipment (telephone terminal blocks, card-dial, push-button telephone components)	3.7
Appliances (hinges, handles, etc.)	2.1
Mill shapes	1.5
Aerosol containers	1.5
Pipe	1.5
Others	1.7

Source: *Modern Plastics,* January, 1966.

The investment cost for a 15.0 million pounds-per-year poly-acetals plant is $15.0 million. A large part of this is due to the need to make a very pure formaldehyde feed to the polymerization plant.

Glycolic acid

In 1965, DuPont was still manufacturing ethylene glycol on a large scale starting from formaldehyde:

$$CH_2O + CO + H_2O \longrightarrow HOCH_2COOH$$

$$HOCH_2COOH + CH_3OH \longrightarrow CH_3OOCCH_2OH$$

$$HOCH_2COOCH_3 + H_2 \longrightarrow HOCH_2CH_2OH + CH_3OH$$

The intermediate, glycolic acid, finds some use in the leather industry, but most of it goes into chelating formulations for iron (in combination with citric acid). The demand in the U.S., other than for ethylene glycol, is around 9.0 million pounds per year.

Textile finishes

Apart from the amino-resins used for this purpose, formaldehyde is an intermediate in the manufacture of a number of reactant textile finishes for cellulosic fibers which act by etherification of the methylol groups with the hydroxyl groups on the cellulose molecule.

Tetrakis (hydroxymethyl) phosphonium chloride (THPC) is the only permanent fire-retardant so far developed for cellulosic fibers; it is also an effective antibacterial and mildew preventive, and is used mainly on institutional linen. It is made by reacting phosphine, formaldehyde and hydrochloric acid:

$$PH_3 + HCl + 4CH_2O \longrightarrow (HOH_2C)_4PCl$$

Sodium sulfoxylate formaldehyde is used extensively in vat dyeing and for preparing oil-in-water emulsions of printing pastes. The domestic demand is around 7.0 million pounds per year. It is made via zinc hydrosulfite.

$$Zn + 2SO_2 \longrightarrow ZnS_2O_4 \xrightarrow{CH_2O} (CH_2OH\ SO_2)_2Zn$$

$$(CH_2OH\ SO_2)_2Zn + ZnO \longrightarrow 2Zn \begin{matrix} \diagup OH \\ \diagdown SO_2CH_2OH \end{matrix} \xrightarrow{NaOH} NaHSO_2CH_2O$$

Triazones are wash-are-wear finishes made from urea, formaldehyde and ethylamines. The tertiary nitrogen molecule acts as a buffer, which makes these finishes suitable for white goods. Their use is fairly static compared with the rapid growth of triazine, DMEU and permanent-crease finishes.

"Uron" finishes are made from urea, formaldehyde and an alcohol. They are nonyellowing and inexpensive.

"Nuactant" is methylolated ethyl carbamate:

$$\begin{matrix} O \\ \parallel \\ C_2H_5OC{-}N(CH_2OH)_2 \end{matrix}$$

Approximately 4.0 million pounds of this finish were used in 1965.

Finally, substantial amounts of formaldehyde are used alone for cotton treating.

3.2 Methylamines

Ammonia and methanol react in the presence of a catalyst to form a mixture of mono-, di-, and trimethylamine (MMA, DMA, and TMA):

$$CH_3OH + NH_3 \longrightarrow CH_3NH_2 + H_2O$$

$$2CH_3OH + NH_3 \longrightarrow (CH_3)_2NH + 2H_2O$$

$$3CH_3OH + NH_3 \longrightarrow (CH_3)_3N + 3H_2O$$

Some methanol decomposes to CO and hydrogen in the process, but yields are approximately 94 percent on methanol and 97 percent on ammonia. Most new plants employ the Leonard process (a flow-sheet is shown in Fig. 2-21). Reactants are first preheated against converter effluent in order to recover some of the exothermic reaction heat. Then the product stream is flashed to remove the noncondensibles and sent to the recovery system. First, ammonia is taken overhead and recycled, together with some trimethylamine. Next water is added in order to break the TMA-ammonia azeotrope, and in the second column pure TMA is taken overhead. The mixture of mono- and dimethylamine is first dehydrated and then fractionated to give pure DMA and MMA. Since the market distribution differs from the equilibrium mixture of the three amines — much more DMA is consumed than either of the other two — the relative amounts produced in the reactor are controlled by recycling the unwanted amines and thus driving the equilibrium towards the ones desired. Carbon steel equipment can be used since amines are corrosion-inhibiting. Special precautions are necessary to avoid leaks of malodorous methylamines.

The demand for methylamines was 93.0 million pounds in 1965, distributed as follows.

End-use		MM lbs, 1965
Monomethylamine		21.0
- "Sevin"	14.0	
- Methyl taurine	4.0	
- Photographic developers	0.5	

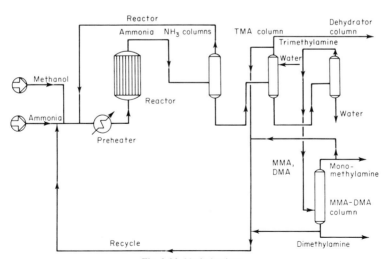

Fig. 2-21 Methylamines.

Continued

End-use		MM lbs, 1965
- Sarcosine	0.2	
- Pharmaceuticals, other pesticides	2.3	
Dimethylamine		58.0
- Dimethylformamide	12.0	
- Dimethylacetamide	7.0	
- Pesticides	10.5	
- Rubber chemicals	7.0	
- Unsymmetrical dimethyl hydrazine	9.0	
- Surface-active agents	9.0	
- Dimethylamine sulfate	0.5	
- Miscellaneous	3.0	
Trimethylamine		14.0
- Choline salts	12.2	
- Betaine, etc.	0.3	
- Ion exchange resins	1.5	
		93.0

Investment for a methylamines plant depends somewhat on the intended product mix, 10.0 million pounds-per-year unit costs around $1.5 million. The following production costs have been calculated for dimethylamine.

	Cost, ¢/lb DMA
Methanol (captive, 3.0 ¢/lb)	4.6
Ammonia (merchant, 4.0 ¢/lb)	1.6
Utilities	1.5
Labor and overhead	1.2
Capital charges	5.0
	13.9

Captive consumption of methylamines is about 60 percent of the total output. The differential between production costs and market prices is high, and the larger consumers have tended to integrate.

Below is a list of the methylamines producers in the U.S.

Producer	Location	Capacity MM lbs/yr	Main captive uses
Commercial Solvents	Terre Haute, Ind.	18.0	Choline salts
DuPont	Belle, W. Va.	50.0	Solvents, rubber chemicals,
	LaPorte, Texas	26.0	pesticides
Escambia*	Pace, Fla.	30.0	
General Aniline	Calvert City, Ky.	10.0	Surfactants, choline
Pennsalt	Wyandotte, Mich.	10.0	Dimethylaminoethanol, rubber chemicals
Rohm & Haas	Bristol, La.	7.0	Quaternaries, ion-exchange
	Philadelphia, Pa.	14.0	resins
		165.0	

*Also, Escambia indicates expansion to 50.0 million by mid-1967.

Source: *Oil, Paint & Drug Reporter*, May 30, 1966.

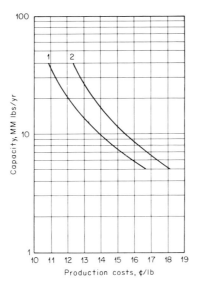

Fig. 2-22 Dimethylamine.   1-Methanol, 3.0¢/lb; 2-Methanol, 4.0¢/lb.

Monomethylamine

The principal outlet for monomethylamine is the manufacture of "Sevin," a carbamate insecticide.

Still in the pesticide field, "Vapam" finds some use as a soil fumigant for vegetables, tobacco and ornamental plants:

$$CH_3NH_2 + CS_2 + NaOH \longrightarrow CH_3NHC\overset{\displaystyle S}{\overset{\displaystyle \|}{}}-S-Na + H_2O$$

The "Igepon" family of surface-active agents is well known in the textile field. Originally developed in Germany, they are made in the U.S. by General Aniline & Film. About 40.0 million pounds per year of these compounds are consumed in the U.S. They are made from fatty acyl chlorides, obtained from fatty acids by reaction with a strong chlorinating agent such as PCl_3, and an N-alkyltaurine:

$$HOCH_2CH_2SO_3Na + CH_3NH_2 \longrightarrow CH_3NHCH_2CH_2SO_3Na$$

$$CH_3NHCH_2CH_2SO_3Na + RCOCl \longrightarrow RCONCH_2CH_2SO_3Na + HCl$$
$$\underset{\displaystyle CH_3}{|}$$

Although other substituted taurines (ethyl, isopropyl, butyl) give more soluble compounds, the monomethylamine derivatives are used most frequently.

Monomethylamine is used to make photographic developers such as "Metol" (N-methyl-p-aminophenol); in the drug field it is an intermediate for pyrazolone analgesics, ephedrin and adrenalin. It is also the raw-material for making sarcosine.

Dimethylamine

Dimethylamine is used largely in the manufacture of spinning solvents for acrylic and polyurethane fibers. Dimethylformamide (DMF), the more widely employed of the solvents, is made from dimethylamine and carbon monoxide:

$$(CH_3)_2NH + CO \longrightarrow (CH_3)_2N\overset{\displaystyle O}{\overset{\displaystyle \|}{C}}H$$

When CO is unavailable, methyl formate can be substituted. Also DMF is used as a solvent in the recovery system of some acetylene plants. Dimethylacetamide, also a spinning solvent, is made from dimethylamine and glacial acetic acid or acetic anhydride:

$$(CH_3)_2NH + CH_3COOH \longrightarrow CH_3\overset{\displaystyle O}{\overset{\displaystyle \|}{C}}N(CH_3)_2 + H_2O$$

In 1965, a total of 30.0 million pounds of these solvents was used; consumption is about 0.2 pound per pound of acrylic fiber. Du Pont uses DMF to produce "Orlon" and "Lycra" fibers. Chemstrand is the largest consumer of dimethylacetamide.

Substituted urea herbicides such as "Diuron" are the largest outlet for dimethylamine as an intermediate in the pesticide field. The DMA salt of 2,4-D is important as a soluble form of this herbicide.

The most important rubber chemicals derived from dimethylamine are tetramethylthiuram disulfide and sodium dimethyldithiocarbamate. The former is an accelerator for low-temperature or rapid high-temperature vulcanization. The latter is used as a short-stop in SBR manufacture.

Unsymmetrical dimethyl hydrazine, or "UMDE" is used as a rocket fuel. It is made by nitrosation and reduction of DMA:

$$NO_2^- + (CH_3)_2NH + H^+ \longrightarrow (CH_3)_2N-NO + H_2O$$

$$(CH_3)_2N-NO + 2H_2 \longrightarrow (CH_3)_2N-NH_2 + H_2O$$

Dimethylaminoethanol is primarily an intermediate for making textile surface-active agents, ion-exchange resins and lubricant additives. It is made from dimethylamine and ethylene oxide:

$$(CH_3)_2NH + CH_2\underset{O}{\overset{}{\diagdown\diagup}}CH_2 \longrightarrow HOCH_2CH_2N(CH_3)_2$$

Its fatty acid esters are nonionic surfactants, and the respective methacrylate polymer is a lube-oil detergent. Reaction with $SOCl_2$ gives dimethylaminoethyl chloride:

$$SOCl_2 + HOCH_2CH_2N(CH_3)_2 \longrightarrow ClCH_2CH_2N(CH_3)_2 + SO_2 + HCl$$

This compound is a major intermediate in making antihistamines, of which about 0.35 million are consumed in the U.S., as well as tranquilizers and spasmolytic drugs.

The sulfuric acid salt of dimethylamine is the most important accelerator for the lime solutions used in unhairing hides for shoe uppers.

Tertiary amines are made by some companies from dimethylamine and an alkyl chloride. This process is used to make lauryl dimethylamine oxide, employed by Procter & Gamble as a foam stabilizer in its liquid detergents instead of alkanolamides.

$$(CH_3)_2NH + RCl \longrightarrow RN(CH_3)_2 + HCl$$

$$RN(CH_3)_2 + H_2O_2 \longrightarrow \underset{\overset{|}{O^-}}{RN^+(CH_3)_2} + H_2O$$

Trimethylamine

The most important outlet for trimethylamine is in the production of choline salts, of which 27.0 million pounds were used in 1965 as additives for high-energy poultry feed. This amounts to an average of 0.6 pound per ton of poultry feed consumed. Choline hydrochloride is made by the following reaction:

$$(CH_3)_3N + HCl \longrightarrow (CH_3)_3N^+HCl^-$$

$$CH_3N^+HCl^- + CH_2\!\!-\!\!CH_2 \longrightarrow (CH_3)_3N^+CH_2CH_2OHCl^-$$
$$\overset{}{O}$$

Commercial Solvents, General Aniline and Hoffman LaRoche are the main U.S. producers of choline salts.

Betaine, obtained by reacting trimethylamine and monochloracetic acid, finds some use in treating liver ailments as well as in the feed industry.

Finally, trimethylamine is used to make anionic ion-exchange resins, usually quaternary ammonium compounds obtained from a chlormethylated styrenic resin and trimethylamine.

3.3 Methyl Chloride

There are two processes used in making methyl chloride. Methanol can be esterified with hydrochloric acid or a mixture of sodium chloride and sulfuric acid:

$$CH_3OH + HCl \longrightarrow CH_3Cl + H_2O$$

$$CH_3OH + \tfrac{1}{2}H_2SO_4 + NaCl \longrightarrow CH_3Cl + H_2O + \tfrac{1}{2}Na_2SO_4$$

or methane chlorinated directly:

$$CH_4 + Cl_2 \longrightarrow CH_3Cl + HCl$$

Despite the ample availability of methane at 0.5 cent per pound or less, approximately 65 percent of all methyl chloride is produced in the U.S. from methanol, available at a cost of around 3.0 cents per pound. Part of the reason lies in the economics of chlorine utilization.   Direct chlorination requires twice as much chlorine, of which half is transformed into hydrochloric acid. The value of the latter has been variously estimated at 30 to 70 percent of that of chlorine.   For example, HCl fed to a Deacon-type chlorine-recovery unit must be valued at 50 percent of the price of captive chlorine if the product of such a unit is to cost the same as fresh chlorine. The higher value of 70 percent applies only to situations in which HCl can be used as such, and does not have to be reoxidized to chlorine. This means that, assuming electrolysis chlorine is worth 2.5 cents per pound, esterification requires no more than 55 percent as much as chlorine in value as does chlorination, on the basis of the 70 percent figure; and if 50 percent is taken, esterification affords chlorine savings of 65 percent over direct chlorination. A further part of the answer consists in the overall structure of the various producing companies. Methyl chloride is primarily an intermediate for other chemicals: silicones, cellulose ethers, tetramethyl lead and others. Therefore, producers with captive requirements often have little interest in producing the higher chlorinated methanes that are byproducts of direct chlorination. Third, in areas where methane is not available, methanol, which is easy to transport, is the only choice.   Finally, making methyl chloride from methanol requires a comparatively small investment.

A flowsheet for methyl chloride production from methanol is shown in Fig. 2-23. Methanol and hydrogen chloride are preheated and sent to catalytic reactor, where conversion takes place in a high-boiling hydrocarbon. Crude methyl chloride leaves the top of the reactor, and is purified by low-temperature fractionation; unreacted methanol is recycled to the reactor feed stream.

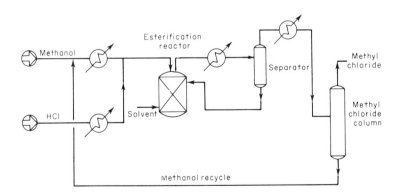

Fig. 2-23 Methyl Chloride.

A 20.0 million pounds-per-year methyl chloride plant via methanol esterification requires an investment of $0.8 million, and production costs can be estimated as shown below.

	Cost, ¢/lb methyl chloride
Methanol (assumed captive, 3.0 ¢/lb)	2.4
Hydrochloric acid ($40/ton)	1.7
Utilities and chemicals	0.2
Labor and overhead	1.2
Capital charges	1.3
	6.8

Producers of methyl chloride in the U.S. and their respective routes, are given below.

Producer	Location	Capacity MM lbs/yr	Raw material
Allied Chemical	Moundsville, W. Va.	12.0	Methanol
Ancon Chemical	Lake Charles, La.	60.0	Methanol
Ansul Chemical	Marinette, Wis.	18.0	Methanol
Diamond Alkali	Belle, W. Va.	8.0	Methane
Dow	Freeport, Texas	35.0	Methane
	Pittsburgh, Calif.	25.0	Methane
Dow-Corning	Midland, Mich.	10.0	Methanol
DuPont	Niagara Falls, N.Y.	25.0	Methanol
Ethyl Corp.	Baton Rouge, La.	25.0	Methanol
General Electric	Waterford, N. Y.	10.0	Methanol
Kolker Chemical	Newark, N. J.	5.0	Methanol
Total		233.0	

Source: *Oil, Paint & Drug Reporter*, October 1, 1964.

In 1964, the demand for methyl chloride was 134.0 million pounds, distributed as follows:

End-use	MM lbs 1964
Silicones	80.0
Catalyst solvent for butyl rubber	16.0
Tetramethyl lead	20.0
Methyl cellulose	9.0
Quaternary ammonium salts and miscellaneous	9.0
	134.0

An additional 35 to 40.0 million pounds of methyl chloride obtained via methanol esterification were converted to other chlorinated methanes.

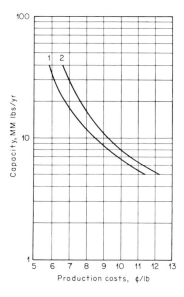

Fig. 2-24 Methyl Chloride. 1-Methanol = 3.0¢/lb; 2-Methanol = 4.0¢/lb.

Silicones

There are two processes for making the raw material for the family of polymers known as polysiloxanes, or silicones. The original Dow process involved a Grignard reaction between methyl chloride and magnesium, followed by combination with silicon tetrachloride:

$$CH_3Cl + Mg \longrightarrow CH_3MgCl$$

$$2CH_3MgCl + SiCl_4 \longrightarrow (CH_3)_2SiCl_2 + 2MgCl_2$$

The present method for making methylchlorosilanes no longer requires use of the hazardous Grignard reaction; it has been found possible to react silicon metal with methyl chloride directly. Numerous byproducts are formed, but the main reactions can be written as

$$2CH_3Cl + Si \longrightarrow (CH_3)_2SiCl_2$$

$$6CH_3Cl + 3Si \longrightarrow 2CH_3SiCl_3 + (CH_3)_4Si$$

$$6CH_3Cl + 3Si \longrightarrow 2(CH_3)_3SiCl + SiCl_4$$

These intermediates polymerize via hydrolysis followed by the splitting off of water:

$$(CH_3)_2SiCl_2 + H_2O \longrightarrow (CH_3)_2Si(OH)_2 + 2HCl$$

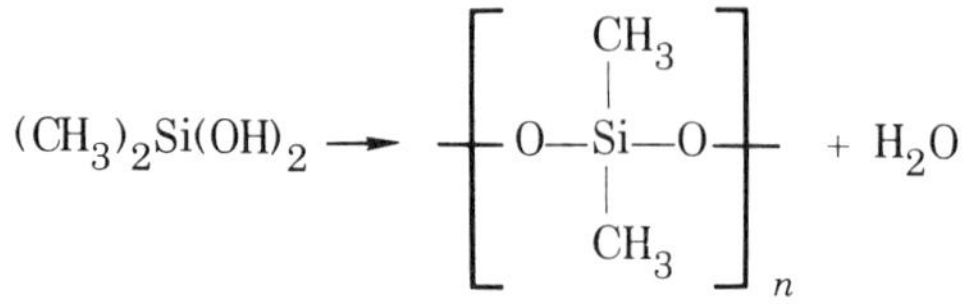

$$(CH_3)_2Si(OH)_2 \longrightarrow \left[\begin{array}{c} CH_3 \\ | \\ O-Si-O \\ | \\ CH_3 \end{array}\right]_n + H_2O$$

Linear polymers are formed from organosilanes containing two chlorine atoms; monoalkyl- (or aryl-) trichlorosilanes give cross-linked resins:

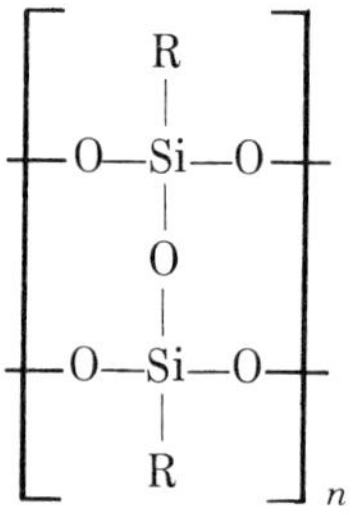

$$\left[\begin{array}{c} R \\ | \\ O-Si-O \\ | \\ O \\ | \\ O-Si-O \\ | \\ R \end{array}\right]_n$$

Trialkyl- (or aryl-) monochlorosilanes are used as chain-length regulators. Although methylchlorosilanes are by far the most important intermediates for silicone production, there are others of commercial significance. Benzene reacts with trichlorosilane to give phenyltrichlorosilane:

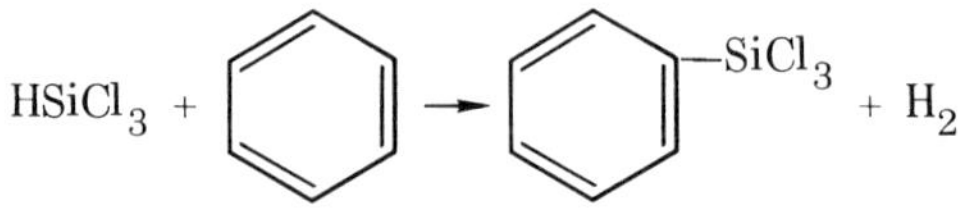

$$HSiCl_3 + \bighexagon \longrightarrow \bighexagon\text{--}SiCl_3 + H_2$$

This compound is used to make silicone resins. Methylphenyldichlorosilane, which improves the low-temperature properties of linear polymers and of peroxide-vulcanized elastomers, is made by the Grignard process:

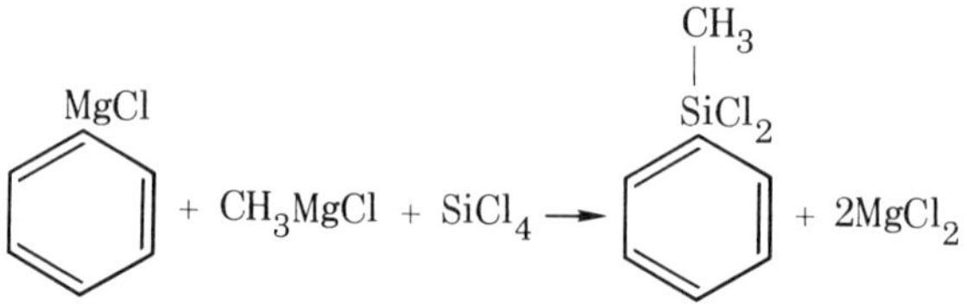

$$\underset{MgCl}{\bighexagon} + CH_3MgCl + SiCl_4 \longrightarrow \overset{CH_3}{\underset{SiCl_2}{\bighexagon}} + 2MgCl_2$$

Alkyltrichlorosilane reacts with alcohols to give alkyltrialkoxy-silanes — monomeric products which have surface-active properties and are extremely stable to hydrolysis. They are also used as urethane-foam stabilizers.

$$RSiCl_3 + 3R'OH \longrightarrow RSi(OR')_3 + 3HCl$$

Vinyltriethoxysiloxane is made from vinyltrichlorosilane (a derivative of vinyl chloride and silicon) and ethanol. It polymerizes to a resin used to prepare glass fibers for conventional dyeing:

$$CH_2\!-\!CHSiCl_3 + 3C_2H_5OH \longrightarrow CH_2\!-\!CHSi(OC_2H_5)_3 + 3HCl$$

The process consists of reacting methyl chloride and 98 percent silicon metal over a copper catalyst to form a mixture of $(CH_3)_2SiCl_2$, the most important product, CH_3HSiCl_2 and CH_3SiCl_3. Dimethyldichlorosilane is separated by fractionation from unreacted methyl chloride, which is recycled, and from the other two products. Methyltrichlorosilane is one of the major raw materials for silicone elastomers.

Dimethyldichlorosilane is hydrolyzed in a stirred reactor, giving off HCl. The latter can be combined with additional silicon to give di- and trichlorosilanes, precursors for other silicone intermediates produced by high-pressure reaction. The polymers are then treated in various ways (emulsified to make textile finishes, compounded with inert materials in the case of elastomers, etc.) to give the desired end-product.

The producers of silicones in the U.S. are given below.

Producer	Location	Capacity MM lbs/yr
Dow-Corning	Elizabethtown, Ky. Midland, Mich. Carrolton, Ky.*	25.0
General Electric	Waterford, N. Y.	15.0
Union Carbide	Sisterville, W. Va.	10.0
Stauffer	Adrian, Mich.	

*Begins operating in 1967.

Source: *Chemical Week*, June 19, 1965.

It is difficult to estimate the relative importance of the enormous number of end-uses. There are 500 or so different products on the market, each with its specific application. The best estimate is that in 1965 the total consumption of silicone materials of all types was around 39.5 million pounds, including monomeric end-products, distributed as follows:

Type of product	MM lbs, 1965
Heat-vulcanized elastomers	8.0
Oils and emulsions	15.0
Resins	12.5
Urethane foam stabilizers	2.0
Room-temperature-cure elastomers	2.0
	39.5

About 38 percent of all silicone heat-vulcanized elastomers go into military applications due to their exceptional high- and low-temperature properties, chemical inertness and resistance to weathering. The remainder goes into household appliance parts (12 percent), wire and cable coatings (14 percent), potting and encapsulation in the electronics field (18 percent), and automotive electrical systems such as spark-plug boots, diesel engine gaskets and other sealing applications (16 percent). A variety of other uses accounts for the rest; for example, their nonadherence makes them suitable for use in the food industry and their lack of volatile components enables them to resist a high vacuum.

Silicone fluids are oil-soluble or, if the presence of hydrocarbons is undesirable, can be emulsified in water. They have a great number of outlets due to their electrical properties (transformer fluids), surface activity (defoamers), chemical inertness (cosmetic), non-toxicity (pharmaceuticals), nonadherence (demolders), water repellency (textile finishes), antistatic properties (polishes), and heat resistance (wire drawing). Their uses can be broken down as follows:

End-use	MM lbs, 1965
Metal polishes	4.6
Defoamers	0.6
Greases	2.4
Textile finishes	0.8
Wire drawing, pharmaceuticals, cosmetics, damping fluids, mold release agents	3.5
Transformer dielectric fluids	3.1
	15.0

Silicone resins also have numerous applications. Alkyd coatings modified with about 25 percent silicones have excellent heat and corrosion resistance, and are used as maintenance paints in exhaust stacks, incinerators, and steam generating equipment. Addition of silicone resins (0.3 pound per cubic yard) to cement improves the strength and increases water repellency. Insulation for heavy-duty electrical generating and transmission equipment (silicones are able to withstand high temperatures for long periods), and other laminating, impregnating and bonding applications in the electrical industry, where silicones compete with epoxy resins, is the most important outlet for silicone resins. They are also used in release papers for tire packaging material, peel-off wall paper, and food wrap for sticky materials. Molding compounds accounted for most of the remainder.

For 1965, the demand breakdown was:

End-use	MM lbs, 1965
Coatings	2.5
Masonry additives	1.5
Electrical insulation	7.0
Release papers	1.0
Others	0.5
	12.5

The most important single application of room-temperature vulcanizable silicone rubbers is in caulking and sealing compounds (about 40 percent of the total); these formulations have the advantage over the cheaper polysulfides of not requiring on-the-job mixing. Other uses are in adhesives and ceramic glazes. Crosslinking is accomplished by the reaction of an organic silicon compound of the form $(RO)_4S$, and a catalyst.

Methyl cellulose

Cellulose can be etherified with methyl chloride to give a water-soluble resin:

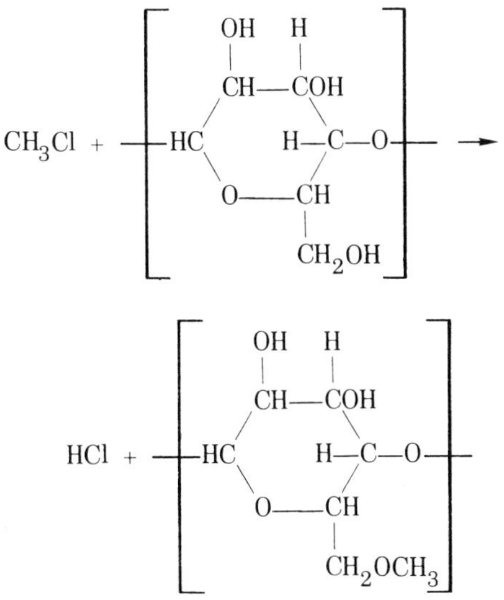

$$CH_3Cl + \left[\begin{array}{c} OH \quad H \\ | \qquad | \\ CH-COH \\ HC \qquad H-C-O \\ O-CH \\ | \\ CH_2OH \end{array} \right] \longrightarrow$$

$$HCl + \left[\begin{array}{c} OH \quad H \\ | \qquad | \\ CH-COH \\ HC \qquad H-C-O \\ O-CH \\ | \\ CH_2OCH_3 \end{array} \right]$$

The degree of methylation varies with the end-use. In its main applications, methyl cellulose competes with other water–soluble resins such as polyvinyl alcohol, carboxymethylcellulose and other cellulosics.   It is used as a paint thickener, a protective colloid for vinyl acetate adhesive, paper processing aid, etc. In 1965, about 20.0 million pounds were consumed.   The only U.S. manufacturer is Dow Chemical.

Arsenicals

Ansul Chemical manufactures a line of arsenical crab-grass killers such as disodium methyl arsonate, made from sodium arsenite and methyl chloride:

$$Na_3AsO_3 + CH_3Cl \longrightarrow CH_3\overset{\displaystyle O}{\overset{\displaystyle \|}{As}}(ONa)_2 + NaCl$$

Reduction, alkaline hydrolysis and reaction with another mol of methyl chloride gives monosodium dimethyl arsonate:

$$CH_3\overset{\displaystyle O}{\overset{\displaystyle \|}{As}}(ONa)_2 + H_2SO_3 \longrightarrow CH_3AsO + Na_2SO_4 + H_2O$$

$$CH_3\overset{\displaystyle O}{\overset{\displaystyle \|}{As}}O + 2NaOH \longrightarrow CH_3AsO_2Na_2 + H_2O$$

$$CH_3AsO_2Na_2 + CH_3Cl \longrightarrow (CH_3)_2\overset{\displaystyle O}{\overset{\displaystyle \|}{As}}ONa + NaCl$$

monosodium dimethyl
arsonate

The plant at Marinette, Wisc. has a capacity for 20.0 million pounds per year of these compounds.

Tetramethyl lead

Two factors are responsible for the increase in the usage of tetramethyl lead as a replacement for tetraethyl lead in gasoline antiknock fluid.   First, it is the lead that provides the antiknock action and use of the methyl compound diminishes requirements per gallon.   Second, TML is more effective in raising the octane numbers of gasolines containing a high proportion of catalytic reformate.   Due to the increase in the percentage of the world's gasoline pool derived from reforming, TML has grown in popularity (although not as fast as was predicted in 1960, when the product first appeared).

Nalco Chemical (Freeport, Texas) is the major producer of TML. It employs a route totally different from the classical TEL process, which is also used by other producers of lead alkyls to manufacture TML. Instead of reacting the alkyl chloride with a lead-sodium amalgam, 3 pounds of lead for each pound converted is recycled to form lead alkyl. The process begins with a Grignard reaction between methyl chloride and magnesium shavings:

$$CH_3Cl + Mg \rightarrow CH_3MgCl$$

Next, the Grignard intermediate is electrolyzed in the presence of excess methyl chloride in a cell where lead pellets constitute the anode. The overall reaction is as follows:

$$2CH_3Cl + 2CH_3MgCl + Pb \rightarrow (CH_3)_4Pb + 2MgCl_2$$

This process has the advantage over the conventional route in that: 1) all the lead is converted to tetraalkyl lead in a single pass; and 2) higher product yields of both lead and methyl chloride are obtained.   The Nalco process is flexible, and can turn out TML, TEL, or, if desired, a lead alkyl containing both methyl and ethyl groups. The byproduct magnesium chloride is recovered in a form suitable for reprocessing by the supplier of magnesium metal.

Approximately 50.0 million pounds of TML were produced in 1965, out of a total of 580.0 million pounds of lead alkyls.

3.4 Dimethylsulfate

Methanol (or the dimethyl ether obtained as a byproduct from methanol manufacture) reacts with oleum to give dimethyl sulfate:

$$CH_3OCH_3 + SO_3 \rightarrow (CH_3)_2SO_4$$

Its main application is as a methylating agent, for example, for making aryl methyl ethers.

3.5 Methyl Glucoside

Glucose (corn sugar) can be etherified with methanol to give methyl glucoside. Its ethylene oxide adducts are used as a polyols for the manufacture of rigid polyurethanes having exceptional thermal and dimensional stability; since the product contains four hydroxyl groups per mol, foams can be made which require less of the more expensive diisocyanate.

A plant producing 20.0 million pounds per year is operated by the Corn Products Refining Co.

3.6 Methyl Bromide

Methanol can be esterified with hydrogen bromide to give methyl bromide:

$$CH_3OH + HBr \rightarrow CH_3Br + H_2O$$

It is a widely used grain and soil fumigant, and is the preferred means of fighting ants in tropical countries. Production in the U.S. reached 17.0 million pounds in 1965, a substantial part of which was exported.

4 OXO CHEMICALS

The oxo reaction is the formation of an aldehyde from an olefin and 1:1 hydrogen-carbon monoxide synthesis gas.

$$RCH{=}CH_2 + CO + H_2 \rightarrow RCH_2CH_2CHO$$

Although any olefin can be used as feedstock to an oxo plant, the most important olefin raw material for oxo chemicals is propylene, as monomer or polymer. The reaction takes place usually in the 3,000 psig range. An olefin reacts with the synthesis gas over a cobalt carbonyl catalyst. The reactor effluent is sent to a phase separator and from there to a cobalt-removal system. The main difference between the various processes lies in the catalyst recovery technique.

In the Kuhlmann version, sodium carbonate is added to the reactor effluent to form a sodium-cobalt carbonyl complex. It is insoluble in the product stream, but soluble in water. Sulfuric acid decomposes the complex into sodium sulfate and cobalt carbonyl. The volatile cobalt carbonyl is recovered from the acid solution by stripping with fresh synthesis gas on its way to the reactor.

In other processes, the cobalt catalyst is decomposed thermally, or tied up as cobalt acetate, which can be removed from the oxo products as an aqueous solution.   Humble has a process in which cobalt acetate is added to the reactor effluent instead of acetic acid.   It forms a cobalt-cobalt carbonyl complex that is more soluble in water than cobalt acetate itself, reduces the amount of water recycled to the reactor, and inhibits corrosion.   There are two types of cobalt recovery: processes such as those of Kuhlmann and Humble, where cobalt is recycled as the desired cobalt carbonyl; and those in which cobalt is sent back to the reactor as a water soluble salt requiring cobalt carbonyl to be formed from the salt in situ by the carbon monoxide contained in the synthesis gas feed.   In all cases, the product stream is sent to a recovery section if the desired end-product is aldehyde, or to a catalytic hydrogenation step if, as is more often the case, the end-product is to be the respective alcohol.   A flow-diagram is shown in Fig. 2-25.

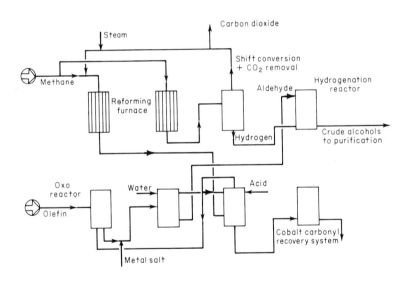

Fig. 2-25 Oxo Alcohols.

Humble developed a modification of the oxo process which is known as "Aldox."   In this process, a zinc catalyst is used to dimerize the oxo aldehyde just before hydrogenation.   Thus, for example, branched C_{16} alcohols can be made from heptene feed. Another modification of the oxo process, developed by Reppe, is being used in Japan to make butanol from propylene directly; it is reported that 85 percent of the total product consists of n-butanol,

with the remainder being isobutanol. The reaction takes place at
only 200-250 psi:

$$CH_3CH—CH_2 + 3CO + 2H_2O \longrightarrow CH_3CH_2CH_2CH_2OH + 2CO_2$$

Theoretically the oxo reaction can place the carbonyl group on
either side of the double bond. For propylene one would expect to
end up with a butyraldehyde-to-isobutyraldehyde ratio of around
2:1. Since the use of n-butanol is of longer commercial standing,
and also since 2-ethylhexanol is made from n-butyraldehyde,
manufacturers of oxo chemicals resort to various methods to raise
the proportion of the n-isomer. This can be accomplished in two
ways: 1) by adjusting the residence time in such a way as to favor
the formation of the unbranched product; and 2) by recycling the
unwanted product to the reactor, thus inhibiting its formation but
slowing down the reaction. Thus, producers are able to raise the
proportion of n-butyraldehyde to 80-85 percent. When higher
olefins are used the number of possible isomers rises, since the
double bond has a tendency to migrate along the carbon chain.

The distribution of oxo products made in the U.S. in 1964 is
shown below.

Product	MM lbs, 1964	Raw material
Propionaldehyde	20.0	Ethylene
n-Butyraldehyde	230.0	Propylene
i-Butyraldehyde	95.0	Propylene
Amyl alcohol	15.0	Isobutylene
Hexyl alcohol	10.0	Pentenes
Isooctyl alcohol	85.0	Heptene
Isodecyl alcohol	70.0	Propylene trimer
Tridecyl alcohol	10.0	Propylene tetramer
Others	10.0	Diisobutylene, etc.

The U.S. producers of oxo chemicals are listed below. Capacities are necessarily approximations, due to the variety of possible feedstocks and hence possible products.

Producer	Location	Capacity
Dow-Badische	Freeport, Texas	30
Enjay	Baton Rouge, La.	120
Gulf Oil	Philadelphia, Pa.	40
Houdry-Tidewater	Delaware City, Del.	30
Oxo Chemicals	Haverhill, Ohio	40
Shell	Houston, Texas	25
Texas Eastman	Longview, Texas	125
Union Carbide	Seadrift, Texas	240
		650

Source: *Chemical Week*, June 6, 1964.

Production costs of iso-octanol and 2-ethylhexanol via the oxo route are estimated below. The investment for a 40.0 million pounds-per-year plant starting from heptene is $4.2 million. Taking heptene at the captive value of 3.5 cents per pound, production costs can be estimated as follows.

Cost, ¢/lb iso-octanol

Heptene	3.8
Synthesis gas	0.6
Catalyst and chemicals	0.8
Utilities	1.6
Labor and overhead	1.2
Capital charges	3.5
	12.5

The investment for a 40.0 2-ethylhexanol-from-propylene plant is $5.7 million. Production costs are:

Cost, ¢/lb 2-ethylhexanol

Propylene	2.1
Synthesis gas	1.5
Catalyst and chemicals	2.4
Utilities	1.6
Labor and overhead	1.2
Capital charges	4.7
	13.5

For comparison, the investment for a plant of the same size starting from acetaldehyde is $3.3 million. Production costs are higher than for the "oxo" process, even if acetaldehyde is taken at 5.6 cents per pound, its internal cost from a large Wacker unit.

Cost, ¢/lb 2-ethylhexanol

Acetaldehyde	9.0
Utilities	1.1
Hydrogen, catalyst, chemicals	0.3
Labor and overhead	0.6
Capital charges	2.7
	13.7

4.1 n-Butyraldehyde

There are two ways of making n-butyraldehyde. Partial hydrogenation of crotonaldehyde, made from acetaldehyde gives a mixture of n-butanol and n-butyraldehyde, which can be separated by fractionation.

$$CH_3CH-CHCHO + H_2 \rightarrow CH_3CH_2CH_2CHO$$

It can also be obtained from propylene by the "oxo" reaction.

2-Ethyl-1, 3-hexanediol

This difunctional alcohol is made by aldolization and subsequent hydrogenation of *n*-butyraldehyde. It is used in the manufacture of polymeric plasticizers, and is well-known as the insect-repellent "6-12."

$$2C_3H_7CHO \longrightarrow C_3H_7C(OH)HCH(C_2H_5)CHO \xrightarrow{H_2} C_3H_7C(OH)HCH(C_2H_5)CHOH$$

Trimethylolpropane

Celanese produces trimethylolpropane by reacting butyraldehyde with three mols of formaldehyde:

$$3CH_2O + C_3H_7CHO \xrightarrow{NaOH} C_2H_5C(CH_2OH)_3 + HCOONa$$

It is used mainly in the manufacture of triols by adduction with propylene oxide, for flexible polyurethane foams, and of synthetic lubricants by esterification with fatty acids. The allyl ethers of TMP are used as crosslinking agents, for example, in acrylic resin systems. The capacity for TMP production is around 12.0 million pounds per year, and the demand about 10.0 million pounds per year, most of it for triols.

Butyric acid

Liquid-phase oxidation of *n*-butyraldehyde yields *n*-butyric acid, of which some 45.0 million pounds per year are used in the U.S.

$$C_3H_7CHO \xrightarrow{(O)} C_3H_7COOH$$

It is used mainly in the manufacture of cellulose acetobutyrate, the preferred cellulose ester for molding powders, especially for automotive applications such as steering wheels. However, other plastics, especially ABS resins, are making inroads into its traditional markets.

n-Butyronitrile, made from butyric acid and ammonia, is one of the raw materials for the widely used coccidiostat "Amprolium."

Butyric acid is also used to make a number of solvent esters of relatively minor importance.

The overall demand for *n*-butyraldehyde in 1964 other than for *n*-butanol and 2-ethylhexanol production, can be broken down as follows:

End-use	MM lbs, 1964
Butyric acid	40.0
Polyvinylbutyral	18.0
Trimethylolpropane	5.0
Other uses	5.0
	68.0

Other uses include a number of aldehyde-amine rubber anti-oxidants, made by condensation of, for example, *n*-butyraldehyde and aniline.

4.2 *i*-Butyraldehyde

Most of the isobutyraldehyde used is a byproduct when making *n*-butyraldehyde from propylene.   It is usually hydrogenated to isobutanol.

Isobutanol

At first, manufacturers of oxo chemicals from propylene found it difficult to eliminate the iso byproduct. However, efforts were made to use isobutanol in place of *n*-butanol at prices low enough to compensate for the difference in reactivity.   This resulted in an excellent growth rate.   While *n*-butanol markets have been rising at a rate of 3.7 percent per year, those for isobutanol have increased at almost 20 percent per year as a result of the price inducements offered by producers.

The demand for isobutanol in the U.S. was around 65.0 million pounds in 1964; the discrepancy between this figure and iso-butyraldehyde (95.0 mm pounds) is due to substantial exports. This demand is broken down below where it can be seen that most of the uses, with the exception of that in lube-oil additives, run parallel to those for *n*-butanol.

End-uses	MM lbs, 1964
Lube-oil additives (dithiophosphates)	22.0
Isobutyl acetate	18.0
Amine resins	9.0
Solvent for coatings	4.0
Plasticizers	1.0
Isobutyl vinyl ether	3.0
Others (acrylate, amines, ether, esters, etc.)	8.0
	65.0

In general, isobutanol esters have a sweeter odor than those of *n*-butanol; thus in addition to the price incentive they have an incentive for use as plasticizers in food packaging films.

Isobutyric acid

Isobutyric acid, made in the same way as butyric acid, is used to make a number of solvents such as isobutyl isobutyrate, a cheap retarding solvent, or plasticizer esters with glycerine, trimethylol-propane and other polyols. One of its important uses is as an intermediate in the manufacture of isobutyronitrile, from which iso-butylamidine hydrochloride, the raw material for the insecticide "Diazinon," is made.

$$(CH_3)_2CHCN + NH_4Cl \longrightarrow (CH_3)_2CHC\overset{\displaystyle NH_2}{=}N^+H_2Cl^-$$

The total demand for isobutyric acid is around 10.0 million pounds per year.

Neopentyl glycol

Isobutyraldehyde reacts with formaldehyde, and the reaction product can be hydrogenated to neopentyl glycol.

$$(CH_3)_2CHCHO + CH_2O \longrightarrow CH_2OHC(CH_3)_2CHO$$

$$CH_2OHC(CH_3)_2CHO + H_2 \longrightarrow CH_2OHC(CH_3)_2CH_2OH$$

Appreciable amounts are used for making polyesters and polymeric plasticizers having exceptional stability due to the alcohol structure.

Pantothenic acid

In the form of its calcium salt, pantothenic acid has become an important enriching compound for bread and animal feeds. The major raw materials are isobutyraldehyde and β-alanine:

$$(CH_3)_2CHCHO + CH_2O \longrightarrow CH_2OHC(CH_3)_2CHO \xrightarrow{HCN; HCl}$$

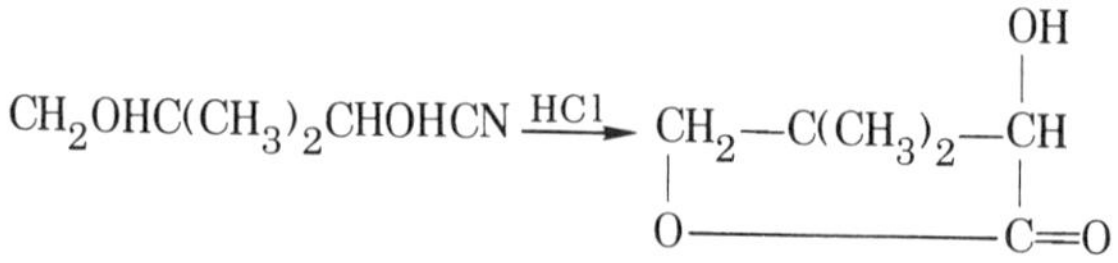

$$CH_2OHC(CH_3)_2CHOHCN \xrightarrow{HCl} \begin{array}{c} OH \\ | \\ CH_2-C(CH_3)_2-CH \\ | | \\ O\text{————————}C=O \end{array}$$

$$2 \begin{array}{c} \text{OH} \\ | \\ \text{CH}_2\text{—C(CH}_3)_2\text{—CH} \\ | \qquad\qquad | \\ \text{O} \text{————————C=O} \end{array} + (\text{NH}_2\text{CH}_2\text{CH}_2\text{COO})_2\text{Ca} \longrightarrow$$

$$[\text{HOCH}_2\text{C(CH}_3)_2\text{CH(OH)}\overset{\overset{\displaystyle O}{||}}{\text{C}}\text{NHCH}_2\text{CH}_2\text{COO}]_2\text{Ca}$$

The consumption of calcium pantothenate in the U.S. in 1965 was close to 2.0 million pounds.

4.3 Octanols

Among the several C_8 alcohols, two are of dominating commercial importance: 2-ethylhexanol, obtained by the aldolization of n-butyraldehyde and subsequent hydrogenation, and so-called "iso-octanol," actually a mixture of several C_8 isomers, made by the oxo reaction from a heptene feedstock.

In the case of 2-ethylhexanol, the n-butyraldehyde can in turn be obtained from two sources: partial hydrogenation of crotonaldehyde from aldolization and from propylene by the oxo process.

There are two other sources of C_8 alcohols. About 15.0 million pounds per year are obtained as a byproduct in the manufacture of sebacic acid from castor oil (ricinoleic acid). It is used to make mixed phthalate ester plasticizers with other C_8 alcohols.

With the decline in relative importance of sebacic as a di-carboxylic acid, this source is not expected to expand. Finally, some normal C_8 alcohol is made by manufacturers of detergent alcohols via fatty acid reduction of the more recent Ziegler-type processes.

While iso-octanol is slightly cheaper than 2-ethylhexanol, by 1965 the latter still accounted for 70 percent of the sum of the two. This is due to several factors: slightly higher volatility of di-iso-octyl phthalate; unavailability of heptene; better odor of di-2-ethylhexyl phthalate; and, finally, that fact that DOP has been available for a longer time, and compounders have become used to it.

The total U.S. demand for these two octanols in 1965 was 285.0 million pounds.

The overall demand for octanols can be broken down as follows:

End-use		MM lbs, 1965
Plasticizers:	Phthalates	210.0
	Phosphates	10.0
	Adipates	10.0
	Azelates	7.0
	Sebacates	3.0
	Epoxidized plasticizers	7.0
Surface-active agents		13.0
Pesticides		10.0
Octoic acid		2.0
2-Ethylhexyl acrylate		15.0
Miscellaneous		8.0
		295.0

Plasticizers are used mainly in PVC compounding, the most important being the phthalates. Esters of aliphatic acids are used in applications where low-temperature flexibility is required, e.g., footwear and household refrigerator parts. Epoxidized plasticizers are made by esterification of oleic acid, followed by epoxidation. Miscellaneous uses are mainly in the lube-oil additives field.

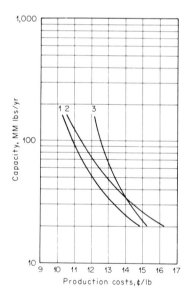

Fig. 2-26 C$_8$ Alcohols. 1-Iso-octanol ("oxo" process); 2-Ethylhexanol ("oxo" process); 3-2-Ethylhexanol (via crotonaldehyde).

Octoic acid

2-Ethylhexanol can be converted to the respective acid by liquid-phase oxidation:

$$C_4H_9CH(C_2H_5)CH_2OH \xrightarrow{(O)} C_4H_9CH(C_2H_5)COOH$$

The main application for octoic acid is in the form of its metal salts, used as odorless paint driers. The development of octoate paint driers represents an attempt on the part of oil-paint makers to retain some of the indoor market since one of the chief selling points of emulsion paints is their less unpleasant odor. In comparison with the cheaper naphthenates, of which some 16.0 million pounds per year are used in the U.S., 3.0 million pounds of octoates are used annually. With the relative decline of caustic treatment as a refinery process, however, the supply of naphthenic acids is not keeping pace with demand. Therefore most of the growth (especially in Europe) is attributed to synthetic long-chain carboxylic acids, among which the heavier "neoacids" also figure. In general, the use of paint driers is decreasing due to the advent of water-based paints. Outside of the coatings field, octoates are used in adhesives.

4.4 Propionaldehyde

Propionaldehyde, as such, has few uses. It is converted to propionic acid or *n*-propanol.

Propionic acid

Since it is very difficult to produce *n*-propyl derivatives from propylene, other routes must be found. Until the 1960's, the major source of propionic acid and *n*-propanol was direct LPG oxidation which produced *n*-propyl derivatives in addition to many other oxygenated compounds. The construction of a number of oxo plants provided a cheap route to propionaldehyde for later oxidation to propionic acid:

$$C_2H_4 + CO + H_2 \longrightarrow CH_3CH_2CHO \xrightarrow{(O)} CH_3CH_2COOH$$

The oxo process yields about 5.0 million pounds annually with diethyl ketone as a byproduct. This is used in miscellaneous solvent applications.   DuPont makes propionic acid by reacting ethyl alcohol, ethyl ether, and carbon monoxide under high pressure:

$$C_2H_5OH + CO \longrightarrow CH_3CH_2COOH$$

$$(C_2H_5)_2O + 2CO + H_2O \longrightarrow 2CH_3CH_2COOH$$

Finally, small amounts are made in the U.S. as a byproduct of hydroxylamine sulfate production and from wood distillation. The major U.S. producers of propionic acid are

Producer	Location	Capacity MM lbs/yr	Process
Celanese	Pampa, Texas	10.0	LPG oxidation
DuPont	Belle, W. Va.	10.0	$(C_2H_5)_2O + H_2O + CO$
Texas Eastman	Longview, Texas	10.0	Oxo (aldehyde only; acid made at Kingsport, Tenn.)
Union Carbide	Seadrift, Texas	10.0	Oxo (aldehyde only; acid made at S. Charleston, W. Va.)

Calcium and sodium propionates, made by the high-pressure reaction between propionic acid and the respective hydroxide, are used almost exclusively as bread preservatives.

$$2CH_3CH_2COOH + Ca(OH)_2 \longrightarrow (CH_3CH_2CO)_2Ca + 2H_2O$$

Boiling acetic anhydride with propionic acid gives propionic anhydride, used in producing cellulose acetopropionate and tripropionate.

Cellulose propionates were used once for making black telephones, but the more popular colored sets are now made of ABS. It is also used in toothbrush handles, pen and pencil housings and toys, but the demand is fairly static.

Several companies make 2-chloro and 2,2-dichloropropionic acid derivatives, which are fairly similar to 2,4D herbicides, but less toxic to farm animals. "Silvex" is the 2-chloropropionic acid analog of 2,4,5T; "Dalapon" is the sodium salt of 2,2-dichloropropionic acid; and "Erbon" is the ester of this acid with ethoxylated 2,4,5 trichlorophenol.

Propionic acid reacts with phenylmercuric acetate to give phenylmercuric propionate, a well-known paint fungicide.

Propionic acid esters, such as ethyl propionate, are solvents for cellulosic resins; ethyl propionate is also used as imitation rum flavoring.   Glycerol tripropionate is a plasticizer for cellulose cigarette filter tips.

Propionyl chloride obtained from propionic acid and a chlorinating agent such as PCl_3 or phosgene, reacts with benzene to give propiophenone

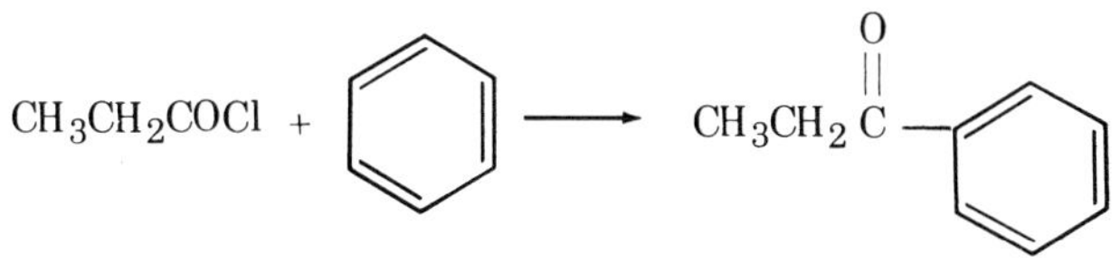

$$CH_3CH_2COCl \; + \; \bigcirc \longrightarrow CH_3CH_2 \overset{\overset{\textstyle O}{\|}}{C} - \bigcirc$$

The reaction of propionyl chloride with phenol produces 4-hydroxy-propiophenone, starting point for making diethylstilbestrol. This compound is the most important anabolic regulator for cattle; it is used to improve the conversion of feed to saleable meat. About 0.3 million pounds per year of this drug are used annually in the U.S. Propionyl chloride reacts with 3,4 dichloroaniline to give "Stam," a post-emergence herbicide used mostly on rice crops.

In 1965, 35.0 million pounds of propionic acid was consumed. The breakdown is as follows:

End-use	MM lbs, 1965
Calcium and sodium propionates	11.0
Cellulose acetopropionate	13.0
Herbicides	5.9
Solvents and plasticizers	3.0
Other uses	3.0

n-Propanol

Propionaldehyde can be hydrogenated to *n*-propyl alcohol. In addition to certain solvent esters, it is used to make *n*-propyl gallate, a food antioxidant usually employed in proprietary mixtures with BHT and BHA. It can be converted to *n*-propyl mercaptan, one of the intermediates for "Tillam," a herbicide of the thiocarbamate family.

The most important thiocarbamate herbicide, "Eptam," is derived from di-*n*-propylamine, also made from *n*-propyl alcohol.

4.5   Heavy Oxo Chemicals

About 70.0 million pounds per year isodecyl alcohol were made from propylene trimer in 1964 by the oxo reaction. The use pattern of this alcohol is shown below; di-isodecyl phthalate is used mainly in PVC wire coatings because of its low volatility. It is also used to make synthetic lubricants and plasticizer esters from acids other than phthalic anhydride.

End-use	MM lbs, 1964
Di-isodecyl phthalate	44.0
Other plasticizers	8.0
Lubricants, etc.	18.0
	70.0

Tridecyl alcohol is made from propylene tetramer. Its main outlet is di-tridecyl phthalate, also used in cases where low volatility is especially desirable, e.g., in PVC wire coating. Some is also used in synthetic lubricants and to make surface-active agents. Consumption in 1964 was 10.0 million pounds. No long-chain alcohols were made by the oxo process in 1964. The first plant for this preparation by Shell began operation at Houston, Texas in 1965. This plant is believed to carry out the oxo reaction on straight chain alpha-olefins. The alpha-olefin is obtained from n-paraffin separated from petroleum, followed by cracking. The chain length of the product depends on the length of the chain of the alpha-olefin. The product is a mixture of 80 percent straight chain primary alcohol and 20 percent straight chain primary alcohol with a methyl group in the alpha position.

A second plant producing the same alcohols is scheduled to be in operation by late 1966 at Geismar, Louisiana, also by Shell. The capacity of the Houston and Geismar plants is respectively 50 million pounds per year and 100 million pounds per year of long-chain fatty alcohol.

5   PHOSGENE

In spite of the notoriety it achieved during World War I as a poison gas, phosgene was hardly more than a laboratory chemical until around 1950. Since then, it has become a heavy chemical, with production reaching 265.0 million pounds in 1964.

Phosgene is made by the reaction between carbon monoxide and chloride:

$$CO + Cl_2 \rightarrow COCl_2$$

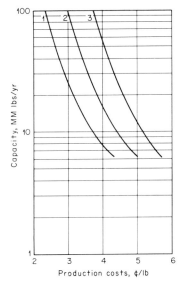

Fig. 2-27 Phosgene. 1-Chlorine = $30/ton;
2-Chlorine = $50/ton; 3-Chlorine = $70/
ton.

The reaction is carried out at essentially atmospheric pressure
in a reactor packed with activated carbon. In the subsequent applications of phosgene, the presence of free chlorine is undesirable;
thus excess carbon monoxide is used in order to ensure complete
conversion.   Purification is accomplished by condensing out the
product, which is a liquid below 47° F. The off-gases are washed
with water to destroy any remaining phosgene, and elaborate precautions are necessary to prevent any loss of product.

Recent phosgene plants constructed in the Gulf Coast area
obtain their CO from synthesis gas by low-temperature fractionation. Early units derived their raw material from coal, directly or
indirectly.   Most phosgene producers have their own caustic
chlorine facilities.

The economics of producing phosgene from captive chlorine
(2.5 cents per pound) and from CO (1.0 cent per pound) are estimated below.   The investment for a 25.0 million pounds-per-year
plant is around 0.8 million.

	Cost, ¢/lb phosgene
Carbon monoxide	0.4
Chlorine	1.8
Utilities	0.2
Labor and overhead	0.2
Capital charges	1.1
	3.7

Since most phosgene is consumed captively, one should assume that the HCl will be recovered in an oxidation unit and chlorine therefore taken at only 30–70 percent of its production cost. Even at the higher figure, captive phosgene can probably be costed in to other units in the plant at only 3.2 cents per pound from a unit of this size.

The ultimate use of phosgene in 1964 can be broken down as follows:

End-uses	MM lbs, 1964
Diisocyanates	165.0
"Sevin"	41.0
Other pesticides	13.0
Cyanuric acid	35.0
Polycarbonates	4.0
Dyes	5.0
Other chemicals	10.0
	265.0

The U.S. manufacturers of phosgene, with their principal captive outlets, are given below.

Producers	Plant location	Capacity MM lbs/yr	Captive chlorine	Main captive uses
Allied Chemical	Moundsville, W.Va	75.0	x	Diisocyanates
Chemetron	Elkton, Md.	6.0	–	Chloroformates, carbonates
	La Porte, Texas	20.0	–	
Corn Products	Muskegon, Mich.	10.0	–	Intermediates
DuPont	Deepwater, N.J.	110.0	x	TDI, MDI, herbicides
FMC	Baltimore, Md.	2.0	x	Chloroformates, carbamates
General Electric	Mt. Vernon, Ind.	20.0	–	Polycarbonates
Kaiser Chemicals	Gramercy, La.	35.0	x	MDI
Mobay	New Martinsville, W.Va.	120.0	x	Diisocyanate, polycarbonates, intermediates, cyanuric acid
Olin Mathieson	Ashtabula, Ohio	30.0	x	TDI
Pittsburgh Plate Glass	Barberton, Ohio	5.0	x	Chloroformates, percarbonates
Stauffer	Richmond, Cal.	8.0	x	Herbicides
Union Carbide	Institute, W.Va.	90.0	–	TDI, methylcarbamates
Upjohn	La Porte, Texas	30.0	–	"PAPI," special diisocyanates
Van de Mark	Lockport, N.Y.	5.0	–	Custom phosgenation
Wyandotte	Geismar, La.	50.0	x	TDI
		616.0		

Source: *Chemical Week*, November 28, 1964.

5.1 Diisocyanates

Primary aliphatic or aromatic amines can be converted to iso-cyanates by reaction with phosgene:

$$
\begin{array}{c}
ArNH_2 \\
+ COCl_2 \\
RNH_2
\end{array}
\quad
\begin{array}{l}
\nearrow Ar{-}N{=}C{=}O + 2HCl \\
\searrow R{-}N{=}C{=}O + 2HCl
\end{array}
$$

Monofunctional isocyanates are raw materials for numerous pesticides, medicinals and textile chemicals. Diisocyanates and polyisocyanates, which for the most are derived from aromatics, are starters for making polyurethane foams.

Toluene can be nitrated to a mixture of 2,4 and 2,6-dinitro-toluene. Nitration is accomplished in a series of countercurrent nitrators, and a mixture of 100 percent nitric acid and nitric acid-water azeotrope is used in order to provide a sufficiently strong acid to dispense with sulfuric acid as a dehydrating agent.

After nitration, the reactor effluent is settled and neutralized with caustic soda. Dinitrotoluene, which also finds some use as an intermediate for sulfur dyes, is next reduced by hydrogenation under pressure; the effluent is fractionated to remove the water and any unreacted or only partially reduced feed, which is recycled.

Finally, toluene diamine is reacted with phosgene in a solvent chosen for its high capacity to dissolve the extremely poisonous gas, usually monochloro- or dichlorobenzene. After passing through a series of phosgenators, the reactants are stripped by inert gas of any remaining phosgene and sent to a purification system, where the solvent is recovered and the product freed of tars.

Toluene diisocyanate is still the main raw material for making flexible foams, which were at first the chief commercial outlet for polyurethanes. With time, however, rigid foams grew in importance for such applications as household and commercial refrigerator insulation, refrigerated trucks and box-cars, etc., where the unusually low conductivity of urethane foams gave them a competitive edge over such earlier foams as polystyrene. Furthermore, in these applications, as well as in construction insulation, there was a great incentive to develop materials that could produce foams "in loco" by mixing diisocyanate and hydroxyl source on the job instead of in a separate operation. For this to be practicable, however, other diisocyanates had to be found, as TDI proved too reactive and thus liberated too much heat to allow formation of a satisfactory foam in a single shot. This is the origin of the fast-growing demand for the isocyanates derived from aniline.

Although one-shot rigid foams can also be made from TDI low in 2,4-isomer content, the major part of these urethane foams is now made from MDI (made by condensing two mols of aniline with one of formaldehyde and phosgenating) and PAPI, a polyisocyanate also made from aniline and formaldehyde. The demand for diisocyanates based on aniline is distributed in roughly equal proportions between PAPI and MDI.

Another advantage of these isocyanates over TDI is that they are more suitable for manufacturing fire-retardant foams. Since much of the growth in demand for these foams is tied to the increase in electrically heated homes — where a higher degree of insulation is required to compete with other methods of space heating that require a higher first cost but use a cheaper source of energy — fire-retardancy is a most important property. Such foams are obtained by reacting the diisocyanate with a chlorine or phosphorus-containing polyol. MDI is also used to make "Spandex" fibers, and both MDI and PAPI are raw materials for urethane elastomers.

The investment for a 20.0 million pounds per year TDI plant, including nitration, reduction and phosgenation facilities, is $11.5

million.   Taking phosgene at a captive cost of 5.2 cents per pound, production costs can be calculated as follows:

	Cost, ¢/lb TDI
Toluene	2.0
Phosgene	6.8
Hydrogen	0.9
Nitric acid	2.3
Labor and overhead	4.8
Utilities	1.2
Capital charges	19.2
	37.0

Isocyanates production can be seen to be highly capital-intensive, as capital charges account for over 50 percent of the total. Major producers of diisocyanates are given below.

Producers	Locations	Capacity MM lbs/yr	Type of diisocyanate
Allied	Moundsville, W.Va.	30.0	TDI
Corn Products	Muskegon, Mich.	5.0	Specialty
DuPont	Deepwater, N.J.	70.0	TDI
Kaiser	Gramercy, La.	20.0*	Aniline-based
Mobay	New Martinsville, W.Va.	70.0	TDI & aniline-based
Olin	Ashtabula, Ohio	20.0	TDI
Rubicon (U.S. Rubber & Imperial Chemical)	Geismar, La.	25.0	TDI
Union Carbide	Institute, W.Va.	25.0	TDI
Upjohn	Houston, Texas	20.0**	TDI & specialty
Wyandotte	Geismar, La.	35.0	TDI

*Plant completed in 1966.
**An additional 50 million lbs/yr of capacity under construction near Houston, Texas.
Source: Modern Plastics, November, 1966.

The first plant to make polyurethanes in the U.S. was that of Mobay, a joint venture undertaken by Monsanto and Bayer using know-how developed by Bayer during World War II. Several of the earlier diisocyanate plants were located in the vicinity of metallurgical sources of carbon monoxide, but more recent plants have been located on the Gulf Coast, and obtain CO from natural gas or as a byproduct from other petrochemical operations. Another significant trend is that two of the three plants erected recently — Rubicon and Kaiser — were designed primarily for aniline-derived isocyanates, reflecting the faster growth-rate of rigid foams.

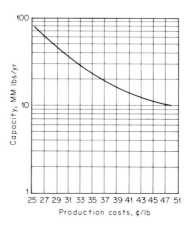

Fig. 2-28 Toluene Diisocyanate (TDI).

Polyurethane foams

Polyurethanes are obtained by reacting a difunctional isocyanate with a polyfunctional hydroxyl source. Polyols and diisocyanates are usually employed.

Polyurethane products are generally classified into four categories: rigid foams; flexible foams; elastomers; and coatings. Of
these, rigid foams are the fastest-growing application, but flexible
foams are still by far the most important. TDI is used only to a
limited extent in making rigid foams, most of the isocyanates for
this application being aniline derivatives. The manufacture of
flexible foams is carried out by one of several slabbing techniques,
in which a thick slab of foam is formed on a continuous conveyor
and then out to size for final use. Since the foam is allowed to
rise freely on the conveyor, it is desirable to mold as thick a

slab as is consistent with proper reaction control in order to minimize trimming losses.

Polyurethane elastomers have the advantage of being castable, or formed by injection and extrusion molding. "Spandex" fibers are usually included in this class.

Polyurethane coatings are mostly oil-modified, and are beginning to compete with alkyds. They are highly resistant to abrasion, corrosion and impact, and their applications are in fields where one or more of these properties are required.

Flexible urethane foams consist of 30 percent isocyanate, 65 percent resin and 5 percent blowing agent, with minor amounts of catalysts and a surface active agent that acts as a cell stabilizer. The demand in the U.S. for flexible urethane foams was 232 million pounds in 1965 requiring 80.0 million pounds of isocyanates. The consumption pattern was as follows:

End-uses	MM lbs, 1965
Furniture	96.0
Bedding	33.0
Transportation	50.0
Textile laminates	30.0
Rug and underlay	3.0
Packaging	5.0
Others	15.0
	232.0

Source: *Modern Plastics*, January, 1966.

In furniture and bedding, flexible urethane foams have taken over much of the market once held by foam rubber. Although costs per unit weight are much higher, polyurethanes have a density of about 1.5 pounds per cubic foot versus 4.0-4.5 pounds per cubic foot for foam rubber, and are cheaper per unit volume. Automotive uses are certain to benefit from demand for safer cars. Among the established applications requiring foams in the 4-6 pounds per cubic foot range, are crash-pads, seating (especially for bucket seats), arm rests and liners. Laminates are made by flame bonding of polyurethanes with textiles. They find wide use in men's garments. Miscellaneous uses include buoyancy products such as the filling of cavities in submarines.

By 1965, roughly 45 percent of the rigid urethane foams were already being made "in place"; this percentage should eventually rise to 60 percent, and the total market for rigid foams is expected to grow at a rate of 15 percent per year. The demand for rigid urethane foams in 1965 can be broken down as follows.

	MM lbs, 1965
Appliances	23
Transportation	20
Building	15
Industrial tanks	3
Flotation	4
Packaging	1
Mining tunnel seats	2
Others	3
	71

Source: *Modern Plastics*, January, 1966.

The best prospects for rigid urethane foams are in construction, e.g., in roofs, panels, shells, and various types of walls. Urethane elastomers are also growing in use. Aside from the 14.0 million pounds of "Spandex" (an elastic polyurethane fiber), about 9.0 million pounds were consumed in 1965.

The use of urethane rubbers in industrial tires is growing. This is because of their excellent resistance to oily surfaces. Industrial and consumer goods components made of cast urethane rubber are extremely durable and chemically resistant. In the sealant field, urethanes are a minor factor, with polysulfides and polyesters accounting for most of the market. Spandex has captured about 50 percent of the 25.0 million pounds per year demand for elastic fibers such as used in girdles and orthopedic garments.

About 6.4 million gallons of urethane coatings were consumed in 1965, the principal outlets being wood finishes for gymnasium floors (and other places where abrasion resistance is desired), and marine finishes. The overall distribution was as follows:

End-uses	MM gallons, 1965
Wood finishes	5.0
Magnet wire coating	0.5
Leather coating	0.6
Other	0.3
	6.4

Source: *Chemical & Engineering News*, October, 1965.

About 1.0 pound of isocyanate is required per gallon of coating.

Finally, polyurethanes are the major constituent of "Corfam," the synthetic leather for shoe uppers developed by DuPont. Ultimate potential for this material can be gauged by the fact that 640 million pairs of shoes are made each year.

Apart from exports, the total demand for isocyanates in polyurethanes in 1965 was 125.0 million pounds, broken down as follows:

End-use	MM lbs, 1965
Flexible foams	80.0
Rigid foams	32.0
Others	13.0
	125.0

Source: *Modern Plastics*, November, 1965.

TDI accounted for approximately 80 percent of this demand.

5.2 Polycarbonates

Polycarbonate resins are made by the reaction between phosgene and sodium salt of bisphenol A or other compounds containing two hydroxyl groups.

These resins are "engineering plastics," that is, thermoplastics that can often replace metals because of their toughness, high strength and softening point. Polycarbonates have not lived up to their very promising future predicted earlier. Apart from high cost, their main disadvantage is poor resistance to ultraviolet light. It is expected that when UV stabilization is satisfactorily solved, polycarbonates will find widespread use in glazing and institutional window panes. Prices are potentially much lower, as can be seen from the raw material costs.

The two U.S. producers are General Electric and Mobay*. Teijin and Edogawa in Japan, and Bayer in Europe round out the list of producers. U.S. capacities are given below.

	Capacity MM lbs/yr
General Electric	
Mt. Vernon, Ind.	20.0
Mobay	
New Martinsville, W.Va.	10.0

Source: *Oil, Paint & Drug Reporter*, August 16, 1965.

*A recent decision has dissolved this partnership. Bayer is now the sole owner.

The resins can be made into film as thin as 1/2 mil., and to a high degree of transparency; glass-reinforced grades have been developed for use in gears and bearings. Major applications are: photographic film (for "Anscochrome"), where it competes with metallized polyester due to the thin gauges that can be attained; safety devices such as goggles, shields and helmets; power tool housings, where their excellent electrical properties are the determining factor, although until now, more stringent safety laws make this application more important in Europe than in the U.S. Their resistance to vandalism is the main attraction for use in street-lighting. Other uses are in making glass-filled molded parts for electronic data-processing machines, household appliance parts, for example, air-conditioner housings and nursing bottles, and finally, in a number of aircraft and military applications. The end-use break-down for 1965 is given below.

End-uses	MM lbs, 1965
Appliances and lighting	5.9
Safety devices	2.6
Molded parts for electrical equipment	2.5
Film and sheet	1.5
Mechanical and transportation	1.2
Other	1.5
	15.2

Source: *Modern Plastics*, January, 1966.

5.3 Chlorinated Isocyanurics

Isocyanuric acid is the starting point for the production of chlorinated isocyanurates, which find considerable application as solid bleaches. The most important member of this group is potassium dichlorocyanurate, made by chlorinating isocyanuric acid and neutralizing with potassium hydroxide. Isocyanuric acid may be made from phosgene and ammonia or from urea.

$$COCl_2 + NH_3 \longrightarrow NH_2COCl + HCl$$

$$3NH_2COCl \longrightarrow 3HCl + $$
or
$$3NH_2CONH_2 \longrightarrow 3NH_3 + $$

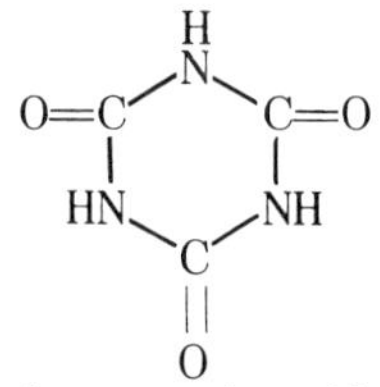

Isocyanuric acid

$$
\begin{array}{ccc}
& \underset{\displaystyle \underset{O=C}{\overset{H}{\big|}}{\overset{N}{\diagdown}}C=O}{} & \\
\text{HN} \diagdown \underset{\displaystyle \underset{O}{\overset{\|}{C}}}{} \diagup \text{NH}
\end{array}
\quad \xrightarrow{\text{Cl}_2;\ \text{KOH}} \quad
\begin{array}{ccc}
& \underset{\displaystyle \underset{O=C}{\overset{K}{\big|}}{\overset{N}{\diagdown}}C=O}{} & \\
\text{ClN} \diagdown \underset{\displaystyle \underset{O}{\overset{\|}{C}}}{} \diagup \text{NCl}
\end{array}
$$

In economies where chlorine is expensive, such as Japan, iso-cyanuric acid is made by pyrolysis of urea, yielding ammonia as a byproduct.

These products have the great advantage of being transportable in concentrated, solid form, instead of in solution as is the case with liquid chlorine bleaches. They are used in household formu-lations, as swimming pool disinfectants, commercial bleaches and scouring compounds, and in dishwashing preparations.

The two U.S. producers are FMC in S. Charleston, W.Va. with a capacity of 6.0 million and Monsanto, with capacity of 30.0 million pounds per year in two plants. The demand for these dry bleaches reached 21.0 million pounds in 1964, with the distribution given in the following.

End-uses	MM lbs, 1964
Household bleaches	10.0
Commercial bleaches	5.0
Dishwashing preparations	3.0
Swimming pools, etc.	3.0
	21.0

5.4 Substituted Urea, Carbamate and Thiocarbamate Pesticides

Substituted urea herbicides are made by the reaction between an aromatic isocyanate and an aliphatic amine.

$$
\text{Ar}-\text{N}=\text{C}=\text{O} + \text{RR}'\text{NH} \longrightarrow \text{R}-\underset{\text{H}}{\overset{\displaystyle}{\text{N}}}-\underset{\text{O}}{\overset{\displaystyle}{\text{C}}}-\text{N}\overset{\text{R}}{\underset{\text{R}'}{\diagup\diagdown}}
$$

Below, we list some compounds which belong to this class.

Product	Producer	Isocyanate parent compound	Amine	Use
"Diuron"	DuPont	3,4-Dichloroaniline	dimethyl	Weed control (preemergence)
"Fenuron"	DuPont	Aniline	dimethyl	Cotton, sugar, fruit trees
"Neburon"	DuPont	3,4-Dichloroaniline	methyl-n-butyl	Ornamentals
"Urox"	Allied	p-Chloroaniline	dimethyl	Nonselective weed control
"Urab"	Allied	Aniline	dimethyl	Nonselective weed control

"Diuron" is the most successful of this group while "Urox" and "Urab" are actually the respective salts of trichloroacetic acid.

Carbamates are made by the reaction between an isocyanate and hydroxy compound, or a chloroformate and an amine.

$$R{-}N{=}C{=}O + R'OH \rightarrow R{-}\underset{\underset{H}{|}}{N}{-}\overset{\overset{O}{||}}{C}{-}OR'$$

or

$$R'OH + COCl_2 \rightarrow R'O\overset{\overset{O}{||}}{C}Cl + HCl$$

$$R'OCOCl + RNH_2 \rightarrow R{-}\underset{\underset{H}{|}}{N}{-}\overset{\overset{O}{||}}{C}{-}OR' + HCl$$

Some examples of carbamate herbicides appear below.

Product	Producer	Hydroxy compound	Amine	Use
"Barban"	Spencer	4-Chloro-2-butynol	*m*-Chloroaniline	Postemergence wild oat control
"CIPC"	Pittsburgh Plate Monsanto, USI	Isopropanol	*m*-Chloroaniline	Preemergence herbicide for cotton and vegetables
"IPC"	Pittsburgh Plate Food Machinery	Isopropanol	Aniline	Annual grass control

The only carbamate that has found successful use as an insecticide is "Sevin." Carbamates are highly selective in their action and have low mammalian toxicity, and "Sevin," for example, is 10 to 100 times less toxic to livestock than most insecticides. At first, their selectivity was considered economically disadvantageous compared with the less discriminating poisons, but the pressure of public opinion is forcing pesticide manufacturers to develop less hazardous materials. Another reason for the spectacular rise in demand for carbamates has been the growing resistance by some of the major pests, for example, boll weevils, to the traditional chlorinated and phosphate insecticides.

The raw materials for "Sevin" are α-naphthol and monomethylamine.

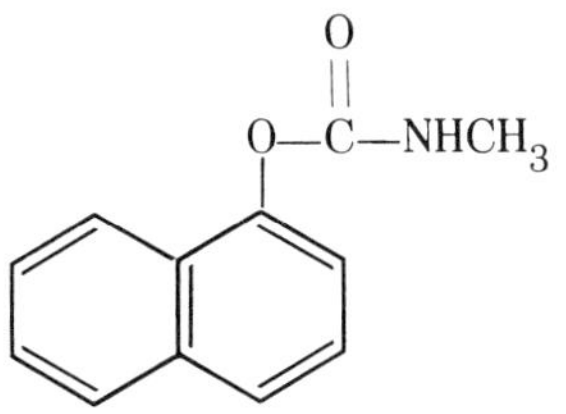

$$O-\overset{\overset{\displaystyle O}{\|}}{C}-NHCH_3$$

Thiolcarbamates differ from carbamates in that a mercaptan replaces the hydroxy compound. Some examples of thiolcarbamate herbicides are given below.

Product	Producer	Mercaptan	Amine	Use
"Avadex"	Monsanto	2,3-Dichloroallyl	Diisopropyl	Wild oat control (pre-emergence)
"Eptam"	Stauffer	n-Propyl	Di-n-propyl	Preemergence herbicide for potatoes, beans, corn
"Tillam"	Stauffer	Ethyl	Ethyl-n-butyl	Preemergence herbicide for sugar beet

5.5 Other Phosgene Derivatives

Diethyl carbonate is made by first converting ethanol to ethyl chloroformate, then reacting with another mol of alcohol:

$$COCl_2 \xrightarrow{C_2H_5OH} C_2H_5O\overset{\overset{\displaystyle O}{\|}}{C}Cl \xrightarrow{C_2H_5OH} (C_2H_5O)_2CO$$

It is a specialty solvent for synthetic resins and an intermediate for certain pharmaceuticals.

Phosgene enters the synthesis of several dyes and dye intermediates. Michler's ketone, for example, is made by the Friedel-Crafts reaction between phosgene and dimethylaniline:

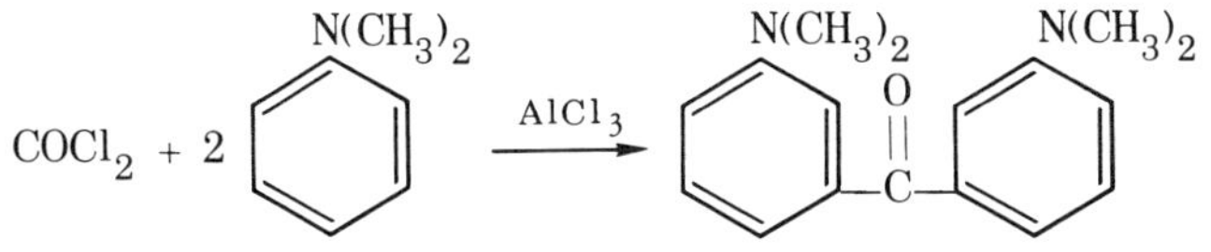

$$COCl_2 + 2$$ $$\xrightarrow{AlCl_3}$$

and is also a coupling agent for many azo dyes.

About 1.0 million pounds per year of isopropyl percarbonate are used in the U.S. as a vinyl polymerization catalyst:

$$(CH_3)_2CH-O-O-\overset{\overset{\displaystyle O}{\|}}{C}-O-O-CH(CH_3)_2$$

Phosgene is one of the best chlorinating agents for making acyl chlorides.

$$2RCOOH + COCl_2 \rightarrow 2RCOCl + CO_2 + H_2O$$

6 FORMIC ACID

Some of the formic acid produced in the U.S. is obtained as a byproduct from the manufacture of pentaerythritol. Most of it, however, is made by absorbing CO either in caustic soda, followed by neutralization:

$$CO + NaOH \rightarrow HCOONa \xrightarrow{H^+} HCOOH$$

or in methanol, at high pressure, followed by conversion to formamide and hydrolysis (DuPont process):

$$CO + CH_3OH \rightarrow H\overset{\overset{\displaystyle O}{\|}}{C}OCH_3$$

$$H\overset{\overset{\displaystyle O}{\|}}{C}OCH_3 + NH_3 \rightarrow HCONH_2 + CH_3OH$$

$$HCONH_2 \xrightarrow[H_2O]{H^+} HCOOH + NH_3$$

Outside the U.S., the main source of formic acid is now that obtained as a byproduct of the Distillers acetic acid process.

Demand for formic acid in 1965 was around 21.0 million pounds. The main application outside the U.S. is as a coagulant for natural rubber latex; domestically, over half of the total is used instead of sulfuric acid in high-temperature acid textile dyeing, and to some extent in leather tanning. About 10.0 million pounds of sodium formate is consumed annually.

Producers of formic acid in the U.S. are given below.

Producers	Location	Capacity MM lbs/yr	Process
DuPont	Philadelphia, Pa.	10.0	CO + CH$_3$OH
Stauffer	Chicago Heights, Ill.	8.0	CO + NaOH
Sunoco	Hartsville, S.C.	1.0	
Tenneco	Fords, N.J.	6.0	By-product

Source: *Oil, Paint & Drug Reporter*, March 29, 1965.

6.1 Oxalic Acid

Sodium formate can be heated to give sodium oxalate and hydrogen.

$$2HCOONa \longrightarrow \begin{matrix} COONa \\ | \\ COONa \end{matrix} + H_2$$

Oxalic acid is recovered by precipitation as calcium oxalate, neutralization and crystallization. Figure 2-29 shows a flow-sheet for a plant producing formic and oxalic acids. Of the three U.S. producers, two use this route. However, only one of these also produces formic acid, the other converting all of it to oxalic. The third obtains oxalic acid by fermentation.   In France, Rhône-Pouleuc has developed an ethylene oxidation route to oxalic acid. The demand for oxalic acid is fairly steady at 22.0 million pounds per year, with the following breakdown.

End-use	MM lbs/yr	%
Rust removal in radiators, other metal cleaning	6.0	27.0
Textile finishing	5.0	23.0
Tanning	1.0	4.0
Oxalates and other intermediate uses	5.5	25.0
Other uses	4.5	21.0
	22.0	100.0

Source: *Oil, Paint & Drug Reporter,* May 18, 1964.

Oxalic acid has several uses as an intermediate, mainly in the pharmaceutical industry; for example: ethyl oxalate is one of the

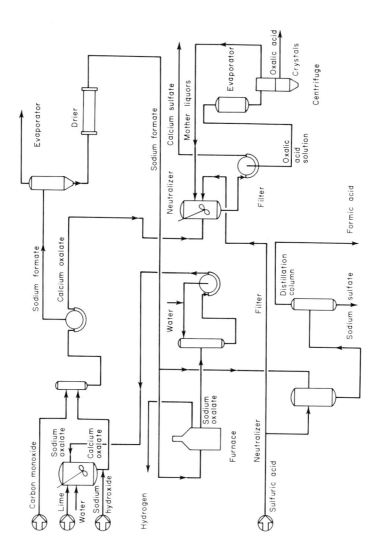

Fig. 2-29 Formic and Oxalic Acids.

raw materials for 4,7-dichloroquinoline, starting point for some of the most effective antimalarial drugs; and for phenobarbital. Its esters are used as high-boiling solvents in printing-ink formulations, and its salts mainly by the metal plating industry. The producers in operation by 1965 were:

Producer	Location	Capacity MM lbs/yr	Process
Allied*	Buffalo, N.Y.	10.0	CO-NaOH
Pfizer	Groton, Conn.	10.0	Fermentation (citric acid by-product)
Stauffer	Chicago, Ill.	5.0	CO-NaOH

*Allied is planning a new oxalic acid plant in N. Claymont, Del. with a capacity of about 10 million lbs. Operation to begin in early 1967.

Source: *Oil, Paint & Drug Reporter*, September 12, 1966.

7 NEO-ACIDS

Olefins react with carbon monoxide and water (the Koch reaction) in a two-step process taking place at around 300 atmos. and in the presence of strong acids to form tertiary carboxylic acids better known as neo-acids or versatic acids. The overall reaction is

$$(CH_3)_2C{=}CH_2 + CO + H_2O \longrightarrow (CH_3)_3C{-}COOH$$
$$\text{Pivalic or neopentanoic acid}$$

Although pivalic acid is the best known member of the series, other olefins are used to produce longer tertiary acids. These can be propylene trimer or tetramer, diisobutylene, and cracked wax in the C_8-C_{10} range.

The peculiar properties of these acids are mild odor and exceptional stability of their esters, due to presence of a tertiary carbon atom. Neopentanoic acid is used to prepare esters of drugs that are intended to be released slowly in the body. It is also an intermediate for t-butyl peroxypivalate, a catalyst rapidly gaining acceptance for low-temperature vinyl chloride polymerization. Higher neo-acids have their most important applications in paint driers, (this may become an important outlet in Europe, where naphthenates are not as abundant as in the U.S.), and in vinyl resins made by reacting these acids with acetylene followed by polymerization.

Enjay (Baton Rouge, La.) operates a 5.0 MM pounds per year plant in the U.S., and Shell uses the same process on a commercial scale at Pernis, Holland, and imports product into the U.S.

8 PURE HYDROGEN

Most pure hydrogen applications require amounts that are very small compared to those involved in making heavy synthesis gas derivatives. In the U.S., even such activities as hydrogenating fats and oils have by now reached such dimensions that plants already find it worthwhile to obtain their hydrogen by steam reforming. In less developed economies, the problem of obtaining small amounts of hydrogen often arises and it is worth discussing some of the usual solutions.

The most common source of pure hydrogen in small amounts is as a byproduct from caustic-chlorine electrolysis plants. A 50 ton-per-day chlorine plant, for example, produces around 500,000 scfd of hydrogen, a quantity far in excess of the needs of a typical fats and oils hydrogenation unit of a fatty alcohols plant. Hydrogen, however, is costly to transport, and the problem of generating it in small amounts frequently arises. Two sources are becoming increasingly important: catalytic decomposition of ammonia

$$2NH_3 \longrightarrow 3H_2 + N_2$$

and steam reforming of methanol

$$CH_3OH + H_2O \longrightarrow 3H_2 + CO_2$$

Hydrogen from ammonia is obtained at 75 percent purity, and concentration must be accomplished by low-temperature fractionation since there is no other convenient way of separating hydrogen and nitrogen. When hydrogen is obtained from methanol, except for small amounts of methane produced during the reforming reaction, the only impurities are CO_2, which can be removed by any of the conventional washing processes, and CO, which can be converted to methane. A third alternative is electrolysis of water. These three methods of hydrogen generation are compared below for a 100,000 scfd plant. Production costs are several times greater than for a steam-reforming unit, but these methods are used in situations where the cost of hydrogen has a slight influence on economics. All three cases assume 99 percent hydrogen or better is being produced.

	Ammonia	Electrolysis	Methanol
Raw material cost	$80/ton	0.6 ¢/kWh	5.0 ¢/lb
Investment, $1,000	120	190	55
Production costs		¢/1000 ft^3	
Raw materials	120	108	145
Less: oxygen credit		(-40)	
Net raw materials	120	68	130
Catalyst, chemicals, utilities	5	–	3
Labor and overhead	25	25	25
Capital charges	115	183	53
	265	276	226

Depending on local conditions of raw material and power supply, small amounts of hydrogen can be generated for somewhere between $2.30 and $3.00/1000 ft^3

The uses of pure hydrogen in the organic industry can be classified into the following five main types of reaction:

1. Double-bond hydrogenation

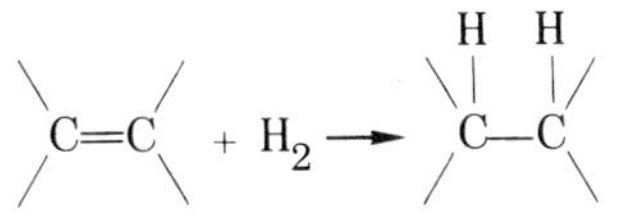

$$\diagdown C = C \diagup + H_2 \longrightarrow \underset{\diagup \quad \diagdown}{\overset{H \quad H}{C - C}}$$

2. Reduction of nitro-groups to amines

$$-NO_2 + 3H_2 \longrightarrow -NH_2 + 2H_2O$$

3. Reduction of aldehydes to alcohols

$$-CHO + H_2 \longrightarrow -CH_2OH$$

4. Reduction of acids, usually as esters, to alcohols

$$-COOH + 2H_2 \longrightarrow -CH_2OH + H_2O$$

5. Hydrogenation of nitriles to amines

$$-CN + 2H_2 \longrightarrow -CH_2NH_2$$

In this section, the discussion will be limited to those applications in which hydrogen is the only link with the petrochemical industry. References to the many others may be found elsewhere.

There are also cases of combinations between two types of reactions for example, 1 and 3.

In addition to these applications in the organic chemical field, there are already numerous uses of hydrogen as a reducing agent in the metallurgical industry. In the HyD process, for example, hydrogen reduces iron ore to a form of synthetic scrap that can be fed directly to steel furnaces. In uranium production, ultrapure hydrogen reduces UO_3 to UO_2; it is also used for decarburizing steel. There are several other nonorganic applications of this kind for hydrogen.

8.1 Hydrogenated Fats and Oils

About 4.0 billion pounds per year of cottonseed and especially soybean oil are hydrogenated in the U.S. for shortening and margarine. These oils contain high percentages of unsaturated fatty chains, responsible for the instability of these natural products to oxidation, which results in rancidity. Cooking oils are also stabilized by hydrogenation. Another 2.0 billion pounds per year of oils are treated in this manner.

Besides the production of edible materials, hydrogenation of natural products is also important in the production of stearic acid from the unsaturated C_{18} acids, oleic and linoleic. In many applications of fatty acid soaps and esters — gel-forming agents in lubricants, PVC stabilizers, plasticizers — it is important that the iodine number (a measure of the degree of unsaturation) be as low as possible to avoid unwanted polymerization and oxidation at the double bond. About 150.0 million pounds of stearic acid are used in the U.S. but part of this is obtained as such by fractional distillation of tallow fatty acids which are rich in this homolog.

8.2 Tetrahydrofuran

Furan, which is obtained by decarboxylating furfural, can be hydrogenated in the presence of a nickel catalyst to give tetrahydrofuran (THF):

$$\begin{array}{ccc}
\mathrm{HC{-\!\!-\!\!-}CH} & \mathrm{HC{-\!\!-\!\!-}CH} & \mathrm{H_2C{-\!\!-\!\!-}CH_2} \\
\| \quad \| & \| \quad \| & | \quad | \\
\mathrm{HC} \quad \mathrm{C{-}CHO} \longrightarrow \mathrm{HC} \quad \mathrm{CH} \; (+\,\mathrm{CO}) \xrightarrow{2\mathrm{H_2}} \mathrm{H_2C} \quad \mathrm{CH_2} \\
\diagdown \diagup & \diagdown \diagup & \diagdown \diagup \\
\mathrm{O} & \mathrm{O} & \mathrm{O}
\end{array}$$

Tetrahydrofuran

In addition to its use in such solvent applications as Dow-Badische acrylates, the Nalco lead alkyls processes, and for making cast PVC film, tetrahydrofuran is the source of hydroxyl groups in the manufacture of "spandex" polyurethane fibers. In Germany, BASF still produces THF from acetylene via Reppe chemistry, the same route that led to butadiene through THF during World War II. In the U.S. DuPont recently began using a very similar process.

$$\mathrm{HOH_2C{-}C{\equiv}C{-}CH_2OH} \xrightarrow[{-H_2O}]{+2H_2} \begin{array}{c} \mathrm{H_2C{-\!\!-\!\!-}CH_2} \\ | \quad\quad | \\ \mathrm{H_2C} \quad \mathrm{CH_2} \\ \diagdown \diagup \\ \mathrm{O} \end{array}$$

Butynediol   Tetrahydrofuran

Furfural is obtained from natural materials, e.g., oat hulls, bagasse and corn cobs.

8.3 Sorbitol

Sorbitol is obtained by reduction of fructose-free glucose, a sugar usually obtained from corn:

$$\begin{array}{cccc}
\mathrm{OH} \; \mathrm{H} & \mathrm{OH} \; \mathrm{OH} & & \mathrm{OH} \; \mathrm{H} \quad \mathrm{OH} \; \mathrm{OH} \\
| \quad | & | \quad | & & | \quad | \quad\; | \quad | \\
\mathrm{HOH_2C{-}C{-}C{-}C{-}C{-}CHO} + \mathrm{H_2} \longrightarrow \mathrm{HOH_2C{-}C{-}C{-}C{-}C{-}CH_2OH} \\
| \quad | \quad | \quad | & & | \quad\; | \quad | \quad | \\
\mathrm{H} \;\; \mathrm{OH} \; \mathrm{H} \;\; \mathrm{H} & & \mathrm{H} \quad \mathrm{OH} \; \mathrm{H} \;\; \mathrm{H}
\end{array}$$

In the older processes, the reaction was carried out at pressures between 1500 and 2000 psi in the presence of a supported nickel hydrogenation catalyst. The use of a promoted Raney-nickel enables operation to take place at pressures in the 300–500 psi range; most new plants use this technique. After hydrogen is flashed off, suspended catalyst is removed by filtration or centrifuging and the product purified by double ion exchange. One of the main problems in sorbitol production is ferrous ion elimination; the presence of this element is due to erosion of the reactor wall by the powdered nickel catalyst; therefore stainless steel equipment is mandatory. A final purification step is carried out by adding activated carbon and filtering. It leaves this step as a 50 percent solution, which is usually consumed captively for oxidation to vitamin C (ascorbic acid),

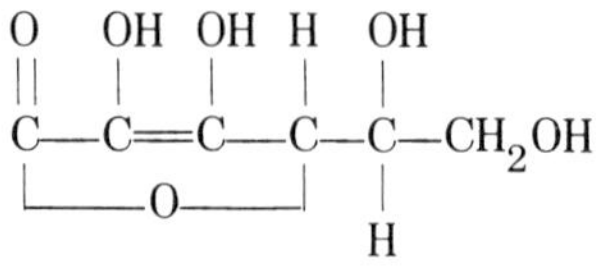

$$
\begin{array}{ccccccc}
O & OH & OH & H & OH & & \\
\| & | & | & | & | & & \\
C & C & C & C & C & CH_2OH \\
\end{array}
$$

or evaporated to a concentration of 70 percent, just short of the point where crystallization begins. This is the prevailing commercial form of sorbitol. In most of its applications other than as a starting material for ascorbic acid, sorbitol competes advantageously with glycerine because of its lower price, better taste and moisture retention. It is used as a vehicle, humectant and sweetener in toothpaste, in proprietary drugs for internal use such as cough syrups, and food processing (in cake fillings, sweets, and diet foods for diabetics). The best growth prospect is polyols used for rigid urethane foams. All these are fields where glycerine is also employed. Nonionic surfactants made from sorbitol find their main outlets in the cosmetic and food-processing field. "Spans" and "Tweens," trademarks of the Atlas Powder Co., are, respectively, fatty acid esters and ethoxylated fatty acid esters of the inner ethers made by dehydrating sorbitol, known as sorbitans. When sucrose (a mixture of glucose and levulose) is hydrogenated, the result is a mixture of sorbitol and mannitol. The latter is separated from sorbitol by crystallization from ethyl alcohol; it is also a surfactant intermediate.

The investment for a 10 million pounds per year sorbitol plant is about $0.8 million, and costs in cents per pound of 70 percent solution as follows

	¢/lb
Corn sugar (7.5 ¢/lb)	5.7
Catalysts and chemicals	1.7
Utilities	1.5
Labor and overhead	2.7
Capital charges	2.6
	14.2

Producers of sorbitol in the U.S. are shown below; the three pharmaceutical companies produce sorbitol for their own consumption in vitamin C manufacture. Atlas is the major manufacturer of polyols and surfactants based on sorbitol. Capacities are expressed in pounds per year of 70 percent solution.

Producer	Location	Capacity MM lbs/yr
Atlas Powder	Atlas Point, Del.	65.0
Baird	Peoria, Ill.	20.0
Hoffman-La Roche	Nutley, N.J.	9.0
Merck	Danville, Va.	17.0
Pfizer	Groton, Conn.	18.0
		129.0

Source: *Chemical Week*, July 18, 1964.

The U.S. demand for 70 percent sorbitol in 1964 was 89.0 million pounds broken down as shown below.

End-use	MM lbs, 1964
Polyols	3.0
Cosmetics (toothpaste, etc.)	18.0
Surface-active agents	13.0
Ascorbic acid	23.0
Pharmaceuticals	11.0
Food processing	13.0
Others (explosives, tobacco humectant, alkyd resin, etc.)	5.0
Export	3.0
Total	89.0

Source: *Chemical Week*, July 18, 1964.

In 1964, 7.4 million pounds of ascorbic acid were consumed in the U.S., of which 4.5 million pounds were in pharmaceuticals and the remainder in various food products.

Sorbitol is a very promising chemical, and especially advantageous for less developed economies where plants as small as 1.0 million pounds per year can operate economically.

8.4 Hydrogen Peroxide

The classical process for making hydrogen peroxide is electrolysis of cold 50 percent sulfuric acid to persulfuric acid. Its reaction with water forms hydrogen peroxide and regenerates sulfuric acid:

$$2H_2SO_4 \longrightarrow H_2 + H_2S_2O_8 \xrightarrow{+2H_2O} 2H_2SO_4 + H_2O_2$$

Sulfuric acid   Persulfuric acid   Hydrogen peroxide

This route, which requires 9.5 kWh per pound of 100 percent hydrogen peroxide as well as large amounts of steam for the hydrolysis step, has been replaced in recently built units by two organic processes, one of them unique. Shell, at Norco, La., produces hydrogen peroxide and acetone by oxidation of isopropanol. The hydrogen peroxide thus produced is used mainly to make glycerine.

$$CH_3CHOHCH_3 + O_2 \longrightarrow CH_3COCH_3 + H_2O_2$$

The more common organic process, known also as the "autoxidation" route, employed in the three largest U.S. plants and others in Europe, actually amounts to indirect oxidation of hydrogen to peroxide. A substituted anthraquinone (for example, 2-ethylanthraquinone) can be reduced to the respective anthraquinol, and then oxidized by air back to the quinone form, giving hydrogen peroxide:

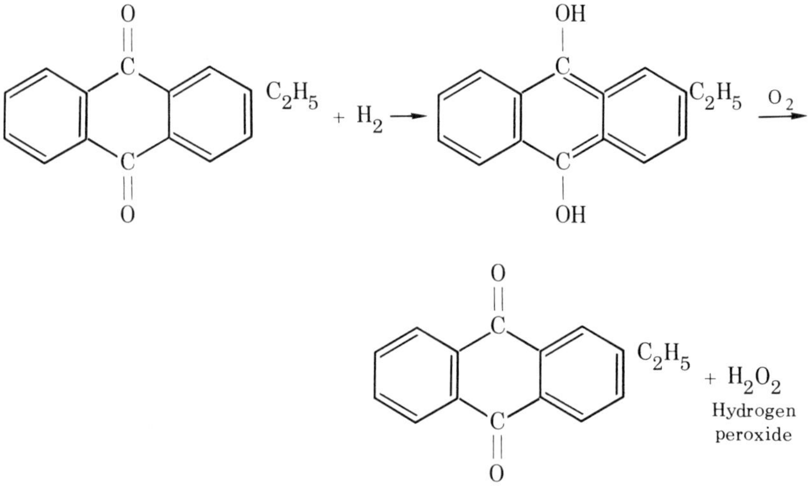

The hydrogenation step is carried out in the presence of a palladium catalyst. After leaving the oxidation reactor, the product stream is extracted with deionized water, which separates the product from a mixture of the anthraquinone, catalyst and a solvent in which the two reactions are carried out, to minimize byproduct formation. The product stream, a 20 percent peroxide solution, can be concentrated by vacuum distillation to the usual commercial forms (35 percent and 90 percent solutions). Figure 2-30 shows a flowsheet of the process. The demand for hydrogen

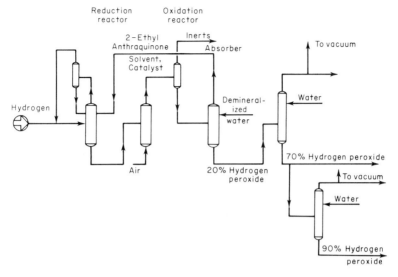

Fig. 2-30 Hydrogen Peroxide Autoxidation Process.

peroxide in 1965 was 106.0 million pounds. Its major application is as an oxidizing agent, and its principal advantages over other oxidants is that it is colorless and relatively safe to handle for bleaching animal and vegetable matter.   The demand can be broken down as shown below.

End-use	MM lbs, 1965
Merchant	
Textile bleaching	32.0
Paper and pulp bleaching	8.0
Plasticizers	7.0
Chemicals	10.0
Miscellaneous	12.0
Captive	
Chemicals	10.0
Glycerine and miscellaneous	27.0
	106.0

Source: *Oil, Paint & Drug Reporter*, June 13, 1966.

About 90 percent of all cotton goods are bleached with peroxide, and minor amounts are used on wool, rayon and polyamide fibers. In the paper field, peroxide bleaching is used mainly on groundwood pulps, which are considerably cheaper than chemical pulps, but impossible to bleach with chlorine. Sodium peroxide (made from metallic sodium and air) is also used for this purpose. Peroxide bleaching has been responsible for the growth in use of groundwood pulp, which allows pulp yields of 100 percent on wood, instead of only 50 percent for chemical pulp, to be obtained. In the cosmetics field, peroxide is used principally for hair-dyeing; whereas hair dyes were formerly based on vegetable compounds and metallic salts, presently they are almost exclusively water-soluble organic dyes that are insolubilized and developed in situ by hydrogen peroxide. A major use for peroxide is the manufacture of sodium perborate, an important component of many household bleach formulations. Among the miscellaneous applications, besides military outlets (as an oxidant for liquid fuel propellant systems), are bleaching of animal glues, wood and marble bleaching, fresh milk preservation, purification of nickel-plating solutions and in several processing techniques involving tungsten, molybdenum and beryllium ores.

The following companies in the U.S. produce hydrogen peroxide.

Producers	Plant location	Capacity MM lbs/yr	Process
DuPont	Memphis, Tenn.	40.0	Autoxidation
FMC	Buffalo, N.Y.	12.0	Electrochemical
	Vancouver, Wash.	12.0	Electrochemical
	S. Charleston, W. Va.	16.0	Autoxidation
Allied Chemical	Syracuse, N.Y.	8.0	Autoxidation
Pittsburgh Plate Glass	Barberton, Ohio	7.5	Electrochemical
Shell	Norco, La.	30.0	Isopropanol oxidation
Pennsalt	Wyandotte, Mich.	3.5	Electrochemical
		118.0	

Source: *Oil, Paint & Drug Reporter,* June 13, 1966.

Epoxides

The term "epoxidation" is applied to the following reaction:

$$\diagup_{\diagdown} C = C_{\diagdown}^{\diagup} + H_2O_2 \longrightarrow \diagup_{\diagdown} C \underset{O}{\overset{}{-}} C_{\diagdown}^{\diagup} + H_2O$$

In actual practice, hydrogen peroxide does not perform the reaction alone. Two methods are employed. Some plants use in situ epoxidation, which involves a peracid — usually peracetic — in the epoxidation reactor, by the action of peroxide on acetic acid:

$$H_2O_2 + CH_3COOH \longrightarrow CH_3COOOH + H_2O$$
$$\text{Acetic acid} \qquad \text{Peracetic acid}$$

Epoxidation is accomplished by the peracid, regenerating the acid:

$$CH_3COOOH + \begin{matrix} \diagdown \\ \diagup \end{matrix}C=C\begin{matrix} \diagup \\ \diagdown \end{matrix} \longrightarrow \begin{matrix} \diagdown \\ \diagup \end{matrix}C—C\begin{matrix} \diagup \\ \diagdown \end{matrix} + CH_3COOH$$
$$O$$

The carboxylic acid is recovered and recycled for reuse. Most producers using in situ epoxidation employ ion-exchange catalyst to carry out the reaction; sulfuric acid can be used instead when there is no danger of harming the product. DuPont, FMC and others have developed processes for carrying out epoxidation continuously.

Although 80 percent of the plants in the U.S. use in situ epoxidation, this technique has its disadvantages. Whereas peracid formation is slow, epoxidation takes place very rapidly. Thus there are increases in productivity to be realized by acquiring the peracid ready-made, and using the available reactor capacity only for epoxidation; this also allows better temperature control of the exothermic epoxidation reaction, and reduces the danger of ring-opening. The carboxylic acid can be recovered and returned to the peracid manufacturer. There are two processes in use for making peracids. Union Carbide employs air oxidation of acetaldehyde, while FMC Corp., the largest user of hydrogen peroxide in the U.S., removes peracetic acid under vacuum from an equilibrium mixture of acetic acid, peroxide and peracetic itself.

The main applications of the epoxidation reaction are listed below, with an estimate of the total quantities produced in each category. Consumption of peroxide naturally varies with molecular weight of the product and the number of double-bonds per mol being epoxidized.

Application	Production, 1965 MM lbs/yr
Epoxy plasticizers	45.0
Dieldrin and endrin	20.0
Epoxidized polyolefins	20.0
Other epoxy resins	16.0
Drugs	1.0

Epoxy plasticizers fall into two categories: epoxidized esters of oleic acid with alcohols such as 2-ethylhexanol; and epoxidized soybean oil. The latter is a mixture of glycerol esters of various fatty acids, among them oleic itself, which has one double-bond per mol, and linoleic, which has two.

Their main function is as synergists for the barium and cadmium soaps used to stabilize nontoxic plasticized PVC. By reducing the concentration of metal stabilizers, the use of epoxy plasticizers prevents haziness and opacity in the finished resin. Being inexpensive, consumption grew rapidly as a cheap method for obtaining thermal and UV stability. Recently, there has been a trend towards more expensive and permanent stabilizers which require no synergist. Almost 95 percent of epoxy plasticizer output is used in vinyl plastics, of which 50 percent is in flooring and the rest in sheeting and film. The remaining 5 percent is used as corrosion inhibitors and as plasticizers for alkyd resins and synthetic rubber. In general, epoxidized soybean oil is preferred for vinyl flooring and sheet, while the alkyl stearate type is used to plasticize vinyl upholstery.

From a total of approximately twenty manufacturers of epoxy plasticizers, the main ones are given below.

Producer	Capacity MM *lbs/yr*
Rohm & Haas	30.0
Swift	10.0
Reichhold	10.0
Union Carbide	10.0
Archer-Daniels-Midland	10.0

Dieldrin and endrin are both epoxidized members of the family of chlorinated insecticides derived from cyclopentadiene.

Epoxidized polyolefins and other materials are used in applications similar to those of the epichlorohydrin-bisphenol epoxy resins. The FMC Corp. produces a family of epoxidized polybutadiene resins known as "Oxirons"; Union Carbide and Ciba make a number of cycloaliphatic epoxy resins said to possess higher deflection temperature and better workability than epichlorohydrin-derived materials. Some of these are derived from cyclopentadiene, others (Union Carbide's "Unox 201") from the Diels-Alder condensation product between crotonaldehyde and butadiene.

Finally, epoxidation of terpenes and steroids is of some importance in the pharmaceutical field.

Organic Peroxides

Production of most organic peroxides involves the use of hydrogen peroxide. Benzoyl and lauroyl peroxides, the two most widely used, for example, are made from the respective acyl chlorides:

$$2RCOCl + H_2O_2 \longrightarrow \overset{\overset{O}{\|}}{R}C\overset{\overset{O}{\|}}{O}OCR + 2HCl$$

Over 12.0 million pounds of all organic peroxides were used in the U.S. in 1965, 90 percent of which was used as free-radical initiators or redox catalysts for polymerization via double-bond addition.

Lauroyl peroxide is used primarily to make PVC. Benzoyl peroxide is the preferred catalyst for styrene polymerization, and for making maleic anhydride-type polyesters. Since ketone peroxides such as MEK peroxide are the safest to handle they are used primarily by small polyester resin makers. The *t*-butyl ester of perpivalic and perbenzoic acids, as well as di-*t*-butyl peroxide, are also employed to catalyze ethylene polymerization. Production of the main peroxides in 1965 is given below.

Peroxide*	Production MM lbs
Benzoyl peroxide	5.0
Lauroyl	1.3
Di-*t*-butyl peroxide	1.0
Decanoyl peroxide	0.3
t-Butyl hydroperoxide	0.1
Methyl ethyl ketone (MEK) peroxide	1.7
Others (est.)	3.0
	12.1

Source: 1965 U.S. Tariff Commission Report on Production of Synthetic Organic Chemicals.

*Cumene hydroperoxide is produced in the hundreds of million pounds for phenol production, but only small quantities for other uses. If for no other reason than this, the above table is open to more than the usual number of questions.

Other hydrogen peroxide derivatives

Among the other uses of hydrogen peroxide as an intermediate, one might mention the manufacture of certain rubber chemicals such as benzothiazyl disulfide from mercaptobenzothiazole. Tertiary amine oxides are used as cationic surfactants; the most common example is lauryl dimethylamine oxide:

$$C_{12}H_{25}N(CH_3)_2 + H_2O_2 \longrightarrow C_{12}H_{25}\overset{\overset{-O}{|}}{N}(CH_3)_2$$

Lauryl dimethylamine
oxide

These compounds were used first by Procter & Gamble in liquid detergents as replacements for alkanolamides; they are unique among cationic materials in that they do not form precipitates with anionic materials of the sodium sulfonate type. Also they act as foam stabilizers, detergency promoters for other surfactants and, above all, are nonirritating. "Omadine" is a bactericide derived from pyridine with peroxides, the use of which has been growing rapidly.

8.5 Furfuryl Alcohol

Furfural can be reduced to furfuryl alcohol by catalytic hydrogenation:

$$\underset{\displaystyle O}{\overset{\displaystyle HC\!\!-\!\!CH}{H_2C \quad C\!\!-\!\!CHO}} \xrightarrow{\;H_2\;} \underset{\displaystyle O}{\overset{\displaystyle HC\!\!-\!\!CH}{H_2C \quad C\!\!-\!\!CH_2OH}}$$

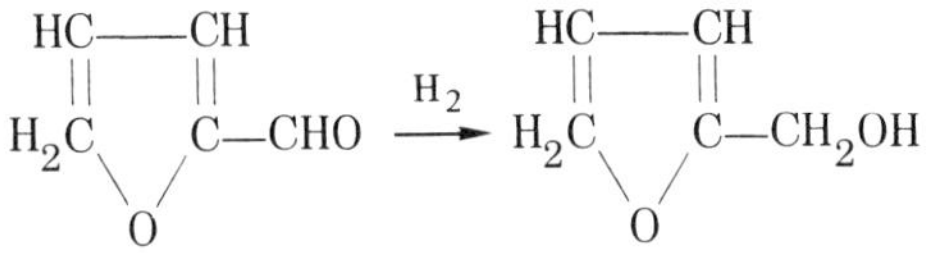

Furfuryl alcohol is used mainly in making modified aminoresins for foundry core binders.   This required 10.0 million pounds of furfuryl alcohol in 1965.

8.6 Fatty Alcohols

Fatty acids were formerly reduced to their respective alcohols by the action of sodium metal.   The disadvantages of this method are that the charge must be bone-dry, and that yields are low.   Also, sodium metal is expensive and it is not easy to find an outlet for the dilute NaOH byproduct.   At present, this can be accomplished by high-pressure catalytic hydrogenation of fatty acids, more usually with their methyl esters.

$$RCOOCH_3 + 2H_2 \longrightarrow RCH_2OH + CH_3OH$$

The use of the acids themselves requires stainless-steel equipment, thereby causing higher investment costs. These are made up in the case of methyl ester reduction by methanol losses and higher utility charges.

Hydrogenation can be carried out either under conditions that favor carboxyl group reduction over double-bond hydrogenation, or in such a way as to produce only saturated alcohols. Reaction is carried out at pressures in the 3000-to-5000 psig range, followed by filtering or centrifuging out the suspended hydrogenation catalyst. Separation of the various alcohols is accomplished by vacuum distillation (around 20 mm Hg), as is usual for fatty derivatives in general. Yields are around 96 percent.

The most important uses for fatty alcohols are in the field of surface active agents. Of the 150.0 million pounds per year of fatty alcohols consumed by the detergent industry, 35.0 million pounds were used to make alcohol ether nonionic surfactants by adduction with ethylene oxide:

$$ROH + n(CH_2\overset{\diagup\diagdown}{\underset{O}{}}CH_2) \longrightarrow RO(CH_2CH_2O)_nH$$

The demand for nonionics of this type was around 90.0 million pounds in 1964. They are used primarily in industrial applications such as pesticide emulsions, textile processing, emulsion polymerization, etc.

Most of the remaining 115.0 million pounds per year were used to make anionic surfactants, mainly in the form of alcohol ether sulfates, made by sulfating the reaction product of an alcohol and an average of three mols ethylene oxide:

$$RO(CH_2CH_2O)_3H \xrightarrow{H_2SO_4} RO(CH_2CH_2O)_3SO_3H \xrightarrow{NaOH} RO(CH_2CH_2O)_3SO_3Na$$

Smaller amounts of fatty alcohols are sulfated directly, either with sulfuric acid when the end-use is in detergents, or with chlorosulfonic or sulfamic acids, which react quantitatively and thus avoid the formation of sodium sulfate upon neutralization:

$$ROH + ClSO_3H \longrightarrow ROSO_3H + HCl$$

$$ROH + NH_2SO_3H \longrightarrow ROSO_3H + NH_3$$

Products of this type are used in cosmetics and toiletries, which consume 20.0 million pounds per year of anionic emulsifiers.

In addition to surfactants, the lower fatty alcohols are used to make lubricant additives, especially the higher polymethacrylates, widely employed as viscosity-index improvers. A total of 190.0 million pounds of viscosity-index improvers, about 25 percent of all motor oil additives, were used in 1965.

Fatty alcohols are used in cutting oils as coupling agents whose function is to modify the interface between metal and lubricant fluid.   The water-soluble oils in which these alcohols are chiefly included are growing at the expense of mineral oil-based straight cutting oils.

Petrochemical materials have begun to compete with fatty alcohols. Primary alcohols can be made using the Ziegler method, as is being done by Ethyl and Conoco. Long-chain olefins, also obtainable by a Ziegler-type process or by cracking n-paraffins, can be converted to a mixture of primary and secondary alcohols by the "oxo" process. Finally, Union Carbide produces secondary alcohols by direct oxidation of n-paraffins.

Producers of long-chain alcohols from fatty acids are given below.

Producers	Capacity MM lbs/yr
Procter & Gamble	110.0
DuPont	40.0
Rohm & Haas	20.0

Procter & Gamble has converted their entire output to surfactant raw materials.   DuPont produces surfactants especially for the cosmetics industry and Rohm & Haas is the dominant producer of methacrylate viscosity index (VI) improvers.

Investment for a 25.0 million pounds per year of fatty alcohols plant is \$1.5 million, and production costs are as follows, assuming coconut oil at 18 cents per pound and byproduct glycerine credited at 15 cents per pound:

	Cost, ¢/lb fatty alcohol
Raw materials (net)	19.3
Hydrogen (250 ¢/1000 ft^3)	1.5
Utilities	0.8
Labor and overhead	0.6
Capital charges	3.3
	25.5

8.7 Fatty Nitriles and Amines

Fatty acids can be converted into the respective nitriles by reaction with excess ammonia:

$$RCOOH + NH_3 \longrightarrow RCN + 2H_2O$$

The reaction is usually carried out in two steps followed by separation and recycling of the excess ammonia and vacuum fractionation of the product stream into compounds of various chain-lengths. Fatty nitriles have some applications as solvents and fiber lubricants, but their main use is as intermediates in the production of fatty amines by catalytic hydrogenation at around 250 psig:

$$RCN + 2H_2 \longrightarrow RCH_2NH_2$$

The reaction takes place in an agitated autoclave, after which the catalyst is separated, yielding the finished product.

The total consumption of fatty amines in the U.S. was 60.0 million pounds in 1965. About 25.0 million pounds were used to make quaternary ammonium salts, and another 9.0 million pounds as flotation agents in potash and phosphate rock processing. The film-forming properties of octadecylamine have determined its widespread use as a corrosion preventive in steam-generating equipment, and for breaking up unwanted emulsions of water in machinery lubricants. Around 1.0 million pounds per year are used in boiler feed water; another 2.0 million pounds per year go into floor polish formulations. Paint and grease thickeners and gelling agents require around 6.0 million pounds per year. Known as "Bentones" they are made by combining fatty amines with Bentonite. Amine phosphates are added to gasoline as anti-stalling agents.

Most quaternary ammonium salts are of the following form:

$$\left[\begin{array}{c} RN(CH_3)_2 \\ | \\ R' \end{array} \right]^{+} Cl^{-}$$

and are made from fatty amines by reaction with two mols of methyl chloride. This is followed by quaternization with another mol of an alkyl chloride, such as benzyl chloride or methyl chloride itself. There are innumerable types of "quats," but these are among the most important. Total consumption of quaternary ammonium salts in the U.S. was around 35.0 million pounds in 1965, most of which went into domestic fabric softeners. These laundry additives are being used increasingly to soften garments and household linen. They are also used as germicides in all kinds of fields: cutting oils (to prevent dermatitis), sanitary paper

products, surgical sheeting and many others, and as dyeing retardants for lighter shades of acrylic fibers. Apparently, new
types of dyes have been developed which dispense with these
auxiliary products. Note: Some companies arrive at these products
via long-chain acyl chlorides and therefore not via fatty amines.

Fatty amines are intermediates for making a number of fatty
diamines widely used for preparing cationic asphalt emulsions.
These have the advantage over the more common anionic emulsions
in that they set faster; asphalt emulsions, of which around 600
million gallons per year are used in the U.S., are safer to handle
and more simple to apply than liquid asphalts.

Octadecyl isocyanates, made by phosgenation of the respective
amine, are intermediates for textile chemicals.

"Dodine," an important fungicide, is derived from dodecyl-
amine and cyanamide.   It is used to combat foliage diseases and
apple scab.

$$C_{12}H_{25}-NH-\overset{\overset{\displaystyle NH}{\|}}{C}-{}^+NH_3$$

$$CH_3COO^-$$

Producers of fatty nitrogen derivatives in the U.S. are given
below.

Producers	Capacity MM lbs/yr
Archer-Daniels-Midland	15.0
Armour	40.0
Cargill	15.0
Foremost Chemicals	15.0
General Mills	15.0
Marcus-Ruth-Jerome	15.0
National Dairy Products	15.0

Source: *Chemical and Engineering News*,
May 25, 1964.

The demand can be broken down as shown in the table below.

End-uses	MM lbs, 1965
Quaternary ammonium compounds	25.0
Gelling agents	6.0
Flotation agents	9.0
Floor polish	2.0
Pesticides	3.0
Asphalt additives	6.0
Corrosion inhibitors	3.0
Lube-oil additives, etc.	6.0
	60.0

The data in the preceding tables are open for many suggestions
and should be taken only as a very rough guide line.

3

METHANE

1 INTRODUCTION

Three important chemicals — carbon disulfide, hydrogen cyanide and chlorinated methanes — are made from relatively pure methane. The presence of heavier hydrocarbons would promote undesirable side reactions, leading to the formation of coke and tar.

Purification of natural gas to yield methane feedstock can be carried out by ambient or low-temperature absorption and low-temperature fractionation. In the former, natural gas is contacted with a heavy hydrocarbon stream at temperature and pressure conditions yielding an overhead of pure methane; the rich stream is stripped of C_2+ hydrocarbons and recycled.

Impurities can be reduced to 5-10 thousand ppm and 1-2 thousand ppm by ambient and low-temperature absorption, respectively. Low-temperature fraction which is more expensive, can produce an overhead containing as little as 100 ppm of hydrocarbons heavier than methane.

2 HYDROGEN CYANIDE

The only method currently in use in the U.S. for making hydrogen cyanide as the major product from hydrocarbons is the Andrussow process based on reaction between ammonia, methane and air.

$$CH_4 + NH_3 + \tfrac{3}{2}O_2 \longrightarrow HCN + 3H_2O$$

A flow-diagram is shown in Fig. 3-1. The three reactants are mixed and fed to a catalytic converter at around 2000°F. The

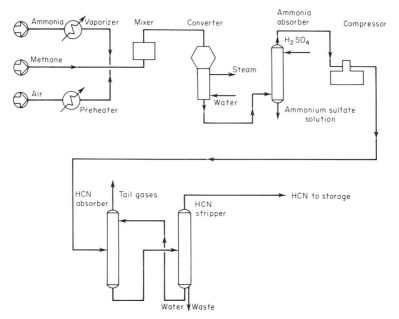

Fig. 3-1 Hydrogen Cyanide (Andrussow Process).

ratios must be adjusted carefully to avoid decomposition of ammonia and methane, which is favored at the high temperatures required to promote the formation of HCN. The gases are cooled in
a waste-heat boiler, where care is taken to avoid the formation of
the so-called "azulmic acids," polymers formed by the reaction
between HCN, ammonia and water which would foul the heat-
transfer surface.   Ammonia conversion per pass is around 65
percent, and unconverted ammonia goes to a scrubbing tower where
it is absorbed in sulfuric acid.   The product is then absorbed in
water, stripped and finally rectified to remove any polymers that
may have formed.   Yields are 70 and 60 percent on methane and
ammonia, respectively.

This process is sensitive to the presence of impurities in the
feed, so the methane feed must be practically pure. Higher hydrocarbons (as well as oil in the ammonia feed) would lead to the formation of soot and poison the catalyst. Desulfurization of the methane
is also necessary.

The Degussa process, used by this firm in Germany, also starts
from methane and ammonia, but requires no air. Platinum is used
as a catalyst in the form of catalyst-lined ceramic tubes. The
endothermic heat of reaction is provided by burning fuel in a combustion chamber which precedes the reactor. The reactor effluent

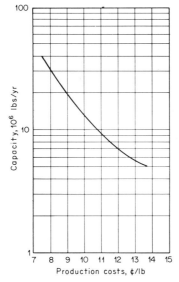

Fig. 3-2 Hydrogen Cyanide (Shawinigan Process).

gases are more concentrated than in the Andrussow process and the remainder is mainly hydrogen instead of nitrogen. Yields are around 83 percent on ammonia, which alone represents a saving of 0.7 cents per pound HCN compound with the Andrussow process.

$$CH_4 + NH_3 \longrightarrow HCN + 3H_2$$

The Shawinigan process permits usage of charges of heavy or light naphtha, and is of this interest in areas where methane is unavailable.   The process was first developed to handle propane as well as methane.   But areas short of natural gas are also short of hydrocarbons in the LPG range.

The reaction between ammonia and the hydrocarbon feedstock occurs in an electrically heated bed of coke particles in the absence of air.   The particles themselves promote uniform heat distribution. Since the bed consists of coke particles, the formation of soot constitutes no problem.

The Shawinigan process has the following advantages: 1) the hydrogen effluent is sufficiently pure to be recycled to the ammonia unit which provides the feed to the process; 2) the purer hydrogen effluent reduces the investment in the recovery section since it eliminates the need for ammonia absorption facilities; and 3) the absence of water vapor and low ammonia concentration also precludes formation of the undesirable "azulmic acids."

To be competitive, the Shawinigan process requires cheap and abundant power.   A 20.0 million pounds per year plant, for example, would draw almost 7000 kW, and power requirements are around 3.0 kWh per pound HCN, depending on the feed. The economics of a plant of this size, employing this process, are estimated below for a 20.0 million pounds per year plant and requires an investment of $2.0 million. Propane is assumed to be available at 1.5 cents per pound and captive ammonia at 2.5 cents per pound. Hydrogen has been credited at 30 cents per cubic foot; the cost of electricity for this particular example was taken at 0.8 cents per kWh.

		Cost, ¢/lb HCN
Raw materials		1.7
Propane	1.0	
Ammonia	1.8	
Less: hydrogen (−1.1)		
Utilities		3.0
Labor and overhead		0.8
Capital charges		3.3
Total		8.8

The DuPont route to HCN consists of indirect reaction between carbon monoxide and ammonia.   First, methyl formate is formed from methanol and carbon monoxide.   Then ammonia is reacted with the methyl formate to produce formamide for dehydration to HCN.

$$CO + CH_3OH \longrightarrow HCOOCH_3 \xrightarrow{NH_3} HCONH_2 + CH_3OH$$

$$HCONH_2 \longrightarrow HCN + H_2O$$

An increasingly important source of hydrogen cyanide is the production of acrylonitrile via the Sohio process, which yields about 0.1 pound HCN per pound of product.   Finally, HCN can be recovered from coke oven gas by washing the gases with a base and then driving off the product by acidification.

The importance of hydrogen cyanide as a building-block is decreasing. Production of acrylates and acrylonitrile is demonstrably cheaper by other routes. As an intermediate for adiponitrile and for hexamethylenediamine for nylon 6/6 only DuPont uses hydrogen cyanide. Since other manufacturers are continually entering the nylon 6/6 field, the relative importance of HCN will probably decline

in this application.   The development of hydrocarbon oxidation
routes to methyl methacrylate, such as the Escambia process,
indicates that the rapid increase in demand for this monomer will
not have a corresponding effect on that for HCN. Only chelates
and amino acids such as EDTA and methionine can be considered
growth areas for hydrogen cyanide.

The demand for HCN in the U.S. was around 285.0 million
pounds in 1966 and is expected to remain constant for a number
of years; it can be broken down as shown in the table below.

End-uses	MM lbs, 1966
Adiponitrile	60.0
Acrylonitrile	75.0
Methacrylates	84.0
Acrylates	10.0
Sodium cyanide	22.0
Chelates	18.0
Merchant and other	16.0
Total	285.0

Source: *Oil, Paint & Drug Reporter,*
May 22, 1966.

It should be pointed out that the amount used to produce acrylonitrile may decline quite suddenly.   This is because the plants
being replaced by Sohio units are actually expected to shut down
permanently.

Producer	Location	Capacity, MM lbs/yr
American Cyanamid	Fortier, La.	13.0
Dow	Freeport, Texas	5.0
DuPont	Memphis, Tenn.	115.0
	Victoria, Texas	60.0
Goodrich	Calvert City, Ky.	5.0
Hercules	Glen Falls, N.Y.	5.0
Monsanto	Alvin, Texas	32.0
	Texas City, Texas	70.0
Rohm & Haas	Houston, Texas	80.0
Sohio	Lima, Ohio	30.0
Union Carbide	Institute, W.Va.	11.0
Total		426.0

Source: *Oil, Paint & Drug Reporter*, May 22, 1967.

2.1   Methacrylates

Except for the one plant making methacrylates by isobutylene
oxidation, all the methyl methacrylate produced in the U.S. is
based on acetone and HCN.

Acetone cyanohydrin is made by the liquid-phase reaction between HCN and acetone in the presence of a strong base. Excess alkali is neutralized with sulfuric acid, which prevents decomposition of the cyanohydrin, then sodium sulfate is filtered out and unreacted HCN and acetone are removed by distillation for reuse. The concentrated cyanohydrin is hydrolyzed with sulfuric acid to form methacrylamide sulfate which reacts with methanol to form methyl methacrylate and ammonium bisulfate. The reacted mixture is then distilled to remove methyl methacrylate and unreacted methanol which is recycled; the last traces of methanol are removed by water extraction, after which the monomer is finally purified in a rerun tower.

$$CH_3COCH_3 + HCN \longrightarrow CH_3 - \overset{\overset{\displaystyle OH}{|}}{\underset{\underset{\displaystyle CH_3}{|}}{C}} - CN \xrightarrow{H_2SO_4}$$

$$(CH_2 = \underset{\underset{\displaystyle CH_3}{|}}{\overset{\overset{\displaystyle O}{\|}}{C}} - C - NH_3^+)HSO_4 \xrightarrow{CH_3OH} CH_2 = \underset{\underset{\displaystyle CH_3}{|}}{\overset{\overset{\displaystyle O}{\|}}{C}} - C - OCH_3 + NH_4HSO_4$$

Figure 3-3 shows a flowsheet of this operation.

In the Escambia process, methyl methacrylate is made by nitric acid oxidation of isobutylene to methacrylic acid and esterification with methanol:

$$CH_2 = \underset{\underset{\displaystyle CH_3}{|}}{C} - CH_3 \xrightarrow{[O]} CH_2 = \underset{\underset{\displaystyle CH_3}{|}}{\overset{\overset{\displaystyle CH_3}{|}}{C}} - COOH \xrightarrow{CH_3OH} CH_2 = \underset{\underset{\displaystyle CH_3}{|}}{C} - \overset{\overset{\displaystyle O}{\|}}{O} - OCH_3$$

This process is also used in Japan.

Methacrylates of long-chain alcohols are made by dehydrating the respective ester of α-hydroxy isobutyric acid, for example, by means of phosphorus pentoxide:

$$CH_3 - \underset{\underset{\displaystyle CH_3}{|}}{\overset{\overset{\displaystyle OH}{|}}{C}} - CN \xrightarrow[H^+]{ROH;} CH_3 - \underset{\underset{\displaystyle CH_3}{|}}{\overset{\overset{\displaystyle OH}{|}}{C}} - COOR \xrightarrow{-H_2O} CH_2 = \underset{\underset{\displaystyle CH_3}{|}}{\overset{\overset{\displaystyle O}{\|}}{C}} - COR$$

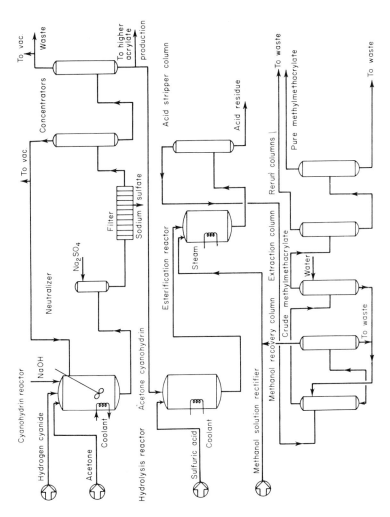

Fig. 3-3 Methyl Methacrylate.

Lower methacrylates can be made by transesterification of methyl methacrylate.

The economics for methyl methacrylate production are evaluated in the following. It is assumed that the producer has captive HCN but receives acetone from an outside source. This is the case for all three U.S. producers using this route. The investment for a 40.0 million pounds per year unit is $3.2 million. Acetone is taken at 8.0 cents per pound, captive HCN at 8.5 cents per pound, and methanol at the market price of 4.5 cents per pound, although producers having captive methanol source can charge in this product at around 3.0 cents per pound.

	Cost, ¢/lb methyl methacrylate
Acetone	5.7
HCN	2.9
Methanol	2.6
Catalyst and chemicals (net)	1.2
Utilities	0.6
Labor and overhead	1.0
Capital charges	2.6
Total	16.6

The isobutylene route seems potentially cheaper. Assuming isobutylene at 6.8 cents per pound, and that captive nitric acid is made from purchased ammonia, exclusive of capital charges, we have the following:

	Cost, ¢/lb methyl methacrylate
Raw materials	9.3
Utilities	1.8
Labor and overhead	1.0
Total	12.1

The total cost would be lower than for the acetone-HCN route provided a 40.0 million pounds per year plant costs less than $5.5 million. Furthermore, a captive isobutylene producer would enjoy an additional advantage of about 1.4 cent per pound of methyl methacrylate.

Producers of methyl methacrylate in the U.S. in 1966 and routes are given in the table below.

Producer	Location	Capacity, MM lbs/yr	Route
Rohm & Haas	Houston, Tex. Louisville, Ky. Bristol, Pa.	240.0	Acetone–HCN

Producer	Location	Capacity, MM lbs/yr	Route
DuPont	Belle, W.Va.	80.0	Acetone—HCN
American Cyanamid	Fortier, La.	40.0	Acetone—HCN
Escambia*	Pensacola, Fla.	20.0	Isobutylene oxidation
Total		380.0	

*Shut down.

Source: *Oil, Paint & Drug Reporter*, March 6, 1967.

The growth in demand for methyl methacrylate has caused several companies to consider entering the field, especially since the adoption by several acrylonitrile producers of the Sohio process has released large amounts of HCN for other uses. Methyl methacrylate demand is growing at a rate of 10 percent per year, reaching 280.0 million pounds in 1966. Recent stimulants have been the ICI process for making extruded sheet directly from methyl methacrylate monomer; and the Swedlow continuous sheet-casting process. The most important use for methyl methacrylate is cast sheet. This is employed in outdoor signs because of its weather resistance, and in construction for panels, windows, skylights and numerous molded products. For use in mercury lamp street lights and automobile tail-lights, copolymers with α-methylstyrene have the advantage of exceptional stability to ultraviolet light and a higher heat-distortion point than pure polymethyl methacrylate. High-impact formulations are made by copolymerization with styrene-butadiene (e.g., Rohm & Haas' "Implex"). Methacrylate competes with styrene for polyesters employed in reinforced plastic paneling, especially for applications requiring weather-resistance. Water-thinned paints based on copolymers of ethyl acrylate and methyl methacrylates have better properties than vinyl emulsions; acrylic paints make up 20 percent of the water-based coatings market. Their use is growing, but may be

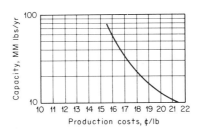

Fig. 3-4 Methyl Methacrylate.

threatened by vinyl acetate–ethylene copolymer emulsions. Methacrylate lacquers are used primarily on automobiles.

Acid-removable floor-polishes are formulated with polymers of diethyl, aminoethyl, or *t*-butyl aminoethyl methacrylate. Lube oil additives are polymers of higher alcohol or amino alcohol methacrylates; around 100.0 million pounds were used as viscosity-index improvers and ash-free detergents in heavy-duty lubricants (for removing cold sludge) and as pour-point depressants.

The pattern of use for methyl methacrylate for 1966 is shown below.

End-uses	MM lbs, 1966
Cast sheet	140.0
Molding and extrusion powders	59.0
Coatings	50.0
Miscellaneous and export	31.0
Total	280.0

Source: *Oil, Paint & Drug Reporter*, March 6, 1967.

Sheet is the main outlet for methacrylate. DuPont is the only monomer producer who does not make sheet.

2.2 Cyanogen Chloride

The reaction between HCN and chlorine gives cyanuric chloride:

$$HCN + Cl_2 \longrightarrow ClCN + HCl$$

And cyanogen chloride is an intermediate in the production of diphenylguanidine, a vulcanization accelerator.

Cyanogen chloride trimerizes readily to cyanuric chloride:

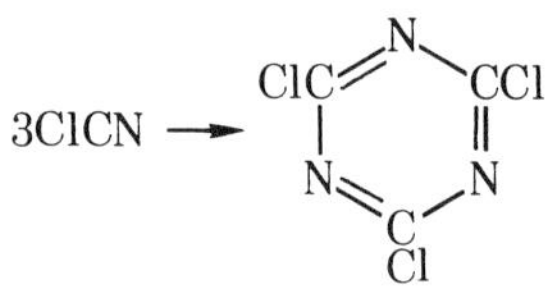

This intermediate has found several important uses; the three chlorine groups have different reactivities which make possible the stepwise introduction of different substituents. The largest outlet is the production of Geigy's family of herbicides, e.g., Atrazine having the general formula where R,R' and R'' can be Cl itself, or a substituted amino group, a methoxy radical, or a thiol group.

Atrazine, the most important of the group with an annual consumption of 10.0 million pounds, is used mainly in sugar cane and industrial weed control. It is made by reacting cyanuric chloride with monoethyl- and monoisopropylamine:

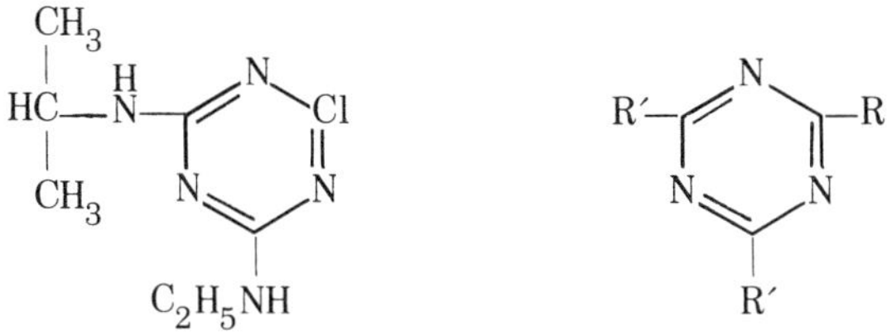

The second largest use for cyanuric chloride is in the manufacture of fluorescent brightening agents, i.e., white dyes. Small quantities of these materials are added to essentially all synthetic detergents sold retail in the U.S.

Triallyl cyanurate, made from cyanuric chloride and allyl alcohol, is used as a comonomer to impart higher temperature resistance to polyesters.

One of the most significant developments in the dye field is the principle of the "reactive dye." As the name implies, the dye actually combines with a reactive group in the fiber chain. The most common reactive dye form is the 1,3,5-triazine ring. Here one or two of the chlorine atoms are free to react with the fiber and the others are attached to a dye molecule. Reactive dyes are being used on polyamide and cellulose acetate fibers, where the reactive chlorine combines with the -NH- and -OH groups on the fibers. So far, they have been considered insufficiently resistant chlorine for use in cotton dyeing. However, indications are that these dyes are becoming increasingly popular as replacements for the disperse dyes previously used in nylon and acetate. Cyanuric chloride is also used to make nonreactive dyes; for example, a green dye can be made by introducing a yellow dye group in one of the reactive positions and a blue dye on another.

Other uses of cyanuric chloride are in the manufacture of hindered-phenol triazines as antioxidants for polyolefin resins.

The total demand for cyanogen chloride and cyanuric chloride in the U.S. was around 21.0 million pounds in 1965. The distribution is as follows:

End-use	MM lbs, 1965
Diphenyl guanidine	2.0
Triazine herbicides	13.0
Reactive dyes	1.0
Fluorescent brightening agents	4.0
Others	1.0
Total	21.0

American Cyanamid, Nilok, and Geigy are the three U.S. producers of cyanuric chloride.

2.3 Sequestrants

The most important application of the Strecker synthesis for making amino acids from HCN, formaldehyde and an amine, is the manufacture of ethylenediamine tetracetic acid (EDTA) and other similar compounds.   The process consists of a one-step reaction between a solution of HCN in formaldehyde and ethylenediamine in NaOH, followed by neutralization if the product is desired in its acid form.

$$
\underset{\mathrm{H_2C-NH_2}}{\overset{\mathrm{H_2C-NH_2}}{|}} + 4HCN + 4CH_2O + 4NaOH \longrightarrow
\begin{array}{c}
\mathrm{H_2C-N}\!\!\begin{array}{l}\diagup \mathrm{CH_2COO^-Na^+}\\ \diagdown \mathrm{CH_2COO^-Na^+}\end{array}\\
|\\
\mathrm{H_2C-N}\!\!\begin{array}{l}\diagup \mathrm{CH_2COO^-Na^+}\\ \diagdown \mathrm{CH_2COO^-Na^+}\end{array}
\end{array}
+ 4NH_3
$$

Similarly, nitrilotriacetic acid (NTA) is made from ammonia, and diethylenetriamine pentacetic acid from diethylenetriamine.

These compounds are used mainly in the form of their sodium salts, as chelating agents in a large number of industries. Chelating agents are used whenever it is desired to tie up as harmless complexes metal ions which otherwise would interfere with product color, catalyze undesirable reactions, affect solubility of other compounds, etc.   The main field of application is in textile wet processing but chelating agents have similar uses in paper processing and water treating.   In agriculture, it has been found that the best way to supply iron-deficient citrus trees with this element is in the form of its EDTA chelate.   As a solubility promoter, it is used to prevent soluble pesticides from precipitating in hard-water solutions, and as a water softener in soap or detergents. In the cold rubber SBR process, EDTA is used to avoid the catalytic effect of metal ions present in the water, the reaction medium.

Demand for ethylenediamine tetracetic acid (EDTA) and other chelating agents, also known as sequestrants, in 1965 was around 36.0 million pounds per year, of which 24.0 million pounds were

EDTA and its sodium salt. Actually, only 14 percent of the total is used as the acid with the remainder consumed in soluble form. The breakdown of the demand is given in the table below.

End-uses	MM lbs, 1965
Textile wet processing	11.0
Soap and cleaning compounds	4.0
Rubber polymerization	2.0
Agricultural	2.0
Miscellaneous	5.0
All others	12.0
Total	36.0

Source: *Oil, Paint & Drug Reporter*,
January 11, 1965.

The miscellaneous uses include various applications as stabilizers, that is, in preventing undesirable side effects due to metal compound traces. Also, mention may be made of vegetable oils, hydrazine, formaldehyde solutions, cellulose esters, etc., as some of the numerous products stabilized with EDTA and related compounds. Since 1966, the use of sodium NTA in detergents has grown remarkably; it is a partial replacement for sodium polyphosphates as a builder. Several plants in the 20 to 50 million pounds per year range were announced. This was made possible by the ample availability of byproduct HCN.

2.4 Lactic Acid

The first commercial method of making lactic acid was by fermentation of carbohydrates. More recently, Monsanto has synthesized lactic acid by the hydrolysis of lactonitrile. Originally, this product was made by reacting with acetaldehyde the HCN off-gas from an acetylene-based acrylonitrile unit. With the change in acrylonitrile technology, the source of HCN for this process is now the byproduct from the Sohio process.

$$CH_3CHO + HCN \longrightarrow CH_3CHOHCN \xrightarrow[H_2O]{H^+} CH_3CHOHCOOH$$

The production of synthetic lactic acid is so inexpensive that most of the older fermentation plants have been forced to shut down. In France, Rhône-Pouleuc has developed a different process involving oxidation of propylene. In 1965, 8.5 million pounds of lactic acid was consumed in the U.S. with 6.8 million pounds being used in food products. About 2.5 million pounds per year are used as a food acidulant. Most of the rest was used to make calcium stearyl lactate, an emulsifier employed in the bakery industry to make the dough more workable, to produce a finer-textured bread. It also keeps the product fresh and tender. Synthetic lactic acid has obtained FDA approval for use in food, thus almost the entire range of uses can now be supplied from this synthetic source.

Outside the food processing field, lactic acid is employed in antiperspirants, leather tanning in competition with other organic acids, as a component of permanent floor-polish formulations, and as a mordant in the textile industry.

Ethyl lactate is a high-boiling solvent for cellulose ethers, of which 2.0 million pounds per year are consumed in the U.S. The main U.S. producer, Kay-Fries, starts from acetaldehyde, HCN and ethyl alcohol.

$$CH_3CHO + HCN \longrightarrow CH_3CHOHCN \xrightarrow{C_2H_5OH} CH_3CHOHCOOC_2H_5$$

Phenyl mercuric lactate, obtained by reacting phenyl mercuric acetate with lactic acid, has some use in slime control of paper stock.

Having two reactive groups, lactic acid can homopolymerize; polylactic acid has been investigated as a material for surgical sutures.

Lactic acid is also made synthetically in Japan, starting from acetaldehyde and HCN.

The only fermentation plant still operating in the U.S. is in Clinton, Iowa, with a capacity of 3.0 million pounds per year. The Monsanto plant can produce 10.0 million pounds per year.

2.5 Sodium Cyanide

Sodium cyanide is formed when hydrogen cyanide is absorbed in NaOH,

$$HCN + NaOH \longrightarrow NaCN + H_2O$$

Absorption is carried out in a 35 percent caustic solution, and recovery of the salt is accomplished by evaporation under vacuum.

In addition to its use in organic synthesis, sodium cyanide is used in electroplating, in the Degussa process for heat treating steel and in gold extraction by oxidation. In electroplating, the largest single application, metal deposition from a solution of cyanide complex ions, gives a much better surface than from a solution of the ions themselves. About fifty million pounds per year are consumed in the U.S. Over 140.0 mm pounds of NaCN are consumed in the U.S. each year, about 120 mm pounds per year for adiponitrile and over 20 mm pounds per year for other uses.

Sodium cyanide use is declining; since it is an expensive source of the -CN group. Users of sodium cyanide for chemical synthesis tend to convert to merchant HCN despite the greater hazards and investment requirements.

2.6 Ferrocyanides

Well known as "Prussian blue," ferric ferrocyanide pigments are still widely used in printing inks and surface coatings despite the competition from phthalocyanines, which have much better alkali fastness and durability. Another important application is in blueprint papers. They are made by first absorbing HCN in a slurry of ferrous hydroxide in caustic soda or caustic potash to form sodium (or potassium) ferrocyanide:

$$7H_2O + 6HCN + FeO + 4NaOH \longrightarrow Na_4[Fe(CN)_6] \cdot 10H_2O$$

The alkali salt reacts with ammonium and ferrous sulfates to give ferrous ferrocyanide, which is then oxidized — for example, by means of sodium chlorate — to the ferric form:

$$2NH_4^+ + Fe^{++} + Na_4[Fe(CN)_6] \longrightarrow Fe(NH_4)_2[Fe(CN)_6] + 4Na^+ \xrightarrow{(O)}$$

$$FeNH_4[Fe(CN)_6]$$

About 12.5 million pounds per year of ferrocyanides are consumed in the U.S. Not all production is based on HCN as a starting material. One process, for example, uses calcium cyanamide as a source of the -CN groups.

2.7 Cyanoacetic Acid

Sodium cyanide reacts with sodium monochloroacetate, and neutralization gives cyanoacetic acid:

$$ClCH_2COONa + HCN \xrightarrow{NaOH} NCCH_2COONa \longrightarrow NCCH_2COOH$$

In 1965, about 2.2 million pounds per year were made in the U.S. by Kay-Fries, the major producer. Its most important use is as an intermediate for diethyl malonate.

$$NCCH_2COOH \xrightarrow[H^+]{C_2H_5OH} CH_2(COOC_2H_5)_2$$

Diethyl malonate is the raw material common to the manufacture of most barbiturates and thiobarbiturates, a family of sedatives of which one million pounds per year are produced in the U.S. The two hydrogen atoms on the middle carbon are very reactive, and derivatives can be made by reaction with alkyl bromides.

Barbiturates are made from these intermediates by reaction with urea, and thiobarbiturates with thiourea.

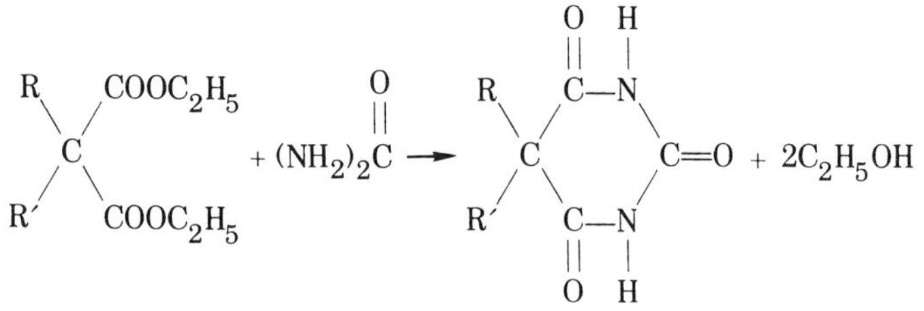

Still in the pharmaceutical field, diethyl malonate is an intermediate for phenylbutazone, an important antirheumatic.

Cyanoacetic acid is also an intermediate for making synthetic caffein and certain polymethine dyes.

2.8 Orthoformic Esters

Kay-Fries, the only U.S. producer of orthoformic esters, starts from HCN instead of using the classical chloroform-sodium alcoholate reaction:

$$HCN + HCl + 3ROH \longrightarrow HC(OR)_3 + NH_4Cl$$

Ethyl orthoformate is an intermediate for acrylic fiber dyes and photographic sensitizers of the polymethine family.

2.9 t-Butylamine

The following net reaction takes place between isobutylene and hydrogen cyanide:

$$(CH_3)_2C{=}CH_2 + HCN + 2H_2O \longrightarrow (CH_3)_3C{-}NH_2 + HCOOH$$

Other tertiary amines are produced similarly.

There are two t-butylamine producers in the U.S. — Rohm & Haas and Monsanto. The former uses this and other tertiary amines to make accelerators and hardeners for epoxy resins (such as dimethyl aminomethyl phenol), and methacrylate polymers used in floor polish formulations. Monsanto makes benzothiazyl t-butylamine sulfonamide, a widely used sulfonamide vulcanization accelerator. It is reputed to be safer, that is, less liable to cause scorching, than other similar compounds.

2.10 Dimethyl Hydantoin

The reaction between acetone, HCN and ammonium carbonate gives dimethyl hydantoin:

$$CH_3COCH_3 + HCN \xrightarrow{(NH_4)_2CO_3} \begin{array}{c} NH \\ (H_3C)_2C \quad C{=}O \\ O{=}C{-}NH \end{array}$$

The original peacetime use of this product was as its di-chloro derivative, the first solid bleach to become commercial. It has since been replaced by the chlorinated isocyanurates, but one million pounds per year are still used in swimming pool chlorination.

Since dimethyl hydantoin is an amine, it can be condensed with formaldehyde to give resins that find some use in hair sprays; in most formulations it acts as an extender of the more expensive polyvinyl pyrrolidone.

2.11 Azobisisobutyronitrile

Sold under the trade-mark "Vazo," azobisisobutyronitrile is manufactured by DuPont and used as a catalyst for vinyl poly-merizations, principally acrylates. Consumption is around one million pounds per year. It is made by reacting acetone, HCN and hydrazine, followed by oxidation. In Europe it is made by Bayer and sold as a blowing agent for making foam rubber ("Porofor").

$$HCN + CH_3COCH_3 + H_2NNH_2 \xrightarrow{OH^-} \underset{\underset{CN}{|}}{(H_3C)_2C}{-}NH{-}NH{-}\underset{\underset{CN}{|}}{C(CH_3)_2} \xrightarrow{(O)}$$

$$\underset{\underset{CN}{|}}{(H_3C)_2C}{-}N{=}N{-}\underset{\underset{CN}{|}}{C(CH_3)_2}$$

3   CARBON DISULFIDE

Carbon disulfide was made traditionally by heating charcoal and sulfur either in gas-fired retorts or in an electric-arc furnace:

$$C + 2S \longrightarrow CS_2$$

At present, 70 percent of U.S. capacity is based on the Thacker process, the reaction of methane and sulfur. Figure 3-5 shows a flowsheet of this process. The reactions are:

$$CH_4 + 4S \longrightarrow CS_2 + 2H_2S$$

The sulfur values in the H_2S are recovered by the Claus process

$$H_2S + \tfrac{3}{2}O_2 \longrightarrow SO_2 + H_2O$$

$$2H_2S + SO_2 \longrightarrow 3S + 2H_2O$$

The net reaction is:

$$CH_4 + 2S + O_2 \longrightarrow CS_2 + 2H_2O$$

Methane, after being processed to remove any traces of C_2^+ material, is mixed with vaporized sulfur and preheated. The mixture enters the converter where the first reaction takes place.

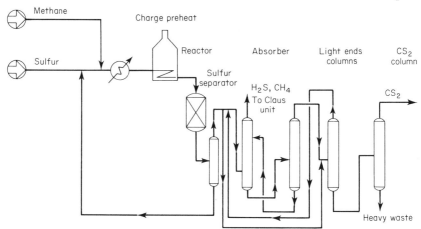

Fig. 3-5   Carbon Disulfide.

The effluent, after heat exchange against the feed, goes to a sulfur-recovery unit where unreacted sulfur is separated and combined with fresh feed. The overhead from this separation goes to an absorption system, from which unreacted methane and H_2S are sent to a multistage Claus unit for sulfur recovery. Crude carbon disulfide is sent to a fractionation system, where the remaining hydrogen sulfide and methane are taken overhead first and recycled to the absorption train, and the product finally separated from the heavy byproducts.

Since this process starts from methane it is limited in its application to areas where either natural gas or coke-oven gas methane are available. A modification has been announced which allows production from any hydrocarbon feed up to and including fuel oil. It is considered to involve lower capital investment and reduced operating costs since heavy hydrocarbon streams have a carbon-to-hydrogen ratio of around 2.0 instead of 4.0 as in methane, forming less H_2S. Therefore much less sulfur need be recovered and vaporized. While the amount of sulfur required per pound product remains the same, recycle is reduced from 1.6 pound per pound to 0.4 pound per pound, and the amount vaporized from 2.5 to 1.25 pound per pound of product. The process had not been employed commercially by 1965. It promises to be attractive to the nonoil-producing areas of the world, although at least two Thacker process plants operate in Europe based on coke-oven gas methane.

The economics of CS_2 production from methane are shown below. Investment for a 100.0 million pounds per year plant is $5.0 million, and production costs are as follows:

	Cost, ¢/lb CS_2
Methane (1.0 ¢/lb)	0.3
Sulfur (1.75 ¢/lb)	1.5
Utilities	0.4
Labor and overhead	0.4
Capital charges	1.7
Total	4.3

Carbon disulfide was sold for 4.4 cents per pound in 1965, which explains why hardly any of the major rayon and cellophane producers have captive facilities.

Carbon disulfide producers in the U.S. for 1964 are listed below, with their respective capacities and processes used.

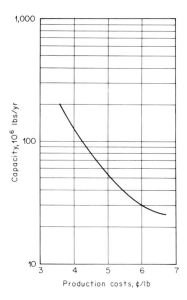

Fig. 3-6   Carbon Disulfide.

Producer	Location	Capacity, MM lbs/yr	Process
FMC–Allied	S. Charleston, W.Va.	220.0	Methane
Baker	Penn Yan, N.Y.	20.0	Charcoal, electric furnace
Pennsalt	Houston, Tex.	10.0	Methane
Pittsburgh Plate Glass	New Martinsville, W.Va.	50.0	Methane
Stauffer	Delaware City, Del.	100.0	Methane
	Le Moyne, Ala.	100.0	Methane
	Lowland, Tenn.	100.0	Charcoal, retort
	Old Hickory, Tenn.	100.0	Charcoal, retort

Source: *Oil, Paint & Drug Reporter*, March 16, 1964.

Domestic demand for carbon disulfide in 1965 was around 754.0 million pounds with the following distribution:

End-use	MM lbs, 1965
Rayon and cellophane	527.0
Carbon tetrachloride	150.0
Other	77.0
	754.0

Source: *Oil, Paint & Drug Reporter*,
September 19, 1966.

3.1 Viscose: Rayon and Cellophane

The term "rayon" applies to fibers and filaments made from certain forms of cellulose that would otherwise be unsuitable for the manufacture of textile fibers. The main sources of cellulose for rayon manufacture are sulfite pulp and cotton linters.

The viscose process consists of three main steps: first, rayon-grade cellulose is reacted with caustic soda, thereby converting the -OH groups on the cellulose unit to -ONa. This step requires pure caustic soda such as is obtained from mercury cells; diaphragm cells produce a grade of soda containing sodium chloride which renders it unusable without further purification. Next, the alkali cellulose, having been left to age for 2 to 3 days, reacts with carbon disulfide in an excess of caustic. Aging reduces the average cellulose chain-length, thus making the cellulose xanthate easier to dissolve in caustic soda.

$$\text{Cell —OH + NaOH + CS}_2 \longrightarrow \text{Cell—O—}\overset{\overset{\textstyle S}{\|}}{\text{C}}\text{—S—Na}$$

An average of 0.7 out of the 3 -OH groups on each glucose unit reacts with carbon disulfide. It is this solution of cellulose xanthate in caustic soda that is known as "viscose."

The regeneration step consists of coagulating the viscose in a bath containing mainly sulfuric acid and sodium sulfate. This decomposes the viscose solution into cellulose fiber, sodium sulfate and a number of sulfur compounds including carbon disulfide, part of which can be recovered by an absorption technique. In the U.S., this step is not warranted due to the low price of CS_2. Rayon and cellophane differ primarily in their assumed forms after regeneration.   Rayon is made by sending the viscose solution through spinnerettes, and cellophane by extrusion through a slot.

Conventional rayon has low wet strength, poor resistance to abrasion and to caustic, and high elongation; it cannot be preshrunk.   Accordingly, rayon began to lose ground to other fibers until the development of high-modulus types.   These are widely used in 50-50 blends with cotton without lowering quality and at considerable raw materials savings, since rayon is the cheapest of all staples.   Apart from apparel, rayon is used in tire-cord for original equipment (although seriously threatened by new types of polyamides, glass fibers and by polyesters), carpets and household goods.   A threat to rayon-cotton blends for apparel applications has been the development of permanent-press resin treatments.   The high percentages (by weight of fiber) in which these resins are used would weaken those blends to the point where they would become unusable. Therefore it is predicted that rayon-cotton will lose a good deal of its market to polyester-cotton mixtures.

The cellophane market has declined considerably in the last few years. It is still used for very high-speed packaging applications where synthetics such as polyethylene cannot compete for lack of adequate machinery. However, it is expected that synthetics soon will assume an even greater part of the cellophane market than they already have. In 1965, rayon consumption was 1080 million pounds, and that of cellophane 405.0 million pounds.

The largest rayon manufacturers are American Viscose, American Enka, and Beaunit Mills.

For a long time, rayon staple has been sold at marginal cost (28 cents per pound) and it is unlikely that any new plants will be built.

3.2 Carbon Tetrachloride

There are two processes currently being used in the U.S. for making carbon tetrachloride. The most important is the chlorination of carbon disulfide by the following reactions:

$$CS_2 + 3Cl_2 \rightarrow CCl_4 + S_2Cl_2$$

$$2S_2Cl_2 + CS_2 \rightarrow CCl_4 + 4S$$

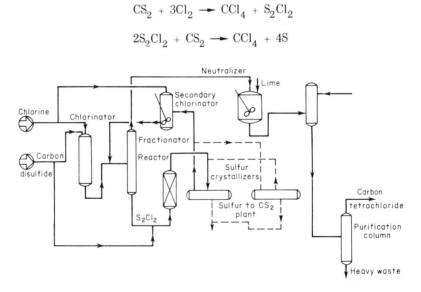

Fig. 3-7 Carbon Tetrachloride (via CS_2).

In one variant carbon disulfide and chlorine are mixed and fed to a cooled reactor where the first reaction takes place. The effluent goes to a column where CCl_4 is taken overhead, and S_2Cl_2 removed from the bottom. This reacts catalytically with more CS_2 giving more carbon tetrachloride and sulfur which is removed by crystallization. The liquid product from the second reactor is fed to the CCl_4 fractionation column. The product is treated with lime to remove any unreacted S_2Cl_2, and finally purified by fractionating out a heart-cut to eliminate light and heavy byproducts. Yields on both raw materials are around 90 percent. Figure 3-7 shows a flowsheet.

A carbon disulfide manufacturer can produce carbon tetrachloride and recycle the byproduct sulfur to the CS_2 plant. This process accounts for 60 percent of the U.S. capacity. The other 40 percent is obtained as a byproduct from perchloroethylene plants that do not start from acetylene, or from plants making chlorinated methanes either from methane or from methanol. Raw materials to these plants are ethylene dichloride, propane, propylene dichloride (a byproduct of propylene oxide manufacture), methane or methyl chloride derived from methanol.

Carbon tetrachloride can also be made by the catalytic reaction between two mols of phosgene:

$$2COCl_2 \longrightarrow CCl_4 + CO_2$$

This process may be attractive to a company capable of producing phosgene at between 3.0 and 3.5 cents per pound, but so far purification costs are such as to preclude its commercial use.

The table below lists the U.S. producers of carbon tetrachloride in 1966, with their respective routes.

Producer	Location	Capacity, MM lbs/yr	Routes
Allied Chemical	Moundsville, W.Va.	5.0	Methyl chloride
Diamond	Painesville, Ohio	30.0	CS_2
Dow	Freeport, Tex.	130.0	
	Pittsburg, Calif.	30.0	Hydrocarbon chlorination
	Plaquemine, La.	20.0	
FMC–Allied	S. Charleston, W.Va.	200.0	CS_2, hydrocarbon chlorination
Vulcan	Wichita, Kan.	25.0	Hydrocarbon chlorination
Stauffer	Le Moyne, Ala.	85.0	Hydrocarbon chlorination
	Louisville, Ky.	70.0	CS_2
	Niagara Falls, N.Y.	125.0	CS_2

Note: Pittsburgh Plate Glass can produce CCl_4 at its Barberton, Ohio, complex.
Source: *Oil, Paint & Drug Reporter*, October 17, 1966.

The demand for carbon tetrachloride is determined almost entirely by that for its fluorinated derivatives. Consumption in 1965 was 590.0 million pounds distributed as follows.

End-use	MM lbs, 1965
Fluorocarbons	530.0
Grain fumigant	24.0
Solvent and export	36.0
Total	590.0

Since carbon tetrachloride is a nonconductor, it is used for fire-fighting in areas where electrical equipment is the main hazard. However, its consumption has been decreasing because of toxic hazards associated with its use. In the field of metal cleaning it is employed primarily on aluminum parts and in machinery maintenance, although less toxic products such as 1,1,1-trichloroethane have replaced it in many applications.

Carbon tetrachloride is also one of the raw materials in the manufacture of nylons 7 and 9, via telomerization with ethylene. It is also a raw material for certain oil additives. Polyamide fibers are being made by this process by ICI, as well as in Japan and in the U.S.S.R.

The reaction between ethylene and CCl_4 takes place at high pressures (2000-3000 psig)

$$CCl_4 + nC_2H_4 \longrightarrow CCl_3(CH_2)_{2n}Cl$$

$$CCl_3(CH_2)_{2n}Cl + 2H_2O \longrightarrow HOOC(CH_2)_{2n}Cl + 3HCl$$

$$HOOC(CH_2)_{2n}Cl + 2NH_3 \longrightarrow HOOC(CH_2)_{2n}NH_2 + NH_4Cl$$

The ω-amino acids autocondense to a polyamide. Nylon 7 corresponds to $n = 3$, and $n - 4$ gives nylon 9.

CCl$_4$ derived Fluorocarbons

Carbon tetrachloride reacts with hydrogen fluoride to form various fluorinated substitution products:

$$CCl_4 + HF \longrightarrow HCl + CCl_3F \text{ (``11'')}$$

$$CCl_4 + 2HF \longrightarrow 2HCl + CCl_2F_2 \text{ (``12'')}$$

The first two products of the series are the most important. The process (see Fig. 3-8) combines carbon tetrachloride with hydrofluoric acid in the presence of antimony chloride. The effluent

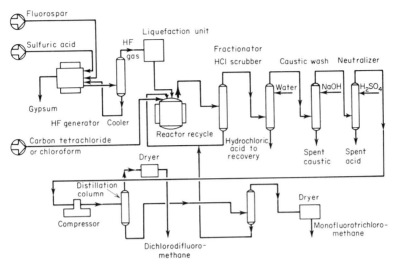

Fig. 3-8 Fluorocarbons.

from this reaction is sent to a column where unreacted CCl_4 is removed as bottoms and recycled. The overhead is treated with caustic to remove excess HF, then with an acid to neutralize any remaining caustic. The product stream goes to a fractionation system, from which CCl_2F_2 and CCl_3F are recovered overhead as products, and additional carbon tetrachloride recycled to the fluorination reactor.

The applications of these gases are due to their nontoxicity, nonflammability, and noncorrosiveness. Hydrocarbons such as butane and propane are satisfactory in most aerosol applications from the viewpoint of physical properties, chemical inertness, and low cost. However, fire laws in most parts of the U.S. require the use of fluorocarbons in personal products, either alone or as mixtures with methyl chloride, methylene chloride, vinyl chloride, or hydrocarbons. Recently, there has been a certain increase in the consumption of aerosols of higher molecular weight fluorocarbons as well as of such gases as CO_2 and nitrous oxide.

Fluorocarbons "11" and "12" still accounted for 85 percent of the aerosol propellant market in 1964. However, it is expected that hydrocarbons will be increasingly used in aerosol formulations.

Next in importance is the use of fluorocarbons as a refrigerant in domestic and industrial cooling, where fluorocarbon "12" is the most common. It requires less power for a given duty, and eliminates the hazard of ammonia or propane leakage.

The most dynamic use for fluorocarbons is as a blowing agent for urethane foams. At first these foams were made by decomposing part of the isocyanate feed by hydrolysis to give CO_2 bubbles. Later, it was discovered that using fluorocarbons was not only cheaper, since it obviates sacrificing part of the expensive isocyanate, but also made it possible to obtain products with a much lower thermal conductivity. Furthermore, flexible urethane foams contain around 6 percent blowing agent, but rigid foams, which are growing in demand at a higher rate, require 15 to 20 percent blowing agent. Fluorocarbon "11" is preferred for this application.

Almost 60 percent of the total CCl_4-derived fluorocarbons consists of "12." However, the uses of "11" are growing faster, and it is expected to increase its share of the market.

The market distribution for the 350.0 pounds per year of carbon tetrachloride-derived fluorocarbons consumed in the U.S. during 1963 can be broken down as follows.

End-use	MM lbs, 1963
Aerosol propellant	200.0
Refrigerant	120.0
Missile and aircraft parts degreasing	10.0
Blowing agent	20.0
	350.0

Source: *Chemical & Engineering News*,
September 7, 1964.

The main manufacturers of fluorocarbons are shown in the table below.

Producer	Location	Capacity, MM lbs/yr	Captive sources	
			Carbon tetrachloride	HF
Allied Chemical	Baton Rouge, La. Elizabeth, N.J. Danville, Ill. El Segundo, Calif.	175.0	—	x
DuPont	Antioch, Calif. Carney's Point, N.J. East Chicago, Ind. Louisville, Ky.	350.0	—	x
Kaiser	Gramercy, La.	35.0	—	x
Pennsalt	Calvert City, Ky.	60.0	—	x
Union Carbide	Institute, W.Va.	70.0	—	—
Racon	Wichita, Kan.	15.0	—	x

Investment for a 25.0 million pounds per year unit is $2.0 million, and production costs are calculated assuming that the producer has captive carbon tetrachloride at 7.0 cents per pound and HF at 15 cents per pound (100 percent). Although carbon tetrachloride is the major cost component, more fluorocarbon producers have captive HF than their own carbon tetrachloride facilities, probably due to the difficulties in handling hydrogen fluoride.

Cost, ¢/lb fluorocarbon "12"

Carbon tetrachloride	9.6	
Hydrogen fluoride	5.6	
Less: Hydrochloric acid	(-0.9)	
Net raw material costs		14.3
Labor and overhead		0.6
Utilities		0.8
Capital charges		2.6
Total		18.3

3.3 Perchloromethyl Mercaptan

Chlorination of carbon disulfide in the presence of iodine gives perchloromethyl mercaptan instead of carbon tetrachloride.

$$2CS_2 + 3Cl_2 \longrightarrow 2Cl_3C\text{—}S\text{—}Cl + S_2Cl_2$$

This compound is an intermediate for making "Captan" and "Folpet," two important fungicides.

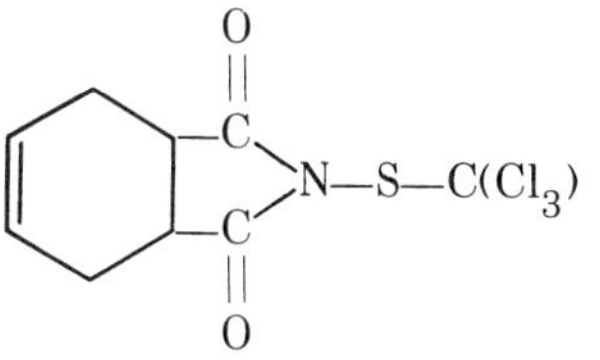

("Captan")

3.4 Xanthates

Alcohols react with CS_2 and a hydroxide to form xanthates. The type of reaction is the same as for viscose rayon.

$$ROH + CS_2 + NaOH \longrightarrow RO\text{—}\overset{\overset{\displaystyle S}{\|}}{C}\text{—}S\text{—}Na + H_2O$$

These compounds are important in the field of ore dressing. Many types of organic chemicals are used in the course of the froth flotation process. These include fatty alcohols and amines, phosphate esters, and others. About 60.0 million pounds per year of flotation aids are used in the U.S., of which approximately 20.0 million pounds are xanthates, principally those derived from isopropanol and ethanol. Zinc xanthates are also used as ultra-accelerators for natural rubber.

3.5 Ammonium Thiocyanate

Ammonia and carbon disulfide react to form ammonium thiocyanate, with ammonium sulfide as a byproduct.

$$CS_2 + 4NH_3 \longrightarrow NH_4SCN + (NH_4)S$$

This product is used mainly as a corrosion inhibitor in ammonia-handling systems. About 4.0 million pounds per year are consumed in the U.S.

4   CHLORINATED METHANES

Methane can be chlorinated to a mixture of all four methane derivatives: methyl chloride; methylene chloride; chloroform; and carbon tetrachloride:

$$CH_4 + Cl_2 \longrightarrow CH_3Cl + HCl$$

$$CH_3Cl + Cl_2 \longrightarrow CH_2Cl_2 + HCl \quad \text{(etc.)}$$

The reaction can be carried out by thermal or photo activation. The former, which requires temperatures in the 700°F range, is preferred commercially since it requires lower investment and maintenance, and allows more complete conversion of chlorine. The reaction is exothermic (about 650 Btu per pound of chlorine reacted) and the temperature in the chlorinator is kept from rising by diluting the feed with recycle methane to prevent decomposition of the reactants to HCl and coke. This process cannot be used to produce pure methyl chloride, since the higher substitutes are necessarily formed; however, it can, as in a few plants, be run so as to produce no methyl chloride at all by recycling all of it to the chlorinator.

The flowsheet in Fig. 3-9 shows a typical chlorinated methane process. Variations are possible with respect to the recycle of unwanted products to further chlorination; with respect to the HCl recovery scheme, the European practice is to refrigerate the reactor effluent to -85° F immediately. This removes the hydrochloric acid, saving on construction materials downstream, but increases utility costs. Preheated reactants are fed to the chlorination furnace and then after exchange against the feed, sent to a column where the chlorinated products are absorbed in a refrigerated mixture of carbon tetrachloride and chloroform; methane and HCl leave overhead and are absorbed in weak hydrochloric acid, which removes the HCl. The remaining methane is first neutralized and then returned to the chlorinator. The product stream is stripped of its light constituents, which are absorbed in water to remove the contained HCl, and then sent to the purification system. The bottoms of this second HCl absorber are fed to the top of the first; the net bottom of the stripper, consisting of the CCl$_4$ and CHCl$_3$ produced in the chlorinator, are also sent to the purification system. The light product stream is neutralized and fractionated into methyl chloride, methylene chloride and bottoms, which go to a liquid-phase mercury-lamp activated chlorinator to convert any remaining methylene chloride to chloroform and carbon tetrachloride. The effluent from this reactor is fractionated in two towers; the first produces a recycle stream to the second chlorinator, and the second produces pure chloroform. The bottoms

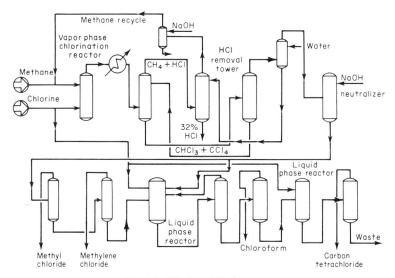

Fig. 3-9 Chlorinated Methanes.

from this last column are chlorinated again and separated into carbon tetrachloride and heavy waste. All four products still contain some dissolved HCl and must be neutralized before going to storage. Although HCl is recovered as a 32 percent solution, some plants still prefer to circulate a water-HCl solution instead of water, which on stripping yields gaseous HCl for use, for example, in vinyl chloride manufacture.

DuPont has developed and piloted a methane oxychlorination process:

$$CH_4 + HCl + \tfrac{1}{2}O_2 \longrightarrow CH_3Cl + H_2O \quad (\text{etc.})$$

The economics of a typical methane chlorination plant are estimated below, assuming a 95 percent yield on chlorine and 75 percent on methane. The investment for a 50.0 million pounds per year plant making all four products is around $3.2 million; production costs, assuming HCl worth 50 percent of the incoming chlorine, valued at 2.5 cents per pound are given in the table that follows.

| | Cost, ¢/lb | | |
	CH_2Cl_2	$CHCl_3$	CCl_4
Pure methane	0.3	0.2	0.2
Chlorine	3.2	3.5	3.6
Utilities	0.5	0.5	0.5
Labor and overhead	0.5	0.5	0.5
Capital charges	2.1	2.1	2.1
Total	6.6	6.8	6.9

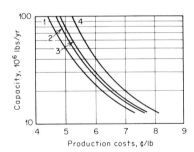

Fig. 3-10A Chlorinated Methanes. 1-Methylene Chloride, HCl = 70% chlorine; 2-carbon tetrachloride, HCl = 70% chlorine; 3-methylene chloride, HCl = 50% chlorine; 4-carbon tetrachloride, HCl = 50% chlorine.

Methane chlorinating plants in the U.S. are listed below. The amounts of carbon tetrachloride indicated are not all derived from methane; part is obtained as a byproduct from the manufacture of perchloroethylene from chlorinated raw materials or from hydrocarbons other than methane. Plants of this type are somewhat flexible, and their product unit can be varied.

Producer	Capacity*, MM lbs/yr			
	CH_3Cl	CH_2Cl_2	$CHCl_3$	CCl_4
Allied	14	50	30	10
Diamond	8	20	10	30
Dow	52	80	45	180
DuPont	25	35	15	–
Vulcan	–	10	11	15
Stauffer	–	55	75	275

*Omitted are those companies producing only methyl chloride which make up an aggregate of 110 million lbs additional capacity, and the 200-million-pound FMC-Allied joint venture at South Charleston, W.Va., for production of carbon tetrachloride.

Source: *Oil, Paint & Drug Reporter*, August 2, 1965.

Allied Chemical originally operated its plant at Moundsville, W.Va., on methane, but has since converted to methanol esterification followed by chlorination of methyl chloride to obtain higher substitutes. It is seen from the table that most methane chlorination plants are run primarily to produce methylene chloride and chloroform.

4.1 Methylene Chloride

The demand for methylene chloride in the U.S. was 175.0 million pounds in 1964. It is almost exclusively a service chemical, having no application of any importance as an intermediate. The end-uses in 1964 can be broken down as shown below.

End-use	MM lbs, 1964
Paint remover	87.0
Solvent degreasing	26.0
Aerosol propellant	18.0
Plastics processing	31.0
Other uses	13.0
Total	175.0

Source: *Oil, Paint & Drug Reporter*, July 13, 1964.

In addition to its wide application as a nonflammable paint-remover, methylene chloride has numerous other applications. In the plastics field it is a solvent for polycarbonates, isocyanates and cellulose diacetate, a urethane foam blowing agent, etc. It is used also in purification of steroids as a reaction medium, e.g., in the manufacture of phosphate insecticides, vegetable oil extraction, etc. In Germany it is important as the spinning solvent for cellulose acetate. Two developments of great potential importance are the use of methylene chloride in aerosol hair sprays and as a polyurethane foam blowing agent.

Apart from the companies making methylene chloride and chloroform via methane chlorination, three other plants employ chlorination of methyl chloride obtained via methanol esterification:

Producer	Location	Capacity, MM lbs/yr	
		Methylene chloride	Chloroform
Allied Chemical	Moundsville, W.Va.	40.0	20.0
DuPont	Niagara Falls, N.Y.	35.0	15.0
Frontier	Newark, N.J.	10.0	3.0

4.2 Chloroform

The consumption of chloroform in 1965 was 150.0 million pounds, distributed as shown below.

End-use	MM lbs, 1965
Fluorocarbon 22: Refrigerant, propellant	75.0
Plastics intermediate	45.0
Pharmaceuticals, etc.	30.0
Total	175.0

Other than as an intermediate, chloroform is used in certain pharmaceutical formulations, for example, cough medicines and rubbing liniments. It is also a solvent for textile degreasing and an extractant for food flavors, steroids, and especially antibiotics.

Fluorocarbon 22

The most important derivative of chloroform is fluorocarbon 22 (dichlorodifluoromethane):

$$CHCl_3 + 2HF \longrightarrow CHClF_2 + 2HCl$$

Most of the 60.0 million pounds of this product consumed in 1965 was used as a refrigerant and aerosol propellant, especially the former.   The trend to smaller and thinner domestic air-conditioning units has brought about an increased use of this refrigerant, which has a boiling point 20°F below that of fluoro-carbon 12.

The second most important use of fluorocarbon 22 is as an intermediate for polymers. Best known by far of all fluorocarbon plastics are the polymers of tetrafluoroethylene (TFE), developed originally as construction materials for handling the highly cor-rosive materials required for concentrating uranium isotopes. The monomer is made by dehydrochlorination of fluorocarbon 22:

$$2CHClF_2 \xrightarrow{\Delta} \begin{array}{c} F \\ \diagdown \\ \diagup \\ F \end{array} C{=}C \begin{array}{c} F \\ \diagup \\ \diagdown \\ F \end{array} + 2HCl$$

Hexafluoropropylene, another important fluorocarbon monomer, is made from TFE:

$$2F_2C{=}CF_2 \xrightarrow{\Delta} CF_3{-}CF{=}CF_2 + F_2$$

Fluorocarbon resins have exceptional chemical resistance and are suitable for applications up to 500°F. In the consumer goods field, tetrafluoroethylene polymers are used to coat numerous cookware items, especially frying pans and ovens, and for compo-nents of other household utilities that have to withstand high temperatures.   They are also used in vending machines, for parts in contact with hot food products.   However, the military and chemical process equipment applications are of greater com-mercial importance.   TFE resins are used for insulation in mili-tary and commercial aircraft, to line corrosion-resistant process vessels and piping, as insulation for electrical steam-tracing cable, for electronic component sockets (TFE maintains its dielectric properties at ultrahigh frequencies), and in all kinds of gaskets, rings, seals, and packagings. In the automotive industry TFE parts are being used in power steering assemblies and piston rings.

Granular TFE resins can be extruded or compression molded; about 75 percent is processed by heat extrusion. Paste grade TFE resins can also be extruded or sintered, whereas the grades sold as aqueous dispersions are processed by impregnation or coating. Granular grades are used to make seals, bearings, wiring devices and such extruded items as piping; paste and dispersion grades are preferred for flexible hose, wire coating, pipe linings, and the well-known nonadherent coatings for frying pans.

Listed below are the producers of TFE resins in the U.S.

Producers	Location	Capacity MM lbs/yr
Allied Chemical	Elizabeth, N.J.	3.0
DuPont	Parkersburg, W. Va.	25.0
Pennsalt	King of Prussia, Pa.	(pilot)
Thiokol	Moss Point, Miss.	2.0

The total demand for fluorocarbon plastics in 1963 was 15.0 million pounds, of which 12.5 million pounds was polytetrafluoroethylene. The market can be broken down as follows:

End-use	MM lbs, 1963	
	Granular	Paste and dispersion
Aerospace industry	1.25	1.00
Electrical and electronics industry	2.00	3.25
Chemical process industry	2.75	1.00
Mechanical goods industry	1.00	—
Miscellaneous industry	—	0.25

Source: *Chemical & Engineering News*, March 9, 1964.

About 60 percent of the remaining 2.5 million pounds of fluorocarbon plastics consisted of fluorinated ethylene-propylene (FEP), a copolymer of TFE and hexafluoropropylene. Its main advantage over TFE is that it can be extruded to film thicknesses of as little

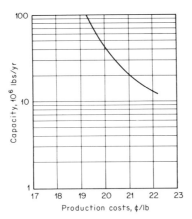

Fig. 3-10B Fluorocarbon "12."

as 1/2 mil; these films are used in the electrical industry and especially in the manufacture of temperature- and corrosion-resistant
process equipment where fluorocarbon must compete with epoxy
resins, polyurethanes and "Hypalon." Polymers of chlorotrifluoroethylene ("Kel-F") and vinylidene fluoride ("Kynar") made
up the rest of the fluorinated plastics market. "Viton" is a
fluorocarbon elastomer useful in the 400 to 500°F range. It is
used in heat- and corrosion-resistant aircraft and industrial parts,
and is a copolymer of hexafluoropropylene and vinylidene fluoride.

Chlorpicrin

Trichloronitromethane, known as chlorpicrin, is used as a
fumigant and as an odorant for methyl bromide.

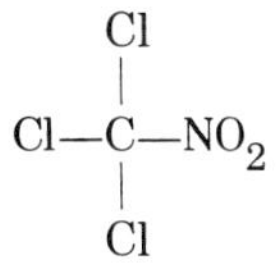

$$\begin{array}{c} Cl \\ | \\ Cl-C-NO_2 \\ | \\ Cl \end{array}$$

4

ACETYLENE

1 INTRODUCTION

Acetylene has become a petrochemical. Some 45 percent of all acetylene capacity in the U.S. uses a hydrocarbon raw material. Petrochemical acetylene is even being produced in Germany, the birthplace of coal-based acetylene chemistry. Until well after World War II, acetylene produced via calcium carbide from coal and electric energy was the foundation of the European chemical industry and of German acyclic chemical manufacture. At one time, even ethylene for styrene production was made in Germany by partial hydrogenation of acetylene.

Acetylene is unique in that it is the only organic building-block which can be derived entirely from inorganic raw materials — lime, coal, water — and stored in the form of a stable solid. The hydrocarbon, acetylene, is generated by simply adding water to the solid calcium carbide. It is also much more reactive than ethylene and therefore easier to convert to other derivatives.

The main inputs for acetylene production via calcium carbide are electricity and coal, the former being the more important. Hence in the U. S., with a few exceptions, carbide is produced in areas such as Niagara Falls, N.Y., and the region served by the Tennessee Valley Authority, where cheap hydroelectric power is available.

Petrochemical acetylene can be made from either natural gas, e.g., methane, which yields only acetylene and synthesis gas, or higher hydrocarbons, which, in addition to hydrogen and acetylene, produce ethylene and heavier hydrocarbons. When only acetylene is required, usually it is best to start from natural gas or from carbide. Acetylene processes starting from higher hydrocarbons become of interest when the producer needs both ethylene and

186

acetylene.   One example is one route for the production of vinyl chloride with total chlorine utilization.

Among the many techniques developed for producing acetylene from hydrocarbons, The Badische Anilin und Soda Fabrik (BASF) or Sachse process, has been most popular in the U.S. It is used in almost half of all hydrocarbon acetylene plants in operation. The Sachse process is a natural gas partial oxidation method, in which sensible and reaction heat are supplied by burning part of the feed.   Methane and oxygen in a molar ratio of 1.5:1 are heated separately to 1200°F in direct-fired heaters, and then the two streams are mixed and cracked in a 2700°F flame resulting from the combustion of more methane.

$$5CH_4 + 3O_2 \longrightarrow C_2H_2 + 3CO + 6H_2 + H_2O$$

Hydrocarbons can be cracked to acetylene only at temperatures where the product readily decomposes to carbon and hydrogen. Thus the residence time in the cracking section must be very short — a typical figure would be 0.0005 sec. The reaction stream is quenched immediately to prevent decomposition of the product. In some processes, quenching is accomplished so that the sensible heat in the product stream can be recovered.   However, in the original BASF process, the cracked gases are quenched by water-spraying, and thus no heat is recovered.   After removal of the carbon-black that inevitably forms, the gases are compressed and sent to the purification section. Acetylene cannot be processed at partial pressures above 1 to 2 atmos., which precludes purification by the low-temperature fractionation techniques used in ethylene plants.   Thus all purification schemes employ some kind of solvent recovery.   The BASF process employs acetone, but other solvents are also used. The Societe Belge de le Azote (SBA) process uses dimethyl formamide; N-methyl pyrrolidone has been suggested by BASF; γ-butyrolactone, ammonia and others have also been proposed.

BASF has announced modifications in its burner (the heart of any acetylene process) thereby making the process applicable to naphtha feeds.   A heat-recovery system has also been developed. It uses an aromatic oil stable at the elevated temperature where the reactor effluent contacts the quench oil. This quench oil also removes soot formed in the reactor, eliminating the coke filters in the earlier version of the process. Filtration is eliminated by centrifuging a slip-stream of quench-oil or by distillation. Losses of the rather costly aromatic oil must be kept as low as possible.

After quenching, the hot oil is sent to a waste heat boiler. Compared with the earlier version, the process is expected to save 14 percent on raw materials. Figure 4-1 shows a flowsheet of the BASF acetylene process starting from natural gas.

There are five other processes being used in the U. S. These include DuPont's rotating electric arc technique whereby very rapid quenching is accomplished using a nonaromatic oil which is cracked to more acetylene. Heat recovery is therefore in the form of reaction heat instead of steam generation. Electric energy consumption is said to be 6.0 kWh per pound acetylene, against 4-5.0 kWh per pound for carbide acetylene. The Kellogg-SBA process, used by Tenneco, is also a partial oxidation route. An all-metal water-cooled reactor is designed to eliminate the refractory problems encountered in other processes. As in the DuPont process, quenching is by means of a crackable, non-aromatic oil. Diamond Alkali employs the Montecatini process. Its distinctive characteristics are that cracking takes place under pressure, and that steam is recovered directly from the hot gases. Union Carbide uses a patented partial oxidation process somewhat similar to the "Sachse" version. Also, it owns all rights to the Wulff process.

Only the Wulff process employs a regenerative, cyclic technique. In the first half of the cycle, a brick checker-work is heated by

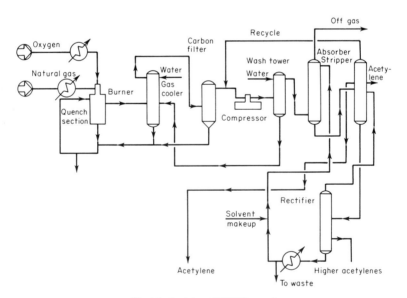

Fig. 4-1 Acetylene (BASF Process).

burning a convenient fuel, part of which is supplied by the process off-gases. In the second half, cracking stocks are introduced under vacuum in the opposite direction and removed from the bricks in the form of reaction heat, the energy previously stored in them. Due to limitations imposed by piping and switch-valve design, a pair of Wulff furnaces can be designed for no more than 10.0 million pounds per year. The main advantages of the process are good heat utilization, and the use of the coke and carbon deposits by burning off during the heating-up part of the cycle. These advantages are offset by poor yields, since long residence times lead to appreciable decomposition of acetylene. The process is very flexible because the automatic coke burn-off permits any feedstock to be employed. As would be expected, this feature makes it more attractive in Europe than in the U.S. where the process was developed. By 1965 three large Wulff plants had been introduced in Europe whereas only one small plant, aside from a pilot plant, existed in the U.S. Since the process is not attractive for methane feeds, Wulff plants will normally produce both ethylene and acetylene. However, recovery of ethylene adds considerably to investment requirements and hence several plants that simply burn the ethylene produced by cracking along with the rest of the off-gases have been built. One of the major advantages of the Wulff process is acetylene and ethylene can be produced without the need to market liquid fuels or to upgrade byproduct synthesis gas.

Union Carbide recovers some acetylene from one of its olefin plants. The usual practice is to eliminate acetylene by catalytic hydrogenation. However, when the olefin plant is large enough, recovery by solvent extraction may become feasible. The increasing size of olefin plants in the U.S. and elsewhere may enhance the importance of this type of acetylene source.

Italy has the largest number of hydrocarbon acetylene plants (five) because of its natural gas reserves. In Germany, coke-oven methane is used in one large plant (Huels), and another plant (Hoechst) uses an unique flame-chemistry process known as HTP (High Temperature Pyrolysis) in which feed is introduced into the hot gases resulting from combustion of a suitable fuel. In France, the largest hydrocarbon acetylene plant is near the Lacq gas-fields, but there is also a unit employing the SBA process on coke-oven methane and a Wulff plant producing acetylene only. In Japan, coke-oven gas is also the most common feed to noncarbide acetylene plants. Of the 20 plants built or announced outside the U.S. up to 1964, 10 used the BASF process, 4 the Montecatini process, 3 the Wulff process, and the remainder accounted for by other techniques. The newest BASF process generates acetylene by submerged combustion. In the Tsutsumi process cracking takes place at 200 mm Hg thereby lowering the temperature to 1750° F.

The economics of acetylene production by oxidation cracking depend on raw material costs. Heavier feeds give better yields and also consume considerably less oxygen. Methane requires 50 percent more oxygen per pound acetylene produced than would ethane, and heavier feeds need even less. However, since on a weight basis, methane costs about 50 percent less than the next-cheapest feedstock (U.S., Gulf Coast conditions), almost all hydrocarbon acetylene plants in the U.S. start from natural gas, which will therefore be considered typical. The reactor effluent stream is about 1.5 times greater for methane feedstock than for heavier hydrocarbons, and plant investment around 25 percent larger. A 100.0 million pounds-per-year partial oxidation plant, excluding the oxygen unit, would cost $11.5 million, starting from methane, compared to $9.0 million from LPG. As can be seen from the table below, the off-gas is almost 90 percent synthesis gas, and acetylene economics thus depend very much on whether or not this gas is used as a raw material or goes to fuel. Most of the U.S. acetylene-from-methane plants use the off-gas for further synthesis. For example, Monsanto and Tenneco make methanol, Rohm & Haas and Dow-BASF produce acrylates, Rohm & Haas make ammonia, Dow-BASF and Union Carbide operate oxo plants, and Monochem (Borden) employs an unique integrated vinyl acetate scheme.

Reactor Effluent: Typical Composition

Compound	Mol %
Hydrogen	55.0
CO	25.8
Methane	5.0
CO_2	5.0
Acetylene	8.5
Heavy acetylenes	0.5
Heavy hydrocarbons	0.2
Total	100.0

Assuming natural gas at 20 cents per 10^6 Btu, oxygen obtainable at $9.00 per ton and crediting the crude synthesis gas at 25 cents per 1000 ft^3, acetylene production costs are as shown in the next table.

A 40.0 million pounds per year Wulff unit producing no ethylene (all off-gas is burned), using gas-oil as raw material, requires an

	Cost, ¢/lb acetylene
Methane	2.50
Oxygen	2.45
Utilities and chemicals	2.34
Labor and overhead	.48
Capital charges	3.80
	11.57
Less: synthesis gas	(−2.87)
Total	8.70

investment of $6.5 million. Production costs are calculated as-suming gas-oil at 0.8 cent per pound, and off-gases credited at fuel value.

	Cost, ¢/lb acetylene
Raw material (net)	1.8
Utilities	3.2
Labor and overhead	1.3
Capital charges	5.4
Total	11.7

Throughout this section, 9.0 cents per pound will be used as a typical cost of acetylene. With ethylene and propylene being sold on the Gulf Coast at less than 4.0 cents per pound, it is evident that acetylene is losing ground as a raw material. This is true even in Europe where the industry had been traditionally acetylene-oriented. In late 1964, however, it was announced that a process based on a better understanding of methane thermal cracking kinetics could produce acetylene for under 5 cents per pound, or even 3 cents per pound if hydrogen were valued as synthesis gas. If these studies lead to a commercial process, acetylene may still recover the uses lost to ethylene and propylene.

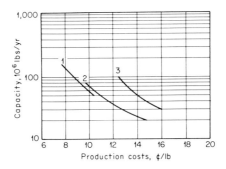

Fig. 4-2 Acetylene. 1-BASF Process; 2-Wulff Process; 3-from Calcium Carbide.

The table on the next page shows the main U.S. producers, capacities, process used and chief chemical uses of the acetylene produced at each plant. Other plants exist, but their output is only intended for such nonchemical uses as metal cutting. The U.S. end-use pattern for acetylene in 1965, aside from the 140.0 million pounds per year used for arc-cutting and welding, was as shown below.

End-use	MM lbs, 1965
Acrylonitrile	200.0
Vinyl chloride	350.0
Neoprene	180.0
Trichloroethylene	70.0
Perchloroethylene	20.0
Vinyl acetate	140.0
Acrylates	30.0
Others	30.0
Total	1,020.0

Source: *Hydrocarbon Processing,*
August 19, 1966.

Very few of these end-uses are still the exclusive domain of acetylene. Ethylene, propylene or butylenes are the preferred feeds for most new projects. In this chapter those derivatives are discussed which are still made predominantly from acetylene, although opposite trends are at work.

2   VINYL ACETATE

Vinyl acetate is made primarily by the vapor-phase reaction between acetylene and acetic acid in the presence of a zinc acetate catalyst:

$$CH_3COOH + C_2H_2 \longrightarrow CH_3COOH{=}CH_2$$

In the typical process (Fig. 4-3), acetylene is mixed with acetic acid in a vaporizer and the combined feed enters an oil-cooled reactor. Effluent is condensed and unreacted acetylene removed and recycled. The remainder of the stream is fractionated into unreacted acetic acid, vinyl acetate product and heavy wastes. Yields are 95 percent on acetylene and 97 percent on acetic acid. The capacities of U.S. plants using this process in 1965 were:

Producer	Location	Capacity, MM lbs/yr	Route	Outlets
Air Reduction	Calvert City, Ky.	250.0	Carbide	Vinyl chloride, vinyl acetate (both captive); Reppe chemicals (General Aniline)
	Louisville, Ky.	100.0	Carbide	Neoprene (DuPont); polyvinyl chloride (Goodrich)*
Allied Chemical	Moundsville, W.Va.	40.0	Carbide	Vinyl chloride
American Cyanamid	Fortier, La.	100.0	BASF	Acrylonitrile
Diamond	Houston, Tex.	40.0	Montecatini	Vinyl chloride; perchloroethylene
Dow	Freeport, Tex.	8.0	Dow	Acrylates; vinyl chloride; chlorinated solvents
DuPont	Montague, Mich.	50.0	Electric arc	Neoprene
Monochem	Geismar, La.	80.0	BASF	Vinyl chloride; vinyl acetate
Monsanto	Texas City, Tex.	125.0	BASF	Vinyl acetate; vinyl chloride; acrylonitrile
Rohm & Haas	Deer Park, Tex.	35.0	BASF	Acrylates
Tenneco	Houston, Tex.	100.0	Kellogg-SBA	Vinyl chloride (captive); vinyl acetate (captive)
Union Carbide	Niagara Falls, N.Y.	150.0	Carbide	Vinyl chloride (Goodrich); trichloroethylene (Hooker, DuPont, captive); vinyl acetate (DuPont, captive)
	Ashtabula, Ohio	40.0	Carbide	Vinyl chloride (General Tire, U.S. Rubber); trichloroethylene (Detrex); acetylene black
	Texas City, Tex.	80.0	UCC	Vinyl chloride
	Institute, W.Va.	10.0	Wulff	
	Seadrift, Tex.	15.0	Olefins plant byproduct	

*Goodrich now uses oxychlorination of ethylene only.
Source: for hydrocarbons, *Oil, Paint & Drug Reporter*, October 12, 1964.

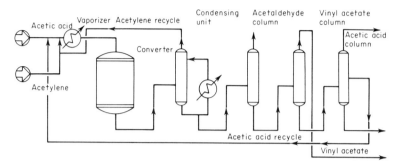

Fig. 4-3 Vinyl Acetate (Wacker Process).

Producer	Location	Capacity, MM lbs/yr	Captive sources	
			Acetylene	Acetic acid
Air Reduction	Calvert City, Ky.	90.0	x	—
Borden Chemical	Geismar, La.	75.0	x	x
DuPont	Niagara Falls, N.Y.	75.0	x	—
Monsanto*	Texas City, Tex.	55.0	x	—
National Starch	Long Mott, Tex.	50.0	—	—
Union Carbide	S. Charleston, W.Va.	50.0	x	x
	Texas City, Tex.	110.0	x	x

*Has acetic source in Canada.
Source: *Oil, Paint & Drug Reporter*, September 12, 1966.

Also, Celanese, the world's largest producer, operates two plants using different processes. The first, at Pampa, Texas, makes use of large amounts of two acetyl products obtained by LPG oxidation to make vinyl acetate from acetaldehyde and acetic anhydride. The overall reaction follows:

$$(CH_3CO)_2O + CH_3CHO \longrightarrow CH_3COOCH{=}CH_2 + CH_3COOH$$

One mol acetic acid is recovered per mol of product and returned to the ketene furnaces. This plant had a capacity of 65.0 million pounds per year in 1966, but is expected to be phased out. The second Celanese plant (at Bay City, Texas) with a capacity of 100.0 million pounds per year, uses the ICI ethylene oxidation route. Reaction takes place over a palladium catalyst.

$$2C_2H_4 + \tfrac{3}{2}O_2 \longrightarrow CH_3COOCH{=}CH_2 + H_2O$$

Several companies have developed liquid phase (Shell, Hoechst, Bayer) or vapor-phase (U. S. I.) techniques for making vinyl acetate from ethylene and acetic acid:

$$CH_3COOH + C_2H_4 + \tfrac{1}{2}O_2 \rightarrow CH_3COOCH=CH_2 + H_2O$$

The U. S. I. process claims yields of 95 percent but uses oxygen instead of air.

Here, mention should be made of the Borden plant because of the unique relation between the various units. Acetylene is produced by the BASF partial oxidation process, and byproduct synthesis gas used to make methanol. This reacts with additional carbon monoxide to give acetic acid:

$$CH_3OH + CO \rightarrow CH_3COOH$$

The acid combines with acetylene in a vinyl acetate plant using a process developed by Borden and Blaw Knox. The acetic acid technology was also developed by BASF.

The vinyl acetate output in the U.S. was 520.0 million pounds in 1965, and has been growing steadily at a rate of 11 percent per year. The end-use distribution, primarily as the polymer, is given below.

End-uses	MM lbs, 1965
Polyvinyl acetate	425.0
Copolymers	60.0
Other uses	35.0
Total	520.0

Source: *Oil, Paint & Drug Reporter*,
September 12, 1966.

The investment required for a 50.0 million pounds per year acetylene based unit is $4.0 million. Production costs can only be considered within the framework of the raw materials position of each individual producer. From the prevailing prices for merchant acetylene and acetic acid it is seen that it would be impossible to produce vinyl acetate at a profit without captive production of at least one raw material. Almost all the companies listed are acetylene producers despite the fact that acetic acid constitutes the major cost component. Costs for a 50.0 million pounds per year plant are estimated below, assuming costs for both raw materials somewhere between captive and merchant values.

	Cost, ¢/lb vinyl acetate
Acetic acid (8.0 ¢/lb)	5.8
Acetylene (9.0 ¢/lb)	2.9
Utilities	0.6
Labor and overhead	0.6
Capital charges	2.6
Total	12.5

Monomer has sold at 12.5 cents per pound and it is not surprising that the plants built since 1963 have either been based on ethylene, or involved integrated production of acetylene and acetic acid. Although sufficient data has not been made available, the cost of making vinyl acetate in a 100.0 million pounds per year unit using the ICI process can be estimated as follows, exclusive of royalties. The investment required is $8.0 million.

	Cost, ¢/lb vinyl acetate
Ethylene (at 4.0 ¢/lb)	1.9
Utilities, catalyst, and chemicals	1.1
Labor and overhead	0.6
Capital charges	2.6
Total	6.2

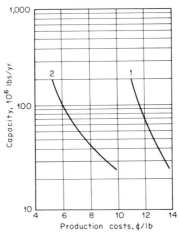

Fig. 4-4 Vinyl Acetate. 1-From C_2H_2 + CH_3COOH; 2-From ethylene (ICI process).

2.1 Polyvinyl Acetate

Vinyl acetate can be polymerized by several methods. The most important are emulsion and suspension polymerization, both using peroxide catalysts. Most polyvinyl acetate is sold in the form of 55 percent solid dispersions.

The use distribution of polyvinyl acetate in 1965 is shown below. Note that the use of a vinyl acetate as an intermediate for polyvinyl alcohol appears under this heading since no direct way of making it has been found.

End-use	MM lbs, 1965
Adhesives	130.0
Latex paints	105.0
Miscellaneous uses	15.0
Paper	25.0
Polyvinyl alcohol	90.0
Polyvinyl butyral	35.0
Textile treating	25.0

Source: *Oil, Paint & Drug Reporter*,
September 12, 1966.

Most synthetic adhesives are phenolics or urea-formaldehyde; polyvinyl acetate represents only 4 percent of the total adhesive consumption, and 10 percent of synthetic adhesives. It is used wherever its freedom from taste, color, odor and toxicity are at a premium. Nonwoven goods include sanitary products where polyvinyl acetate is the preferred binder. However, in paint, polyvinyl acetate latex has succeeded in capturing much of the market formerly held by rubber paints. This was due to the abundant supplies of butadiene and styrene readily available after World War II. Polyvinyl acetate has the advantage of being suited to production in the simple type of equipment used for making alkyd resins, which brings it within reach of a great number of coatings producers. Styrene-butadiene and acrylic resins are made by only a limited number of large producers. Being stable to ultraviolet light, vinyl acetate latices are expected eventually to find wide application as outdoor coatings, where alkyds hold the largest share of the market but where rubber latex cannot be used because of its instability to oxidation. Vinyl acetate and its copolymers account for over 50 percent of the latex paint market, compared to 25 percent each for SBR and acrylics; the proportion of vinyl acetate is even higher in those parts of the world where butadiene is not as readily available. Other uses include inks, where compatibility with other resins is an advantage, and as a base for chewing gum.

2.2 Copolymers

The most important vinyl acetate (VAc) copolymer is with vinyl chloride (VC). Most LP-records are made from a 15-85

VAc-VC copolymer; this outlet consumes about 13.0 million pounds per year of VAc. Beer, vegetable and fruit-can lining include a VC-VAc top coating, which accounts for 8.0 million pounds per year of vinyl acetate. Another 4.0 million pounds per year are used in VAc-VC calendered sheeting used in outdoor signs. Other outlets are numerous bonding and coating applications, as well as "Vinyon," a fiber consisting of 40 percent vinyl acetate. VC-VAc copolymers are also used in flooring and textile coating.

Of more recent interest are ethylene-vinyl acetate copolymers first produced by DuPont under the name "Elvax," and later also by U.S.I., Union Carbide and Air Reduction. They are used in textile and paper coatings; in book-binding and hot-melt adhesives, where they compete with straight polyvinyl acetate products; as wax extenders in floor polish formulations; and especially in surface coatings. Emulsion coatings made from ethylene-VAc copolymers are considered to be as good as acrylics for both interior and exterior use, and also a great deal cheaper. The use of these copolymers in all fields amounted to 20.0 million pounds in 1964, and excellent growth is expected — partly, however, at the expense of overall vinyl acetate use. Total use of vinyl acetate in all copolymers for 1964 was 70.0 million pounds.

2.3 Polyvinyl Alcohol

Since it has proved impossible so far to produce monomeric vinyl alcohol, polyvinyl alcohol is made from polyvinyl acetate by hydrolysis, mostly with methanol.

$$\left[\begin{matrix} -CH_2-CH- \\ | \\ O \\ | \\ CH_3-C=O \end{matrix} \right] \xrightarrow{CH_3OH} \left[\begin{matrix} -CH_2-CH- \\ | \\ OH \end{matrix} \right] + CH_3COOCH_3$$

The resulting methyl acetate is hydrolyzed, with methanol being recycled and acetic acid sent back to the vinyl acetate plant. Since polyvinyl acetate is a solid, saponification usually takes place in large kneaders, although more recent developments are in the direction of continuous processes, with polymerization and hydrolysis both taking place in solution. Figure 4-5 shows a flowsheet.

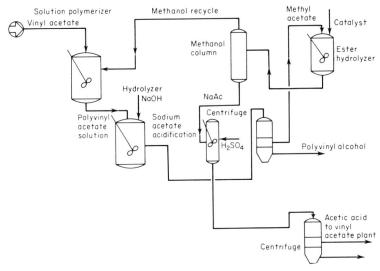

Fig. 4-5 Polyvinyl Alcohol.

Numerous grades are made, with properties varying with viscosity (i.e., molecular weight) of the original polymer and with
degree of saponification. The most widely used grades were about
90 percent saponified. When losses in yield are considered it takes
1.85 pounds of vinyl acetate to make 1.0 pound of polyvinyl alcohol.
As seen from the table below, all except one producer also manufactures vinyl acetate. A 10.0 polyvinyl alcohol plant, including
vinyl acetate polymerization facilities, costs $5.0 million.

Producer	Location	Capacity, MM lbs/yr
Air Reduction Company	Calvert City, Ky.	20.0
	Cleveland, Ohio	2.0
American Aniline & Extract	Calvert City, Ky.	1.0
Borden	Leominster, Mass.	8.0
DuPont	Niagara Falls, N.Y.	25.0
Franklin Chemical Company	Columbus, Ohio	2.0
Monsanto	Springfield, Mass.	10.0

In 1964, 38.0 million pounds were used in the U.S., distributed as
follows. (No use for polyvinyl butyral is accounted for below.)

End-use	MM lbs, 1964
Vinyl emulsifier and protective colloid	4.5
Adhesives	17.1

Film	3.8
Paper and carton coating	6.1
Textile sizing, finishing, and nonwoven textile binder	4.5
Other uses	2.0
Total	38.0

Source: *Oil, Paint & Drug Reporter*, August 25, 1965.

At a concentration of about 5 percent, polyvinyl alcohol (PVAl) is used as a thickener, viscosity regulator and suspending agent in many vinyl acetate adhesives. It acts as a humectant, retarding water loss and increasing shelf life. It also is used to avoid particle agglomeration in suspension-polymerized PVC. Another application, which requires higher concentrations, is in remoistenable polyvinyl acetate adhesives for envelopes and labels. Water solubility is also what determines its use as a film; being solvent-resistant but water-soluble, these films are finding some use in packaging single doses of such things as household items as bleaches. PVAl is used to impart scuff-resistance to returnable cartons, and as a borax substitute in paper making. Borden produces a group of "internal" plasticizers — that is, plasticizers that are themselves polymerizable, known as "Lemoflex" — by ethoxylating polyvinyl alcohol. A copolymer with polyvinyl pyrrolidone is used in hairspray formulations. When used in paper sizing, it gives a stronger, lighter product. Particularly noteworthy are the polyvinyl alcohol fibers made by Kurashiki in Japan. In addition to excellent elasticity and dyeability, these fibers are unique among synthetics in that they have a degree of water absorbency comparable to that of cotton, and are spun from hot water. Since polyvinyl alcohol is water-soluble, it is insolubilized after spinning with formaldehyde.

In the Kurashiki process, vinyl acetate is polymerized continuously in methanol. Next, continuous methanolysis takes place in the presence of sodium hydroxide; methyl acetate is hydrolyzed to acetic acid and methanol which are reused. Some sodium acetate also forms and is reacted with sulfuric acid giving more acetic acid and Glauber's salt, which is used during the spinning process in the coagulant solution.

Polyvinyl acetals

Polyvinyl acetals are made from polyvinyl acetate in two steps: 1) hydrolysis to polyvinyl alcohol; and 2) insolubilization of the

alcohol by combining two mols of alcohol with one mol of an aldehyde, usually butyraldehyde or formaldehyde:

$$\left[-CH_2-CH-CH_2-CH-\right] \quad \xrightarrow{RCOH} \quad \left[-CH_2-CH-CH_2-CH-\right] + H_2O$$

(with OH, OH groups on the left structure; on the right, the two O atoms bridge to a central C bearing H and R)

About 40.0 million pounds per year of polyvinyl butyral are
consumed in the U.S., much of it as the inner layer for automobile
safety glass, and most of the rest in wash-primers. The question
of whether or not safety glass should be required other than in
automobile windshields is a topic of much discussion. Since the
trend in cars is towards greater safety, polyvinyl butyral is still
growing in demand.

The capacities of U.S. producers are given in the table below.

Producer	Capacity, MM lbs/yr
DuPont	15.0
Shawinigan	30.0
Union Carbide	5.0

Polyvinyl formal is a tough, heat-resistant resin used as wire
enamel. About 12.0 million pounds per year are consumed in the
U.S., the main producer also being Shawinigan.

2.4 Other Uses

Vinyl acetate is the intermediate from which other vinyl esters,
e.g., vinyl stearate and crotonate, are produced. It is also one of
the intermediates for the production of aminothiazole, a bactericide
and one of the intermediates for sulfathiazole, a well-known sulfa
drug. Finally, vinyl acetate is the main component of "Darvan," a
fiber developed by B.F. Goodrich, manufactured at first by Celanese
and now by Hoechst in Germany. The other component is vinylidene
cyanide.

3   NEOPRENE

Acetylene can be dimerized to vinyl acetylene and then hydrochlorinated.

$$2C_2H_2 \longrightarrow H-CC-CH=CH_2 \xrightarrow{HCl} H_2C=\overset{\overset{\text{Cl}}{|}}{C}-CH=CH_2$$

This monomer, known as chloroprene, can be polymerized to a rubber that has outstanding resistance to oil, solvents and ozonolysis cracking. The presence of chlorine in the molecule imparts flame resistance, which permits its use in wire and cable coatings and in making conveyors that handle inflammable materials, notably coal. Its resilience is responsible for its use in engine mountings (automotive as well as others), in flexible couplings, hose, etc. Developed by DuPont, this company is still the only producer in the U.S. with two plants, one at Montague, Mich. (60.0 million pounds per year) and one at Louisville, Ky. (260.0 million pounds per year).

Distillers (G.B.) have developed a cheaper route from butadiene. The key step is production of 3,4-dichlorobutene by high-temperature chlorination and subsequent isomerization; this can be dehydrochlorinated to neoprene:

$$C_4H_6 + 4Cl_2 \longrightarrow C_4H_6Cl_2 \xrightarrow{cat.} CH_2=CH-CHCl-CH_2Cl \xrightarrow{-HCl}$$

$$CH_2=CH-\overset{\overset{\text{Cl}}{|}}{C}=CH_2$$

The first plant using this process was built in France at Champagnier, and has a capacity of 45.0 million pounds per year. In the U.S., Petrotex has announced a plant of the same size to be built at Houston, Tex.

This process allows a considerable reduction in the production cost of neoprene. Investment for a 45.0 million pounds per year plant is $14.0 million, compared to $22.0 million for one of the same size based on acetylene. Monomer production costs are estimated in the table below.

	Cost, ¢/lb chloroprene	
	From butylene	From acetylene
Raw materials	6.6	7.2
Labor, utilities, etc.	2.5	2.5
Capital charges	10.3	16.1
Total	19.4	25.8

Consumption in the U.S. is fairly steady at about 190.0 million pounds per year, in addition to which large quantities are exported. The demand pattern is shown below. A large variety of goods are made from neoprene latex — gloves, coated fabrics, high wet-strength papers, dipped metal coatings, and fuel pump diaphragms. It is also used in combination with "Hypalon" to make industrial tank linings. There are more than 40 types of neoprene on the market.

End-use	1963 Consumption MM lbs/yr
Wire and cable coating	45.0
Tires	8.0
Adhesives	11.0
Mechanical goods	105.0
Shoes and heels	11.0
Miscellaneous	10.0
Total	190.0

Source: *Oil, Paint & Drug Reporter,*
December 30, 1963.

4   TRICHLOROETHYLENE

Trichloroethylene is made by dehydrochlorinating tetrachloroethane. Formerly this was accomplished with lime, usually available from carbide acetylene production. Today most producers have developed cracking techniques which yield anhydrous HCl instead of calcium chloride. Although there are installations making tetrachlorethane by chlorination or oxychlorination of ethylene-derived ethylene dichloride, as of 1965, 75 percent of the trichloroethylene capacity was still based on acetylene. New projects underway in 1965 will eventually lower this to about 65 percent. A flowsheet of trichloroethylene production is shown in Fig. 4-6.

$$C_2H_2 + 2Cl_2 \rightarrow CHCl_2CHCl_2 \xrightarrow{-HCl} CHCl{=}CCl_2$$

Chlorinated solvents have gained wide acceptance due to their quick drying and low flammability. This in turn means lower insurance rates and improved safety. Both are used for extraction and for a new type of quick drying, one-coat paint. Perchloroethylene and trichloroethylene each is used almost entirely in one application. Perchloroethylene is primarily a dry-cleaning solvent, being the more stable of the two and therefore less likely to ruin

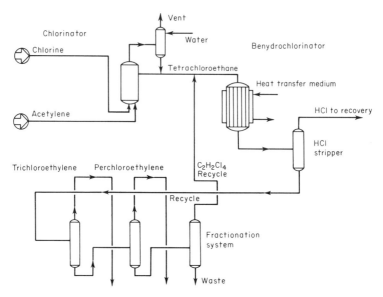

Fig. 4-6 Trichloroethylene.

clothing by decomposition to HCl. Almost all trichloroethylene is used in metal degreasing. Trichloroethylene must be stabilized. At first this was done with alkylamines, which are simply neutralized by the HCl given off until there is no more amine and the stabilizing effect destroyed. Later, compounds that actually inhibit decomposition were developed; although their exact nature is considered secret, acetylenic alcohols and certain pyrroles are known.

The U.S. producers, capacities and routes in 1964 are given below.

Producer	Location	1964 Capacity, MM lbs/yr	Route
Detrex	Ashtabula, Ohio	70.0	Acetylene
Dow	Freeport, Tex.	30.0	EDC (oxychlorination)
	Plaquemine, La.	35.0	EDC (oxychlorination)
DuPont	Niagara Falls, N.Y.	225.0	Acetylene
Hooker	Tacoma, Wash.	25.0	Acetylene
	Niagara Falls, N.Y.	45.0	Acetylene
Pittsburgh Plate	Lake Charles, La.	80.0	EDC (oxychlorination)
	Barberton, Ohio	45.0	Acetylene

Source: *Oil, Paint & Drug Reporter*, June 15, 1964.

The demand for trichloroethylene in 1965 was over 500.0 million pounds, 90 percent of which was used in metal degreasing and the rest for solvent, extraction and other applications. Exports are small; however, the imports during that year accounted for 80.0 million pounds. The purity requirements, principally of the defense industries, have made specifications in the U.S. more stringent than abroad. As a result, chlorinated solvents not meeting specifications, but nevertheless suitable for many other purposes, can be acquired more cheaply outside the U.S.

Over the years there has been a shift away from market-oriented, high-cost raw material operations in the Northern part of the U.S., towards raw material-oriented plants situated in the Gulf Coast area. This trend is responsible for the gradual shift from acetylene to ethylene as the starting point for trichloroethylene.

5   ACRYLATES

Several processes for making acrylates and acrylic acid co-exist in the U.S. However, those starting from acetylene account for over 50 percent of the total capacity.

Rohm & Haas uses the older version of the two acetylene-based (or Reppe) routes, where the nickel carbonyl catalyst supplied 20 percent of the CO requirements, and pure CO the rest. The reaction can be represented as

$$20\,C_2H_2 + 20\,ROH + Ni(CO)_4 + 16\,CO + 2\,HCl \longrightarrow$$

$$20\,CH_2{=}CH{-}COOR + H_2 + NiCl_2$$

After the reaction which takes place in the esterification alcohol, unreacted acetylene is stripped off and sent back to the reactor. Nickel chloride in aqueous solution is sent to a unit where nickel carbonyl is regenerated with carbon monoxide at a partial pressure of 100 atmos. The remaining mixture is neutralized to remove HCl and any acrylic acid present. Then the product is purified by vacuum distillation.

Since nickel carbonyl is one of the most toxic products handled anywhere in industry, a large part of plant investment is attributable solely to the elaborate safety precautions required. The process can be termed semicatalytic, because nickel acts both as a catalyst for the reaction between CO and acetylene and as a CO carrier. The described process is used to make methyl and ethyl acrylates, which represent almost 90 percent of the U.S. market; higher acrylic esters are made by batchwise transesterification.

A more elegant version of this process is used by Dow-Badische. Acetylene is absorbed in tetrahydrofuran at 6 atmos. and the solution pumped up to 80 atmos. and sent to the reactor where acrylic acid is produced in the presence of a nickel salt catalyst:

$$-C\equiv C- + CO + H_2O \xrightarrow{Ni} CH_2{=}CH{-}COOH \xrightarrow{ROH} CH_2{=}CH{-}COOR$$

Unreacted acetylene leaves the top of the reactor and is absorbed in tetrahydrofuran, the solution joining the product stream and entering a degasser. Overhead from this tower is unreacted acetylene, which is washed free of acrylic acid and recycled; bottoms, consisting of acrylic acid and dissolved in tetrahydrofuran, are stripped of the solvent which is recycled, and the purified acid stored as a product or sent on to esterification. The Dow-Badische plant employs an ion-exchange medium to effect continuous esterification with ethanol or methanol, kettle-type reactors and acid catalysts which are employed when making the less common higher acrylates. Investment costs are substantially below those for the indirect route. A flowsheet appears in Fig. 4-7.

Two other routes are used in the U.S. In the B.F. Goodrich process used by Celanese and Goodrich, ketene (obtained by cracking acetic acid) reacts with formaldehyde to give β-propiolactone, which combines with water or an alcohol in the presence of a dehydrating catalyst to form acrylic acid or the respective ester.

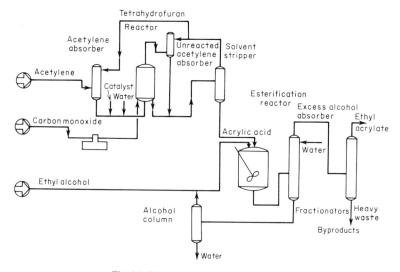

Fig. 4-7 Ethyl Acrylate (Reppe Process).

B.F. Goodrich produces only glacial acrylic acid, while Celanese turns out the whole range of esters:

$$H_2C{=}C{=}O + ROH \rightarrow CH_2{=}CH{-}COOR$$

In a process now abandoned, Union Carbide started from ethylene oxide and hydrogen cyanide; the resulting ethylene cyanohydrin is submitted to simultaneous dehydration and alcoholysis:

$$\underset{O}{CH_2{-}CH_2} + HCN \rightarrow HOCH_2CH_2CN \xrightarrow[ROH]{-H_2O} CH_2{=}CHCOOR$$

The Distillers route, employed in Great Britain by Border Chemicals, is a different approach. Propylene, mixed with 5 parts air and 4 parts steam, is oxidized to acrylic acid and acrolein. Conversion is 22.5 percent of feed per pass, of which 12.5 percent is to acrylic acid and 10 percent to acrolein, which can be further oxidized to more acrylic acid. The overall yield is around 80 percent. Nippon Shokubai (Japan) has announced a similar process. The low cost of the main raw material is offset partly by high investment requirements, but overall costs are about 35 percent lower than for the acetylene-based routes.

$$CH_2{=}CHCH_3 \xrightarrow{(O)} CH_2{=}CH{-}COOH$$

In 1966, Union Carbide announced plans to build a 200.0 million pounds per year acrylates plant at Taft, La., using the Distillers process. The investment for a 30.0 million pounds per year plant using this process was reported to be $8.5 million, and it may be estimated that Union Carbide, being also an ethanol producer, will be able to produce ethyl acrylate for a net-back of 15.0 cents per pound.

It is somewhat surprising that the three processes used in the U.S. could at one time coexist despite their differences in raw materials costs. Assuming overall yields for the β-propiolactone process are 90 percent, for the Union Carbide route 80 percent, and for the Dow-Badische plant 55 percent, and the following raw material internal costs

Raw material	Cost, ¢/lb
Acetic acid	8.0
Formaldehyde (100%)	5.5
HCN	8.5
Acetylene	9.0
Ethylene oxide	8.0
CO	0.8

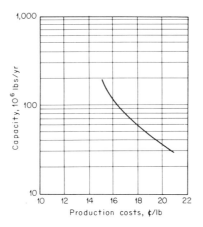

Fig. 4-8 Ethyl acrylate (Distillers process estimate).

the following raw material costs, in cents per pound of acrylic acid are obtained.

Route	Raw material cost, ¢/lb acrylic acid
Ethylene cyanohydrin	10.2
β-Propiolactone	10.0
Acetylene	5.7

This indicates the power of vertical integration in determining the process chosen by a given producer when several alternatives are present.

The capacities of the five producers in 1964 are given below.

Producer	Location	Capacity, MM lbs/yr	Route
Celanese	Pampa, Tex.	35.0	β-Propiolactone
Dow-Badische	Freeport, Tex.	15.0	Reppe
B. F. Goodrich	Calvert City, Ky.	15.0	β-Propiolactone
Rohm & Haas	Houston, Tex.	80.0	Reppe
Union Carbide	Institute, W.Va.	45.0	Ethylene cyanohydrin

Source: *Oil, Paint & Drug Reporter*, May 11, 1964.

In 1964, the U.S. consumption of acrylates was 147.0 million pounds. Below is a breakdown by type of ester and of end-use.

Product	MM lbs, 1964
Acrylic salts	7.0
Methyl acrylate	15.0

Ethyl acrylate	81.0
Butyl acrylate	7.0
2-Ethylhexylacrylate	21.0
Other esters	16.0
Total	147.0

End-uses	MM lbs, 1964
Latex coatings	56.0
Thermosetting finishes	11.0
Floor polish	8.0
Paper coatings	7.0
Leather finishes	10.0
Textile finishes	18.0
Acrylic fiber modifier	8.0
Elastomers	3.0
Latex thickeners	4.0
Adhesives	3.0
Plastics	4.0
Water-soluble resins	7.0
Miscellaneous	8.0
Total	147.0

Acrylates are used mainly in protective coatings. In 1964, about 30 million gallons of acrylic coatings were produced in the U.S., containing 67.0 million pounds of acrylates, mainly the ethyl ester, primarily for both interior and exterior trade coatings. Acrylic coatings are based mainly on copolymers of ethyl acrylate and methyl methacrylate or glacial methacrylic acid, although there has been a notable increase in demand for vinyl acetate copolymer latices with butyl and especially 2-ethylhexyl acrylate. Usually, in acrylate-methacrylate copolymers, the function of the latter is to strengthen the more flexible acrylates. Generally, thermosetting finishes are made by solution polymerization. Although acrylics represent only 5 percent of the entire U.S. market for protective coatings, rapid growth is expected despite their high price (compared with straight vinyl acetate emulsions).

Leather finishes are mainly made from ethyl and butyl acrylates, which provide crack-proof, adherent base coatings.

Acrylic finishes compete with vinyl acetate resins in wash-and-wear textile coatings; the better properties of acrylics are offset by higher prices. Another important outlet in the textile field are the fluorinated acrylate oil and water-repellent finishes developed by Minnesota Mining and Manufacturing.

Specialty paper finishes represent about 7.0 million pounds per year; while this is only 25 percent of the market for the cheaper butadiene-styrene binders, the improvement in printability obtained with acrylics gives these finishes a solid position despite higher prices.

Plastic uses are in the form of copolymers with methyl methacrylate and other monomers. Methyl acrylate is used as a comonomer with acrylonitrile in some acrylic fibers to improve dyeability.

Most polyacrylate elastomers are copolymers of 95 percent ethyl acrylate and 5 percent chloroethyl vinyl ether. A total of 3.0 million pounds was used in 1965. They are employed where high-temperature resistance (up to 350° F) is required, for example, in truck and automobile power system seals; their use has been enhanced by the trend towards smaller differentials and transmissions, which in turn implies a smaller lubricant inventory operating at higher temperatures. They compete with nitrile rubber in this application, which accounts for 70 percent of the demand for these elastomers; molded parts, hose and tubing make up the remainder.

Other major uses of importance are in self-polishing floor polishes, adhesives, thickening agents for SBR latices, and in shampoo formulations. Copolymers with vinyl ethers are common in floor polishes, and those with ethylene are used as hot-melt adhesives.

In general, higher esters are preferred when the main requirement is flexibility; ethyl acrylate is the usual compromise between hardness and flexibility, which is why its consumption is the largest.

Polymers of acrylic acid salts are water-soluble resins of which some 7.0 million pounds per year are consumed in tile cements, grout, mortar, and ceramic glazes.

Acrylic acid itself has been used for graft-polymerization to provide dye-sites on polypropylene fibers, and copolymers with acrylamide are offered by Dow for increasing paper wet-strength.

Finally, carboxylated styrene-butadiene resins are made by adding a certain proportion of acrylic or methacrylic acid to the recipe. They are used widely as carpet-backing; also their market has been growing at the expense of conventional SB resins.

6   REPPE CHEMICALS

The name "Reppe Chemistry" is used to designate a family of processes in which acetylene is handled at pressures above its explosive limit. Developed by Germany between World War I and II by Wilhelm Reppe, these processes, despite their hazardous nature, were the backbone of the German chemical war-effort.

They were used to make butadiene, synthetic fibers and other heavy chemicals. At present, Reppe chemistry is used to make a number of very interesting but much less important products; the two main producers in the world are Badische Anilin und Soda Fabrik (Germany) and General Aniline and Film at Calvert City, Ky., formerly a subsidiary of I.G. Farben in the U.S. and a buyer of much of Germany's peculiar acetylene-based technology. In 1966, GAF announced a new Reppe chemicals plant.

Acetylene can react with one or two mols of formaldehyde, giving first, propargyl alcohol and, finally, butynediol.

$$C_2H_2 + CH_2O \longrightarrow -C\equiv C-CH_2OH \xrightarrow{CH_2O} HOH_2C-C\equiv C-CH_2OH$$

Butynediol can be hydrogenated to butanediol, which as a difunctional alcohol finds some application in alkyd resins and polyesters in competition with compounds such as propylene glycol.

Butanediol can be dehydrogenated to γ-butyrolactone:

$$HOH_2C-CH_2CH_2-CH_2OH \xrightarrow{-2H_2} \begin{matrix} H_2C-\!\!-\!\!-CH_2 \\ | \qquad | \\ O=C \quad CH_2 \\ \diagdown \diagup \\ O \end{matrix}$$

This finds considerable use as a solvent, and is also an intermediate in the production of two herbicides, "Tropotox" and "Embutox." γ-Butyrolactone reacts with monomethylamine to give N-methyl pyrrolidone. This product has been promoted as an extractant for acetylene (modified BASF process) and butadiene, with the advantage that the recovered product is pure enough for polybutadiene manufacture.

$$\begin{matrix} H_2C-\!\!-\!\!-CH_2 \\ | \qquad | \\ O=C \quad CH_2 \\ \diagdown \diagup \\ O \end{matrix} + CH_3NH_2 \longrightarrow \begin{matrix} H_2C-\!\!-\!\!-CH_2 \\ | \qquad | \\ O=C \quad CH_2 \\ \diagdown \diagup \\ NCH_3 \end{matrix} + H_2O$$

γ-Butyrolactone reacts with ammonia to give pyrrolidone.

Polyvinylpyrrolidone is made by polymerizing the reaction product between pyrrolidone and acetylene. It can be used as a leveling agent for floor polishes, in ball-point and typewriter inks as a dye solubilizer and to control viscosity, in detergents as a suspending agent (as a minor competitor of carboxymethyl cellulose), and as a binder in paper napkins. Its best known applications are as an extender for human plasma and in aerosol hair sprays, where it acts as the stiffening agent. Polyvinylpyrrolidone is also being developed as a clarifier for beer, that is, as a substitute for the proteolytic enzymes now used for this purpose.

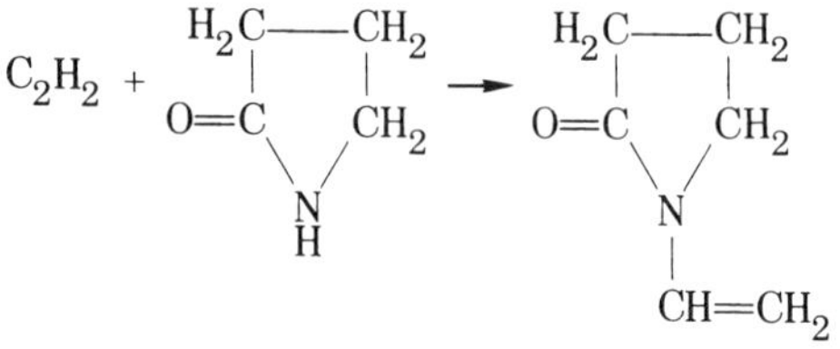

$$C_2H_2 + \begin{array}{c} H_2C{-}{-}CH_2 \\ | \quad\quad | \\ O{=}C \quad CH_2 \\ \diagdown N \diagup \\ | \\ H \end{array} \longrightarrow \begin{array}{c} H_2C{-}{-}CH_2 \\ | \quad\quad | \\ O{=}C \quad CH_2 \\ \diagdown N \diagup \\ | \\ CH{=}CH_2 \end{array}$$

Acetylene reacts with phenolic compounds to give alkylaryl vinyl ethers.   These ethers can be polymerized to a family of resins known as "Koresins," used as rubber tackifiers.

$$RArOH + C_2H_2 \longrightarrow RArO{-}CH{=}CH_2$$

Ketones react with acetylene to form acetylenic alcohols. The most important are the results of the reaction between acetylene and acetone or MEK.

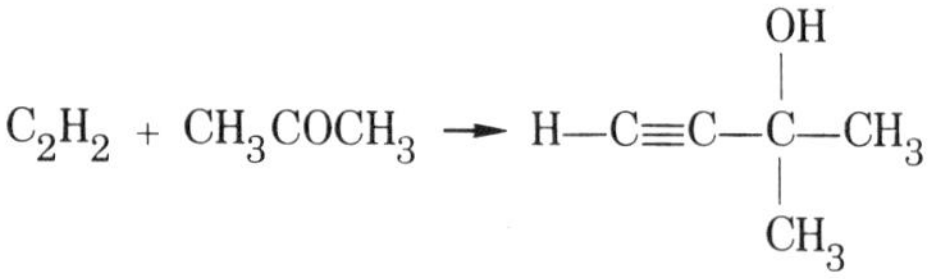

$$C_2H_2 + CH_3COCH_3 \longrightarrow \begin{array}{c} OH \\ | \\ H{-}C{\equiv}C{-}C{-}CH_3 \\ | \\ CH_3 \end{array}$$

In addition to its use as a stabilizer for trichloroethylene and perchloroethylene, and other acid-inhibiting applications, methylbutynol is the intermediate from which vitamin A and several perfumery chemicals are made. It is also the starting point for the manufacture of alethrin, a synthetic substitute for pyrethrum.

Propargyl alcohol and butynediol are used as acid inhibitors in chlorinated solvents and acid baths for metal cleaning or plating. The latter is also an intermediate for "Barban," a carbamate postemergence herbicide used to combat wild oats.

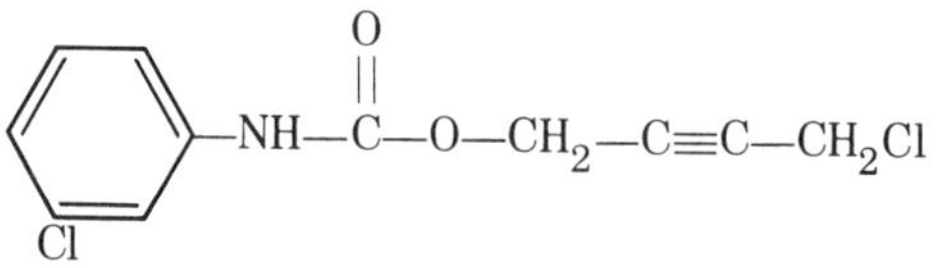

$$\overset{Cl}{\underset{}{\bigcirc}}{-}NH{-}\overset{\overset{O}{\|}}{C}{-}O{-}CH_2{-}C{\equiv}C{-}CH_2Cl$$

Propargyl alcohol is also an intermediate for certain herbicides, and propargyl bromide the starting point for the Eastman vitamin A process.

Acetylene reacts with alcohols to give vinyl ethers:

$$ROH + C_2H_2 \longrightarrow ROCH{=}CH_2$$

Since polyvinyl ethers have excellent adhesion properties, they find their main use in adhesives, for leather and pressure-sensitive tapes, chiefly as copolymers with other monomers. Higher alcohol ethers are used in floor polish formulations; they are also used, in competition with castor oil, in leather processing. Some vinyl ethers are intermediates in the production of lube oil additives. A copolymer of PVC and polyisobutyvinyl ether is employed in the formulation of metal-protecting varnishes. GAF markets a family of synthetic starch substitutes under the trade-name "Gantrez," which are copolymers of methyl vinyl ether and maleic anhydride.

ICI has developed a process for making vinyl ethers by reacting the respective alcohol (methyl, ethyl, isobutyl) with ethylene, in the presence of a palladium catalyst:

$$ROH + C_2H_4 + \tfrac{1}{2}O_2 \longrightarrow ROCH{=}CH_2 + H_2O$$

It is difficult to estimate the total production of Reppe chemicals since those that account for most of the volume in the U.S. are made by a single manufacturer. However, some idea can be obtained from the fact that in 1965, GAF decided to build a 45.0 million pounds per year (37 percent) formaldehyde plant, which would correspond to a capacity of about 25.0 to 30.0 million pounds per year in 1965.

7   HIGHER ACETYLENES

Higher homologs of acetylene are produced as byproducts in the partial oxidation acetylene process to the extent of 5 percent of the main product. They are separated during purification, and are essentially a mixture of propadiene and methylacetylene. Dow sells this mixture under the name "MAPP," a gas that is said to handle like propane, but on burning reaches the flame temperature of acetylene.

5

ETHYLENE

1   INTRODUCTION

The sources of ethylene in a given economy depend largely on the way in which petroleum is utilized. Generally speaking, the objective of a petroleum refinery operator in the U.S. is to maximize the production of specification gasoline, subject to the constraints that other products must also meet specifications and that investments must show a certain minimum profitability. Hence, refiners in the U.S. have been led to transform ever heavier fractions of the original crude into the more profitable light fuels. This transformation is accomplished mainly by such operations as catalytic cracking, hydrocracking, and delayed coking. In addition to producing the desired gasoline, some of these operations generate a considerable amount of light olefins. It has been estimated that a typical U.S. refinery has an ethylene potential of 125 million pounds per year for every 100,000 barrels per day of capacity, assuming ethane and propane in the off-gases are also cracked to ethylene — roughly twice that of a refinery not operated to maximize gasoline production.

The percentage of all ethylene produced in the U.S. recovered from refinery gases has been steadily declining. At present, 26 percent of the total ethylene demand is produced from refinery off-gases, 65 percent is made by cracking ethane or propane derived from natural gas, and 9 percent is made by cracking natural gasoline and other liquid feeds. In comparison to other feedstocks, ethane yields almost no byproducts. In a refinery off-gas recovery plant, the net feed enters the unit via the purification section, where the originally contained olefins are removed and the ethane and propane returned to the cracking section. When cracking virgin stock, feedstock enters the plant via the cracking furnaces.

In the oil-producing regions of the U.S., a typical ethylene plant receives a variety of feedstocks from neighboring refineries as well as natural gas constitutents. In Europe and Japan, the usual process is steam-cracking of naphtha with recycle of recovered ethane and propane to the cracking section, and with upgrading of the various byproducts to recover such valuable materials as butadiene and aromatics. Due to the shortage of naphtha in the U.S., chemical companies desiring to crack naphtha and thus base their operations on the concept of a "chemical refinery" have had to import their raw materials. They have also had to face the opposition of the oil companies, who oppose the idea of large amounts of high-octane gasoline being turned out by companies outside the refining business. Figure 5-1 shows a typical ethylene-plant flow-diagram. Cracking is accomplished in tubular furnaces, and steam is added to the feed to reduce the hydrocarbon partial pressure, since high pressures favor coke formation. Reducing coke laydown in the tubes lengthens the period between shutdowns for burning off the coke, thereby lowering operative costs. The heavier and the less paraffinic the feedstock, the higher the rate of coke deposition. This explains the interest in nontubular cracking sections in Europe, where light paraffinic feedstocks such as natural gas derivatives are in short supply.

After cracking, the gases are quenched immediately with water. This prevents further coke formation and removes tarry material heavier than about C_{10}. Gaseous effluent from the water quench is compressed in the main feed compressor (which differs from the refrigeration compressors in the purification section) to the pressure at which fractionation is to take place. Earlier plants operated at about 450 psig, but the trend has been to lower pressures. These require lower temperatures in the fractionation towers (around -205°F at the top of the demethanizer) which increases material of construction costs, but at the same time imply higher relative volatilities and lower reflux ratios thereby reducing overall utility requirements. Before purification, three preparatory operations take place: 1) removal of CO_2 and other acid gases by alkaline washing; 2) acetylene removal; and 3) thorough drying. Acetylene cannot be tolerated in high-pressure polyethylene manufacture; it can be recovered if the amount present justifies the extra investment, or as is usually the case, catalytically hydrogenated to ethylene. The dry hydrocarbons are sent to a deethanizer, the overhead of which is demethanized leaving a fraction containing only ethane and ethylene; this is fractionated into pure ethylene and ethane, the latter being recycled to the cracking section. The

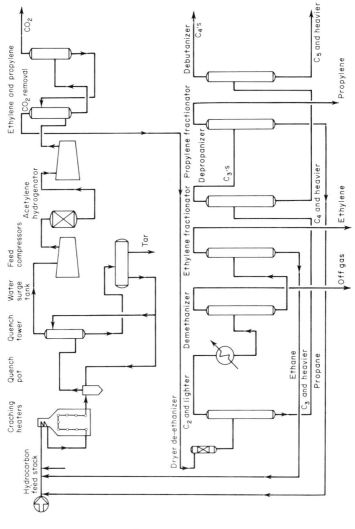

Fig. 5-1 Ethylene Plant Flow Diagram.

deethanizer bottoms are depropanized, the overhead from this operation being split into propylene and propane, which is also recycled to the furnaces. Finally, the depropanizer bottoms are split into a C_4 fraction that may constitute the feed to a butadiene extraction unit and a high-octane gasoline bottoms stream containing around 50 percent aromatics.

In Europe, at least until the mid-1960's, most oil refineries were designed to produce the minimum amount of gasoline compatible with average octane requirements. By proper choice of crude, this may mean no catalytic cracking is needed. Octane requirements are met by catalytic reforming of straight-run naphtha, and the rest of the barrel fractionated into the various distillate fuels. Hence, European refineries produce much less ethylene per barrel of crude oil than those in the U.S. Furthermore, the recent natural gas discoveries in the North Sea area contain only small amounts of hydrocarbons in the LPG range. Refinery off-gases and ethane furnish only 6 percent of the ethylene produced in Europe; 89 percent comes from cracking naphtha and heavier feeds, and the remaining 5 percent are recovered from coke-oven gas. A similar proportion holds true for Japan. The effect of the conversion to natural-gas space heating as a result of the discovery of the Slochteren, North Sea, and Lacq gas deposits is difficult to predict. The less intense use of automobiles in Europe probably means that much of their fuel oil will not be converted into gasoline, as in the U.S., but rather that the current balance between light and heavy petrochemical feedstocks will be maintained.

There also has been considerable interest in Europe in cyclical or continuous nontubular processes for olefin production, which could increase the range of feedstock choice. Figure 5-2 shows flowsheets for three of these. By 1967, none of these processes had reached full-scale commercial utilization because economics still favored steam-cracking of naphtha.

In the ONIA-GEGI cyclical process, naphtha is cracked in a previously heated brick-lined reactor, with the exit gases generating the 8 mols of steam per mol of naphtha required for dilution. When the endothermic cracking reaction has proceeded far enough and coke build-up reaches a certain level, the reactor is reheated first by means of hot flue gases and finally by burning off the coke deposits, thereby preparing the reactor for another cycle. The advantage of the ONIA and other nontubular processes is that they are flexible as to feedstock and product distribution, and that a much higher temperature can be attained than in a metal-tube furnace. The ONIA process also produces little ethane

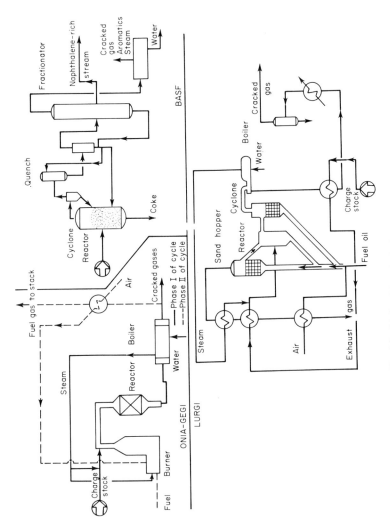

Fig. 5-2 Olefins (Nontubular Processes).

and propane. Pyrolysis gasoline production may be as low as 7 percent depending on feed and cracking conditions. Reported yields on naphtha feed are as high as 45 percent total olefins, and the ethylene-to-propylene ratio can be varied between 5:4 and 4:1. Erected cost for a plant producing 100.0 million pounds per year of ethylene is reported around $8.1 million.

In the LURGI process, feedstock and steam are injected into a fluidized bed of hot sand, which performs the double function of transferring heat to the feed and acting as a carrier for the deposited coke. Sand is continually removed from the reactor to a regenerator, where the coke is burned off and the hot sand returned to the cracker.

The BASF process consists of cracking whole crude oil in a fluidized bed of coke particles. Process heat is generated by the partial oxidation of some of the feed by the addition of around 4.0 pounds of oxygen per pound of crude, in addition to dilution steam. After the removal of the coke particles entrained from the fluidized bed by a series of cyclone separators, cracked hydrocarbons are separated into recycle oil and a main product stream which is fractionated into three parts. The lightest is further fractionated into an olefin-rich stream, which is treated to recover the desired products, and an aromatics-rich fraction from which benzene can be extracted. The middle stream from the main fractionator is rich in naphthalene, and the bottom stream is also recycled to the reactor. This is a high-investment process since it requires an oxygen plant as well as a complicated solids recovery system. The economics thus seem to depend on the full utilization of all the various byproducts. On the other hand, it is designed to use the cheapest of all hydrocarbon feedstocks.

The development of greatest interest to those areas where liquid feedstocks are widely used for olefin production has been the short residence time naphtha cracking furnace. This concept is based on the observation that coke is formed at a slower rate than the desired olefins. By reducing the residence time in the furnace coil to about one-third of the former practice, and compensating for this by increasing the tube-wall temperature to around 2000°F the average heat-flux through the tube walls is multiplied by around 2.5, excellent ethylene yields were obtained from naphtha while a reasonable period between shut-down for coke burn-off was maintained. Dilution steam requirements were also reduced compared with previous naphtha cracking processes. As cracking severity increases, so does the ethylene yield, and consequently that of the various byproducts decreases.

Oil companies in the U.S. hold a comparatively minor but increasing position in the chemicals derived from olefins. Therefore roughly 75 percent of the ethylene produced from refinery off-gas is sold. This quantity amounts to almost a third of the total ethylene produced. The distribution of ethylene sources between refinery off-gases and natural gas constituents roughly approximates that of total ethylene output between oil refineries and chemical companies. Despite a certain trend towards market-oriented location of olefin plants, the U.S. raw materials structure has kept about two-thirds of the ethylene capacity in the Gulf Coast region. On the other hand, outside the U.S., because of the almost total reliance on liquid feeds, olefin plants tend to locate near the centers of ultimate consumption. The table below shows the location of the U.S. ethylene plants, with their respective capacities, as of 1966.

Producer	Location	Capacity, MM lbs/yr
Atlantic	Watson, Calif.	100
Dow	Bay City, Mich.	70
	Freeport, Tex.	1,300*
	Midland, Mich.	150
	Plaquemine, La.	400
DuPont	Orange, Tex.	800*
El Paso-Rexall	Odessa, Tex.	290*
Enjay	Baton Rouge, La.	1,000*
	Baytown, Tex.	90
	Linden, N.J.	175
Goodrich	Calvert City, Ky.	250
Gulf	Cedar Bayou, Tex.	400
	Port Arthur, Tex.	425
Jefferson	Port Neches, Tex.	450*
Koppers	Kobuta, Pa.	35
Mobil	Beaumont, Tex.	450
Monsanto	Alvin, Tex.	500
	Texas City, Tex.	100
National Distillers	Tuscola, Ill.	320
Olin Mathieson	Brandenburg, Ky.	90
Petroleum Chemicals	Lake Charles, La.	360
Phillips	Sweeny, Tex.	550
Shell	Deer Park, Tex.	275
	Norco, La.	500*
	Torrance, Calif.	70
Sinclair-Koppers	Houston, Tex.	500*
Sun Olin	N. Claymont, Del.	125*
Texas Eastman	Longview, Tex.	450

(Continued)

Producer	Location	Capacity, MM lbs/yr
Union Carbide	Institute, W.Va.	350
	Seadrift, Tex.	900
	S. Charleston, W.Va.	400
	Taft, La.	600*
	Texas City, Tex.	750
	Torrance, Calif.	150
	Whiting, Ind.	275
Total, end of 1966		13,650
Probable capacity to be added in 1967		
Allied-Wyandotte	Geismar, La.	500
American Can-Skelly Oil	Clinton, Iowa	400
Continental Oil	Lake Charles, La.	400
Phillips-Houston Nat. Gas	Sweeny, Tex.	500
Union Carbide	Gulf Coast	550**
Total capacity to be added in 1967		2,350
Total U.S. capacity, end 1967		16,000

Source: *Hydrocarbon Processing*, October 1966.

*Part or all of this capacity completed or to be completed during 1966; total capacity completed or to be completed during 1966 = 3,600 million pounds.

**New plant of 1,300 million pounds will replace existing 550 million pounds installation.

The greatest recent change in ethylene plants has been the use of centrifugal compressors. When using reciprocal compressors, the practical ethylene production limit for a single main compressor was 150 million pounds per year and plants were usually built to multiples of that. Commercial designs for a capacity of one billion pounds per year for a single centrifugal compressor are known to be available. Thus, the capacity of future plants will be much larger than those built up to now. Compressors represent about 35 percent of the total investment for an olefins plant. Cracking and quenching sections make up 25-30 percent of the erected cost, the higher figure applying to a naphtha-cracking plant; recovery facilities constitute the remaining 35-40 percent.

Ethylene production costs depend very much on the criteria for attributing credits to the various byproducts. Propylene, whether valued as alkylation feed, petrochemical feedstock or for inclusion in LPG, is worth around 2.0-2.3 cents per pound. Butadiene, on the other hand, varies in value with the amounts actually available. For example, the 8.0 million pounds per year which could be recovered from a 400.0 million pounds per year ethane-cracking plant would not be worth much more than fuel value. However, the 50.0 million pounds per year available from a naphtha cracking plant of equal rating could be valued at 5-6 cents per pound. Butane is worth more as LPG than as fuel if it can be sold as such. The pyrolysis gasoline stream, containing as much as 50 percent aromatics, under high severity cracking conditions, can be hydrogen treated and fed to an aromatics extraction unit. This treatment converts the unstable diolefins in the cracked gasoline to paraffins.

If aromatics are to be extracted, then the mono-olefins are also hydrogenated under conditions that avoid saturation of the aromatic rings. Demethanizer overhead is usually employed as the source of hydrogen. Here again the same stream would be worth more from a naphtha cracking plant because the amount of aromatics involved is so much larger than when cracking ethane. Thus, an evaluation of the net raw material cost of a pound of ethylene depends on the size of the unit as well as on the nature and yields of the various byproducts. In the case of naphtha-cracking, yields depend considerably on the raw material itself; no olefins plant based on naphtha is ever designed without prior de-termination of the yield structure in a pilot cracking unit. The four cases below were selected at random; however, countless variations are possible. In general, high-severity naphtha crack-ing presents the most favorable economics provided byproducts can be valued as chemical feedstocks. Low-severity naphtha crack-ing, under such assumptions, appears even more favorable. In this case, however, these assumptions are less than realistic because the demand for the various byproducts whose yield rises at lower severities, is seldom enough to justify valuing them at maximum levels. This has led some engineering companies to advocate designing ethylene plant recovery sections with enough flexibility to handle the furnace effluent from both high- and low-severity steam-cracking. The four cases are:

I) ethane cracking, 200.0 million pounds per year, all by-products credited as fuel;
II) propane cracking, 600 million pounds per year, byproducts credited as in table;
III) high-severity naphtha cracking, 200.0 million pounds per year, all byproducts except propylene credited as fuel;
IV) high-severity naphtha cracking, 600.0 million pounds per year, byproducts credited as in table

Case	Investment, $MM
I	10.5
II	19.5
III	12.5
IV	21.0

These figures include a propylene recovery section, but no other downstream processing units.

Net raw material costs are calculated below; in all cases it is assumed that ethane and propane are recycled to extinction.

	Lbs consumed or produced per lb ethylene	Value, ¢/lb	¢/lb +	Ethylene −
Case I				
Ethane	1.25	1.0	1.25	
Byproducts	0.18	1.0		0.18
Net feedstock cost			1.07	
Case II				
Propane	2.3	0.9	2.07	
Propylene	0.32	2.1		0.67
Butadiene	0.08	5.5		0.44
Other products	0.92	0.5		0.46
Net feedstock cost			0.50	
Case III				
Naphtha	2.8	1.1	3.08	
Propylene	0.51	2.1		1.07
Other byproducts	1.22	0.8		0.98
Net feedstock cost			1.03	
Case IV				
Naphtha	2.8	1.1	3.08	
Propylene	0.51	2.1		1.07
Butadiene	0.12	5.5		0.66
Butane	0.15	2.0		0.30
Aromatic oils (pyrolysis gasoline)	0.40	1.4		0.56
Other byproducts	0.55	0.5		0.28
Net feedstock cost			0.21	

Thus, naphtha can be seen to be the most favorable feedstock provided the olefin plant is large enough and byproducts can indeed be disposed of. Therefore, it is not surprising that the great interest on the part of highly diversified U.S. chemical companies such as Dow and Union Carbide in cracking imported naphtha, instead of ethane or LPG coincided with the construction of the first ethylene plants in the 500.0 million pounds per year range.

For the purpose of economic evaluation, the following net feedstock costs will be assumed:

Case	Cost, ¢/lb net feedstock
I	1.1
II	0.5
III	1.0
IV	0.2

Utility costs vary inversely with the ethylene content of the furnace effluent, and are thus lowest for ethane and highest for naphtha. Production costs can be estimated as shown below.

	Cases			
	I	II	III	IV
Net feedstock	1.1	0.5	1.0	0.20
Utilities and chemicals	0.8	0.9	1.1	1.10
Labor and overhead	0.3	0.1	0.3	0.10
Capital charges	1.7	1.1	2.1	1.15
Total	3.9	2.6	4.5	2.55

Naphtha is the most desirable olefin-plant feedstock in an economy where the demand for ethylene is large enough to justify crediting the byproducts at the figures assumed for Case IV. A country short of ethane and propane therefore has a strong incentive to encourage the construction of large petrochemical complexes, where several plants all receive their feedstock from a single common cracking plant. The structure of the entire Japanese chemical industry already reflects this principle. In Europe the chemical industry is dominated by a few large and integrated companies that can utilize the full range of products from a naphtha

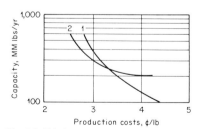

Fig. 5-3 Ethylene. 1-From ethane; 2-from naphtha [steam cracking] (high severity; byproducts valued as chemical feedstocks).

cracker.   The entry of U.S. companies into the European market, however, may transform this picture and make it resemble more closely the Japanese model.

The distribution of ethylene demand in the U.S. during 1965 was as follows:

End-use	MM lbs, 1965
Polyethylene and copolymers	3,300
Ethylene oxide	2,250
Ethanol	1,300
Styrene	840
Vinyl chloride	530
Miscellaneous	1,130
Total	9,350

Source: *Hydrocarbon Refining*, October 1966.

2   POLYETHYLENE

Polyethylene was discovered accidentally at ICI in 1933. It was the first polyolefin plastic to be commercialized, and in the U.S. and Great Britain, it is by far the most important polymer in terms of amount consumed.   It is often said that the sign of a mature, consumer-oriented economy is the consumption of poly-ethylene in greater amounts than PVC, which tends to be the most important plastic in less developed economies.

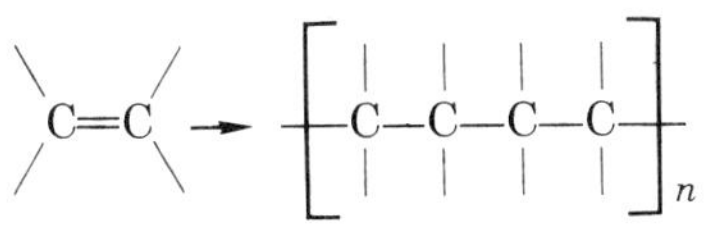

The lower molecular-weight polymers of ethylene — below 10.000 — are waxy materials that compete in their applications with petroleum and natural waxes.   The properties of higher molecular-weight resins depend on their degree of crystallinity; the predominantly amorphous grades, products of high-pressure technology, are at the lower end of the density scale (about 0.92); the more crystal-line grades are made by completely different, low-pressure processes, and have higher density as well as certain properties which distinguish them from high-pressure polyethylene.

Both high- and low-pressure polyethylene are thermoplastic. The more characteristic properties of the former are a softening point, below the boiling point of water, which permits the easiest processing of all plastics. Their high impact strength and flexibility make them suitable for such applications as film and injection molding of items where constant dimensions are not required or even undesirable, for example, squeeze bottles or snap-cap closures. Their good transparency makes them highly desirable in packaging films. Low-pressure polyethylenes have softening points above 212°F, higher yield and tensile strength, lower flexibility, and poor transparency. Low-pressure polyethylene is also more expensive to make than high-pressure polyethylene; it requires both higher investments and higher operating costs.

Despite the appearance of several independent processes, most high-pressure polyethylene plants use some variant of the original ICI mass-polymerization technology.

A typical flowsheet for high-pressure mass polymerization of ethylene is shown in Fig. 5-4. The feed, 99.5+ percent ethylene, is compressed to around 1500 atmos. and heated to 380°F. The higher the pressure, the higher the molecular weight of the polymer. After introduction of the 0.06 percent oxygen or organic peroxide that acts as a catalyst, ethylene enters the polymerization reactor (where conversion is 25 percent per pass). Monomer and polymer are separated, the unconverted ethylene being recycled. An ethylene slip stream is returned to the low-temperature fraction section of the ethylene plant, or sent to other process units that can be fed by ethylene of less than polymerization grade. The polymer is extruded, chilled and finally chopped into granular finished product. In the DuPont high-pressure process, water containing oxygen acts as the reaction initiator.

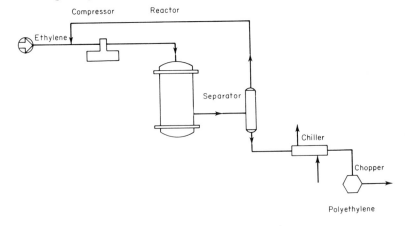

Fig. 5-4 Polyethylene (High Pressure).

Two types of process are used to make high-density poly-ethylene — those using metallic oxides as catalyst and operating at around 400 psig, and those employing Ziegler catalysts (a mixture of a titanium halide and a trialkyl aluminum) which operate at practically atmospheric pressures.   The metal oxide processes were slow in gaining acceptance because the catalyst had a way of forming extremely malodorous complexes, thereby making the product practically unusable.   This disadvantage was overcome, but the section of the plant where the water-insoluble catalyst is stripped from the polymer represents a major part of the total investment.

The most widely used medium-pressure process in the U.S. employs the Phillips Cr_2O_3-alumina catalyst.   Ethylene is poly-merized in a solvent such as cyclohexane, which has the additional function of moderating the reaction.   The polymer is precipitated by running the reactor mass into water, where it comes down in fairly large pellets.   These are filtered, after which solvent and catalyst are separated from the product and the latter sent on to finishing.   Figure 5-5 shows a flowsheet of this operation. Another variant of the medium pressure technique is that of Standard Oil of Indiana.   In 1965, this process was already being used in Japan and Sardinia, although not yet in the U.S.

Although the ICI or Phillips processes are licensed in the form of fully engineered plants, the low-pressure Ziegler process

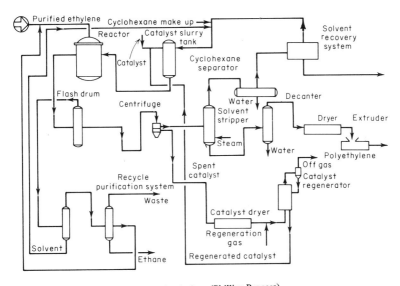

Fig. 5-5 Polyethylene (Phillips Process).

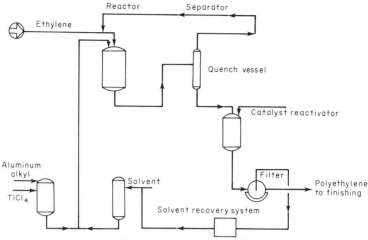

Fig. 5-6 Polyethylene (Low-Pressure Ziegler).

licensees acquire the know-how on the Ziegler catalyst, but must develop their own technology. The diagram in Fig. 5-6 describes a typical sequence of operations. First, the catalyst is prepared and then introduced together with high-purity ethylene into the polymerization reactor in the presence of a solvent such as heptane. The temperature is low enough (120° - 170°F) to cause the polymer to precipitate in the solvent itself. The precipitate is slurried with water to dissolve the catalyst and then the usual filtration and finishing operations take place. The Ziegler catalyst can also be used to produce low-density polyethylene, and DuPont is reported to be producing grades covering the full range of densities between .925 and .965 at a plant in Sarnia, Ont. The Ziegler and Phillips processes together account for most of the U.S. high-density polyethylene plants.

Process	High-pressure	Low-pressure
Investment ($ MM)	21.8	25.5
	Production costs, ¢/lb polyethylene	
Ethylene	4.3	4.8
Utilities	0.6	2.2
Catalyst and chemicals	0.9	1.4
Labor and overhead	1.1	1.1
Capital charges	7.2	8.4
Total	14.1	17.9

The economics of polyethylene production are compared above. Low-pressure polyethylene requires a somewhat higher investment, but the main differences appear in utilities, chemicals consumption and yields.   Ethylene is assumed available at 4.0 cents per pound, and capacities in either case are 100.0 million pounds per year.   The capacities of U.S. polyethylene plants as of 1966 are shown below.

Producer	Location	1966 Capacity, MM lbs/yr
A-B Chemicals	Houston, Tex.	250
Dow	Freeport, Tex.	120
	Plaquemine, La.	110
DuPont	Orange, Tex.	300
	Victoria, Tex.	80
El Paso-Rexall	Odessa, Tex.	300
Gulf (Spencer)	Orange, Tex.	200
Monsanto	Texas City, Tex.	130
National Distillers	Tuscola, Ill.	120
Sinclair-Koppers	Port Arthur, Tex.	120
Texas Eastman	Longview, Tex.	150
Union Carbide	Seadrift, Tex.	200
	S. Charleston, W.Va.	130
	Texas City, Tex.	250
	Torrance, Calif.	80
	Whiting, Ind.	160
Total		2,700
High density		
Allied	Baton Rouge, La.	125
	Orange, Tex.	25
Celanese	Pasadena, Tex.	120
Dow	Plaquemine, La.	40
DuPont	Orange, Tex.	90
Hercules	Parlin, N.J.	80
Monsanto	Texas City, Tex.	50
National Petrochemicals (National Distillers & Owens-Illinois Glass)	Houston, Tex.	130
Phillips	Pasadena, Tex.	120
Sinclair-Koppers	Port Reading, N.J.	35
Union Carbide	Seadrift, Tex.	125
Total		940

Source: *Hydrocarbon Processing*, October 1966.

2.1 High-pressure Polyethylene

Domestic use of high-pressure polyethylene in 1965 reached 2300 million pounds distributed as shown in the following table.

End-use	MM lbs, 1965
Film and sheeting	890
Injection molding	370
Extrusion coating	290
Wire and cable coating	230
Pipe and other extrusion	65
Blow-molding	50
Export	340
Miscellaneous	65
Total	2,300

Source: *Modern Plastics*, January 1966.

High-pressure polyethylene has captured a substantial part of the film applications once the exclusive domain of cellophane. The trend towards pre-packaging, especially of produce and processed food, ensures sustained growth of the polyethylene film market despite competition from other plastic films. Wrapping machinery originally designed for high-speed cellophane packing at first were only able to handle polyethylene at reduced speeds, which offset the price advantage of around 0.4 cents per thousand square feet offered by polyethylene. Later, however, high-speed machines were designed to handle polyethylene film which has since captured many markets such as bread-wrapping. Cellophane still is used in such extremely high-volume applications as cigarettes and candy. Polyethylene film is also used in heavy-duty bags, especially for fertilizers. Polyethylene accounts for about 5 percent of the 2.5 billion industrial bags used annually in the U.S. Nonpackaging uses include wrapping synthetic rubber bales, in civil and highway construction, and for mulching in agriculture.

The table below breaks down the demand for polyethylene film.

End-use		1965 MM lbs/yr
Packaging		660.0
Rack and counter	40.0	
Bakery goods	80.0	
Produce	160.0	
Frozen foods	20.0	
Meat and poultry	20.0	
Other foods	40.0	
Dry cleaning and laundry	60.0	
Industrial bags	80.0	
Soft goods and paper products	60.0	
Miscellaneous	100.0	
Nonpackaging		220.0
Construction	50.0	
Agriculture	40.0	
Household uses	50.0	
Rubber industry	30.0	
Other nonpackaging	50.0	

Injection-molded polyethylene has numerous applications, such as closures, containers and housewares.

The main application of extrusion coating is in milk containers, where low molecular-weight high-pressure polyethylene has almost entirely replaced paraffin wax, despite a recent trend towards blends of the two.   Due to its high electrical resistivity, polyethylene is used to insulate and protect electrical wire and cables. About 65 percent of the cable and wire insulated with polyethylene is used in residential construction, where it shares the market equally with polyvinyl chloride despite the disadvantage of a low degree of flame-retardancy.   Crosslinked polyethylene is widely used on wires carrying power from the main lines to residences; it is admitted in lines up to 15 kilovolts.

Despite its use restriction to water, and even then only in low temperature and pressure service, high-density flexible polyethylene pipe accounts for 60 percent of the total plastic pipe market.   This, however, is only temporary; the fairly conservative building codes still prevailing in most U.S. cities are gradually being altered.   Polyethylene is used mainly in drainage and irrigation — markets already close to having reached their full potential.   But the growth in use of other plastics that can serve plumbing applications will reduce this share to about 40 percent by 1970.   This trend is already apparent in countries like Japan. Polyethylene pipe is more expensive per linear foot than steel, but has the advantage that it can be laid continuously from a coil, greatly reducing labor requirements.

Blow-molding applications are mainly jars, carboys, toys and housewares.   The demand for polyethylene foam, which in 1963 represented only 3 percent of all types of plastic foam, is expected by some to grow at a rate of 45 percent per year, thanks to such applications as packaging of electronic components and wire coating.   Powdered polyethylene resins, made by grinding extruded pellets, have largely replaced rubber latices as backing for automobile carpets; they are cheaper and allow the material to be molded to fit the contour of the car interior. However, up to now, mechanical shortcomings and their tendency to stick have limited their use in domestic carpets.   They also compete with vinyl plastisols for rotational molding outlets such as toys. Emulsifiable polyethylene waxes are low molecular-weight products of which 20.0 million pounds went mainly into floor polish and textile softeners.   The use of emulsifiable polyethylene softeners is expected to grow with that of wash-and-wear finishes, their main function being to offset the deleterious effect of these finishes on fiber strength.

Nonemulsifiable waxes are used in hot-melt adhesives, paper coatings and a number of packaging outlets; they are also used as replacements for natural wax in printing inks.

2.2 Low-pressure Polyethylene

Low-pressure polyethylene, compared with high-pressure products, is definitely a premium plastic. In 1964 prices of most grades stood between 20 and 24 cents per pound compared to 12.5-14.5 cents per pound for high-pressure resins. However, it is difficult to generalize since there are hundreds of different grades of each type representing differences in density (often obtained by blending), molecular weight distribution, degree of branching and of copolymerization with other olefins, or in the additives blended into the resin. The table below lists the main applications for low-pressure polyethylene in 1965.

End-use	MM lbs, 1965
Export	80
Blow-molded containers	330
Injection molding	170
Pipe and profiles	45
Film and sheet	35
Wire and cable coating	30
Extrusion coating	10
Miscellaneous	40
Total	740

Source: *Modern Plastics*, January 1966.

Among blow-molded articles the most important applications are household bleach and liquid-detergent bottles. This application is due to the low permeability of low-pressure polyethylene. The detergent and bleach bottle markets are essentially saturated and growth will have to be in other areas. On the other hand, part of this market may be captured by blow-molded PVC.

Most of the use in injection-molding is actually in the form of blends with low-density material, especially for housewares. Injection-molded bottle carrying-cases made of high-density polyethylene far outlast wood, and are expected to consume 10 million pounds per year when the full potential of this outlet is realized.

Extrusion products — film, coating and pipe — are of higher quality than those made from most thermoplastics (except the engineering plastics), but are also comparatively expensive. This can often be partly compensated for by using less material, especially in very thin films (used for food packaging in competition with polypropylene film), wire and cable coating, and pipe. All of these are growing markets, although low-pressure polyethylene pipe suffers from cold-flow and stress-cracking. Of special interest are carbon black-reinforced and peroxide cross-linked grades, used increasingly for cable coating and pipe. In Japan, both types of polyethylene began to be produced at about the same time. Low-pressure polyethylene injection molding, blown packaging film and filament for rope and fish-netting are much more developed than in the U.S.

About 12 percent of all polyethylene produced in 1965 was in the form of its copolymers. High-density polyethylene for making blow-molded containers, for example, is actually a copolymer with about 4 percent 1-butene.

A word of caution applies to all discussions of plastics applications: minor changes in the combination of price-properties-fabrication costs can have the effect of completely displacing a product from a given application within as little as 3 years; good judgement has to be exercised in extrapolating data, and in employing U.S. figures to draw conclusions about other economies.

2.3 "Hypalon"

DuPont produces this elastomeric material, which has properties similar to neoprene, by reacting polyethylene with chlorine and sulfur dioxide. The final product contains 30 to 40 percent chlorine and about 1.5 percent SO_2. Its advantages are its electrical properties and resistance to heat, and especially to abrasion; its main shortcoming is poor adherence to metals. The demand in 1965 was around 20.0 million pounds per year and may grow considerably if a type of high-impact polyvinyl chloride made by blending PVC and "Hypalon" becomes an important outlet.

In 1964, the major markets for "Hypalon" were automotive, wire and cable coating, shoes, and tires.

2.4 Chlorinated Polyethylene

Medium-pressure polyethylene can be chlorinated to 25-35 percent chlorine, giving an elastomeric material, or 68-73 percent where the result is a rigid compound. It is used in acid or fireproof coatings, flooring, roofing and wire coatings. Allied Chemical is the main producer in the U.S.

3   ETHANOL

Although fermentation plants still operate in the U.S., three-quarters of the ethanol produced in the country comes from ethylene. There are two processes used for conversion of ethylene to ethanol. In the first, originally used by Union Carbide, ethylene in concentrations as low as 35 percent passes through a series of sulfuric acid absorbers at pressures between 200 and 600 psig. The following reaction takes place:

$$C_2H_4 + H_2SO_4 \rightarrow C_2H_5HSO_4$$

Ethyl hydrogen sulfate and ethyl sulfate, formed as a byproduct, are hydrolyzed with steam in another reactor, the effluent being sent to a steam stripper from which alcohol, water and byproduct ethyl ether are sent on to fractionation:

$$C_2H_5{-}HSO_4 + H_2O \rightarrow C_2H_5OH + H_2SO_4$$

$$(C_2H_5)_2SO_4 + H_2O \rightarrow (C_2H_5)_2O + H_2SO_4$$

Spent acid is reconcentrated and sent back to the absorbers. This process has the advantage of handling a feed stream such as the deethanizer overhead of an ethylene plant without further purification, as conversion to sulfate is practically 100 percent and no recycle of unconverted feed is required. Ethyl alcohol yields are 90 percent, with formation of ethyl ether being responsible for most of the losses. Ethanol plants employing this process account for most of the ethyl ether produced in the U.S. A flowsheet is shown in Fig. 5-7.

In the second process, catalytic hydration, used in the U.S. by Shell, high-purity ethylene is heated to 570°F, compressed to 1000 psi, mixed with condensate and introduced into a reactor containing a phosphoric acid catalyst. Since conversion per pass is only 4.2 percent, the recycle-to-feed ratio is around 23:1. Gas and liquid are separated and the unconverted ethylene is recycled after inerts contained in the feed are bled off. Ethanol is removed by steam-stripping. Before going to the purification section it is catalytically hydrogenated to remove any traces of acetaldehyde by conversion to alcohol. Yields are reported around 97 percent. The process is advantageous since lower investment is required because the expensive acid reconcentration facilities are not needed. Maintenance costs are also lower than in the sulfation process. No sulfuric acid is consumed, but this is offset by the power required to recirculate unreacted feed.

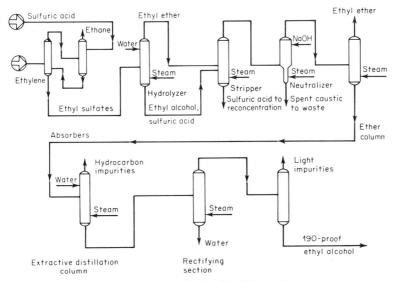

Fig. 5-7 Ethanol (Union Carbide Process).

The main disadvantages are high utility costs and the fact that the feed must be polymerization-grade ethylene. Figure 5-8 shows a flowsheet of the Shell process.

Some European countries still produce ethanol by hydrogenation of acetaldehyde, which in turn was all made from coal or natural

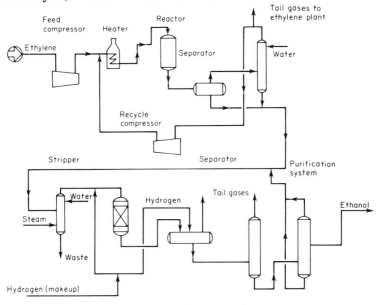

Fig. 5-8 Ethanol (Shell Process).

gas-derived acetylene.   Today some is made by oxidation of ethylene (Wacker process).

Synthetic ethanol capacities in the U.S. are given in the table below.

Producer	Location	1965 Capacity, MM lbs/yr
Union Carbide	S. Charleston, W.Va.	335
	Texas City, Tex.	600
Humble	Baton Rouge, La.	370
U.S. Industries	Tuscola, Ill.	350
Shell	Houston, Tex.	210
Texas-Eastman	Longview, Tex.	150
Total		2,015

Below, the uses for alcohol in fiscal 1965 are given. Solvent outlets are expected to grow no faster than the population, and industrial uses other than acetaldehyde at 6-7 percent per year. Acetaldehyde as an outlet for ethanol will at best remain static. The more economical production process for direct conversion of ethylene to acetaldehyde (first introduced in 1962), should account for all future expansion. During World War II large amounts of ethanol were converted into butadiene, but this process is no longer used. The end-uses are shown in the following.

End-use	Fiscal 1965, MM lbs/yr	
Chemical		
Acetaldehyde	955.0	
Ethyl amines	20.0	
Ethyl acetate	42.0	
Ethyl chloride	43.0	
Glycol ethers	77.0	
Acetic acid	35.0	
Vinegar	76.0	
Other ethyl esters	81.0	
Other chemical uses	50.0	
Total chemical uses		1,380.0
Solvent		
Cellulose and other resins	195.0	
Hair and scalp preparations	151.0	
Other cosmetic uses	72.0	
External pharmaceutical	32.0	
Detergents, disinfectants, flavors, etc.	76.0	
Food and drug processing	70.0	
Other solvent uses	14.0	
Total solvent uses		610.0
Total		1,990.0

Source: "Uses of Specially Denatured Alcohol—Fiscal 1965," U.S. Treasury Department.

Although until recently only fermentation alcohol was acceptable for cosmetic use, Union Carbide announced the construction of a 200.0 million pounds per year plant capable of turning out synthetic cosmetic-grade ethanol. Only fermentation alcohol is used in alcoholic beverages, and if this ever changes it will mean an enormous increase in demand for the synthetic product.

The process unit for a 200 million pounds per year ethanol plant using the sulfation process costs about $5.5 million. Production costs for ethanol from ethylene are:

	Cost, ¢/lb ethanol
Ethylene	1.9
Chemicals	0.2
Utilities	0.5
Labor and overhead	0.2
Capital charges	0.9
Total	3.7

Ethylene has been assumed at 3.0 cents per pound, to account for the fact that an impure stream can be fed to the unit.

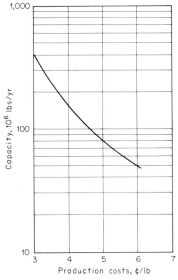

Fig. 5-9 Ethanol.

3.1 Ethylamines

Most modern alkylamines plants are designed to convert alcohols ranging from methanol to isopropanol or from ethanol to cyclohexanol, into the respective amines by catalytic reaction with ammonia:

$$ROH + NH_3 \rightarrow RNH_2 + H_2O$$

$$2ROH + NH_3 \rightarrow R_2NH + H_2O$$

$$3ROH + NH_3 \rightarrow R_3N + 3H_2O$$

The flowsheet in Fig. 5-10 applies to alkylamine production in general. Alcohol, ammonia and recycled amines are vaporized, superheated and sent on to the reactor, with hydrogen acting as a diluent. After the hydrogen has been separated and recycled the product stream is sent to the fractionation section. Water formed in the reaction is removed from the phase-separator between the secondary and tertiary amine columns. The ratio between the three amines is controlled by recycling the unwanted product and thus inhibits its formation.

The producers of higher alkylamines in the U.S. are:

Producer	Location	1964 Capacity, MM lbs/yr
Pennsalt	Wyandotte, Mich.	28.0
Union Carbide	Taft, La.	25.0
DuPont	La Porte, Tex.	7.0
Escambia	Pace, Fla.	10.0
Total		70.0

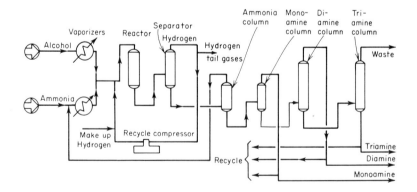

Fig. 5-10 Alkylamines.

The consumption of ethylamines in the U.S. reached 32.0 million pounds in 1965, and is growing at a rate of 8 percent per year. Consumption was distributed as shown in the table below.

End-use	MM lbs, 1962
Rubber chemicals	16.0
Agricultural chemicals	9.0
Pharmaceuticals	2.0
Chemical intermediates	3.4
Surfactants	1.0
Others	4.6
Total	36.0

Source: *Chemical Engineering Progress*, October 1963.

Rubber chemicals

The most important rubber chemical made from diethylamine is zinc diethyldithiocarbamate:

$$2(C_2H_5)_2NH + 2CS_2 + ZnO \rightarrow (C_2H_5)_2N\overset{\displaystyle S}{\overset{\|}{C}}-S-Zn-S-\overset{\displaystyle S}{\overset{\|}{C}}-N(C_2H_5)_2 + H_2O$$

Its main application is as a vulcanization accelerator in solid rubber, the more soluble sodium compound being used for curing rubber latex.

Agricultural chemicals

Diethylamine is a raw-material for "Vegadex," a herbicide made by Monsanto for use against wild oats. Monoethylamine is an intermediate for "Simazine," a triazine herbicide developed by Geigy.   Triethylamine finds some use in solubilizing 2,4,*D* acid.

$$(C_2H_5)_2N\overset{\displaystyle S}{\overset{\|}{C}}-S-CH_2-\overset{\displaystyle Cl}{\overset{|}{C}}=CH_2 \quad (\text{"Vegadex"})$$

Triazone textile finishes

The most important member of the triazone family of reactant cotton finishes is made by reacting formaldehyde, dimethylol urea and monoethylamine:

$$(C_2H_5)NH_2 + (NHCH_2OH)_2\overset{O}{\overset{||}{C}} + 2CH_2O \longrightarrow \underset{HOH_2C-N}{\overset{C_2H_5}{\underset{}{\underset{}{\overset{|}{\overset{N}{\underset{H_2C}{}\underset{}{}CH_2}}}}}}\underset{\overset{||}{O}}{\overset{}{\underset{C}{}}}N-CH_2OH + 2H_2O$$

Their main application is in wash-and-wear finishes for white goods, since the tertiary nitrogen atom acts as an effective buffer against the HCl produced from the chlorine bleaching compounds remaining in the fiber upon ironing. They have been suffering from increasing competition from other types of wrinkle-resistant finishes.

Other ethylamine derivatives

Ethylamines are used as corrosion inhibitors, for example, to stabilize trichloroethylene.   They do not provide permanent stability, since when the HCl produced on degradation of the solvent has finally neutralized all the amine originally present, the stabilizing effect disappears.   For users who recycle the solvent many times over, more effective stabilizers are necessary. They have been found to be activators for methylene chloride paint removers in concentrations around 5 percent.

Diethylaminoethyl chloride, made by reacting ethylene oxide and diethylamine followed by chlorination, is an intermediate for numerous pharmaceutical products, antihistamines, spasmolytics, antimalarials and tranquilizers.

3.2 Ethyl Ether

Although ethyl ether can be made from ethanol via sulfuric acid dehydration, in countries where ethylene is the main building block for acetyl chemicals, most of it is obtained as a byproduct from synthetic ethanol units.   About 93.0 million pounds per year were consumed in the U.S. in 1965, chiefly as an extraction solvent in a variety of applications, as an anaesthetic and in automobile starting fluids for cold weather.

4   ACETALDEHYDE

Until the introduction of the Wacker process, acetaldehyde was produced almost entirely via ethyl alcohol in the U.S. In

Germany, one-third of all acetaldehyde is still being made from acetylene, and as late as 1957, a plant based on hydrocarbon acetylene was built at Lacq, France. In Italy acetaldehyde was also predominantly acetylene-derived, at least as long as the Po Valley natural gas reserves held out.

$$C_2H_2 + H_2O \longrightarrow CH_3CHO$$

The two processes for making acetaldehyde from ethanol are dehydrogenation and oxidation:

$$C_2H_5OH \longrightarrow CH_3CHO + H_2$$

$$C_2H_5OH + \tfrac{1}{2}O_2 \longrightarrow CH_3CHO + H_2O$$

Oxidation has the advantages of allowing higher conversion per pass, and therefore involves lower utility costs. Dehydrogenation is preferred when pure hydrogen is required, for example, if the producer also intends to make butanol or 2-ethylhexanol. A flowsheet of an oxidation variant is shown in Fig. 5-11. Acetaldehyde and air enter the oxidation furnace and the exit gases,

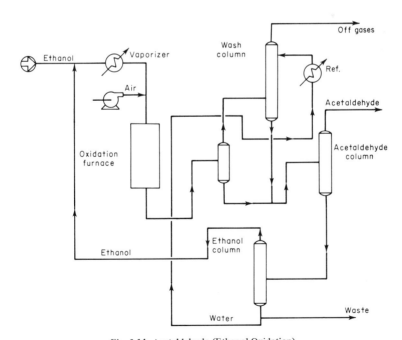

Fig. 5-11 Acetaldehyde (Ethanol Oxidation).

after passing through a condenser, go to a phase separator. The vapor phase is absorbed in refrigerated water, and the wash is combined with the liquid phase. The combined stream is fractionated into acetaldehyde and a water-alcohol mixture which is further separated into alcohol and waste water. Conversion per pass is 50 percent, and acetaldehyde yields 95 percent of theoretical. No air is used in the dehydrogenation process. Ethanol is passed over a dehydrogenation catalyst, yielding acetaldehyde and approximately 1 mol H_2/mol product which after removal of organic impurities can be used elsewhere in the plant. The disadvantage of this alternative is that for a reasonable catalyst life, conversion per pass must be kept down to 35 percent, thus utility costs are about 50 percent higher than for oxidation. Investment costs are also higher.

The Wacker process, used for the first time in the U.S. by Celanese, consists of oxidizing ethylene to acetaldehyde directly over a cupric chloride-palladium chloride catalyst. Two variants exist: in the one-stage process, the oxidant is pure oxygen and the catalyst is regenerated *in situ*; in the two-stage process, spent catalyst is regenerated in a separate reactor and the oxidant is air. The two-stage process forms more byproducts, operates at a higher pressure, involves a 20-25 percent higher investment cost for the process unit itself but requires no oxygen or air-plant. The choice between the two alternatives depends mainly on whether the producer already has oxygen at his disposal, or needs the byproduct nitrogen; yields are reported around 95 percent in either case. A flowsheet for the one-stage process is shown in Fig. 5-12. Materials of construction requirements are stringent: titanium equipment throughout, and a rubber-lined reactor further protected by an acid-resistant brick wall; this explains the large incremental investment costs when the second reactor is added. The economics depend on the raw material structure and accounting practices of the individual company. There seems to be little difference between the two alternatives, the higher investment cost for the two-stage process being roughly offset by the difference in cost between air and oxygen. Nevertheless, the two-stage version has proved more popular. There seems little doubt that most additional acetaldehyde plants will be built using this process, except in those situations where a high value can be attributed to the hydrogen obtained by dehydrogenation. So far, Celanese and Texas Eastman use the two-stage version of this process in the U.S., and Shawinigan Chemical operates a one-stage plant in Canada. Sicedison in

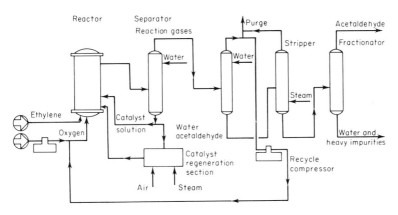

Fig. 5-12 Acetaldehyde (Single-Stage Wacker Process).

Italy, Rhône-Poulenc in France, Pemex in Mexico, Hoechst, Wacker and Knapsack-Griesheim in Germany and at least four producers in Japan also use the process. It is also possible to use this technology for making acetone, as is already being done in Japan. The net reaction is:

$$C_2H_4 + \tfrac{1}{2}O_2 \xrightarrow[\text{CuCl}_2]{\text{PdCl}_2} CH_3CHO$$

In the U.S., the LPG oxidation processes employed by Celanese at its two plants in Bishop and Pampa, Texas, are a major source of acetaldehyde, as well as of acetic acid, propionic acid, methyl ethyl ketone, acetone, glycols, methanol, and formaldehyde. Celanese is the largest producer of acetyl derivatives in the world, and until the above-mentioned Wacker plant was built, its entire output was based on this process. The plants consist essentially of a battery of liquid or vapor-phase oxidation reactors, followed by separation of the numerous products in an extremely complex fractionation plant. Process conditions can be varied to produce a wide range of product mixes.

Acetaldehyde capacities in the U.S., by producer, are listed below.

Producer	Location	1966 Capacity, MM lbs	Feedstock
Celanese	Bay City, Tex.	210	Ethylene
	Bishop, Tex.	200	LPG
	Pampa, Tex.	10	Byproduct

Commercial Solvents	Agnew, Calif.	1	Ethanol
DuPont	Louisville, Ky.	10	Byproduct
Eastman	Kingsport, Tenn.	200*	Ethanol
	Longview, Tex.	250	Ethylene
Goodrich	Calvert City, Ky.	1	Byproduct
Hercules	Parlin, N.J.	35	Ethanol
Monsanto	Texas City, Tex.	5	Byproduct
Publicker	Philadelphia, Pa.	80*	Ethanol
Union Carbide	Institute, W.Va.		
	S. Charleston, W.Va.	650	Ethanol
	Texas City, Tex.		
Total		1,652	

*Major portions of the acetaldehyde produced here are not isolated.
Source: *Oil, Paint & Drug Reporter*, April 3, 1967.

The principal uses for acetaldehyde are as an intermediate for acetic acid and for higher alcohols via crotonaldehyde. In the breakdown for 1966, the acetic acid figure is necessarily an approximation since the Celanese LPG oxidation processes yield both acetaldehyde for conversion to acetic acid and the acid directly.

End-use	MM lbs/yr, 1966
Acetic acid and anhydride	600.0
Peracetic acid	15.0
n-Butanol	300.0
Pentaerythritol	30.0
Chloral	25.0
2-Ethyl hexanol	280.0
Pyridines and picolines	30.0
Vinyl acetate	30.0
Trimethylolpropane	15.0
1,3 Butylene glycol	20.0
Miscellaneous	55.0
Total	1,400.0

Source: *Oil, Paint & Drug Reporter*, April 3, 1967.

The investment of a 50.0 million pounds per year acetaldehyde plant via ethanol dehydrogenation is $1.0 million. The same capacity using the one-stage Wacker process would cost $2.4 million, and the two-stage version — $2.9 million. The two processes are compared below. The assumptions are: purchased ethylene at 4.0 cents per pound, and $9 per ton captive oxygen for the Wacker process, captive alcohol from purchased ethylene for the ethanol process. Hydrogen is credited at 30 cents per thousand cubic feet.

	Production cost, ¢/lb		Ethanol dehydrogenation, ¢/lb
	One-stage Wacker	Two-stage Wacker	
Ethylene	2.7	2.7	—
Oxygen	0.4	—	—
Ethanol	—	—	4.1
Utilities	0.3	0.4	0.6
Labor and overhead	0.6	0.6	0.3
Capital charges	1.6	1.9	0.7
	5.6	5.6	5.7
Hydrogen credit	—	—	-0.2
Total	5.6	5.6	5.5

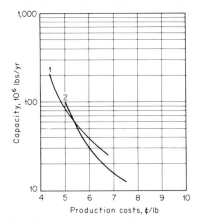

Fig. 5-13 Acetaldehyde. 1-Wacker process (two-stage); 2-ethanol dehydrogenation.

4.1 Acetic Acid

In all countries except the U.S., the most important process for making acetic acid is liquid-phase oxidation of acetaldehyde, using either oxygen or air:

$$CH_3CHO + \tfrac{1}{2}O_2 \longrightarrow CH_3COOH$$

In the U.S., LPG oxidation accounts for about 50 percent of all acetic acid produced. The Distillers' naphtha oxidation process, used for the first time in the early 1960's, is reported to produce acids almost exclusively — formic, and propionic in addition to acetic, instead of the wide range of oxygenated chemicals produced by Celanese — and also depends on recovery of these byproducts for economic operation.

Figure 5-14 shows the flowsheet for an acetaldehyde oxidation plant. Oxygen or air and acetaldehyde are bubbled through liquid acetic acid. The liquid effluent from this reactor is stripped of unreacted acetaldehyde by the main air stream, which carries it back to the reactor, and then enters an acid stripper where the incoming air is saturated with acetic acid. Before entering the aldehyde stripper, the air stream passes through a condenser where most of the net product condenses out and is sent to a fractionation tower. The bottoms from this saturator wash the aldehyde in the gaseous effluent from the reactor, to which they are returned. Meanwhile the gaseous effluent from the reactor is washed first with the liquid bottoms from the acid stripper and then with refrigerated water; liquid from the first wash returns to the reactor and the water stream is also sent to the dehydration column. Yields are 95 percent.

In 1964, acetic acid consumption in the U.S. reached 1,070 million pounds. The table below shows the distribution.

End-use	MM lbs, 1964
Cellulose acetate	480.0
Vinyl acetate	270.0
Esters	160.0
Chloroacetic acid	42.0
Miscellaneous	118.0
Total	1,070.0

Source: *Oil, Paint & Drug Reporter*,
October 5, 1964.

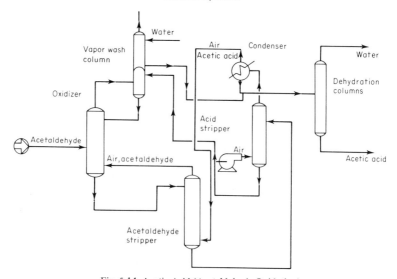

Fig. 5-14 Acetic Acid (Acetaldehyde Oxidation).

The overall growth-rate for the acetic acid demand in the U.S. is expected to be 7 percent per year; the most dynamic markets are vinyl acetate and chloroacetic acid. Producers and their respective capacities are:

Producer	Location	1966 Capacity, MM lbs/yr	Process
Celanese	Bishop, Tex.	200	LPG oxidation
	Pampa, Tex.	500	LPG oxidation
Union Carbide	Brownsville, Tex.	310	LPG oxidation
	S. Charleston, W.Va.	140	Acetaldehyde
	Texas City, Tex.	100	Acetaldehyde
Hercules	Parlin, N.J.	40	Acetaldehyde
Eastman	Kingsport, Tenn.	325	Acetaldehyde
Total		1,615	

Source: *Oil, Paint & Drug Reporter*, February 7, 1966.

The main captive applications for Celanese are cellulose acetate, vinyl acetate and acrylates; Union Carbide makes vinyl acetate as well as the complete line of esters and salts, of which it is the dominant producer; the company is also a producer of merchant peracetic acid from acetaldehyde.

Eastman is mainly a cellulose acetate manufacturer. Hercules originally entered the field as a manufacturer of cellulose acetate and RDX, an explosive. Monsanto, a large producer of vinyl acetate and monochloroacetic acid, has its source of supply in Canada.

Investment for a 50.0 million pounds per year acetic acid plant from captive ethanol taken at 3.7 cents per pound, via acetaldehyde, is $2.2 million. Production costs, assuming captive ethanol production, can be estimated as shown below.

	Cost, ¢/lb acetic acid
Ethanol	3.0
Utilities	0.9
Labor and overhead	0.6
Capital charges	1.4
Total	5.9

An estimate of the economics of the BASF process is given below. The only U.S. plant is said to have cost $7.5 million for a capacity of 100.0 million pounds per year, and captive methanol from a very large plant can be transferred at 2.5 cents per pound.

	Cost, ¢/lb acetic acid
Methanol	1.5
Carbon monoxide	0.2
Labor and overhead	0.3
Utilities (est.)	0.9
Capital charges	2.5
Total	5.4

Due to its high investment and low raw material costs, this process seems attractive at high capacities.

A liquid hydrocarbon oxidation plant for 80.0 million pounds per year of oxygenated products requires an investment of $12.5 million. About 0.2 pound per pound of byproduct is produced per pound of acetic acid, which contributes around 2.8 cents per pound credit to the cost of cracking acetic acid. Production costs for the Distillers' process can be estimated as shown below.

Naphtha (at 1.0 ¢/lb)	1.7	
Catalyst, chemicals, utilities	2.2	
Labor and overhead	0.4	
Capital charges	5.2	
		9.5
Byproduct	-2.8	
Total		6.7

At 1500 million pounds per year this cost falls to 5.2 cents per pound. The process thus appears competitive at high capacities provided the byproducts can still be disposed of at the same price.

Investment for an acetic-acid-from-acetylene plant is approximately the same as from ethanol; costs for the two routes would be equal for acetylene at 6.5 cents per pound. If it ever becomes possible to make acetylene for 3 to 5 cents per pound, acetylene may become the preferred building block for acetyl chemicals. See Fig. 5-15, on next page, for relative costs.

Ketene and diketene

The most important immediate derivative of acetic acid is ketene, formed by pyrolysis of acetic acid in a tubular furnace:

$$CH_3COOH \rightarrow CH_2{=}C{=}O + H_2O$$

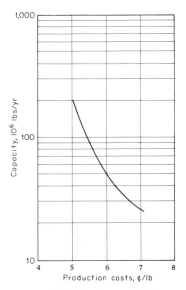

Fig. 5-15   Acetic Acid (via Acetaldehyde).

Ketene can also be made from acetone

$$CH_3COCH_3 \rightarrow CH_2=C=O + CH_4$$

Ketene is extremely reactive. It is the starting point for several important intermediates; reactions take place at the carbon-to-carbon double bond or the carbonyl group. Examples of the two reaction types are:

$$CH_2=C=O + CH_3COOH \rightarrow (CH_3CO)_2O \quad \text{(Acetic anhydride)}$$

$$2CH_2=C=O \rightarrow \begin{array}{c} CH_2=C\!\!-\!\!O \\ | \quad\quad | \\ CH_2\!\!-\!\!C=O \end{array} \quad \text{(Diketene)}$$

Most ketene is made from acetic acid; however, sometimes when there is an oversupply of acetone, some of it still is transformed to acetic anhydride to keep availability in line with requirements. In the few instances when ketene producers have no captive source of acetic acid, acetone turns out to be the cheaper raw material of the two. There are several instances of this in the U.S. and Europe. Cracking of acetone has the added advantage of yielding methane instead of water; this can be burned to produce enough heat for the cracking step.

Acetic anhydride production consumes 95 percent of all ketene produced, the remainder going to ethyl acetoacetate, β-propiolactone, vitamin A, sorbic acid, perfumery products, etc.

Ketene dimerizes to diketene, a highly reactive intermediate with several important uses in the dye and pharmaceutical industries. It reacts with ethyl alcohol to form ethyl acetoacetate:

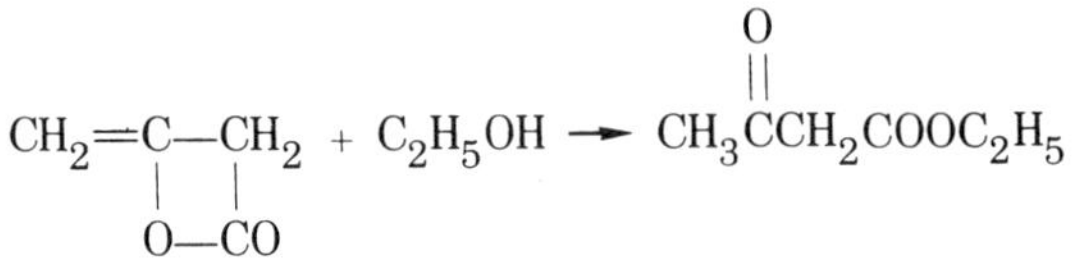

$$\text{CH}_2\!\!=\!\!\overset{\displaystyle |}{\underset{\displaystyle |}{\text{C}}}\!\!-\!\!\overset{\displaystyle |}{\underset{\displaystyle |}{\text{CH}_2}} + \text{C}_2\text{H}_5\text{OH} \longrightarrow \text{CH}_3\overset{\displaystyle \text{O}}{\overset{\displaystyle \|}{\text{C}}}\text{CH}_2\text{COOC}_2\text{H}_5$$
$$\text{O}\!\!-\!\!\text{CO}$$

This compound, of which around 6.0 million pounds are used annually in the U.S., is an intermediate for most yellow azo pigments, such as Benzidine Yellow, the various Hansa Yellows, Vulcan Fast Yellow 5G, and others, of which some 7.0 million pounds per year are made. Ethyl acetoacetate is also an intermediate for the production of 1-phenyl-3-methyl pyrazolone, an intermediate for the production of tartrazine, a yellow certified color of which 1.2 million pounds per year are used, and for the well known antipyretics, analgesine and aminopyrine. Ethyl acetoacetic ester is an intermediate in the production of thiamine, a constituent of the vitamin B complex. An important outlet is the production of "Diazinon," a phosphate insecticide developed by Geigy for use against livestock insects:

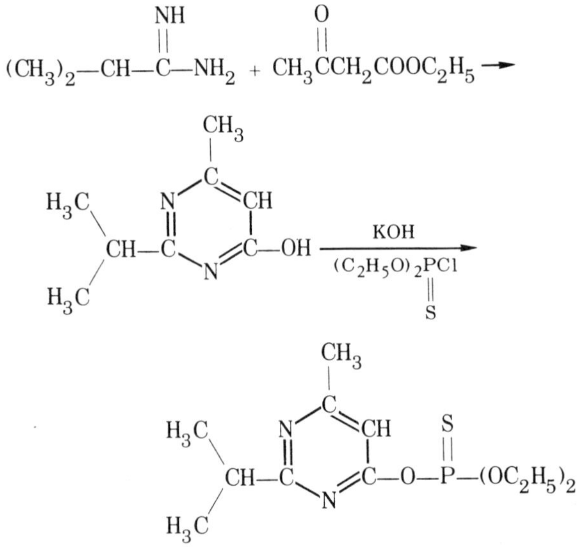

$$(\text{CH}_3)_2\!\!-\!\!\text{CH}\!\!-\!\!\overset{\displaystyle \text{NH}}{\overset{\displaystyle \|}{\text{C}}}\!\!-\!\!\text{NH}_2 + \text{CH}_3\overset{\displaystyle \text{O}}{\overset{\displaystyle \|}{\text{C}}}\text{CH}_2\text{COOC}_2\text{H}_5 \longrightarrow$$

Most ethyl acetoacetate derivatives can also be obtained directly from diketene; since this product is too unstable to be stored and transported, only captive producers can bypass ethyl acetoacetate, which is, in effect, a way of shipping diketene. The market distribution of ethyl acetoacetate is:

Yellow pigments	2.8
Tartrazine	0.6
Other dyes	0.4
"Diazinon" and others	2.2
	6.0

Ketene reacts with paraformaldehyde to form β-propiolactone:

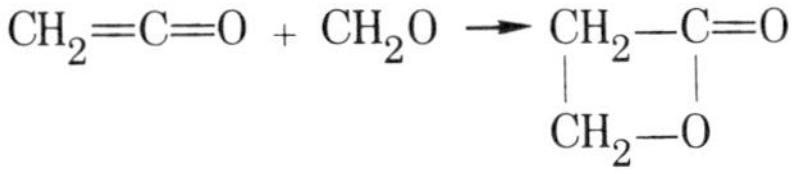

$$CH_2{=}C{=}O + CH_2O \rightarrow \begin{array}{c} CH_2{-}C{=}O \\ | \qquad | \\ CH_2{-}O \end{array}$$

which is also a highly reactive intermediate. Its most important use is the B. F. Goodrich process for making acrylic acid and acrylates employed by the company itself and by Celanese.

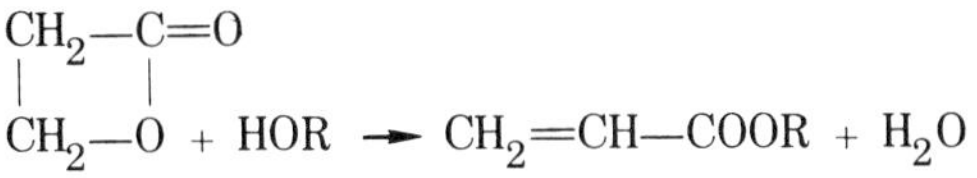

$$\begin{array}{c} CH_2{-}C{=}O \\ | \qquad | \\ CH_2{-}O \end{array} + HOR \rightarrow CH_2{=}CH{-}COOR + H_2O$$

β-Propiolactone reacts with ammonia to give β-alanine, one of the starting points for making calcium pantothenate:

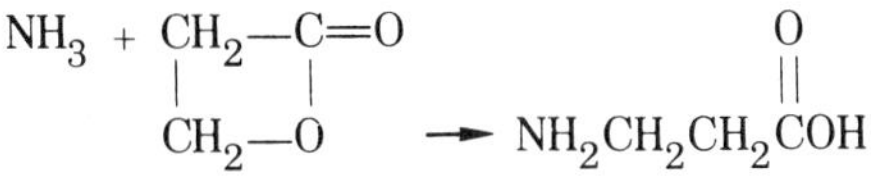

$$NH_3 + \begin{array}{c} CH_2{-}C{=}O \\ | \qquad | \\ CH_2{-}O \end{array} \rightarrow \begin{array}{c} O \\ || \\ NH_2CH_2CH_2COH \end{array}$$

B. F. Goodrich, the major producer, uses this route at Avon Lake, Ohio. Some is still produced from acrylonitrile by a two-step synthesis:

$$CH_2{=}CH{-}CN + NH_3 \rightarrow NH_2CH_2CH_2CN \xrightarrow{H^+} NH_2CH_2CH_2COOH$$

This route is used by Abbott and Nopco, the other two producers of β-alanine.

Acetic anhydride

Although acetic acid is reactive enough for esterification of aliphatic alcohols and metal hydroxides, acetic anhydride is required to acetylate cellulose and aromatic hydroxyl groups. Acetic anhydride is formed by absorbing ketene in acetic acid as it leaves the cracking furnace.   A simplified flowsheet of the Wacker process is shown in Fig. 5-16. After the entire feed has been cracked to ketene, water is removed by low-temperature condensation, at which point some ketene hydrolyzes and is either sent to the central acetic acid recovery unit or to ethyl acetate manufacture.   Ketene is then absorbed in a concentrated solution of acetic anhydride in acetic acid.   In the Celanese process acetic acid is cracked, at a lower temperature than in the Wacker process, to an equimolar mixture of ketene and acetic acid. After water is removed by flashing with benzene, the two reaction products combine to give acetic anhydride.   The yield by either process is around 94 percent.   While this route is by far the most important commercially, some acetaldehyde is still transformed into acetic anhydride using the peracetic acid process:

$$CH_3CHO + O_2 \longrightarrow CH_3COOOH$$

$$CH_3COOOH + CH_3CHO \longrightarrow (CH_3CO)_2O + H_2O$$

In all acetylation reactions involving acetic anhydride, acetic acid is formed as a byproduct.   This means that for every mol of acid used in the form of acetic anhydride, a mol of acid must be recovered and is usually recycled to the ketene furnace.   The demand for acetic anhydride can be broken down as shown below.

End-use	MM lbs/yr, 1965
Cellulose esters	1,315.0
Acetylsalicylic acid	29.0
Vinyl acetate	87.0
Other uses	29.0
Total	1,460.0

Source: *Oil, Paint & Drug Reporter*, October 25, 1965.

Apart from the manufacture of cellulose esters, acetic anhydride is used in numerous other acetylations, some of which are discussed under the main raw material involved. Most of these uses are in the pharmaceutical and perfumery fields. Nalco Chemical has developed a process for refining petroleum distillate fuels with acetic anhydride. Finally, during wartime, large quantities of R.D.X., a violent explosive, are produced

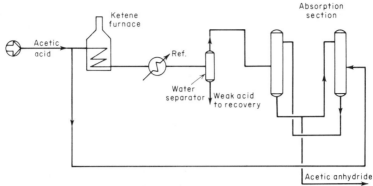

Fig. 5-16 Acetic Anhydride.

from hexamethylene tetramine, acetic anhydride and nitric acid. The U.S. producers of acetic anhydride have the following capacities:

Producer	Location	1965 Capacity, MM lbs/yr
Celanese	Cumberland, Md. Narrows, Va. Pampa, Tex. Rock Hill, N.C.	800.0
Eastman	Kingsport, Tenn.	400.0
Hercules	Parlin, N.J.	50.0
Union Carbide	Brownsville, Tex. S. Charleston, W.Va. Texas City, Tex.	350.0
Total		1,600.0

Source: *Oil, Paint & Drug Reporter*, October 25, 1965.

A 50.0 million pounds per year acetic anhydride plant requires an investment of $1.6 million, and production costs are

	Cost, ¢/lb acetic anhydride
Acetic acid (at 8.0 ¢/lb)	10.0
Utilities	0.8
Labor and overhead	0.8
Capital charges	1.1
Total	12.7

Cellulose acetate

Cotton linters or wood pulp can be acetylated to form a substance that is either spun into fibers from an acetone solution, or compounded with plasticizers and other additives to give molding powders.   Compared to rayon, which is simply regenerated cellulose, acetate manufacture involves chemical modification of the cellulose molecule.   Each glucose unit in cellulose contains three hydroxyl groups.   The material generally used for fibers corresponds to cellulose diacetate and contains about 2.2 acetylated hydroxyl groups per glucose unit.   Cellulose triacetate is a 92 percent acetylated product.   In the usual process for cellulose diacetate, cellulose is first completely acetylated and then partly saponified with sulfuric acid back to diacetate. The product is then separated from the acetic acid by precipitation in water, filtered and dried to give cellulose acetate flake. Acetylation takes place in a large excess of acetic acid; this results in recirculation via an acetic acid recovery unit of 7.5 pounds of acid per pound of fresh acid consumed.   Triacetate is also used commercially, both as a fiber that is shinier, silkier but less resistant than cellulose acetate proper, and for photographic and X-ray film; it was once the main raw material for magnetic tape, now usually made of polyester film.   Triacetate fiber (for example, "Arnel") is cheaper to make than diacetate: the saponification step is eliminated, which reduces capital costs; most important, the fiber can be spun directly from the acetic acid solution in which acetylation takes place.   Apart from fibers, cellulose acetate has two other uses — cigarette filter tips and molding powders.

Cellulose acetate is relatively expensive as a raw material for fibers.   The cost advantage of cellulose acetate over most synthetic fibers appears in the spinning process; cellulose acetate is solvent-spun from a "dope" of acetate in acetone or methylene chloride, a process that affords considerable savings over melt-spinning.   The investment for a 50.0 million pounds per year cellulose acetate plant, including anhydride manufacture and acid recovery, is $14.5 million, and production costs are given below.

	Cost, ¢/lb cellulose acetate
Cellulose ($140/ton)	4.7
Fresh acid (6.5 ¢/lb)	5.2
Labor and overhead	1.9
Utilities	4.8
Capital charges	9.5
Total	26.1

U.S. cellulose acetate producers are:

Producer	Location	1964 Capacity, MM lbs/yr	Captive acetic acid
FMC	Meadville, Pa.	25.0	—
Celanese	Cumberland, Md. ⎫ Narrows, Va. ⎬ Rock Hill, S.C. ⎭	400.0	x
DuPont	Waynesboro, Va.	60.0	—
Tennessee Eastman	Kingsport, Tenn.	215.0	x
Total		700.0	

Source: *Oil, Paint & Drug Reporter*, June 29, 1964.

Cellulose acetate consumption in the U.S. during 1965 was:

End-use	MM lbs, 1964
Yarn	320.0
Staple and tow	190.0
Plastics	145.0
Total	655.0

Source: *Oil, Paint & Drug Reporter*,
June 29, 1964.

Consumption of fibers and filter tips is growing at rates of 5 and 7 percent per year, respectively.

Cellulosic molding powders, however, are expensive compared with polypropylene or high-impact polystyrene, and are expected to lose ground over the next few years. These powders include not only acetate but also, and chiefly, cellulose acetopropionate and acetobutyrate. These are made by acylating cellulose with a mixture of acetic and propionic, or acetic and butyric anhydrides. Whereas cotton linters are the raw material for acetate fibers, cellulose for molding powder is usually sulfite pulp. Higher cellulose esters are preferred for making molding powders, and straight acetate for film.

Propionate, of which some 50.0 million pounds per year are consumed, is easier to mold than butyrate, and is the more important of the two. It is also odorless and more resistant to abrasion. A breakdown of cellulosic plastic end-uses is given below. Sheet under 3 mils is declining in importance because of competition from cheaper substitutes whereas that over 3 mils is showing some growth in blister-packaging. Other important applications are outdoor signs, where weatherability and toughness are at a premium, and cellulosic automobile steering wheels, for which the higher esters are preferred. The estimated market in 1965 for cellulosic plastics can be broken down by application as follows:

End-use	MM lbs/yr, 1965
Automotive	7.0
Blister packaging	30.0
Optical goods	19.0
Toothbrushes, hairbrushes, pens, pencils	24.0
Packing containers	13.0
Electrical appliances	9.0
Tools	9.0
Tubing	5.0
Toys	8.0
Heavy sheeting	11.0
Signs	5.0
All others	17.0
Total	157.0

Source: *Modern Plastics*, January 1966.

Propionic and butyric anhydrides needed for higher cellulose ester manufacture are made by the reaction:

$$(CH_3CO)_2O + 2RCOOH \rightarrow (RCO)_2O + 2CH_3COOH$$

Theoretically, no acetic acid is consumed, since in effect a mol of water is lost by the higher acid and hydrolyzes acetic anhydride back to acetic acid; only a small amount is needed for make-up.

Acetic esters

Acetic acid is esterified in the presence of an acid catalyst, usually sulfuric acid, to form the respective acetate:

$$CH_3COOH \rightarrow CH_3COOR + H_2O$$

Yields are around 95 percent on both reactants, depending somewhat on the alcohol. Ethyl acetate, for example, is made by reacting acetic acid and ethanol; a ternary azeotrope of ethanol, water and ethyl acetate is taken overhead, and is washed with water to remove the unreacted alcohol. The wet ester is purified by fractionation, and the alcohol, after removal of water, returned to the esterification step. Since water is formed in the reaction anyway, dilute acetic acid recovered in other parts of an acetyl products plant often is sent directly to the ethyl acetate unit. One major U.S. producer obtains ethyl acetate as a byproduct from polyvinyl alcohol production.   In Europe, where the acetyl radical is still based chiefly on acetylene, ethyl acetate is made from acetaldehyde by the Cannizzaro reaction:

$$2CH_3CHO \longrightarrow CH_3COOC_2H_5$$

The main U.S. producers of ethyl acetate are:

Producer	Location	1965 Capacity, MM lbs/yr
Publicker	Philadelphia, Pa.	5.0
Shawinigan	Springfield, Mass. } Trenton, Mich.	20.0
Eastman	Kingsport, Tenn.	20.0
Union Carbide	S. Charleston, W.Va. } Texas City, Tex.	60.0
Celanese	Bishop, Tex.	5.0
Hercules	Parlin, N.J.	18.0
Total		128.0

Source: *Oil, Paint & Drug Reporter*, January 25, 1965.

Acetate esters are used mainly as solvents.   The shorter the alcohol, the faster-boiling the solvent. Methyl acetate is obtained by the two main polyvinyl alcohol producers as an 80:20 mixture with methanol.   It is used mainly as a paint remover but, being flammable, is less desirable than methylene chloride.   Of total consumption, 6.0 million pounds (100 percent methyl acetate) were used in paint-remover formulations. Ethyl acetate is a low-boiling solvent for lacquers.   The various butyl acetates are popular medium-boiling solvents also used mainly in surface coatings.   "Cellosolve" acetate is important as a high-boiling solvent.

Ketones such as MEK, MIBK and MIAK have taken over most of the market once held by ethyl, butyl and heavier acetates. They are more expensive per unit weight but their solutions have a lower viscosity for a given solids content. Other acetate solvents are methoxybutyl acetate, widely used in Europe but less so in the U.S., and ethylene-glycol monomethylether acetate, somewhat similar to "Cellosolve" acetate but not as important. Triacetin, the acetic acid ester of glycerine, is nontoxic and used specifically as a plasticizer for cellulose acetate filter-tips. The U.S. consumption of the various acetic esters in 1964 was:

Esters	MM lbs, 1964
Methyl acetate (100% basis)	9.0
Ethyl acetate (100% basis)	95.0
Isopropyl acetate	45.0
n-Butyl acetate	90.0
Other butyl acetates	45.0
"Cellosolve" acetate	62.0
Amyl acetates	8.0
Triacetin	9.0
Total	363.0

Source: Primarily from U.S. Tariff Commission Annual Report, "Organic Chemical Production."

Acetic salts

The most important salt of acetic acid is sodium acetate, obtained by the action of acetic acid on caustic soda. Approximately 16.0 million pounds per year are used in the U.S., mainly in the textile industry as a buffer for dyeing baths. Lead acetate is the only soluble lead salt, and is used as a mordant in the textile industry, in lotions and other pharmaceutical applications, and wherever else lead is desired in solution.

Zinc acetate is best known as the catalyst for the vapor-phase vinyl acetate process.

Mercuric acetate is an important intermediate for the production of numerous organometallic compounds. Among these are the well-known household antiseptics, Mercurochrome, Merthiolate and Metaphen, but in terms of tonnage the most important application is in fungicides.

Chloroacetic acids

The usual process for making monochloroacetic acid is chlorination of acetic acid. This can be done in the presence of acetic anhydride, or by using SO_2Cl_2 as the chlorinating agent:

$$SO_2 + Cl_2 \rightarrow SO_2Cl_2$$

$$SO_2Cl_2 + 2CH_3COOH \rightarrow 2ClCH_2COOH + SO_2$$

Companies having acetic anhydride and acetic acid recovery units tend to prefer straight chlorination in the presence of some anhydride.

$$(CH_3CO)_2O + Cl_2 \rightarrow ClCH_2COOH + CH_3COCl$$

$$CH_3COCl + CH_3COOH \rightarrow (CH_3CO)_2O + HCl$$

Producers receiving acetic acid from an outside source are forced to use the SO_2 route. Yields on acetic acid are around 95 percent. Since the products are very corrosive, glass-lined equipment is used in most plants; this is also necessary to meet food-grade specifications.

Monochloroacetic acid can also be made by sulfuric acid hydrolysis of trichlorethylene:

$$HCl{=}CCl_2 + 2H_2O \rightarrow ClCH_2COOH + 2HCl$$

While at least two large European producers still employ this route, only one U.S. plant still finds it competitive with the acetic acid process. In the U.S., di- and trichloroacetic acids are obtained similarly by chlorinating acetic acid.

It has been reported that trichloroacetic acid is obtained in Eastern Europe by oxidizing chloral:

$$CCl_3CHO \xrightarrow{(O)} CCl_3COOH$$

The capacities of U.S. producers of monochloroacetic acid as of 1966 are given below. Note that consumption is above domestic capacity, the difference being made up by imports.

Producer	Location	1966 Capacity, MM lbs/yr	Major captive end-use
Hercules	Hopewell, Va.	10.0	Carboxymethylcellulose
Buckeye Cellulose	Memphis, Tenn.	3.0	Carboxymethylcellulose
Dow	Midland, Mich.	40.0	Phenoxy herbicides
Diamond	Newark, N.J.	3.0	Phenoxy herbicides
Monsanto	Monsanto, Ill.	15.0	Herbicides

Source: *Oil, Paint & Drug Reporter*, May 16, 1966.

The demand for monochloroacetic acid reached 80.0 million pounds in 1965, distributed as follows:

End-use	MM lbs, 1965
2,4-D acid and derivatives ⎫ 2,4,5-T acid and derivatives ⎬	52.0
Carboxymethyl cellulose and other cellulose ethers	18.0
Thioglycolic acid	2.0
Glycine	2.0
Other uses	6.0
Total	80.0

Source: *Oil, Paint & Drug Reporter*, May 16, 1966.

The use of post-emergence herbicides of the phenoxyacetic type is fairly static, since the trend is toward the much safer preemergence products.

The investment for a 20.0 million pounds per year monochloroacetic acid plant via direct chlorination is $850,000. Production costs assume captive chlorine at 2.7 cents per pound and that HCl is worth 50 percent of fresh chlorine; captive acetic acid is assumed available at 8.0 cents per pound.

	Cost, ¢/lb monochloroacetic acid
Chlorine	1.7
Acetic acid and anhydride	6.7
Utilities	0.4
Labor and overhead	1.2
Capital charges	2.8
Total	12.8

The trichlorethylene route becomes competitive if internal raw materials costs are around 5.0 cents per pound.

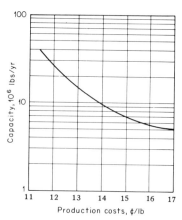

Fig. 5-17 Monochloroacetic Acid (Trichlorethylene).

Numerous herbicides are derived from chloroacetic acid. In the U.S. 2,4-*D* and 2,4,5-*T* dominate the field, whereas in Europe MCPA is preferred. "Randox," a preemergence product, is used mainly on corn.

In the form of its salts with sodium or certain arylamines, trichloroacetic acid is used to make a number of nonselective weed killers used mainly along highways and under utility lines. Dichloroacetic acid is an intermediate in the synthesis of chloramphenicol.

Carboxymethyl cellulose

Monochloroacetic acid and caustic soda react with cellulose to form a water-soluble resin known as sodium carboxymethyl cellulose.

$$Cell—O—Na + ClCH_2COONa \longrightarrow NaCl + Cell—O—CH_2COONa$$

The average degree of carboxymethylation of the grades used commercially is 0.66 carboxymethyl groups per glucose unit, or 22 percent of the cellulose hydroxyl groups. The larger plants use continuous processes in which caustic soda and chloroacetic are sprayed onto powdered wood, bleached sulfite pulp, cotton linters, or any other form of cellulose in a rotary reactor. After it has been aged and broken up, the product is flash-dried and bagged. An elaborate solids recovery system is one of the major components of plant investment. In the older batch process, sodium monochloroacetate is mixed with shredded cellulose. U.S. producers and their capacities are shown below.

Producer	Location	1965 Capacity, MM lbs/yr
Buckeye Cellulose	Memphis, Tenn.	4.5
DuPont	Carney's Point, N.J.	15.0
Hercules	Hopewell, Va.	30.0
Kohnstamm	Camden, N.J.	2.5
	Clearing, Ill.	0.5
Union Carbide	Niagara Falls, N.Y.	5.0
Wyandotte	Wyandotte, Mich.	4.0
Warner Chemical	Muncie, Ind.	0.5
Total		62.0

Source: *Oil, Paint & Drug Reporter,* January 11, 1965.

Many different grades exist — varying in viscosity, degree of acetylation and purity. The purification process consists of removing sodium chloride and sodium glycolate, which is toxic, by dissolving in alcohol and filtering off the pure product. Production in the U.S. was 45.0 million pounds in 1965, and expected growth is 2.5 percent per year. In most of its numerous applications it must compete with both natural and other synthetic water-soluble resins, with the exception of detergents, where it is by far the suspending agent in use. CMC goes mainly into solid detergents, its functions being to prevent suspended dirt from redepositing and to increase the proportion of inorganic builders to active detergent. CMC is widely used as a viscosity improver in drilling muds. This use has been decreasing largely due to competition from lignosulfonates recovered from the kraft paper process. It is used in ice-cream, where its job is to prevent the formation of large ice particles and make it easier to increase the volume by beating air bubbles into the product. Food uses, apart from ice-cream, include noncaloric soft drinks and weight-reducing wafers.

End-use	MM lbs, 1965
Detergents	17.0
Drilling muds	6.0
Food	6.5
Textiles	5.0
Paper size	3.5
Pharmaceuticals	3.5
Paint thickener	1.3
Other uses	2.2
Total	45.0

Source: *Oil, Paint & Drug Reporter*,
January 11, 1965.

Investment costs for a 5.0 million pounds per year continuous plant are $1.5 million, and approximate production costs are as follows:

	Cost, ¢/lb CMC
Raw materials	13.6
Utilities	0.6
Labor and overhead	6.4
Capital costs	9.9
Total	30.5

4.2 Crotonaldehyde

Crotonaldehyde is obtained by condensation of two mols of acetaldehyde and subsequent dehydration:

$$2CH_2CHO \xrightarrow{\text{OH}^-} CH_3CH(OH)CH_2CHO \rightarrow CH_3CH=CHCHO + H_2O$$

The two reactions are carried out in separate reactors after which a fractionation system removes unreacted acetaldehyde and heavy condensation products.

Production of crotonaldehyde in the U.S. was estimated at 285.0 million pounds in 1965. The economics of this product are linked with those of the oxo process; both routes are used to produce n-butyraldehyde, from which in turn n-butanol and 2-ethylhexyl alcohol are produced. Economics still seem slightly in favor of crotonaldehyde as a source of butanol but the oxo route accounts for the major portion of 2-ethylhexanol output. By 1965, crotonaldehyde was still the starting point for over 60 percent of the U.S. butanol output, but only 12 percent of 2-ethylhexanol; the advent of the Wacker acetaldehyde process has given the crotonaldehyde route to butanol new life and may counteract the noticeable trend in favor of the oxo process as a source of n-butanol as well as of C_8 alcohols.

Methoxybutanol

Crotonaldehyde reacts with methanol, and the product can be hydrogenated to 3-methoxybutanol:

$$CH_3-CH=CHCHO + CH_3OH \rightarrow CH_3\overset{\overset{\displaystyle OCH_3}{|}}{C}HCH_2CHO \xrightarrow{H_2} CH_3\overset{\overset{\displaystyle OCH_3}{|}}{C}HCH_2CH_2OH$$

It is only important as a paint solvent, in the form of its acetic ester. While its use is widespread in Europe, especially in Germany, it is of secondary importance in the U.S.

Crotonic acid

The chief outlet for crotonic acid, obtained by the liquid phase oxidation of crotonaldehyde, is in polyvinyl acetate coatings, where addition of around 1 percent greatly improves adherence to the substratum.   About 0.5 million pounds per year go into this application.   Other uses are as an intermediate (via crotonyl chloride) for "Karathane," a fungicide developed by Rohm & Haas; vinyl crotonate, a crosslinkable monomer for surface coatings; and "Bidrin," made via crotonamide, a compound developed by Shell to fight Dutch elm disease. The total demand for crotonic acid is 1.0 million pounds per year.

Sorbic acid

Crotonaldehyde reacts with ketene, giving sorbic acid:

$$CH_2{=}C{=}O + CH_3{-}CH{=}CH_2{-}CHO \longrightarrow CH_2{=}CH{-}CH{=}CH{-}CH_2{-}COOH$$

It is being used in rapidly increasing amounts as a fungistat for foods such as cheese, packed meat products and margarine, and in food packaging materials; consumption in 1965 was around 2.5 million pounds.   In the U.S. this process is used by Hostachem, a subsidiary of Farbwerke Hoechst. Union Carbide obtains sorbic acid by oxidizing a trimer of acetaldehyde:

$$CH_2{=}CH{-}CH{=}CH{-}CH_2{-}CHO \xrightarrow{\text{[o]}} CH_2{=}CH{-}CH{=}CH{-}CH_2COOH$$

4.3 *n*-Butanol

Hydrogenation of crotonaldehyde gives *n*-butanol:

$$CH_3CH{=}CHCHO + 2H_2 \longrightarrow CH_3CH_2CH_2CH_2OH$$

Note that no hydrogen production is necessary when acetaldehyde is obtained by ethanol dehydrogenation, since the hydrogen given off is exactly equal to that needed to convert an equivalent amount of crotonaldehyde to butanol. In a unit based on Wacker acetaldehyde, butanol production requires an outside hydrogen source. Butanol is also produced from propylene via the oxo reaction:

$$CH_3CH{=}CH_2 + CO + H_2 \longrightarrow CH_3CH_2CH_2CHO \xrightarrow{H_2} CH_3CH_2CH_2CH_2OH$$

The major problem in this process is the large quantities of isobutanol obtained as a coproduct.

In Japan, n-butanol is made by the Reppe process, which amounts to simultaneous hydroformylation and CO shift-conversion. The reaction takes place at 200 psi and yields almost 6 times more n-butanol than i-butanol.

$$CH_3—CH{=}CH_2 + 3CO + 2H_2O \rightarrow CH_3CH_2CH_2CH_2OH + 2CO_2$$

In the acetaldehyde route, vaporized crotonaldehyde and recycled hydrogen enter the hydrogenation furnace. The effluent is sent to a dehydration column from which a crotonaldehyde-water azeotrope leaves overhead and is decanted into an organic phase, which returns to the reactor, and an aqueous phase which is stripped of the remaining unreacted crotonaldehyde. Bottoms from the dehydration column are split into n-butanol product and heavy waste. A flowsheet for the process, starting from acetaldehyde, is shown in Fig. 5-17. U.S. producers of butanol are:

Producer	Location	1965 Capacity, MM lbs/yr	Route
Celanese	Bishop, Tex.	125.0	Crotonaldehyde
Dow-Badisch	Freeport, Tex.	15.0	Oxidation
Texas Eastman	Longview, Tex.	60.0	Oxidation
Humble	Baton Rouge, La.	10.0	Oxidation
Shell	Houston, Tex.	50.0	Oxidation
Union Carbide	Seadrift, Tex.	180.0	Oxidation
	S. Charleston, W.Va.	60.0	Crotonaldehyde
	Puerto Rico	40.0	Crotonaldehyde
Publicker	Philadelphia, Pa.	35.0	Fermentation
Total		575.0	

Source: Based on *Oil, Paint & Drug Reporter*, May 1, 1967 and June 3, 1963.

Production of n-butanol in 1965 was around 429.0 million pounds per year with the following distribution.

End-use	MM lbs, 1965
Esters and others	55.0
n-Butyl acetate	55.0
Solvent (surface coatings)	86.0
Glycol ethers and amino resins	100.0
Plasticizers	30.0
Exports	73.0
Total	399.0

Source: Derived from *Oil, Paint & Drug Reporter*, June 3, 1965 and later data.

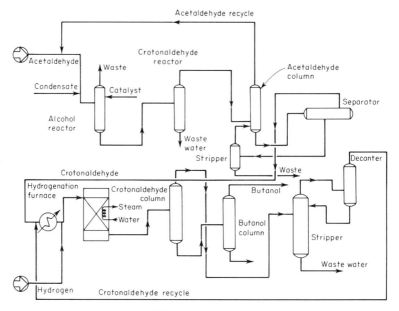

Fig. 5-18 Butanol.

The investment for a 50.0 million pound butanol plant starting from ethanol is $3.5 million. Production costs have been calculated assuming captive ethanol.

Ethanol	8.1
Utilities	2.1
Labor and overhead	0.7
Capital costs	1.1
Total	12.0

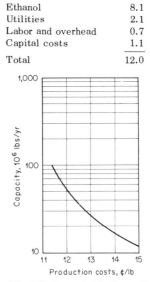

Fig. 5-19 Butanol (via Crotonaldehyde).

Modified amino-resins

Melamine and urea-formaldehyde resins are made compatible with alkyds by etherification with a C_4 alcohol. Water formed in the reaction is removed as an azeotrope with the alcohol. There is a trend towards modified alkyd coatings at the expense of straight alkyd formulations, hence this outlet is expanding.

n-Butylamines

Butylamines are made by a process analogous to ethylamine production.   About 4.0 million pounds per year, mostly di-*n*-butylamine, were used in 1965. Of this total, the largest quantity went into the production of zinc dibutyl dithiocarbamate, an accelerator used principally for vulcanizing transparent rubber goods, and also as a stabilizer for polyurethane foams, which is where most of the growth potential lies.

Fatty acid salts of butylamines are used as corrosion inhibitors and in protective coatings. A well-known derivative of mono-*n*-butylamine is tolbutamide, better known as the oral antidiabetic "Orinase."

Miscellaneous uses make up the remainder. For example, methyl-*n*-butylamine is an intermediate for "Neburon," a carbamate-type herbicide used in the control of woody plants and in commercial nurseries.

4.4 Pyridine and Picolines

Traditionally derived from coal-tar, pyridine bases are now, for the most part, made synthetically since coal-tar sources have not kept up with the growing demand for these compounds and their derivatives.

All processes used at present are based on the reaction between acetaldehyde, or paraldehyde, its cyclic trimer, and ammonia.

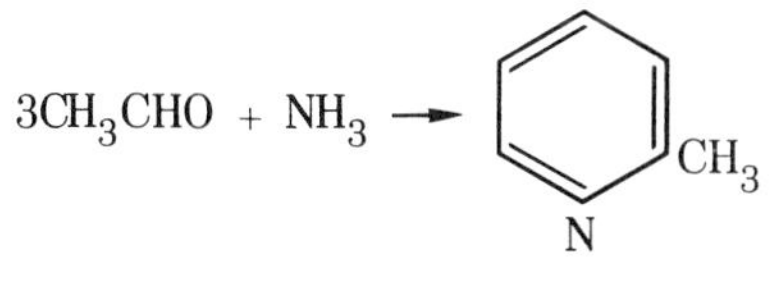

$$3CH_3CHO + NH_3 \longrightarrow$$

When pyridine itself is to be produced, formaldehyde is mixed with acetaldehyde in the feed to form odd-numbered carbon compounds. The reaction mechanism is thought to be *in situ* formation of acrolein, which reacts with more formaldehyde and with NH_3.

$$CH_3CHO + CH_2O \xrightarrow{-H_2O} CH_2{=}CH{-}CHO$$

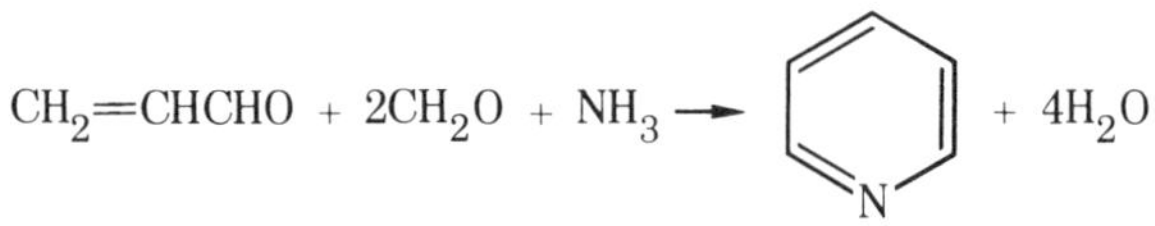

$$CH_2{=}CHCHO + 2CH_2O + NH_3 \longrightarrow \text{(pyridine)} + 4H_2O$$

It has been reported that the largest U.S. plant, operated by Reilly Tar & Chemical, can turn out either 50 percent pyridine and 50 percent β-picoline, or 50 percent α-picoline, 25 percent γ-picoline and 25 percent pyridine. 2-Methyl-5-ethyl pyridine is made by a similar process:

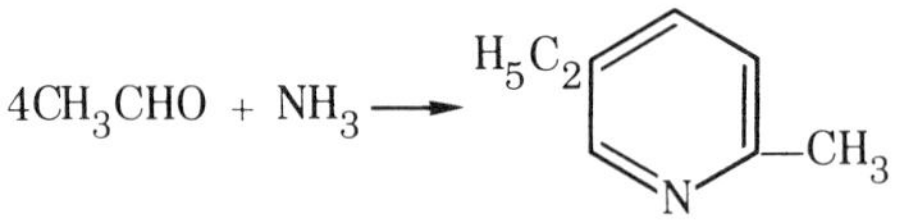

$$4CH_3CHO + NH_3 \longrightarrow \text{(2-methyl-5-ethyl pyridine)}$$

U.S. producers of synthetic pyridine bases and their capacities are as follows

Producer	1965 Capacity, MM lbs/yr	Product
Phillips	5.0	2-Methyl-5-ethyl pyridine
Union Carbide	2.0	2-Methyl-5-ethyl pyridine
Reilly Tar & Chemical	12.0	Pyridine, picolines
Nepera	5.0	Picolines
Total	24.0	

Source: Based on *Oil, Paint & Drug Reporter*, January 10, 1966; February 15, 1965.

The U.S. production in 1965 was 6.5 million pounds of pyridine and 4.8 million pounds of picolines. About 1.5 million pounds per year of pyridine were used as a solvent in numerous applications, especially in the dye, plastics and pharmaceutical industries, wherever a basic and fairly high-boiling medium is required. Another 4.0 million pounds were consumed as an intermediate. Being a tertiary amine, pyridine can be quaternized to form such compounds as lauryl pyridinium chloride, a spinning aid for rayon, or stearamidopyridinium chloride, employed as a waterproofing compound for tarpaulins and other nonapparel outlets. In 1965, about 2.0 million pounds of these textile aids were consumed.

Several agricultural chemicals are derived from pyridine: "Reglone," a nonarsenical potato-vine killer developed by ICI;

"Raticate," a rodenticide, as well as fungicides or bactericides such as zinc pyridinethione and l-hydroxy pyridinethione, sold as "Vancide SP" and "Omadin," respectively.

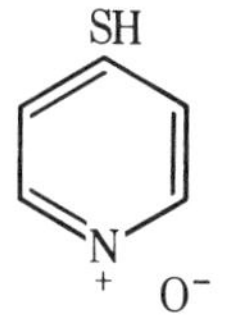

"Omadin"

These fungicides are responsible for much of the recent growth in the demand for pyridine.

Pyridine reacts with sodamide to give 2-aminopyridine, an intermediate for drugs such as sulfapyridine and pyrilamine maleate, the latter a widely used antihistamine base. Pyridyl mercuric acetate, made by reacting pyridine with mercuric acetate, has been introduced as a slime control agent in paper stock. Sulfonation of pyridine and subsequent reaction with sodium cyanide gives β-cyanopyridine which can be hydrolyzed to nicotinamide

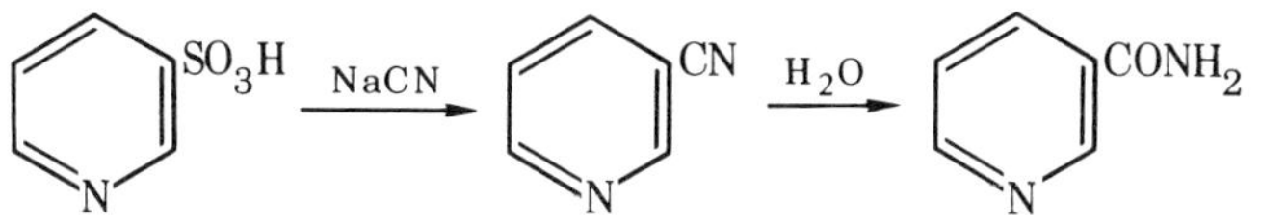

Pyridine can be hydrogenated to piperidine:

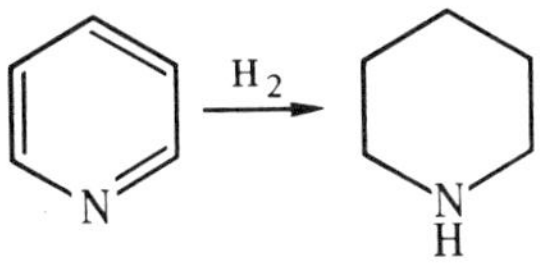

This is used to make a dithiocarbamate rubber accelerator known as "pip-pip," and is also an intermediate in the manufacture of "Artane," an anti-Parkinsonian drug.

The use of α-picoline in the U.S. is about 5.5 million pounds per year. Its main application is in the manufacture of vinyl pyridine. Some 3.5 million pounds per year of this compound go into butadiene-styrene-vinyl pyridine terpolymers; 18.0 million pounds per year of these resins are used to make tire-cord adhesives. Vinyl pyridine is produced from α-picoline by reacting with formaldehyde, followed by dehydration.

α-picoline is also used in the production of "Tordon," a herbicide developed by Dow for use along highways; a side-chain chlorinated derivative, 2-chloro-6-(trichloromethyl) pyridine, made by Dow under the name "N-Serve," which functions in the control of nitrogen-destroying bacteria in soil; and as one of the raw materials for "Amprolium," a widely used coccidiostat developed by Merck.

With the advent of new processes which can produce β-picoline without producing the γ-isomer, β-picoline has dropped in cost to the point where it is competitive with methyl-ethyl-pyridine as the raw material for making niacin (nicotinic acid). As long as coal-tar was the only source of β-picoline, the two isomers were difficult to separate because they have nearly the same boiling point and the preferred raw material for making nicotinic acid was quinoline; later, 2-methyl-5-ethyl pyridine became available as the starting point for 2-methyl-5-vinyl pyridine, and proved a convenient alternate source for niacin. At present, Nepera and Reilly make niacin from captive β-picoline, and producers having no captive raw material source such as Nopco, Merck and Abbott prefer to oxidize MEP. Quinoline oxidation has been abandoned.

As the acid itself, or the respective amide, some 5.0 million pounds per year of niacin are consumed in the U.S. in feed supplements and pharmaceutical products; a deficiency of this member of the vitamin B complex is the cause of pellagra.

Nicotinamide can be made from niacin as well as from cyanopyridine. The main outlet for γ-picoline is isonicotinic acid, the intermediate for "Isoniazid," a very effective anti-tubercular drug.

Methyl vinyl pyridine (MVP) is produced by high-temperature dehydrogenation of methyl ethyl pyridine. Its largest outlet is as a 3 percent constituent in "Acrilan," an acrylic fiber produced by Chemstrand, where its function is to improve dyeability. This application requires about 2.0 million pounds per year. Great interest has developed in the possibility of using MVP for a similar purpose in polypropylene fibers whose poor dyeability has so far restricted their use. It is reported that Montecatini uses this technique at Terni, Italy. Phillips is the leading MVP producer in the U.S.

4.5 Chloral

Chloral is presently produced exclusively by chlorination of acetaldehyde:

$$3Cl_2 + CH_3CHO \longrightarrow Cl_3CCHO + 3HCl$$

The only important use for chloral is in the production of D.D.T. (dichlorodiphenyltrichloroethane), an insecticide developed by Geigy in 1942, and which since has become and remains the most widely used pesticide in the world.

Chloral is also an intermediate for other insecticides: "Methoxyclor," analogous to D.D.T. but with anisole replacing chlorobenzene, and DDVP, a phosphate insecticide, both of which are used mainly in aerosol sprays. D.D.T. is both cheaper and more effective than many of the newer carbamates and phosphates; the main selling point of carbamates is their safety and that of the phosphate poisons their ready disappearance by hydrolysis.

Monochlorobenzene reacts with chloral in the presence of sulfuric acid to form D.D.T.:

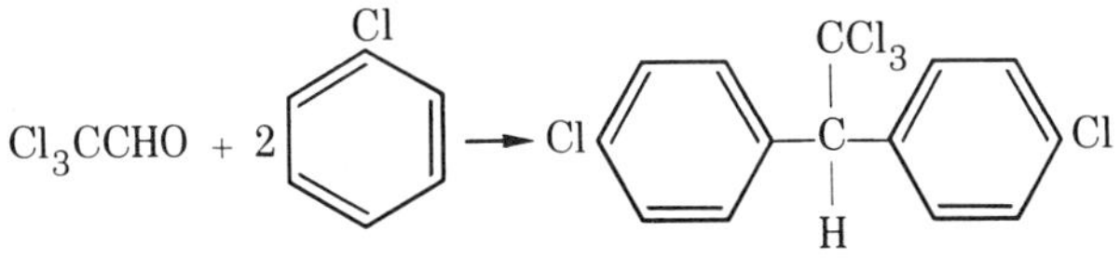

Like most chlorinated poisons, D.D.T. has the disadvantage of being chemically stable. Whereas, for example, phosphorodithiocarbamates hydrolyze easily, and thus disappear after having performed their function, chlorinated insecticides remain active for a long time afterwards. The popularity of D.D.T. in agriculture has been declining and now at least 60 percent of the D.D.T. produced in the U.S. is bought by the G.S.A. for allocation to the World Health Organization or UNICEF for malaria eradication in underdeveloped countries. The use of D.D.T., especially in the form of aerial sprays, is already forbidden in several states; also, immunities have developed in some of the insects against which D.D.T. is directed. Apart from its use as an agricultural or public health pesticide, D.D.T. finds limited use in the textile industry as a moth-proofing agent for wool.

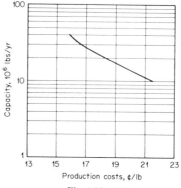

Fig. 5-20 DDT.

Investment for a 20.0 million pounds per year D.D.T. plant, including chloral and chlorobenzene facilities, is $3.6 million, and production costs are as shown below.

Raw materials	8.0
Chemicals	1.8
Utilities	1.4
Labor and overhead	1.2
Capital charges	6.0
	18.4

The selling price of D.D.T. has at times fallen below this figure, with the result that such major producers as DuPont and Monsanto have abandoned the field.

The domestic demand is steady at around 70.0 million pounds per year, most of it used against the boll weevils and the corn-borer. U.S. producers of D.D.T. are

Producer	Location	1965 Capacity, MM lbs/yr
Diamond	Greens Bayou, Tex.	30.0
Montrose	Torrance, Calif.	85.0
Olin	Huntsville, Ala.	25.0
Allied Chemical	Marcus Hook, Pa.	20.0
Geigy	McIntosh, Ala.	20.0
Lebanon Chemical	Lebanon, Pa.	15.0
Total		195.0

Source: *Chemical Week*, June 24, 1965.

4.6 Peracetic Acid

Union Carbide and Celanese have developed several important processes employing peracetic acid as an oxidant. The acid is made by air or oxygen oxidation of acetaldehyde to acetaldehyde monoperacetate and decomposition of the latter. The process takes place in an ethyl acetate medium, and ozone acts as the catalyst.

$$2CH_3CHO \longrightarrow O_2 \longrightarrow CH_3COOOC(OH)CH_3 \overset{\triangle}{\longrightarrow} CH_3COOOH + CH_3CHO$$

Peracetic acid is a powerful oxidant; its action is accompanied by reduction to acetic acid. This sequence therefore constitutes a way of making acetic acid from acetaldehyde obtaining simultaneously a valuable service. Peracetic acid can also be made from acetic and hydrogen peroxide; this procedure is much simpler, and less hazardous, but involves the use of an expensive oxidant.

Union Carbide has built a 120.0 million pounds per year peracetic plant at Taft, La. It is aimed at the market for epoxidation reactions. Epoxidation is still often carried out by oxidizing acetic acid to peracetic *in situ* via hydrogen peroxide.

$$CH_3COOH \xrightarrow{H_2O_2} CH_3COOOH$$

$$CH_3COOOH + \overset{\diagdown}{\underset{\diagup}{C}}=\overset{\diagup}{\underset{\diagdown}{C}} \longrightarrow \overset{\diagdown}{\underset{\diagup}{C}}\underset{\underset{O}{\diagdown\diagup}}{\text{—}}\overset{\diagup}{\underset{\diagdown}{C}} + CH_3COOH$$

Celanese has announced a propylene oxide process that bypasses propylene chlorohydrin:

$$CH_3CH{=}CH_2 + CH_3COOOH \longrightarrow CH_3\underset{\underset{O}{\diagdown\diagup}}{CH{-}CH_2} + CH_3COOH$$

This company also employs peracetic acid to make nylon 66 intermediates, via caprolactone and 1,6 hexanediol.

Union Carbide developed a caprolactam process that has the unique feature of producing no ammonium sulfate. The key step is oxidation by peracetic acid of cyclohexanone to caprolactone. The latter is also an intermediate for polymers and copolymers with ethylene oxide used in elastomers and polyurethane foams. In general, peracetic acid oxidation is a very interesting technique for companies that already produce acetyl chemicals; other producers would not know how to handle the byproduct acetic acid, which cannot be reoxidized to peracetic except by hydrogen peroxide. In the U.S., processes of this type are most likely to be used by Celanese, Union Carbide and Eastman.

5   ETHYLENE OXIDE

The two processes for manufacturing ethylene oxide are: 1) first, by reacting ethylene with hypochlorous acid, and dehydrochlorinating the resulting chlorohydrin with lime to form calcium chloride and ethylene oxide:

$$C_2H_4 + HOCl \longrightarrow HOCH_2CH_2Cl$$

$$2HOCH_2CH_2Cl + Ca(OH)_2 \longrightarrow 2CH_2CH_2 + CaCl_2 + 2H_2O$$
$$\underset{O}{\diagdown\diagup}$$

this is the older of the two processes; and 2) by direct ethylene oxidation:

$$C_2H_4 + \tfrac{1}{2}O_2 \longrightarrow CH_2{-}CH_2$$
$$\underset{O}{\diagdown\diagup}$$

The chlorination process has the advantage of an 85 percent yield on ethylene and investment requirements 30 percent lower than those for an equivalent oxidation plant. But these advantages are outweighed by a number of factors, which have completely displaced this process from new plant construction since direct oxidation became feasible. First, chlorine requirements are 2.0 pounds per pound of product; any firm choosing this route must consequently have a captive source of chlorine. Second, the process yields 1.3 pound of calcium chloride per pound product, 1.2 pound per pound of ethylene dichloride and .08 pound per pound of "Chlorex" (dichlorodiethyl ether). Whereas an outlet for ethylene dichloride can always be found, the only market for "Chlorex" was as an extraction solvent for lubricating oils. Calcium chloride actually represents a waste disposal problem. Third, the need to handle solids makes the process rather inconvenient.

The chlorination process received new impetus when polyethers based on propylene oxide began to be used to make polyurethane foams. Since no process for making propylene oxide by direct oxidation had been commercialized by 1965, most existing chlorohydrin plants were converted profitably to propylene oxide just when it seemed that the oxidation routes to ethylene oxide would force them to shut down altogether.

There are three major oxidation processes for ethylene oxide. All three use a silver catalyst; two of the three use air oxidation, the Shell process being the only one to employ oxygen. Yields for all three are around 70 percent. The Union Carbide process is used only by the company itself, the world's first and still largest producer of ethylene oxide and of its main derivatives; most other manufacturers use either Scientific Design or Shell technology. Shell uses oxygen, which has the advantage of reducing the size of the entire plant, although due to explosion hazards, the carbon dioxide formed by combustion of ethylene (the main side-reaction) is allowed to build up in the system to an appreciable concentration to reduce the ethylene partial pressure; on the other hand, an air plant or an external source of oxygen is required. In the Scientific Design process, which until 1964 was the more successful of the two, oxidation is with air, with nitrogen acting as the diluent. In either case, reaction temperature control is of extreme importance and is achieved using an organic coolant, mostly a close-boiling petroleum cut. In general, catalysts contain only 10 percent silver; if more were to be used yields would be higher, but these would be offset by higher catalyst losses due to more frequent servicing. In 1965, Deutsche Erdoel announced the development of a new catalyst that solves this problem. A flowsheet of the Shell process appears in Fig. 5-21. U.S. producers of ethylene oxide are given below.

Producer	Location	1965 Capacity, MM lbs/yr	Process
Allied Chem.	Orange, Tex.	45	SD
Calcasieu Chem.	Lake Charles, La.	90	Shell
Dow	Freeport, Tex.	425	UCC air oxidation
	Plaquemine, La.	150	Chlorohydrin
General Aniline	Linden, N.J.	65	SD
Houston Chem.	Beaumont, Tex.	80	SD
Jefferson	Port Neches, Tex.	290	SD
Olin	Brandenburg, Ky.	90	Shell
Sun-Olin	Claymont, Del.	65	Shell
Texas Eastman*	Longview, Tex.	60	Shell
Union Carbide	Institute, W.Va.	250	UCC air oxidation
	Seadrift, Tex.	400	UCC air oxidation
	S. Charleston, W.Va.	60	
	Ponce, Puerto Rico	80	UCC air oxidation
	Texas City, Tex.	220	UCC air oxidation
	Torrance, Calif.	50	UCC air oxidation
	Whiting, Ind.	150	UCC air oxidation
Wyandotte*	Geismar, La.	185	Shell
Total		2,755	

*In operation by 1966.
Source: Based on *Chemical Week*, April 5, 1966; *Oil, Paint & Drug Reporter*, January 31, 1966.

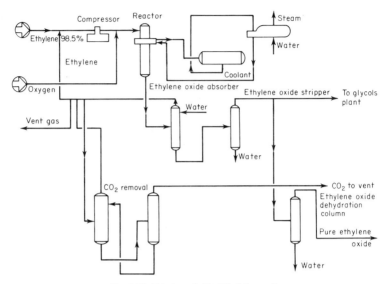

Fig. 5-21 Ethylene Oxide (Shell Process).

The 1965 U.S. demand distribution for ethylene oxide can be estimated as follows:

End-use	MM lbs, 1965
Ethylene glycol	1,310.0
Polyglycols	175.0
Glycol ethers	165.0
Surface active agents	220.0
Ethanolamines	150.0
All other	160.0
Total	2,180.0

Source: Based on *Chemical Week*, April 5, 1966;
Oil, Paint & Drug Reporter, January 31, 1966.

This use pattern, with ethylene glycol predominating so markedly, is peculiar to the U.S. where glycol automotive antifreeze has gained widespread acceptance.   In countries with less severe winters, ethylene glycol represents a much smaller share of the total ethylene oxide market. This has even inhibited the building of plants in temperate-climate countries, where the producer would be forced from the start to develop a complete line of derivatives instead of counting on a single product as his major outlet as is the case with several U.S. producers.

The overall growth rate for ethylene oxide up to 1970 is estimated to be 3.5 percent per year. This modest rate is due to the fact that most of the markets are saturated, except perhaps for surfactants based on ethylene oxide which are expected to grow at the expense of sulfonated alkyl aryl compounds.

Investment for a 60.0 million pounds per year oxidation plant using the Shell process is $2.5 million, exclusive of the oxygen plant. An SD-process plant of the same capacity would cost $4.8 million. Production costs, based on ethylene at 4.0 cents per pound and captive oxygen at $15 per ton, can be estimated as follows for the two processes.

	Process:	Cost, ¢/lb ethylene oxide	
		Shell	SD
Ethylene		3.8	3.8
Oxygen		1.0	—
Utilities, catalyst and chemicals		0.8	0.8
Labor and overhead		0.7	0.7
Capital charges		1.4	2.6
Total		7.7	7.9

Thus the two processes can be seen to be practically equivalent, the difference in capital charges in favor of the Shell process (exclusive of an eventual air plant) compensated by the additional cost of buying or producing oxygen. Economics at higher capacities, however, appear to favor air oxidation.

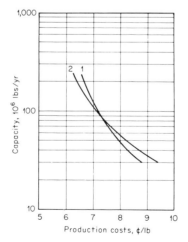

Fig. 5-22 Ethylene Oxide. 1-Shell Process; 2-SD Process.

5.1   Ethylene Glycol

Ethylene glycols are produced by the high-pressure, high-temperature hydration of ethylene glycol:

$$CH_2\!\!-\!\!CH_2 + H_2O \longrightarrow HO(CH_2CH_2O)_nH$$

Figure 5-22 shows a flowsheet for a glycol plant. Higher glycols obtained represent 8 to 9 percent of the monoethylene glycol output, but, since the combined higher glycol market is about 20 percent of that for ethylene glycol, producers actually have an incentive to increase the oxide-to-water molar ratio and produce more of the higher homologs. The capacities of U.S. producers expressed in terms of monoethylene glycol are given below.

Producer	Location	1965 Capacity, MM lbs/yr
Allied Chem.	Orange, Tex.	35
Atlas	Atlas Point, Del.	10
Calcasieu Chem.	Lake Charles, La.	110
Dow	Freeport, Tex.	350
	Plaquemine, La.	75
DuPont	Belle, W.Va.	150
Eastman	Longview, Tex.	50*
GAF	Linden, N.J.	35
Houston Chem.	Beaumont, Tex.	100
Jefferson	Port Neches, Tex.	360
Olin	Brandenburg, Ky.	110
Union Carbide	Institute, W.Va. Ponce, Puerto Rico Seadrift, Tex. S. Charleston, W.Va. Texas City, Tex. Torrance, Calif. Whiting, Ind.	880
Wyandotte	Geismar, La.	90**
Total		2,355

*On stream early 1966.
**30 tons/yr capacity on stream early 1966.
Source: *Oil, Paint & Drug Reporter*, February 7, 1966.

In the U.S. the bulk of ethylene glycol goes into permanent-type automotive antifreeze. Until recently, producers tried to induce the public to change back and forth from water to glycol antifreeze every year. Increased competition, however, prompted some

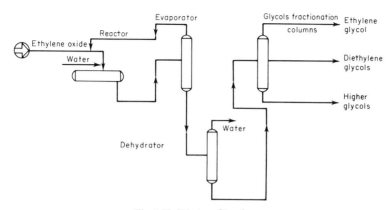

Fig. 5-23 Ethylene Glycols.

producers to promote the use of a product that was both dilute and sufficiently noncorrosive to be used all year. Despite a natural interest on the part of producers to convince motorists to purchase new antifreeze every year, it can be foreseen that the demand for ethylene glycol in the U.S. will soon taper off or even decline. Other trends are at work in the same direction, for example, smaller cars, air-cooled engines, research in the design of automobiles driven by air-cooled gas turbines, etc. On the other hand, ethylene glycol is being promoted increasingly as a coolant for nonautomotive applications, e.g., construction and farm equipment, stationary engines, air-conditioning, and, in general, applications where the much more corrosive brine cooling systems have traditionally been used.

The Food and Drug Administration has ruled that ethylene glycol is too toxic as a cellophane plasticizer wherever the film comes into contact with food. Most of this market has been taken over by glycerine and propylene glycols, which are nontoxic.

Ethylene glycol dinitrate is used in low-freezing dynamite. But much of this market has been taken over by ammonium nitrate.

Finally, the most dynamic market for ethylene glycol is terephthalate polyester fibers and film. The use pattern for ethylene glycol in 1965 is given below.

End-use	MM lbs, 1965
Antifreeze and other coolants	1,310.0
Polyester fibers	110.0
Export	200.0
Other uses	220.0
Total	1,840.0

Source: *Oil, Paint & Drug Reporter*, February 2, 1966.

The investment for an 80.0 million pounds per year glycol plant is $0.8 million, and production costs for ethylene glycol, assuming ethylene oxide is available at 8.5 cents per pound, are as shown below.

	Cost, ¢/lb ethylene glycol
Ethylene oxide	6.3
Utilities	0.2
Labor and overhead	0.2
Capital charges	0.3
Total	7.0

Glyoxal

Union Carbide, the only U.S. producer, makes glyoxal by catalytic oxidation of ethylene glycol.

$$HOCH_2CH_2OH \xrightarrow{[O]} \underset{\substack{\| \quad \| \\ O \quad O}}{HC-CH}$$

The demand for glyoxal is almost entirely for one use in permanent press fabrics. This developed from 500,000 pounds in 1964 to 12,000,000 pounds in 1965. The original application of glyoxal was for shrink-proofing rayon, the "Sanforset" process. It is also used to impart wet-strength to paper towels, in the manufacture of photographic gelatine, leather processing and other minor uses. The spectacular rise in demand was due to the role of glyoxal in the manufacture of certain reactant permanent-press textile resins. These resins are used to impregnate the material which is then cut and sewn into the finished garment before the resin is cured. The main derivative of glyoxal is dihydroxydimethylol urea (DHEU), used in the process widely licensed by Koratron:

$$2CH_2O + \underset{\substack{\| \quad \| \\ O \quad O}}{HC-CH} + (NH_2)_2CO \longrightarrow \begin{array}{c} HOH_2-C-N\text{————}CHOH \\ \diagdown \qquad | \\ C=O \\ \diagup \qquad | \\ HOH_2C-N\text{————}CHOH \end{array}$$

The expected growth for this product is such that in 1966, Union Carbide expanded its capacity to 150.0 million pounds per year.

5.2 Higher Glycols

Diethylene glycol finds its main applications in natural gas dehydration and in the Udex process for extracting from catalytic reformate, which is responsible for most of the petroleum aromatics capacity in the U.S. Actually, mono- and triethylene glycols are also used in natural gas drying; diethylene glycol is a compromise between the lower oil solubility of ethylene glycol and the lower volatility of triethylene glycol. There is a trend towards lower temperatures in gas processing, which in turn means more stringent dryness requirements and thus a more intensive use of glycols per unit of gas throughout. As natural gas continues to replace fuel oil for space heating, and as the incremental demand for aromatics is met from petrochemical sources, these uses will continue to expand.

Diethylene glycol is also used as a binder solvent in printing inks and as a solvent for several classes of textile dyes.

Triethylene glycol is employed in vapor disinfecting systems for public buildings. Fatty acid esters of polyethylene glycols are wetting agents for textiles; consumption of these is around 6.0 million pounds per year. Higher glycols are used as rubber, PVC lubricants and demolding agents, in hair cream, hydraulic fluids and as replacements for natural gums. Those with molecular weights of 400 and 1500 are the most common.

Higher ethylene glycols, with their two hydroxyl groups, can be used in unsaturated polyester resin recipes; they impart better flexibility than propylene glycol. The various homologs are also used in a large number of applications, e.g., solvents, coupling agents, humectants, leveling agents, etc. The production of higher ethylene glycols in 1965 is given below.

Compound	MM lbs, 1965
Diethylene glycol	160.0
Triethylene glycol	50.0
Polyethylene glycol	40.0
Total	250.0

The expected growth rate for these products until 1970 is about 7 percent per year.

5.3 Glycol Ethers

Glycol ethers are the reaction product of one or more mols ethylene oxide with an alcohol. The two examples below are typical for the whole product family.

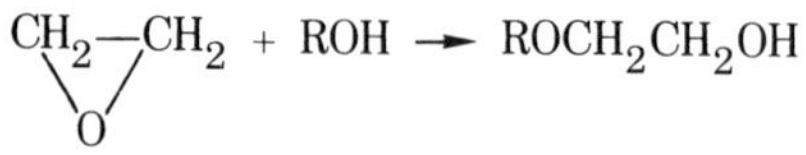

$$CH_2-CH_2 + ROH \rightarrow ROCH_2CH_2OH$$
$$\diagdown O \diagup$$

$$ROCH_2H_2OH + C_2H_4O \rightarrow ROCH_2CH_2OCH_2CH_2OH$$

Yields are around 95 percent.

The capacities of U.S. producers are shown below. It should be pointed out that the names "Cellosolve" and "Carbitol" by which ethylene glycol and diethylene glycol ethers, respectively, are universally known, are actually Union Carbide trademarks; the dominant position of this firm in ethylene oxide carries through to its immediate derivatives.

Producer	1964 Capacity, MM lbs/yr
Union Carbide	250.0
Dow	80.0
Olin	20.0
Jefferson	30.0
Total	380.0

Source: *Oil, Paint & Drug Reporter,*
May 4, 1964.

In 1965, production of all glycol ethers reached 230.0 million pounds, distributed as follows

Product	MM lbs, 1965
Ethylene glycol monoethyl ether (EGMEE)	74.0
Ethylene glycol monomethyl ether (EGMME)	74.0
Ethylene glycol monobutyl ether (EGMBE)	31.0
Diethylene glycol ethers	45.0
Triethylene glycol ethers	6.0
Total	230.0

Source: Based on U.S. Tariff Commission, "Synthetic Organic Chemistry."

The principal use for glycol ethers as a group is in thinners and coatings, and is due to their excellent solvent powers, coupling action and lack of odor.

Roughly 50 percent of all EGMEE ends up as "Cellosolve" acetate, a popular high-boiling solvent for surface coatings. Two-thirds of the EGMME is used in an 87:13 mixture with glycerol as an additive for military jet fuel, which has the double function of preventing ice formation and fighting bacterial corrosion.

More recently, this ratio was 98:2 of monomethyl ether and glycerine.   Dimethoxyethyl phthalate is a plasticizer for vinyl acetate emulsion, which consumes 5.0 million pounds per year EGMEE.   EGMBE is an intermediate for tri-(butoxyethyl) phosphate, a plasticizer for polystyrene-based floor-polish formulations.

Diethylene glycol monobutyl ether is an intermediate for piperonyl butoxide, a synergist for pyrethrum and allethrin preparations, and also for "Lethane," an insecticide developed by Rohm & Haas for household and dairy space-sprays.   The higher glycol ethers function as diluents for viscosity control in brake fluids and constitute about 50 percent of the 95.0 million pounds per year of these fluids consumed in the U.S.   They are also used in paint and floor-polish removers, as heat-stabilizers for PVC, and have captured most of the market for printing ink solvents that was once held by ethylene glycols.   The total domestic demand for glycol ethers can be broken down as follows:

End-use	MM lbs, 1965
Surface coatings	78.0
Brake fluids	48.0
Military jet-fuel additive	43.0
Plasticizers	8.0
Printing inks	15.0
Cleaning compounds	8.0
Cutting oils	2.0
Miscellaneous uses	28.0
Total	230.0

5.4 Ethanolamines

Ethanolamines are produced by reacting one, two or three mols of ethylene oxide with ammonia:

$$CH_2\!-\!CH_2 + NH_3 \longrightarrow HOCH_2CH_2NH_2$$
$$\diagdown\!\diagup$$
$$O$$

$$2CH_2\!-\!CH_2 + NH_3 \longrightarrow (HOCH_2CH_2)_2NH$$
$$\diagdown\!\diagup$$
$$O$$

$$3CH_2\!-\!CH_2 + NH_3 \longrightarrow (HOCH_2CH_2)_3N$$
$$\diagdown\!\diagup$$
$$O$$

A typical flowsheet for an ethanolamines plant is shown in Fig. 5-24.

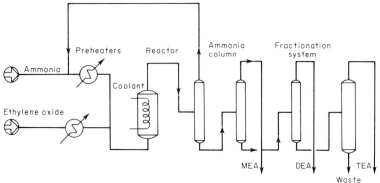

Fig. 5-24 Ethanolamines.

Liquid ethylene oxide reacts with aqua-ammonia in a cooled reactor at 85° F-100° F and a pressure that varies, according to the process employed, from essentially atmospheric to 1500 psi.   Reactor effluent is sent first to a stripping section where ammonia and water are removed and the former recycled. Then the product stream is sent to a vacuum rectifying system, where MEA, DEA, and TEA are fractioned into pure products. The distribution of the three amines is controlled by varying the mol-ratio of the two raw materials in the reactor. Since the relative demand for the three homologs has changed more than once within the last few years, the fractionation system must have sufficient flexibility to handle varying amounts of the three products. U.S. plants and capacities are given below.

Producer	Location	1965 Capacity, MM lbs/yr
Allied Chem.	Orange, Tex.	15.0
Dow	Midland, Mich.	40.0
	Freeport, Tex.	20.0
Jefferson	Port Neches, Tex.	60.0
Union Carbide	Seadrift, Tex.	100.0
	S. Charleston, W.Va.	20.0
Total		255.0

Source: *Oil, Paint & Drug Reporter*, May 23, 1966.

In 1965, production of the three ethanolamines was just over 200.0 million pounds distributed as follows:

Product	MM lbs/yr, 1965
Monoethanolamine (MEA)	67.0
Diethanolamine (DEA)	78.0
Triethanolamine (TEA)	56.0
Total	201.0

Source: U.S. Tariff Commission, "Synthetic Organic Chemicals."

This demand has been growing at 11 percent per year.

Monoethanolamine

The demand for monoethanolamine (MEA) in the U.S. reached 67.0 million pounds in 1965. MEA is still the most widely used scrubbing liquid for removing acid constituents from gas streams, despite increased competition from compounds such as sulfolane, hot potassium carbonate, or propylene carbonate, which require less steam and have other advantages.

Alkanolamides made from MEA and coconut fatty acids are effective foam-builders and stabilizers for alkylarylsulfonates.

Both MEA and DEA are intermediates for emulsifiers used in the textile industries as antistatic and mothproofing agents, as cleaners and in bleaching baths.

Jefferson Chemical produces piperazine from monoethanolamine from the hydrochloride of MEA.

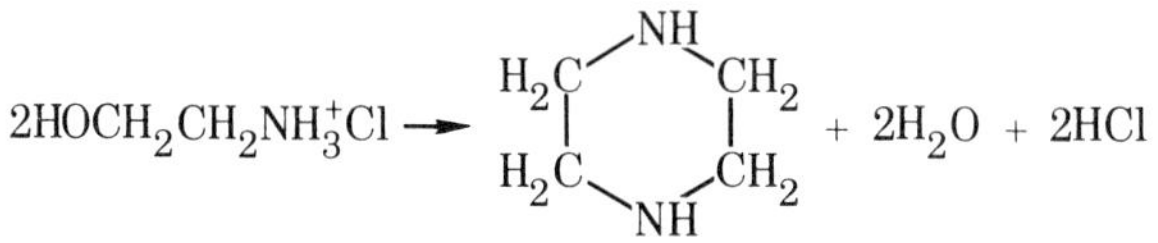

$$2HOCH_2CH_2NH_3^+Cl \rightarrow \underset{\substack{}}{\overset{NH}{\underset{NH}{\bigcirc}}} + 2H_2O + 2HCl$$

It can also be produced from ethylene dichloride and ammonia, or as a byproduct in the production of ethylene diamine from ethylene dichloride and ammonia.

$$\left.\begin{array}{l} 2CH_2ClCH_2Cl + 2NH_3 \\ NH_2CH_2CH_2NH_2 + CH_2ClCH_2Cl \end{array}\right\} \rightarrow \underset{\substack{}}{\overset{NH}{\underset{NH}{\bigcirc}}} + 2HCl$$

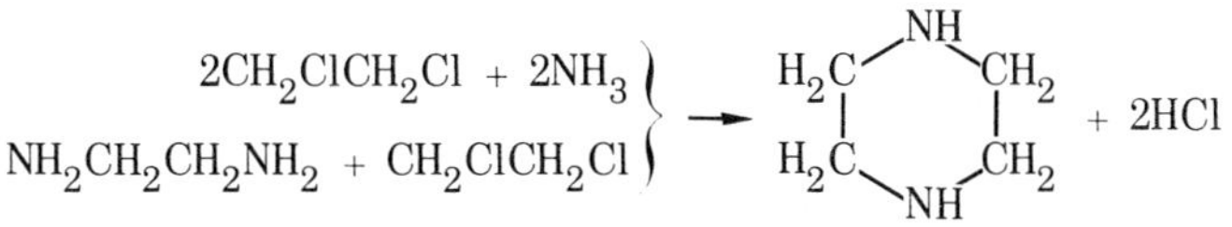

In the form of its citrate or other salts, piperazine has widely replaced the formerly used dialkyl tins as a swine and poultry antihelminthic. The demand in the U.S. is around 3.0 million pounds per year. Jefferson, Union Carbide and Dow are the largest U.S. producers.

In Japan, a new type of polyazide fiber, a copolymer of piperazine, caprolactam and terephthalic acid, has been developed. It is reported to have better water absorption and dyeing characteristics than either nylon or polyester fibers.

Two interesting derivatives are MEA sulfite, developed as a crease-proofing treatment for wool, and ethoxyethyl amines, intermediates for petroleum emulsion breakers used in API separators. The latter are made by ethoxylating MEA in such a way that reaction takes place at the -OH instead of at the -NH group.

Diethanolamine

The demand for DEA in 1964, was 78.0 million pounds.

Lauric acid diethanolamide is a foam-building constituent of liquid detergents. Despite the fact that tertiary amine oxides have already replaced alkanolamides in some of Procter & Gamble's formations and that low-foaming liquids are the fastest-growing type of detergent, total consumption of alkanolamides in 1965 was 85.0 million pounds. The major users are Lever and Procter & Gamble. DEA is used in the gas-treating of streams rich in sulfur compounds, since MEA is degraded by carbonyl sulfide. DEA can also be dehydrated to morpholine and is the major source of this material.

$$HN(CH_2CH_2OH)_2 \rightarrow \begin{array}{c} H_2C-CH_2 \\ HN \qquad O \\ H_2C-CH_2 \end{array}$$

The fatty acid salts of morpholine are used as corrosion inhibitors, especially in boiler feed water and, increasingly as emulsifiers in floor-polish formulations. 2-Benzothiazyl morpholine sulfonamide is a rubber accelerator marketed among others by American Cyanamid under the trade name "NOBS." The morpholine demand was around 20.0 million pounds per year in 1966. 2,4-D acid is solubilized by salt formation with amines, e.g., DEA. DEA, as morpholine, is also an intermediate for certain optical bleaches used in the textile industry.

Triethanolamine

Domestic consumption of TEA in 1964, was 56.0 million pounds. TEA is used in cosmetic formulations both as such and in the form of its fatty acid salt. These are also used in textile surfactants. TEA is also used in many other applications, such as corrosion inhibitors, water-proofing agents, etc. The combined end-use pattern for ethanolamines is shown below.

End-use	MM lbs, 1965
Surfactants	54
Gas treating	37
Textile chemicals	20
Cosmetics	17
Miscellaneous (includes morpholine)	40
Export	32
Total	200

Source: *Oil, Paint & Drug Reporter*, May 23, 1966; September 28, 1964.

The main differences in the consumption pattern of ethanolamines in U.S. and other countries are, first, that use of MEA is unusually high because of the importance of natural gas within the energy resource structure of the country, and second, that in other economies the use of liquid detergents is still not as widespread as in the U.S., thus relatively less DEA is used. In less-developed countries, TEA tends to be the most important of the three ethanolamines; this was actually also the case in the U.S. until around 1950.

A 12.0 million pounds per year ethanolamines plant requires an investment of $1.8 million. Production costs for DEA were cal-

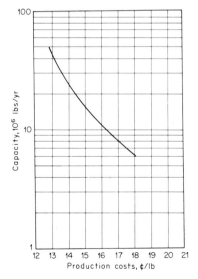

Fig. 5-25 Diethanolamine.

culated assuming the producer has captive ethylene oxide at 8.5 cents per pound, and buys ammonia for 4.0 cents per pound.

	Cost, ¢/lb DEA
Ethylene oxide	7.4
Ammonia	1.1
Utilities	2.0
Labor and overhead	1.0
Capital charges	4.5
Total	16.0

5.5 Nonionic Detergents

The reaction products of several mols of ethylene oxide with compounds containing an active hydrogen molecule constitute an important class of surface active agents:

$$
\left.\begin{array}{l} RNH_2 \\ ROH \\ RSH \\ RArOH \end{array}\right\} + n \underset{O}{CH_2-CH_2} \longrightarrow \left\{\begin{array}{l} RNH(CH_2CH_2O)_nH \\ RO(CH_2CH_2O)_nH \\ RS(CH_2CH_2O)_nH \\ RArO(CH_2CH_2O)_nH \end{array}\right.
$$

Since the early 1960's, a number of factors have caused the demand for nonionic surfactants to expand rapidly, especially the derivatives of fatty or synthetic long-chain alcohols. With the increased preference for liquid detergents, the demand for nonionic surfactants began to grow faster than that for detergents as a whole. Most important, pollution and biodegradability became such a political issue that detergent manufacturers were forced to look for materials that would break down faster under the action of oxygen than conventional alkylaryl sulfonates or nonionic detergents based on alkylphenols. Alcohol-based nonionics have the double advantage of producing little foam, which constitutes visible evidence of the presence of detergents in water streams, and of being more biodegradable than most major detergent raw-materials.

Because of their low foaming properties, nonionics are used widely in industrial applications: for example, textile aids, pesticide emulsifiers, emulsion polymerization, cosmetics, etc. Industrial outlets account for 60 percent of the demand for nonionics. The total consumption of nonionic surface active agents in 1965 was 531.0 million pounds broken down as follows by type of compound:

Type of surfactant	MM lbs, 1965
Nonethylene oxide-derived:	
Alkanolamides	86.0
Ethylene oxide-derived:	
Alkylphenol adducts	184.0
Alcohol adducts	171.0
Tall oil and fatty esters	30.0
Sulfur- and nitrogen-containing adducts, etc.	70.0
Total	541.0

Source: U.S. Tariff Commission and estimate.

5.6 Other Ethylene Oxide Derivatives

Ethylene oxide can be polymerized to form a water-soluble film sold by Union Carbide under the trade-name "Hylox." In 1965, use of this film was estimated at 0.5 million pounds, mainly in prepackaged single doses of bleaches.

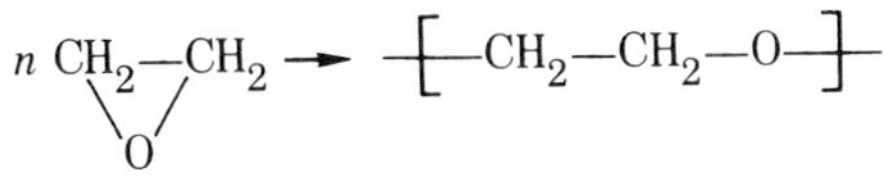

$$n\ CH_2\!-\!CH_2 \longrightarrow \left[\!-CH_2\!-\!CH_2\!-\!O\!-\!\right]$$

Reaction under high pressure between ethylene oxide and carbon dioxide yields ethylene carbonate:

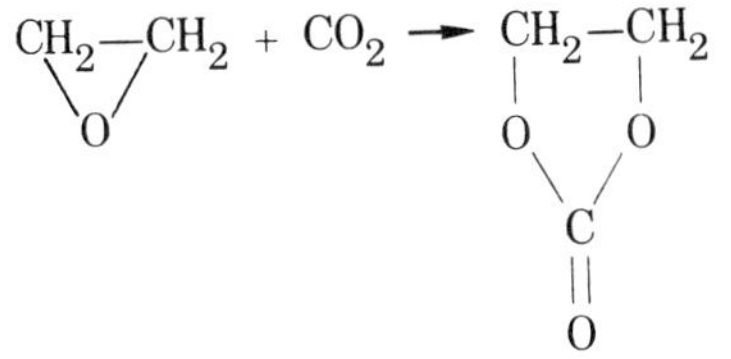

$$CH_2\!-\!CH_2 + CO_2 \longrightarrow \begin{array}{c} CH_2\!-\!CH_2 \\ |\quad\ | \\ O\quad O \\ \diagdown\ \diagup \\ C \\ \| \\ O \end{array}$$

Celanese has reported a process for direct esterification of terephthalic acid using ethylene carbonate as a combination solvent and hydroxyl-group source. It is also a solvent for general use.

Ethylene oxide reacts with sodium bisulfite to form sodium isethionate, an intermediate for the production of n-alkyl taurines. About 5.0 million pounds per year of ethylene oxide are used in this application.

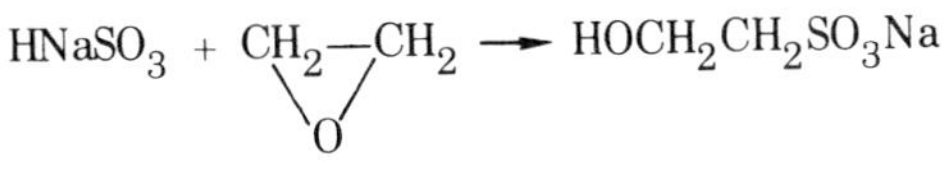

$$HNaSO_3 + CH_2\!-\!CH_2 \longrightarrow HOCH_2CH_2SO_3Na$$

Trichloroethyl phosphate and higher homologs are made by reacting ethylene oxide with $POCl_3$. They are used as polyols for making flame-retardant rigid polyurethane foams. General research is now being conducted for reactant fire retardants to make rigid urethanes sufficiently fire-proof for most U.S. building codes, and trichloroethyl phosphate alone has been estimated to have a potential market of 6.0-8.0 million pounds per year.

Monsanto makes a family of flame-retardant plasticizers known as "Phosgards" by reacting ethylene oxide or propylene oxide with PCl_3 and an aldehyde or ketone. Their main application is also in rigid urethane foams.

Hydroxyethyl cellulose is a synthetic water-soluble resin, made by ethoxylating cellulose to an ethylene oxide content of about 75 percent. Producers and their capacities are as follows:

Producer	Capacity, MM lbs/yr
Hercules	6.0
Union Carbide	2.0

In 1965, about 5.0 million pounds were used in a variety of applications, for example, paint thickeners, emulsifiers, adhesives, paper sizing, ceramics, etc.

Phenylethyl alcohol is made by the Friedel-Crafts reaction between ethylene oxide and benzene:

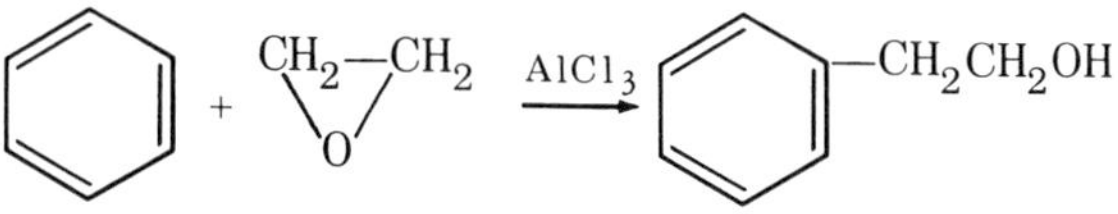

About 1.8 million pounds per year are consumed. It is used in perfume formulations, mainly in the form of its acetic ester, as a relatively cheap source of rose odor; di-(phenylethyl-ether), a plasticizer for polyvinylidene chloride, is obtained as a byproduct.

Ethylene chlorohydrin is the starting point for making dichloroethyl formal, one of the raw materials for polysulfide elastomers

$$2ClCH_2CH_2OH + CH_2O \longrightarrow CH_2(OCH_2CH_2Cl)_2 + H_2O$$

and for chloroethyl vinyl ether

$$ClCH_2CH_2OH + C_2H_2 \longrightarrow ClCH_2CH_2-O-CH=CH_2$$

a constituent, together with ethyl acrylate, of polyacrylate elastomers.

Ethylene oxide is used to ethoxylate reactive hydrogen atoms on raw material for disperse dyes that once dominated the field for cellulose acetate dyeing. However, this field is being taken over by reactive dyes. At least two producers are promoting the use of ethylene oxide as a fumigant; it has the advantages of leaving no odor, high penetrating power and low cost. Copolymers with propylene oxide find use as synthetic lubricants, heat-transfer agents, brake fluids, textile antistatic agents and fiber lubricants, and rubber processing aids. Ethoxylated polyvinyl alcohol derivatives are used as plasticizers.

Ethylenc oxide is used in the manufacture of viscose modifiers for high-modulus rayon; it is used also as a modifier for making polyacetal resins of the copolymer type, for example, "Celcon" (Celanese).

6   ETHYLENE DICHLORIDE

Ethylene dichloride (1,2 dichloroethane) can be made from ethylene ether by chlorination:

$$Cl_2 + C_2H_4 \longrightarrow CH_2ClCH_2Cl$$

or by oxychlorination:

$$2HCl + \tfrac{1}{2}O_2 + C_2H_4 \longrightarrow CH_2ClCH_2Cl + H_2O$$

The investment for a 170.0 million pounds per year ethylene chlorination plant, sufficient to feed a 100.0 million pounds per year vinyl chloride unit, is $1.0 million. Production costs are as follows, assuming ethylene at 4.0 cents per pound and captive chlorine at 2.50 cents per pound

	Cost, ¢/lb EDC
Ethylene	1.2
Chlorine	1.8
Utilities	0.1
Labor and overhead	0.1
Capital charges	0.2
Total	3.4

U.S. producers and their 1965 capacities are:

Producer	Location	1965 Capacity, MM lbs/yr	Oxychlorination unit
American Chem.	Watson, Calif.	150	
B.F. Goodrich	Calvert City, Ky.	600	x
Diamond Alkali	Deer Park, Tex.	90	
Dow	Freeport, Tex.	300	x
	Plaquemine, La.	200	x
Ethyl Corp.	Baton Rouge, La.	350	
	Houston, Tex.	140	
Monsanto	Texas City, Tex.	165	
Olin	Brandenburg, Ky.	100	
Pittsburgh Plate Glass	Lake Charles, La.	200	x
Union Carbide	S. Charleston, W.Va.	125	x
	Texas City, Tex.	150	
Jefferson	Port Neches, Tex.	70	
Total		2,640*	

*Ethylene dichloride plants, primarily using oxychlorination are being added very rapidly. Thus, these numbers, although correct, may appear low.
Source: *Oil, Paint & Drug Reporter*, November 9, 1964.

Domestic production of ethylene dichloride in 1964 was 2,200 million pounds, broken down as follows:

End-use	MM lbs/yr, 1964
Vinyl chloride	1,540.0
Lead scavenger (antiknock fluid)	220.0
Ethylene amines	90.0
Chlorinated solvents and others	220.0
Exports	130.0
Total	2,200.0

Source: *Oil, Paint & Drug Reporter*, November 9, 1964.

As a lead scavenger, ethylene dichloride is not as effective as ethylene dibromide, but much cheaper. Its use is limited to automotive fuel, dibromide being used alone in aviation gasoline; the practical disappearance of the aviation gas market has therefore not affected the demand for ethylene dichloride as an antiknock fluids constituent. On the other hand, the average tetraethyl lead concentration in the U.S. gasoline pool has been declining steadily, as refinery economics have shifted towards obtaining incremental octane numbers in other ways.

Most trichloroethylene is still made from acetylene; however, ethylene, via ethylene dichloride is an important starting point for making perchloroethylene.   EDC is also used as a solvent and in grain fumigation formulations.

6.1 Vinyl Chloride

Although polyethylene has tended to become the most important plastic in highly industrialized nations, in less-advanced economies this position is held by polyvinyl chloride. As noted earlier, the polyethylene demand in U.S. is largely a result of the trend toward prepackaging commodities and goods that, in less developed countries, are still merchandized by more labor-intensive methods. On the other hand, PVC goes into more essential products such as household goods, flooring, piping or corrugated roofing. Another factor that has an influence on the relative positions of polyethylene and polyvinyl chloride is the degree of maturity of the nation in question as distinct from simple per-capita income considerations. Although the applications of plasticized PVC usually compete with natural raw materials — flooring with wood, automobile upholstery with leather, household goods with textiles — rigid PVC competes with traditional industrial products such as steel, brick and tile, solidly established in the developed countries since the Industrial Revolution. The considerable interests vested in these activities of long standing have somewhat inhibited expansion of the demand for PVC. These interests, for example, are reflected in building codes that restrict the use of PVC piping, or in labor-union pressure against the application of PVC roofing or siding; and it is therefore not unexpected that the four areas in the table below should appear listed with respect to demand distribution between rigid and plasticized PVC approximately in order of the time in history when industrialization began.

Country	% Rigid PVC
U.S.	10
Great Britain	21
Common Market	32
France	
Germany	
Italy	
Japan	48

Source: Olivier in talk before Chemical Marketing Research Association, May 1966.

Historically, vinyl chloride is definitely an acetylene derivative, and as late as 1965, around 45 percent of all vinyl chloride produced in the U.S. was still based on acetylene, derived from either hydrocarbons or calcium carbide. In Europe and Japan, acetylene is still the most important starting point for vinyl chloride, although there is a trend toward ethylene; but in the U.S., ethylene is already the major raw material.

The earliest route to vinyl chloride was from acetylene and hydrochloric acid. Acid was obtained either by combining hydrogen and chlorine or as a byproduct from other organic chlorinations. Acetylene was at first exclusively carbide-derived but later also obtained from hydrocarbon cracking:

$$C_2H_2 + HCl \longrightarrow CH_2{=}CHCl$$

As ethylene became more abundant, the next scheme to gain importance was cracking ethylene dichloride to vinyl chloride, recovering byproduct HCl and using it to make more vinyl chloride from acetylene:

$$CH_2ClCH_2Cl \xrightarrow{\Delta} CH_2{=}CHCl + HCl$$

$$C_2H_2 + HCl \longrightarrow CH_2{=}CHCl$$

With economics continually shifting away from acetylene, two schemes were devised that started entirely from ethylene and yet did not terminate with large amounts of byproduct hydrochloric acid. One is to oxidize the byproduct acid back to chlorine, electrolytically or by air; the other, stoichiometrically speaking, a variant of the first, is to react hydrochloric acid with ethylene directly in the presence of oxygen to obtain more ethylene dichloride. This last process is known as "oxychlorination."

$$C_2H_4 + 2HCl + \tfrac{1}{2}O_2 \longrightarrow CH_2ClCH_2Cl + H_2O$$

Finally, two ingenious processes based on naphtha and involving total chlorine utilization have been devised. In both, naphtha is cracked to an equimolar mixture of acetylene and ethylene. Then the mixture of the two reacts with hydrochloric acid which forms vinyl chloride with the acetylene contained in the cracked gases. In the Societé Belge de l'Azote process, this vinyl chloride, made from the acetylene in the cracked gas stream, is separated from ethylene by an independent solvent-extraction system. In the Kureha process, vinyl chloride is absorbed in ethylene dichloride, which is produced further downstream from the ethylene remaining in the gases and the net chlorine being fed to the system. As electric power becomes more expensive and chlorine becomes the principal instead of the secondary product in the caustic-chlorine balance, and with no natural gas to supply cheap hydrocarbon acetylene, these or similar naphtha-to-vinyl chloride routes may be able to compete with oxychlorination of ethylene for the incremental capacity, not only in Europe and Japan, but also in developing countries.

In the acetylene-based process, acetylene and hydrogen chloride react in a cooled tubular furnace containing a mercuric chloride catalyst. A stripping column removes the small amount of unreacted raw material (conversion is almost complete) and the stripper bottoms are fractionated into vinyl chloride and heavy byproducts. Overall yields are around 95 percent.

When starting with ethylene dichloride, the feed must be preheated since the cracking reaction is endothermic. After flashing off the hydrogen chloride, the reactor effluent is fractionated to heavy ends save for a bleed-stream being recycled to the cracking furnace since conversion per pass is around 55-60 percent. Yield is 92 percent on ethylene dichloride, and around 85 percent on ethylene since the ethylene-EDC step also gives yields of 92 percent. So-called "balanced" vinyl chloride operations are juxtapositions of these two processes.

Oxychlorination began receiving a great deal of attention in the early 1960's. Basically, three alternatives are in use. The Shell process consists of regenerating chlorine from HCl in a separate catalytic fluidized bed reactor containing a mixture of copper and other chlorides, at temperatures between 630-800°F which are moderate compared to the 1000°F or more required by the original Deacon process. This enables roughly the same "balanced" operation to be employed.

A second alternative is that used by most "oxychlorinators." It is employed to react ethylene, hydrogen chloride and air to produce ethylene dichloride without separate regeneration of chlorine.

A third alternative is that which would bypass both chlorine production and ethylene dichloride production. This is the oxychlorination of ethylene directly to vinyl chloride.

$$2CH_2 = CH_2 + 2HCl + O_2 \rightarrow 2H_2O + 2CH_2 = CHCl$$

This process has been the subject of numerous patents, and it was discussed extensively in the literature in 1964. Recent references omit this version of oxychlorination and thus it will not be discussed further here.

The higher yields on ethylene and chlorine obtained by separate regeneration of chlorine must be compared with the additional investment. The lower investment costs implied by direct oxychlorination are offset by the fact that some of the ethylene feed is burned to carbon dioxide, thus lowering yields. Numerous oxychlorination processes exist, varying around catalysts, reactor design, yields, ways of keeping the catalyst from volatilizing, materials of construction, etc. Reported ethylene dichloride yields are usually 95 percent on ethylene, and 90 percent on chlorine.   Stauffer, Goodrich, Monsanto, PPG and Dow in the U.S., ICI and Toyo Koatsu are among the vinyl chloride producers employing oxychlorination processes.

In addition to the two processes in which HCl reacts with unseparated acetylene, another possible "balanced" route could be used on an acetylene-ethylene plant, for example, one using the Wulff process; this would have the disadvantage with respect to the Kureha and SBA processes of requiring separation of ethylene from acetylene, which adds considerably to the cost of both.

Figure 5-26 shows block-flow diagrams for various routes to vinyl chloride monomer. The economics of vinyl chloride production depend on how the unit is to fit within the plant as a whole. Ethylene costs can range from 3.0 cents per pound for a very large captive producer to 4.5 cents per pound for ethylene acquired over-the-fence.   Acetylene from natural gas can cost a captive producer as little as 9 cents per pound; from a merchant supplier or from purchased calcium carbide it would cost 12 cents per pound or more.   Chlorine can cost between 2.0 and 3.0 cents per pound; and hydrochloric acid economics are even more complicated because in some plants it actually represents a waste-disposal problem.

The economics of various routes are compared below. Five cases are considered, all assuming a capacity of 100.0 million pounds per year:

1. from captive acetylene (10 cents per pound) and HCl (1.75 cent per pound);
2. from ethylene (4.0 cents per pound) and captive chlorine (2.5 cents per pound), with recovery of HCl (valued at 1.25 cent per pound);
3. "balanced" process, raw materials as in 1 and 2;
4. oxychlorination, all chlorine being used up;
5. Kureha process from naphtha (1.0 cent per pound) and captive chlorine.

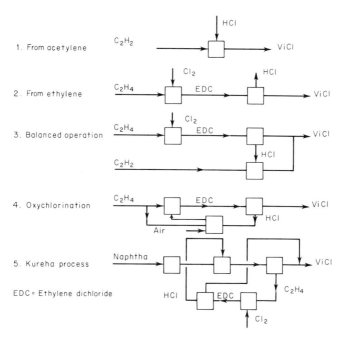

Fig. 5-26 Vinyl Chloride. Block-flow Diagrams.

			Case		
	1	2	3	4	5
			Investment, $MM		
	2.1	3.1	3.6	3.1	5.5
		Cost, ¢/lb vinyl chloride			
Naphtha					0.9
Ethylene		1.9	1.1	1.9	
Acetylene	4.2		2.3		
Chlorine		3.0	1.7	1.7	1.7
HCl	1.1	(-0.7)			
Oxygen					0.4
Catalyst and chemicals	0.3	0.6	0.6	0.4	0.4
Total raw materials	5.6	4.8	5.7	4.0	3.4
Utilities	0.4	0.4	0.4	0.4	1.0
Labor and overhead	0.3	0.4	0.5	0.5	0.6
Capital charges	0.7	1.0	1.2	1.0	1.8
Total	7.0	6.6	7.8	5.9	6.8

The advantage seems to lie in the exclusive use of ethylene as a raw material, and transforming all the chlorine to vinyl chloride

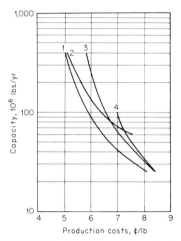

Fig. 5-27 Vinyl Chloride. 1-Oxychlorina-
tion; 2-Kureha process (from naphtha); 3-
ethylene (HCl recovery); 4-acetylene+HCl.

by oxychlorination, although at capacities around 500.0 million
pounds per year the more capital-intensive Kureha process may
become more attractive. U.S. vinyl chloride monomer producers,
capacities and routes for 1965 are given below.

Producer	Location	1965 Capacity, MM lbs/yr	Raw material	Captive chlorine or HCl
Allied Chem.	Moundsville, W.Va.	100	AC-C	x
American Chem.	Watson, Calif.	170	E-C	–
Air Reduction	Calvert City, Ky.	60	AC-C	–
Diamond Alkali	Houston, Tex.	100	AH-C; E:M	x
Dow	Freeport, Tex.	150	AH-C; E:M	x
	Plaquemine, La.	200	E-C	x
Ethyl Corp.	Baton Rouge, La.	190	E-M	x
	Houston, Tex.	110	E-M	x
General Tire	Ashtabula, Ohio	75	AC-M	–
Goodrich Chem.	Calvert City, Ky.	440	E-C	–
	Niagara Falls, N.Y.	40	AC-M	–
Goodyear Tire & Rubber	Niagara Falls, N.Y.	70	AC-M	–
Monochem	Geismar, La.	250	AH-C	–
Monsanto	Texas City, Tex.	150	AH-C; E-C	x
Tenneco	Houston, Tex.	220	AH-C	–
Union Carbide	S. Charleston, W.Va.	120	AC-C; E-C	–
	Texas City, Tex.	180	AH-C; E-C	–
Total		2,625		

AC = carbide acetylene; AH = hydrocarbon acetylene; E = ethylene; C = captive;
M = merchant.

Source: Based on *Oil, Paint & Drug Reporter*, October 11, 1965; *Chemical Week*, January 7, 1967.

Producers of carbide acetylene are located in the regions served by cheap TVA or Niagara Falls power. Ethylene and hydrocarbon acetylene users are mostly on the Gulf Coast. The proportion of manufacturers who have captive hydrocarbon sources is considerably larger than those who make their own chlorine. This is mainly because the latter can be transported whereas acetylene can only be shipped as calcium carbide and ethylene only at very high cost. With the exception of a few small miscellaneous uses the total vinyl chloride monomer demand in the U.S. of 2,000 million pounds in 1965 was utilized in polymer and copolymer production. These miscellaneous uses include application as an aerosol propellent and as an intermediate for the chlorinated insecticides "Isodrin" and "Endrin."

Vinyl Chloride Polymers and Copolymers

Most polyvinyl chloride is made either by emulsion or suspension polymerization. Emulsion polymerization consists of carrying out the reaction in an aqueous medium in the presence of a surface-active agent, the product being a latex of dispersed polymer. Suspension polymerization consists of dispersing the monomer in water by means of violent mechanical agitation, the polymer being obtained in the form of beads. Emulsion-polymerized PVC is recovered as a solid by drying in spray, rotary or flash driers, depending on the desired particle size; the beads obtained by suspension polymerization are centrifuged, washed, dried and finally ground to a powder. Suspension polymerization is carried out discontinuously; its advantages are that the polymer beads are practically free of auxiliary products that would interfere with electrical and other properties, lower chemicals and utility requirements and lower investment costs. Emulsion polymerization, being easier to carry out continuously, involves lower labor costs. The product is easier to plasticize and process due to the presence of emulsifying agents that act as lubricants.

More recently mass polymerization of vinyl chloride has begun to be used. In the U.S., for example, Hooker and Dow employ the Pechiney-St. Gobain two-stage process. It has the advantage of yielding a pure clear product and using no auxiliary chemicals. A comparison of the economics of emulsion and suspension polymerization costs is shown below for a 50.0 million pounds per year operation. Monomer is assumed at 6.5 cents per pound.

	Process	
	Suspension	Emulsion
	Investment, $MM	
	4.0	5.0
	Production costs, ¢/lb PVC	
Monomer	7.1	7.1
Utilities, catalyst and chemicals	0.8	1.4
Labor and overhead	1.2	0.7
Capital charges	2.6	3.3
Total	11.7	12.5

In the U.S., all but 10 percent of the total amount of polyvinyl chloride is used in plasticized form. In other countries, such unplasticized applications as pipe, corrugated roofing and residential siding make up a substantial part of the market — suffice it to say that in Japan consumption of plastic pipe is 2.5 times that of the U.S. The advantages of PVC are its flame-retardancy, good electrical properties and chemical resistance. On the negative side, PVC is susceptible to heat and ultraviolet light degradation, and therefore can be stabilized. Heat stabilizers can be of several types, the most important being lead, cadmium and salts of organic or inorganic acids for plasticized PVC and organotins for rigid PVC. Benzotriazoles and benzophenones are among the more satisfactory UV stabilizers. U.S. polyvinyl chloride producers and their capacities are listed below.

Producer	Location	1966 Capacity, MM lbs/yr
Allied Chem.	Painesville, Ohio	125.0
Air Reduction	Calvert City, Ky.	115.0
American Chem.	Long Beach, Calif.	30.0
B.F. Goodrich	Niagara Falls, N.Y. Louisville, Ky. Watson, Calif. Calvert City, Ky. Avon Lake, Ohio	460.0
Borden Chem.	Leominster, Mass. Illiopolis, Ill.	160.0
Tenneco	Flemington, N.J. Burlington, N.J.	190.0
Diamond Alkali	Deer Park, Tex.	100.0
Diamond	Delaware City, Del.	170.0
Dow Chem.	Midland, Mich.	85.0

Ethyl Corp.	Baton Rouge, La.	80.0
Firestone Tire & Rubber	Pottstown, Pa.	125.0
General Tire	Ashtabula, Ohio	60.0
Goodyear	Niagara Falls, N.Y.	80.0
Monsanto	Springfield, Mass.	150.0
Pantasote	Passaic, N.J.	105.0
Stauffer-Hoechst	Delaware City, Del.	45.0
Thompson Chem.	Aberdeen, Miss. Assonet Village, Mass.	250.0
Union Carbide Plastics	S. Charleston, W.Va. Texas City, Tex.	320.0
U.S. Rubber	Painesville, Ohio	135.0
Others		230.0
Total		3,015.0

Source: Based on *Chemical Week*, September 11, 1965.

Most end-uses of PVC are due to its flame resistance and chemical inertness. The most common means of processing PVC are calendering, extrusion, and molding. Blow-molding of PVC has just gained importance. The introduction of PVC into the disposable packaging materials market around 1964-65 touched off a wave of plant expansions. It is considered by some that the domestic demand for PVC will surpass that for low-density polyethylene and become the most important plastic in the U.S., as it already is in Japan and in many Western European countries. The main disadvantage of PVC in the field of disposable packaging is its high density, which makes it expensive on a volume or surface basis despite its low cost per unit weight.

The end-use statistics given below amalgamate pure polymer with copolymer (mainly with vinyl acetate) applications. In many PVC applications, use of vinyl acetate-vinyl chloride copolymers is strictly a matter of economics. Vinyl acetate is almost twice as expensive as vinyl chloride, but in copolymers it replaces a substantial proportion of the even more expensive plasticizers. The usual vinyl acetate concentration is 15 percent, thus the polymer retains its fire-retardant properties.

The end-use breakdown for vinyl chloride polymer and copolymer consumption in 1965 is given below. Calendering applications include most vinyl flooring (which is beginning to compete with carpeting in public buildings), vinyl-to-fabric laminates for use in upholstery, wall covering, automobile seat-covers and upholstery and numerous household applications such as wall covering, curtains, place-mats, etc. Extrusion is used to coat wire and cables,

to make garden hose and in numerous other fields such as medical
tubing, film and sheet and the various extruded rigid products —
pipe, siding, roofing, panels and other construction accessories,
the market for which is only just beginning to grow rapidly in the
U.S. Most records made of PVC are molded, although some are
made of extruded sheet.   Other molding applications are pipe-
fittings (rigid), slush and rotation-molded toys of various types,
and footwear; more recently such applications as blow-molded
rigid disposable containers have been noted.   The main outlets
for vinyl-coated paper, fabrics and other materials are furniture
and automobile upholstery, luggage, footwear and precoated metals.
A small amount of vinyl chloride goes into fibers (for example,
"Dynel").

The approximate breakdown among these various methods of
fabrication can be estimated as follows:

Method	% of total
Calendering	35
Coating and bonding	15
Extrusion	28
Molding	10
Other	17
Export	5
Total	100

By type of application, the demand can be classified as follows:

End-use		MM lbs, 1965
Flooring		320.0
Vinyl-asbestos	169.0	
Vinyl-felt	88.0	
Solid vinyl	63.0	
House furnishings		317.0
Furniture upholstery	140.0	
Wall covering	50.0	
Shower curtains	20.0	
Curtains, table cloths, place mats	30.0	
Closet accessories	18.0	
Garden hose	30.0	
Appliances	15.0	
Other	14.0	
Wire and cable coating		220.0
Construction	150.0	
Communications	40.0	
Automotive	24.0	
Other	6.0	
Construction accessories		191.0
Weather stripping	23.0	

Pipe and fittings	94.0	
Waterstop	20.0	
Swimming pool liners	20.0	
Windows	10.0	
Lighting	7.0	
Siding, panels	6.0	
Other	11.0	
Phonograph records	95.0	95.0
Transportation accessories		164.0
Automobile upholstery	89.0	
Automobile seat covers	40.0	
Automobile mats	35.0	
Packaging		73.0
Food wrap	38.0	
Nonfood products	30.0	
Blow-molded containers	5.0	
Coatings (except textiles)	70.0	70.0
Clothing		63.0
Outerwear (raincoats, etc.)	35.0	
Baby pants	18.0	
Other	10.0	
Toys		56.0
Dolls	14.0	
Inflatables	16.0	
Balls	12.0	
Other	14.0	
Footwear		30.0
Rainwear	18.0	
Shoes	12.0	
Miscellaneous		233.0
Laminates (industrial, military, etc.)	25.0	
Stationery supplies	21.0	
Medical tubing	19.0	
Sporting goods	16.0	
Tools and hardware	12.0	
Printing plates	7.0	
Credit cards	5.0	
Novelties	8.0	
Agriculture	10.0	
Other	110.0	
		1,830.0

Source: *Modern Plastics*, January 1966.

PVC floor covering accounts for 80 percent of all nonfibrous plastic flooring. The rest is primarily hydrocarbon resin-bound asphalt tile.

There has been a certain mortality among calendered sheet and film applications, as householders tend to switch to more expensive materials for curtains and bedspreads. Sheeting is used for thermoformed products such as packaging materials, signs and lighting fixtures.   Most thermoforming stock is made of PVC-PVAc copolymers.

The demand for vinyl flooring adhesives has grown proportionally to that for vinyl flooring itself.   Fabric coating is about equally distributed between calendered, laminated and knife-coated applications.   In civil construction PVC wire coating shares the market almost equally with polyethylene.

In the U.S., in contrast to many other developed countries, PVC pipe is still discriminated against in building codes and union agreements, although it has the approval of the Underwriters Laboratory.   It has the best all-around properties of any plastic pipe, although for such uses as irrigation polyethylene is preferred because it is cheaper; only 50 percent as much PVC as polyethylene pipe is used in the U.S.   Construction accessories such as window parts, doors, panels, awnings, shutters, siding, corrugated roofing and other outlets for rigid PVC are also hampered by building codes, but also to a large extent by the lack of heat and UV-stabilizers sufficiently effective to give PVC the aging and weatherability characteristics needed in construction. They have only recently become available.

Packaging films, mostly extruded and only slightly plasticized, have begun to compete with polypropylene for that part of the plastic film market not yet taken over from cellophane by polyethylene.   PVC film is supposed to have the machinability properties that have kept polyethylene out of such very high-speed applications as cigarette packing.   Also, now that the Food and Drug Administration has approved nontoxic stabilizers (thiodipropionates) and plasticizers (thiodiglycols), the use of PVC film will also increase in food-packaging.   About 12.5 percent of all PVC is used in the form of disperse resins; their main applications are slush and rotational moldings, floor covering and certain types of fabric coating.

Blow-molded bottles of rigid PVC are a promising application, since they can be made as clear as glass and much lighter, as well as unbreakable.   However, in the U.S. high-density polyethylene has the greater part of the market despite poorer optical qualities.   Vinyl chloride is used in copolymer-type industrial fibers such as "Dynel."   Most of the polymer included in the miscellaneous category is off-grade or reprocessed material.

Pure vinyl chloride can be spun from a benzene-acetone solution into either staple or filament.   This fiber has a rather low softening point, but feels like wool and is practically free of wrinkle problems.   It is made by Teijin, in Japan, and Rhône-Poulenc in

France. A newer version of this fiber, having a considerably higher melting point, is made from a copolymer of vinyl chloride and vinylidene chloride.

Polyvinyl dichloride

B.F. Goodrich makes polyvinyl dichloride by chlorinating a slurry of PVC particles in a chloroform-HCl slurry, in the presence of light. Pipe made from polyvinyl dichloride softens at 220°F and can therefore be used for hot-water lines in construction. Depending on the revision of building codes, consumption of this material may reach 10.0 million pounds per year by 1969.

6.2 Ethylene Amines

Ethylene dichloride reacts with ammonia to form ethylene diamine (EDA) and its homologs:

$$2NH_3 + CH_2ClCH_2Cl \rightarrow 2HCl + NH_2CH_2CH_2NH_2$$

$$NH_2CH_2CH_2NH_2 + CH_2ClCH_2Cl + NH_3 \rightarrow NH_2CH_2CH_2NHCH_2CH_2NH_2 + 2HCl$$

Mention has been made of a commercial process for making EDA via

$$NH_3 + HCN + CH_2O \rightarrow H_2N-CH_2CN \xrightarrow{H_2} H_2N-CH_2CH_2NH$$

This process is apparently being used by one producer in Europe. It is reported that in Japan, ethylene diamine is being produced from ethanolamines instead of via ethylene dichloride. This process also produces homologs of ethylene diamine (10 percent) and piperazine (15 percent). The producers and their capacities in 1965 are given below.

Producer	Location	1965 Capacity	1968 Capacity
		MM lbs	
Dow	Freeport, Tex.	30	65
Union Carbide	Texas City, Tex.	50	50
	Taft, La.	−	50
Jefferson	Port Neches, Tex.	−	50
Total		80	215

Source: *Oil, Paint & Drug Reporter*, September 5, 1966.

The investment for a 15 million pounds per year ethylene amine plant, producing about 12 million pounds ethylene diamine and 3 million pounds higher amines, is \$2.5 million, and production costs are estimated as follows:

(Per pound pure ethylene diamine)

Ethylene dichloride (at 4.5 ¢/lb)	12.0
Ammonia (\$50/ton)	1.5
Caustic soda (\$50/ton), other chemicals	3.7
Total	17.2
Less: Higher amines at 20 ¢/lb credit	-5.0
Sodium chloride at \$4/ton	-0.5
Net raw materials	11.7
Utilities	4.1
Labor and overhead	2.7
Capital charges	6.9
Total	25.4

The largest single outlet for ethylene diamine is in the manufacture of ethylenebisdithiocarbamate fungicides.

$$\begin{matrix} CH_2NH_2 \\ | \\ CH_2NH_2 \end{matrix} + 2CS_2 + 2NaOH \longrightarrow 2H_2O + \begin{matrix} CH_2NHCSSNa \\ | \\ CH_2NHCSSNa \end{matrix} \xrightarrow{ZnCl_2}$$

$$\begin{matrix} CH_2NHCSS \\ \diagdown \\ \diagup \\ CH_2NHCSS \end{matrix} Zn + 2NaCl$$

The metal can, for example, be manganese ("Maneb," consumption around 25.0 million pounds per year), zinc ("Zineb," 5.0 million pounds per year) or sodium ("Nabam," 2.5 million pounds per year). These are the three most important compounds of this type. They are used especially to combat leaf-blights on potatoes and tomatoes.

Ethylene diamine reacts with salicylaldehyde to form N,N′ disalicylal ethylene diamine, of which 2.0 million pounds per year were at one time consumed as a metal deactivator for motor gasoline. However, in recent years, the propylene diamine homolog has gained favor because of its superior oil solubility. A number of polyether manufacturers produce polyols for rigid urethane foams from ethylene diamine, by combining it with propylene glycol. In 1965, nearly 20.0 million pounds of ethylene diamine tetracetic acid were made from ethylene diamine by reaction with formaldehyde and hydrogen cyanide. It is the most important of a total of about 36.0 million pounds of all chelating or sequestering agents produced in the U.S.

Both ethylene diamine and its higher homologs form polyamides with dimerized fatty acids, used especially in the paper industry, in flexographic and rotogravure inks, and to modify alkyd resins. Dimethylol ethylene urea (DMEU) was for a while the most important of all reactant cotton textile crease-proofing agents, but its use is no longer increasing due to growing competition from glyoxal.

Its main disadvantage is its tendency to yellow. In 1965, use of DMEU in wash-and-wear and permanent-crease finishes still reached around 30.0 million pounds.

$$\begin{array}{c} H_2C-NH_2 \\ | \\ H_2C-NH_2 \end{array} + 2CH_2OH + (NH_2)_2CO \rightarrow HOH_2C-N \underset{\underset{O}{\overset{\|}{C}}}{\overset{H_2C-CH_2}{\overbrace{}}} N-CH_2OH + 2NH_3$$

Imidazoline and imidazole derivatives made from ethylene diamine and fatty acids are important in the surface active agent field either as such or as intermediates for cationic textile softeners of the quaternary ammonium type.

$$RCOOH + \begin{array}{c} CH_2NH_2 \\ | \\ CH_2NH_2 \end{array} \rightarrow R-C \overset{N-CH_2}{\underset{\underset{H}{N-CH_2}}{}} \rightarrow R-C \overset{N-CH}{\underset{\underset{H}{N-CH}}{}} + H_2$$

These surfactants are also made by some European manufacturers from glyoxal, ammonia and an aldehyde. "Glyodin," made from ethylene diamine and stearic acid, is an important foliage fungicide. The demand for ethylene diamine in 1965 was about 80.0 million pounds. This includes the 15 million pounds that was exported. Most uses of the higher ethylene amines are analogous to those for ethylene diamine. Diethylene triamine gives diethylene triamine pentaacetic acid, also used as a sequestrant; its propylene glycol adducts are also used as polyols for rigid urethane foams; polyamides are formed by reaction with difunctional acids, etc. Ethylene amines are intermediates for polyamine resins used as textile antistatic agents. Triethylene tetramine and tetraethylene pentamine are additives for petroleum products. About 2.5 million pounds per year of diethylene triamine and triethylene tetramine go into SBR manufacturing recipes as shortstops. These various uses brought the consumption of higher ethylene amines in the U.S. during 1965 to 7.0 million pounds per year.

Triethylene diamine or diazabicyclo-2,2,-octaine is used as a curing agent for urethane foams. Relatively more amine curing agents are used in Europe than in the U.S., where the tendency is to use more fluorocarbon blowing agents and consequently less amines, which catalyze the decomposition of isocyanates to CO_2. The total use of urethane curing agents in the U.S. was around 1.4 million pounds in 1965, of which 1.1 million pounds was triethylene diamine.

$$3ClCH_2CH_2Cl + 2NH_3 \longrightarrow$$

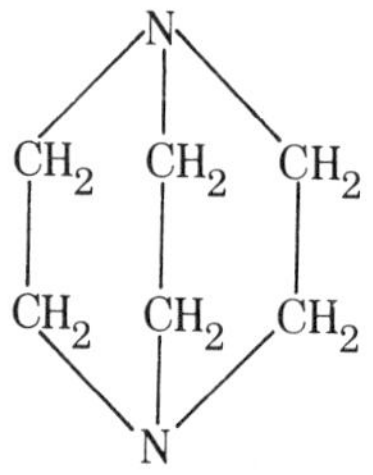

6.3 Polysulfide Rubbers

The most common types of polysulfide rubbers, also known as "Thiokols," are made by reacting ethylene dichloride or dichloroethylformal with sodium polysulfide:

$$n ClCH_2CH_2OCH_2OCH_2CH_2Cl + n Na_2S_4 \longrightarrow$$

$$\left[CH_2CH_2OCH_2OCH_2CH_2SS \right]_n + 2n NaCl$$

The main attraction of these elastomers is their excellent oil and solvent resistance. Their main drawbacks are poor heat resistance and an unpleasant odor. Nonmilitary uses are around 10.0

million pounds per year. Of this amount, 6.0 million pounds are used as caulking and sealing materials in the airplane and construction industries; other applications are as liners in paint, spray hose, printing roll coating, the manufacture of molded gasketing and sealing components, and airport runway construction. In addition, large amounts are used as a binder in solid-fuel rockets.

Thiokol Chemical and Diamond Alkali are the two U.S. producers of polysulfides.

6.4 Vinylidene Chloride

Ethylene dichloride can be chlorinated to 1,1,2-trichloroethane and dehydrochlorinated by cracking to vinylidene chloride:

$$CH_2ClCH_2Cl + Cl_2 \longrightarrow CH_2ClCHCl_2 \xrightarrow{-HCl} CH_2{=}CCl_2$$

Some U.S. producers obtain part of their 1,1,2-trichloroethane requirements by oxychlorination:

$$CH_2ClCH_2Cl + HCl + \tfrac{1}{2}O_2 \longrightarrow CH_2ClCHCl_2 + H_2O$$

Little data has been published on the production of polyvinylidene chloride. However, well over 100 million pounds per year of polyvinylidene homopolymers and copolymers were produced in 1965. The bulk of the resin was for film production (Cryovac and Saran) with very large quantities being used for filament manufacture, latex production, production of plastic and plastic-lined pipe. By far the fastest rate of growth is that of latex production.

1,1,1-*Trichlorethane*

Vinylidene chloride can be hydrochlorinated catalytically to 1,1,1-trichlorethane:

$$CH_2{=}CCl_2 + HCl \longrightarrow CH_3{-}CCl_3$$

The three U.S. producers are as follows:

Producer	Location	Capacity, MM lbs/yr
Dow	Freeport, Tex.	100.0
Pittsburgh Plate Glass	Lake Charles, La.	40.0
Ethyl Corp.	Baton Rouge, La.	40.0
Total		180.0

Its main use is as a cleaning fluid for metal degreasing. Consumption in 1965 was 80.0 million pounds per year, making it a major competitor for trichloroethylene. It has replaced carbon tetrachloride in most of its metal degreasing and equipment cleaning applications, mainly because it is only one-twentieth as toxic. About 50 percent of the total demand is for cold cleaning, the rest for vapor degreasing.

1,1,1-trichlorethane can be hydrofluorinated and dehydrochlorinated to vinylidene fluoride, a monomer for fluorocarbon resins of relatively minor importance compared with tetrafluoroethylene and chlorotrifluoroethylene.

$$CH_3-CCl_3 \xrightarrow{HF} CH_3-CF_3 \xrightarrow{-HF} CH_2=CF_2$$

6.5 Ethyleneimine

The process used in the U.S. by Dow Chemical to produce ethyleneimine starts from ethylene dichloride:

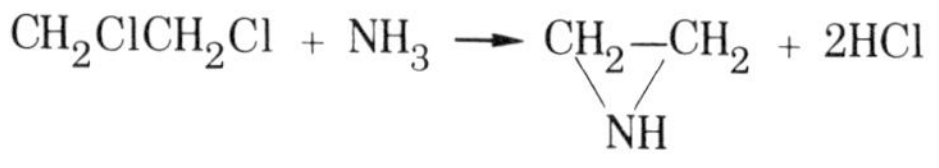

$$CH_2ClCH_2Cl + NH_3 \longrightarrow \underset{\underset{NH}{\diagdown \diagup}}{CH_2-CH_2} + 2HCl$$

In Europe, BASF obtains ethyleneimine by sulfuric acid dehydration of monoethanolamine. Its main outlets are as intermediates for certain textile finishes, flame-proofing compounds, and as a cationic flocculant for use in water. Other derivatives impart wet and nub strength additives for paper, which increase the speed of water removal from the stock as it is dried on the paper machine. Some of these compounds are polymers of ethyleneimine, others its reaction products with a fatty isocyanate of an alkyl chloride.

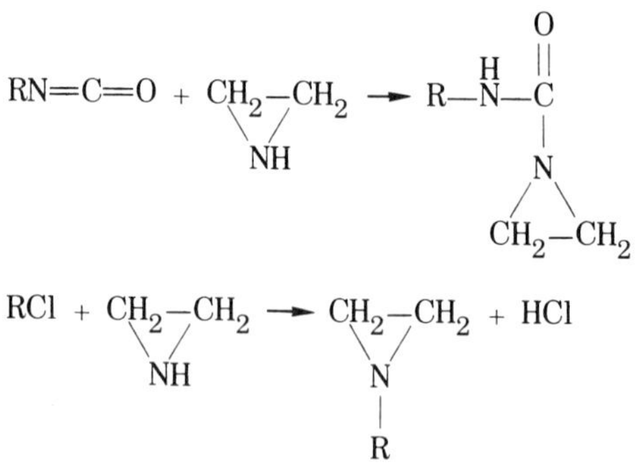

$$RN=C=O + \underset{\underset{NH}{\diagdown \diagup}}{CH_2-CH_2} \longrightarrow R-\overset{H}{\underset{\underset{\underset{CH_2-CH_2}{\diagup \diagdown}}{N}}{N}}-\overset{\overset{O}{\|}}{C}$$

$$RCl + \underset{\underset{NH}{\diagdown \diagup}}{CH_2-CH_2} \longrightarrow \underset{\underset{\underset{R}{|}}{N}}{CH_2-CH_2} + HCl$$

In 1965, about 2.0 million pounds of ethyleneimine were consumed in U.S., and some expect the demand for 1965 to be around 7.0 million pounds. Dow has a capacity of 10.0 million pounds per year.

7   ETHYLENE-PROPYLENE RUBBER

Ethylene and propylene can be copolymerized by methods similar to the manufacture of stereospecific polybutadiene and polyisoprene to give elastomers whose outstanding properties are excellent abrasion, ozonolysis and oxidation resistance and a potentially low price. These rubbers, known as EPR, are more oil-extendable than SBR, and have a lower density which would make EPR even cheaper with respect to SBR per unit volume than per unit weight. If certain obstacles to the use of EPR as a general-purpose rubber are overcome, it may eventually account for about one-third of all the rubber consumed in the U.S.

Neither ethylene nor propylene have more than a single double-bond. Thus EPR can only be cured by strong oxidizing agents, such as peroxides. This has several disadvantages, one being that the crosslinked product has a most unpleasant odor. The trend has been towards the production of sulfur-curable EPR by incorporation of a third monomer containing two double-bonds. These elastomers are known as "EPT," for "Ethylene-Propylene Terpolymer." The third monomer can be cyclic or acyclic. Cyclooctadiene and dicyclopentadiene are examples of possible cyclic comonomers, and 1,4-hexadiene of an acyclic third constituent.

Until now, the most important uses of EPT have been in mechanical goods such as automobile weather stripping, body-mounts and bumpers. These parts were once made mainly of butyl rubber, the most ozone-resistant elastomer before EPT was developed. By 1965 about 2.5 pounds of EPT per car were being consumed in these applications, 80 percent of which was weather stripping. The other field in which EPT has found wide applications is high-voltage cable coatings. In 1965, the total EPT use in the U.S. was 33.0 million pounds, of which 20.0 million pounds was used in automobile components. Tire-building with EPT has been limited so far by the lack of a suitable tire-cord adhesive, poor compatibility with other elastomers, and a slow curing rate that severely cuts down productivity. Nevertheless, it has been predicted that these problems will be overcome and that EPT will eventually achieve general-purpose status. One forecast predicted that by 1970, 82 percent of all EPT will be used in tires, 10 percent in automotive components, and the rest in cable coatings, carpet underlay, hose, and in chemical goods. By then, EPT may already represent 15-20 percent of all rubber consumption.

Major producers and their capacities are given below; in addition to these plants, however, EPT could be produced in existing butyl rubber or polybutadiene plants with only minor alterations. Montecatini is the dominant process licensor, having developed the prevailing catalyst system.

Producer	Location	1965 Capacity, MM lbs/yr
DuPont	Beaumont, Tex.	33.0
Enjay	Baton Rouge, La.	30.0
U.S. Rubber	Geismar, La.	33.0
		96.0

Source: *Oil, Paint & Drug Reporter*, May 3, 1965.

The investment for a 50.0 million pounds per year EPT plant is $10.0 million, and the production costs are similar to those for other stereospecific elastomers.

8   PERCHLOROETHYLENE

Previously, the most important route to perchloroethylene started with acetylene, via trichloroethylene:

$$C_2H_2 + 2Cl_2 \longrightarrow Cl_2CHCHCl_2 \xrightarrow{-HCl} ClCH{=}CCl_2 \xrightarrow{Cl_2} Cl_2HC{-}CCl_3$$

$$Cl_2HC{=}CCl_3 \xrightarrow{-HCl} Cl_2C{=}CCl_2$$

At present, less than 20 percent of the U.S. plants still in operation employ this route. Several market-oriented plants were forced to shut down as natural gas and ethylene became cheaper and competitors began to produce chlorinated solvents from these raw materials on the Gulf Coast. At first the preferred route was chlorination of methane in an atmosphere of carbon tetrachloride at temperature high enough to produce the overall reaction:

$$6Cl_2 + 2CH_4 \longrightarrow Cl_2C{=}CCl_2 + 8HCl$$

More recently, ethylene and propylene dichloride have become the favored raw materials; the latter is a byproduct of propylene oxide production, which has been rising in conjunction with that for urethane foams.   High-temperature chlorination produces mainly perchloroethylene and carbon tetrachloride, most of the other chlorination products being recycled to the reaction furnace. By recycling the undesired byproducts, the end-product mix can be adjusted within broad limits; carbon tetrachloride, as well as other chlorinated compounds, can be recycled to extinction, or recovered as byproducts.

Propane is also used as a raw material in one U.S. installation, although the rising cost of LPG and natural gas is pushing economics towards ethylene. A typical flowsheet appears in Fig. 5-28.

Most of the demand for perchloroethylene is for dry-cleaning. In less advanced countries, petroleum or "Stoddard" solvents are still used to a greater extent than chlorinated compounds, especially since most dry cleaning in these countries still is done in large central plants where the emphasis is on low operating cost. In the U.S., however, by 1964 perchloroethylene already accounted for 50 percent of the market and its share was still growing as the trend continued towards neighborhood dry cleaning plants where nonflammability of the solvent is of prime concern.

With the advent of coin-operated dry-cleaning machines, the demand for perchloroethylene showed a sudden increase since these

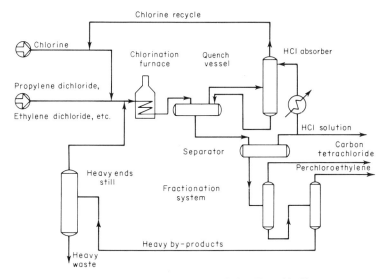

Fig. 5-28   Perchloroethylene and Carbon Tetrachloride.

machines had to receive their initial charge of solvent. For a time, faster-drying solvents such as "Valclene" (1,1,2-trichloro-1,2,2-trifluorethane), itself derived from perchloroethylene, were mentioned as candidates, but eventually the machine manufacturers were able to shorten the cleaning cycle by mechanical means, so the considerably cheaper perchloroethylene has prevailed. The demand for perchloroethylene in 1965 was 490.0 million pounds, of which 70.0 million were, however, imported. Dry-cleaning consumed 85 percent of this total, the rest going into general solvent services (for example, it is the preferred solvent for ethylene-propylene elastomers) and as an intermediate for fluorocarbons 113 and 114. U.S. producers and routes are as shown below.

Producer	Location	Capacity, MM lbs/yr	Raw materials
Detrex	Ashtabula, Ohio	25.0	Acetylene
Diamond	Deer Park, Tex.	50.0	Ethylene, propylene dichloride
Dow	Freeport, Tex.	70.0	Ethylene, propylene dichloride
	Pittsburgh, Calif.	30.0	Methane
	Plaquemine, La.	70.0	Ethylene, propylene dichloride
DuPont	Niagara Falls, N.Y.	60.0	Acetylene
Frontier	Wichita, Kans.	20.0	Methane
Hooker	Tacoma, Wash.	5.0	Acetylene
Pittsburgh Plate Glass	Barberton, Ohio	45.0	Propane, propylene dichloride
	Lake Charles, La.	80.0	Ethylene, propylene dichloride
Stauffer	Louisville, Ky.	55.0	Methane
Total		510.0	

Source: *Oil, Paint & Drug Reporter*, May 25, 1964; February 1, 1965.

Economics for a plant producing 20.0 million pounds per year of carbon tetrachloride and 30.0 million pounds per year of perchloroethylene from propane are shown below. Investment requirements are $3.2 million. Raw materials are assumed to be valued as follows.

Raw material	Cost, ¢/lb
Propane	1.5
Chlorine	2.5
HCl	1.25

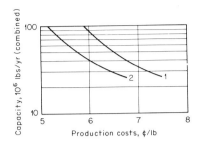

Fig. 5-29 Perchloroethylene and Carbon Tetrachloride.   1-Perchloroethylene; 2-Carbon Tetrachloride.

Production costs are as follows:

	Cost, ¢/lb perchloroethylene		Cost, ¢/lb carbon tetrachloride	
Propane	0.3		0.15	
Chlorine	5.4		5.15	
		5.7		5.30
HCl	-1.5		-1.8	
Net raw materials		4.2		3.5
Utilities	0.4		0.4	
Labor and overhead	0.6		0.6	
Capital charges	2.1		2.1	
Total		7.3		6.6

8.1 Fluorocarbons 113 and 114

Perchloroethylene can be chlorinated to hexachloroethane;
hydrofluorination of hexachloroethane yields various fluorocarbons:

$$CCl_3CCl_3 + 3HF \longrightarrow CCl_2FCClF_2 + 3HCl \quad (\text{``113''})$$

$$CCl_3CCl_3 + 4HF \longrightarrow CClF_2CClF_2 + 4HCl \quad (\text{``114''})$$

Fluorocarbon ''114'' is used as an aerosol propellant in combination with F-12; the demand in 1965 was 15.0 million pounds.
Fluorocarbon ''113'' is the starting point for making chlorotrifluoroethylene (CTFE), the monomer for several fluorocarbon resins,
for example, ''Kel-F.''

$$CCl_2FCClF_2 \xrightarrow{\Delta} ClFC{=}CCl_2 + Cl_2$$

Fluorocarbon "114" also is used as a dry cleaning solvent, and is undergoing rapid growth. The demand in 1965 was estimated at 30.0 million pounds.

9 ETHYLENE DIBROMIDE

Ethylene dibromide is made from ethylene and bromine:

$$C_2H_4 + Br_2 \rightarrow CH_2BrCH_2Br$$

Its main application is as a lead scavenger in antiknock fluid. Since ethylene dibromide is more expensive than ethylene dichloride, it is partly replaced by the latter in automotive gasoline, but is used to the exclusion of the dichloride in aviation gasoline; with the advent of jet aircraft, this market has practically vanished. Some 200.0 million pounds were used in antiknock fluids during 1965. In addition, about 5.0 million pounds per year are used to control nematodes that stunt the growth of tobacco leaves. Dow and Great Lakes Chemical are the sole U.S. producers.

10 ETHYL BROMIDE

Dow employs a process to produce ethyl bromide from ethylene and hydrogen bromide in which the ethylene is activated by means of radiation. The demand is about 1.0 million pounds per year, primarily for ethylations in the pharmaceutical and fine chemical fields.

11 ETHYL CHLORIDE

Three processes are used to produce ethyl chloride. Shell, Ethyl Corp., and National Distillers start from ethylene and HCl:

$$C_2H_4 + HCl \rightarrow CH_3CH_2Cl$$

This process accounts for 70 percent of U.S. capacity. Hydrochlorination takes place in the liquid phase in the presence of a chlorinated solvent, which may be ethyl chloride itself. Aluminum chloride acts as a catalyst. Reactor effluent is heated and flashed, net production being taken overhead and sent on first to a scrubber, where the remaining HCl is washed out, and then to a purification train where lights and heavies are removed by fractionation. The need is not for polymerization-grade ethylene; several U.S. ethyl chloride plants use purge gas from a neighboring polyethylene unit. Reported yields on ethylene are 90 percent.

DuPont and Hercules start from ethanol, or a mixture of ethanol and some byproduct ethyl ether, and HCl:

$$C_2H_5OH + HCl \longrightarrow C_2H_5Cl + H_2O$$

Finally, Dow makes part of its ethyl chloride requirements by chlorinating ethane:

$$C_2H_6 + Cl_2 \longrightarrow C_2H_5Cl + HCl$$

The investment for a 50.0 million pounds per year unit using the ethylene-HCl route is \$3.3 million, with the following production costs:

	Cost, ¢/lb ethyl chloride
Ethylene (4.0 ¢/lb)	2.0
HCl (1.75 ¢/lb)	1.2
Utilities	0.2
Labor and overhead	0.3
Capital charges	2.2
Total	5.9

The capacities are:

Producer	Location	Capacity, MM lbs/yr
Ethyl Corp.	Baton Rouge, La.	270.0
	Houston, Tex.	100.0
Dow	Freeport, Tex.	100.0
DuPont	Deepwater, N.J.	110.0
Hercules	Hopewell, Va.	10.0
National Distillers	Tuscola, Ill.	60.0
Shell	Houston, Tex.	85.0
Total		735.0

Source: *Oil, Paint & Drug Reporter*, October 19, 1964.

The production of ethyl chloride was around 690.0 pounds per year, in 1965, of which 640.0 million pounds per year were used to make tetraethyl lead.

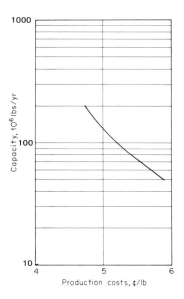

Fig. 5-30 Ethyl Chloride (from Ethylene and HCl.

The most important use for ethyl chloride apart from TEL manufacture is as a Friedel-Crafts reaction promoter, chiefly in ethylbenzene manufacture. A small amount is used to make ethyl cellulose, a water-soluble cellulose ether having most of the same uses as hydroxy ethyl cellulose, of which some 5.0 million pounds per year are produced in the U.S. by Dow and Hercules. Ethyl chloride also has some use as a refrigerant and anaesthetic.

11.1 Tetraethyl Lead

In the classical process for making TEL, ethyl chloride reacts with an amalgam of sodium and lead.

$$4C_2H_5Cl + 4PbNa \rightarrow (C_2H_5)_4Pb + 4NaCl + 3Pb$$

The lead is recovered and recycled.

This is still the most widely used process, and accounts for the consumption of some 90,000 tons per year of sodium metal. Ethyl Corp. reportedly can use triethyl aluminum in the presence of a cadmium salt. Nalco employs a Grignard reaction with magnesium, followed by electrolysis during which the alkyl ions migrate towards a sacrificial lead anode where the lead alkyl is formed. Actually, this company only produces tetramethyl lead, although the process may be employed to make either TEL or TML.

The domestic consumption of lead alkyls in 1964 was 690.0 million pounds, of which 550.0 was TEL and the rest TML. Several long-term trends are at work against further expansion of the market for antiknock fluids. Refinery overcapacity often makes it cheaper to obtain extra octane numbers by refining techniques alone; the advent of jet planes has eliminated most of the aviation gasoline market; for a time, compact cars were in demand, requiring both less and lower-octane gasoline. In the future, perhaps, turbine engines requiring nongasoline fuels may achieve maturity. The average antiknock fluid content of the U.S. gasoline pool is now 2.2 cubic centimeters per gallon, and has been declining steadily over the years. The U.S. producers of lead alkyls and their capacities are given below.

Producer	Location	Capacity, MM lb/yr
DuPont	Antioch, Calif. Deepwater, N.J.	340.0
Ethyl Corp.	Baton Rouge, La. Houston, Tex.	370.0
Pittsburgh Plate Glass	Houston, Tex.	100.0
Nalco	Houston, Tex.	40.0
Total		850.0

Source: *Oil, Paint & Drug Reporter*, August 3, 1964.

By 1965, only about 30.0 million pounds per year of TEL were being used in aviation gas. The rest went into automotive anti-knock fluids and exports.

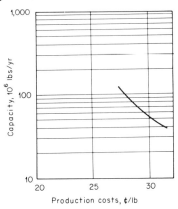

Fig. 5-31 Tetraethyl Lead.

Investment costs for a 40.0 million pounds per year TEL plant are \$10.0 million. This capacity corresponds to 65.0 million pounds per year of antiknock fluid, the rest being lead scavengers— in the case of automotive gasoline, a roughly 50-50 mixture of ethylene dichloride and dibromide. Production costs per pound of pure TEL are calculated assuming captive metallic sodium at 12.0 cents per pound and ethyl chloride at 5.5 cents per pound.

	¢/lb TEL
Ethyl chloride	4.9
Sodium	3.8
Lead (17 ¢/lb)	11.4
Utilities	1.5
Labor and overhead	1.6
Capital charges	8.3
Total	31.5

12 LONG-CHAIN ALCOHOLS

Two plants in the U.S. use processes based on Ziegler catalysts for making straight-chain alcohols from ethylene. One is at Lake Charles, La. (Continental Oil) the other at Houston, Texas (Ethyl Corp.) with capacities of 100.0 and 50.0 million pounds per year, respectively. They begin by reacting aluminum metal with ethylene and hydrogen to form triethyl aluminum. Ethylene adds to this compound at high pressures to give trialkyl aluminum compounds, which are then oxidized with bone-dry air to aluminum trialkoxides. These are hydrolyzed by sulfuric acid to primary alcohols, having an even number of carbon atoms.

$$3C_2H_4 + 1\tfrac{1}{2}H_2 + Al \longrightarrow (C_2H_5)_3Al$$

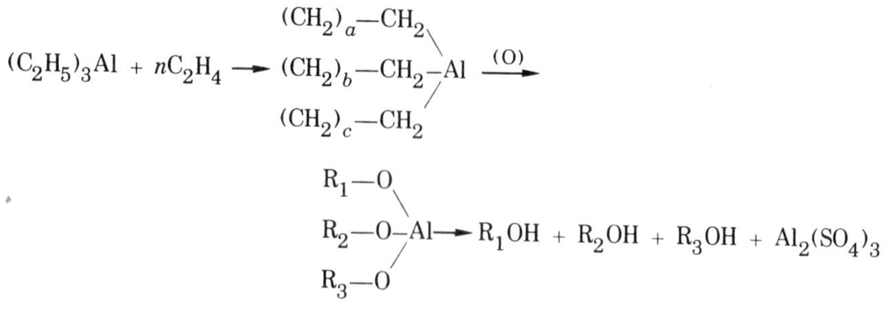

$$(C_2H_5)_3Al + nC_2H_4 \longrightarrow \underset{(CH_2)_c-CH_2}{\overset{(CH_2)_a-CH_2}{(CH_2)_b-CH_2}}\!\!\!\!Al \xrightarrow{(O)}$$

$$\underset{R_3-O}{\overset{R_1-O}{R_2-O}}\!\!\!\!Al \longrightarrow R_1OH + R_2OH + R_3OH + Al_2(SO_4)_3$$

Straight-chain fatty alcohols are expensive due to the high and unstable prices of natural raw materials, and most "oxo" alcohols derived from low polymers of propylene and butylene are highly branched.   However, Ziegler alcohols have straight chains and therefore better biodegradability characteristics than other synthetic alcohols. On the other hand, the process involves extremely pyrophoric intermediates which in turn require heavy investments in the appropriate safety precautions. The chain-length distribution among the various alcohols has a Poisson ratio and can be varied somewhat to suit market requirements by changing process conditions.   However, plasticizer-range homologs make up around 45 percent of the total output. Those in the C_8-C_{10} range are used to make plasticizers, which have better low-temperature properties and lower volatility than those made with branched alcohols. C_{12}-C_{16} is the preferred range for detergents and C_{16}-C_{18} alcohols are modifiers for wash-and-wear resins. Heavier homologs are used as evaporation retarders (C_{18}) and lubricants, mold-release agents, etc. (C_{20}-C_{26}).

Straight-chain detergent-range alcohols are also obtained by the oxo reaction starting from straight-chain α-olefins and by direct oxidation of normal paraffins.   Suitable techniques limit chain branching to 20 percent α methyl derivatives. Producers of long-chain synthetic alcohols in the U.S. are as follows:

Producer	Location	1965 Capacity, MM lbs/yr	Type of alcohol	Process	Raw material
Continental	Lake Charles, La.	100.0	Primary	Ziegler	Ethylene
Ethyl	Houston, Tex.	50.0	Primary	Ziegler	Ethylene
Shell	Houston, Tex.	50.0	80% Primary 20% Secondary	Oxo	Cracked wax
Shell*	Geismar, La.	100.0	80% Primary 20% Secondary	Oxo	Cracked wax
Union Carbide	Texas City, Tex.	40.0	Secondary	Oxidation	n-paraffins

*Due on stream in 1966.
Source: *Oil, Paint & Drug Reporter*, August 26, 1965.

6

PROPYLENE

1 INTRODUCTION

Propylene requirements in the U.S. as a petrochemical feedstock reached 3300 million pounds in 1965. This demand may remain static for a few years, since the growth in certain end-uses is compensated somewhat by the switch away from propylene tetramer for alkyl benzene sulfonates (ABS) to linear alkyl sulfonates (LAS) as the main raw material for household detergents. Countering this expected decline in the use of propylene as tetramer are such dynamic fields as acrylonitrile, polypropylene, propylene oxide-based polyols and ethylene-propylene rubber. Thus, the growth curve for propylene as a whole can be expected to swing upwards again after replacement of alkylbenzene sulfonate has gone far enough.

A shortage of propylene in the U.S. is currently shaping up. Consumption in 1965 was estimated to be 18.0 billion pounds, of which 15.8 billion pounds were from refineries and the remaining 2.2 billion pounds from ethylene plants. Of the 13.4 billion pounds per year not used in the chemical industry, over 60 percent is consumed in catpoly units to make polymer gasoline. However, increasing amounts are being fed to alkylation units instead of the more usual butylenes. On the other hand, individual producers may find themselves with an excess of propylene. Phillips, for example, consumes some of its excess propylene by promoting the reaction

---

NOTE: All data must be considered approximate. Both production and consumption of propylene can vary widely dependent upon operation decisions.

$$2CH_3{=}CH{-}CH_2 \longrightarrow C_2H_4 + n{-}C_4H_8$$

The n-butylenes can be fed to a butadiene unit.

The propanizer off-gas of a typical ethylene plant contains 50-60 percent propylene, the rest being mostly propane, plus some C_4's and C_2's. The cost of producing polymerization grade (99.5 percent) propylene from such a feed is shown below. Separation is by low-temperature fractionation; the relative volatility of propane and propylene is such that about 180 plates are required to achieve the desired purity. The investment for a 50.0 million pounds per year propylene purification unit is around $660,000. Production costs are:

	Costs, ¢/lb propylene
Raw propylene	2.3
Utilities and labor	0.3
Capital charges	0.4
Total	3.0

The distribution of the various uses of propylene for 1965 is shown below.

End-use	*MM lbs, 1965*
Isopropanol	1420.0
Propylene oxide	570.0
Acrylonitrile	550.0
Polypropylene	390.0
Cumene	260.0
Miscellaneous	1410.0
Total	4600.0

Source: *Hydrocarbon Processing,*
January, 1967.

The U.S. producers of propylene used in chemicals and their capacities are given in the table below.

Producer	Location	Source of propylene**	Approximate capacity MM lbs/yr*
Ashland	Catlettsburg, Ky.	G	130.0
Atlantic	Watson, Cal.	G	230.0
Amoco	Whiting, Ind.	G	400.0
	Wood River, Ill.	G	110.0
Chevron	El Segundo, Cal.	G	160.0
Cities Service	Lake Charles, La.	G	320.0
Clark	Blue Island, Ill.	G	70.0
Dow	Bay City, Mich.	E & G	60.0
	Freeport, Texas	E	270.0
	Midland, Mich.	E	50.0
	Plaquemine, La.	E	60.0

(Continued on next page)

(Continued from preceding page)

Producer	Location	Source of propylene**	Approximate capacity MM lbs/yr*
DuPont	Orange, Texas	E	200.0
El Paso	Odessa, Texas	E	60.0
Enjay	Baton Rouge, La.	E & G	1000.0
	Baytown, Texas	G	440.0
Goodrich	Calvert City, Ky.	E	90.0
Gulf	Cedar Bayou, Texas	E	130.0
	Philadelphia, Pa.	G	260.0
	Port Arthur, Texas	E & G	440.0
Jefferson	Port Neches, Texas	E	150.0
Marathon	Texas City, Texas	G	70.0
Mobil	Beaumont, Texas	E & G	520.0
Monsanto	Alvin, Texas	E	300.0
	Texas City, Texas	E	60.0
Petroleum Chemicals (Cities Service)	Lake Charles, La.	E	60.0
Phillips	Sweeney, Texas	E & G	180.0
Shell	Deer Park, Texas	E & G	280.0
	Norco, La.	E & G	320.0
	Wilmington, Cal.	G	130.0
Signal	Houston, Texas	G	90.0
Sinclair	Houston, Texas	G	220.0
	Marcus Hook, Pa.	G	160.0
Sinclair-Koppers	Houston, Texas	E	100.0
Skelly	El Dorado, Kan.	G	90.0
Sohio	Lima, Ohio	G	110.0
Sun	Marcus Hook, Pa.	G	300.0
Suntide (Sunray D-X)	Corpus Christi, Texas	G	70.0
Texaco	Westville, N.J.	G	140.0
Texas City Ref.	Texas City, Texas	G	100.0
Texas Eastman (Eastman Kodak)	Longview, Texas	E	100.0
Tidewater- Air Prod.	Delaware City, Del.	G	250.0
Union Carbide	Institute, W.Va	E	50.0
	Seadrift, Texas	E	100.0
	S. Charleston, W.Va.	E	50.0
	Taft, La.	E	200.0
	Texas City, Texas	E	100.0
	Whiting, Ind.	E	50.0
Union Oil	Los Angeles, Cal.	G	140.0

*This table must be used with extreme care since the "approximate capacities" are not necessarily on a consistent basis, e.g., part of the propylene capacity is used in "poly" gas or alkylate and some for chemicals. Plants where propylene is used solely for non-chemical purposes are not listed. Because of these problems, no total capacity for chemicals has been determined.

**E = From ethylene manufacture; G = From gasoline production.

Source: *Hydrocarbon Processing*, January 1967.

2 POLYPROPYLENE

Whereas polyethylene can be polymerized at high or low pressures, using two entirely different types of catalyst and technology, polypropylene is manufactured exclusively using Ziegler-type catalysts. One original process was developed by Montecatini, who have licensed most of the world's manufacturers.

The term "polypropylene" is applied to a polymer having a molecular weight around 40,000, and a degree of crystallinity below that of linear polyethylene. It is not to be confused with the low polymers of propylene. The outstanding properties of polypropylene are, first, the lowest specific gravity of any commercial plastic, which makes it one of the cheapest plastics per unit volume; second, a melting point of 170°C for the isotactic polymer, and even higher for sindiotactic polypropylene; third, good dielectric properties, low creep and chemical inertness.

Polypropylene is produced at low pressures in a hydrocarbon solvent, using an aluminum alkyl and titanium chloride (Ziegler-type) catalyst. After polymerization has reached the desired stage, the reactor is flashed to remove unreacted monomer, and the polymer mix centrifuged and washed with carbon tetrachloride to inactivate the remaining catalyst. After centrifuging, the polymer is washed with acetone and then with water to remove the last traces of metal salts, and finally extruded and chopped into pellets. Higher pressures favor the polymerization reaction but low pressures are required to produce the steric arrangements needed for commercially desirable properties.

For most applications, polypropylene competes with low-pressure polyethylene. While this is considerably cheaper per unit weight and volume, polypropylene is often preferred on the basis of its superior physical properties. The most important single application of polypropylene is in injection molding, where it has taken over a portion of the market that once belonged to cellulose acetate. Polypropylene film can be clearer and stiffer than polyethylene. It can be produced in thinner gauges, and thus come closest to competing with cellophane for high-speed packaging applications. Biaxially oriented film combines toughness, strength, and low price, and is potentially the material best suited for shrink film in such applications as poultry packaging; film oriented in only one direction also finds numerous packaging applications such as flat wrapping, or as a laminate with cellophane, e.g., for packaging potato chips. Other applications are for pipe, which has limited prospects because of its poor low-temperature properties;

monofilament, which may eventually find large-scale applications in baler twine (in competition with sisal) despite the objection that cattle cannot digest it; and fibers. The idea of a fiber made of a raw material as cheap as propylene is naturally very appealing, but there are two very important obstacles to overcome. The first is its softening point, which is about 100° F lower than that required for apparel that must be ironed; attempts to raise the softening point have included additives, radiation crosslinking and graft polymerization. The most important objection, however, is that polypropylene, having no polar group that could act as a dye-site, is difficult to dye except by spin-dyeing. As a solution, two techniques are finding favor. The first is graft-polymerization with other monomers containing polar groups, which has the ancillary advantage of raising the softening point. The second is to form tiny interconnected cracks in which the dye can be received, by blowing air through the melt. The first large-scale production of polypropylene fibers other than for carpeting and similar less noble uses is the Montecatini operation at Terni, Italy, which has a capacity of 60.0 million pounds per year; the technique used is said to be graft polymerization with around 5 percent of a 50-50 mixture of isopropenyl pyridine and methyl methacrylate, which is said to render the fiber dyeable by numerous conventional dyeing techniques. Polypropylene fibers have an enormous potential. They have 65 percent more coverage per unit weight than cotton, and are already about 25 percent cheaper than nylon with further reductions in price to be expected.

The development of polypropylene was a long and costly process. Controls were complicated and expensive, making it difficult to obtain uniformity in the product. The most recent plants use continuous instead of batch polymerization. The rate of crystallization must be controlled to limit crystal size and avoid brittleness. Polypropylene is unstable to light and must therefore be stabilized immediately. Finally, polypropylene is more difficult to process than most other thermoplastics. Even so, growth has been extremely rapid, and estimated use in all applications for 1965 was 365.0 million pounds against 172.0 million pounds in 1963. Distribution of domestic end-uses in 1965 is given below.

End-use	MM lbs, 1965
Blow-molding profiles	5.0
Injection molding	155.0
Film and sheeting	55.0
Pipe and profiles	10.0
Fibers (mono, multifilament)	90.0

Wire coating	2.0
Export	30.0
Miscellaneous	18.0
Total	365.0

Source: *Modern Plastics*, January, 1966.

The principal molding applications are closures, where polypropylene is competing successfully with thermosetting plastics (phenolics and urea-formaldehyde) because it gives superior threads, and automotive applications taken mainly from cellulosics, such as steering wheels. Copolymers are used for blow-molding, in competition with polyethylene and polyvinyl chloride. Monofilament is used as rope, furniture webbing and to make industrial fabrics, apart from the recent development of plastic baler twine. Among sheet products, the most important applications are luggage and strapping. The earliest U.S. producer was Hercules, using a process developed jointly with Farbwerke Hoechst. This was followed by very rapid capacity expansion, and in mid-1966 the following producers were operating:

Producer	Location	MM lbs/yr
Hercules	Lake Charles, La.	120.0
Dow	Torrance, Cal.	30.0
Enjay	Baytown, Texas	80.0
Novamont (Montecatini)	Neal, W.Va.	40.0
Shell	Woodbury, N.J.	80.0
Texas Eastman	Longview, Texas	30.0
Alamo Polymer	Houston, Texas	70.0
Rexall-El Paso Natural Gas	Odessa, Texas	60.0
Total		610.0

Source: *Petroleum Processing*, January, 1967.

Hercules was the only resin producer to go into fiber production. The other fiber manufacturers depend on outside sources of raw materials. Capacities in 1964 were as shown below.

Producer	Capacity, MM lbs/yr
Hercules	30.0
Alamo Industries, Inc. (Reeves & Gerfil)	17.0
Uniroyal, Inc. (U.S. Rubber)	2.0
Vectra Co. (National Plastic Products)	6.0
Total	55.0

Source: *Chemical Week*, September 5, 1964.

The major tire producers were experimenting with polypropylene for tire cord applications, but no commercial production had been announced by 1965.

3 PROPYLENE TRIMER AND TETRAMER

The low polymers of propylene are produced in refinery units designed to make polymer gasoline from C_3 and C_4 olefins. Cat-poly gasoline, being highly olefinic, is the most unstable component of the average gasoline pool. Since the introduction of jet aircraft has freed large amounts of alkylation gasoline for the motor fuel pool, this type of high-octane gasoline is now preferred to polymer gasoline because of its greater stability. At the same time, the main application of propylene tetramer is losing ground rapidly as a result of objections against the slow rate of biodegradation of synthetic detergents derived from benzene and propylene.

The manner in which, in the space of a few years, alkylbenzene sulfonates (ABS) were almost completely replaced by linear alkyl sulfonates (LAS) is an example of how domestic affairs can affect a whole sector of the chemical industry. When attention was first called to the pollution and foaming caused by the mounting concentration of detergent in the U.S. water resources, it was pointed out that detergents only accounted for 10 percent of the total pollution level. Furthermore, if certain other pollutants were present even as little as 1 ppm of detergent residue would cause plentiful and lasting foaming. Pressure from all sides became so strong, however, that detergent manufacturers were forced to look for raw materials that would be degraded more rapidly by bacteria in the presence of air. By 1965, conversion of detergent alkylate plants to LAS (linear alkyl sulfonate) manufacture had become almost complete. Trimer and tetramer are still important raw materials for making industrial detergents and as intermediates for other alkylations and "oxo" derivatives.

A typical tetramer plant flowsheet is shown in Fig. 6-1. Fresh C_3 feed is combined with recycle, which serves to influence the distribution of various products by suppressing the formation of the recycled fractions. The feed is preheated and introduced into the polymerization reactor, which in this case consists of several fixed beds of solid phosphoric acid. The products stream is cooled between reactor beds by injecting cold propane. The polymer stream is fed to a depropanizer where the net propane entering the system is removed and sent to the LPG plant. The remainder is recycled.

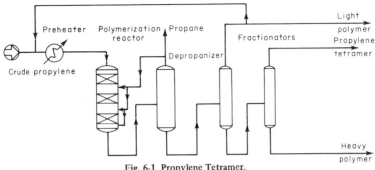

Fig. 6-1 Propylene Tetramer.

Two other towers first remove the light polymers, which are fractionated again if trimer is also to be produced, and second, the heavy polymer leaving a tetramer stream which is actually a mixture of numerous higher isomers and homologs. Careful distillation is required to ensure a reproducible product. Propylene tetramer has a boiling range of 350°F to 480°F, and if desired, the yield of this particular fraction can be made as high as 95 percent by recycling light polymer to the polymerization reactor. The production of propylene tetramer in the U.S. approached 460.0 million pounds in 1965, with the largest portion being consumed in industrial detergents. The production of propylene trimer was about 240.0 million pounds, the greatest percentage being used for nonyl phenol and isodecyl alcohol. All manufacturers of propylene trimer and tetramer are oil companies, since a cat-poly gasoline unit can easily be adapted for their production.

4 ACRYLONITRILE

Although acrylonitrile is still being made from HCN and acetylene, it should be considered a derivative of propylene. Since the commercialization of the Sohio process in the early 1960's, all acrylonitrile plants built in the world have been based on the reaction between propylene, ammonia and air. Most of these plants employ the Sohio process, the exceptions being the plants based on the Distillers process, which is based on the same reaction.

$$CH_3CH{=}CH_2 + NH_3 + O_2 \rightarrow NC{-}CH{=}CH_2 + 3H_2O$$

A flowsheet of an acrylonitrile from propylene process is shown in Fig. 6-2. Approximately stoichiometric proportions of air, ammonia and propylene are introduced into a fluid bed reactor at 15

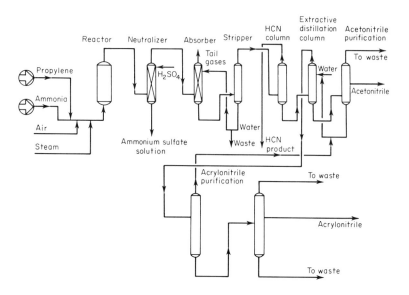

Fig. 6-2 Acrylonitrile from Propylene and Ammonia.

psig and 750-925° F.   One attractive feature of the process is that
it is not necessary to use polymerization-grade propylene.   The
contact time is of the order of several seconds. Once-through
flow is used since conversion of propylene is practically complete.
The reaction is exothermic, so heat-removal must be provided.
The reactor effluent is neutralized to remove unconverted ammonia. It is next washed with water, to yield an unabsorbed stream
of inert gases and a solution of acetonitrile, acrylonitrile and
HCN.   This solution is stripped of the dissolved products, which
are fractionated to remove pure HCN and then sent to the main
purification section.   First, a main fractionator yields an overhead
consisting of wet acrylonitrile, and a bottom stream of wet
acetonitrile.   The overhead is treated first by extractive distillation, and then by conventional fractionation to remove undesirable
lights and heavies and thus yield pure acrylonitrile.   The bottom
from the main fractionator can be fractionated in a two-column
system to yield pure acetonitrile.

Approximately 0.1 pound each of HCN and acetonitrile are
produced per pound acrylonitrile output. While there is a market
for pure HCN, the combined Sohio-process plants operating in the
U.S. as of 1965 can produce about 10 times as much acetonitrile
as can be consumed in the country.   In practice, therefore, it is

incinerated, although some producers may eventually find it more profitable to hydrolyze it to acetic acid. Recently, Sohio has begun to produce methacrylonitrile by substituting isobutylene for propylene. Their plant at Lima, Ohio, has a capacity of 10.0 million pounds per year.

In the acetylene-HCN process, the reactants combine in the presence of cuprous chloride at essentially ambient conditions; a 6:1 excess of acetylene is employed. The product is recovered as a very dilute solution (2 percent) in a water absorber-stripper system, and the unreacted acetylene, after removal of higher homologs, is recycled to the reactor. Crude acrylonitrile is separated from light and heavy ends. The heavy ends are rerun to remove any remaining acrylonitrile, and the overhead from this final column recycled to the light ends column. Reported yields are 80 percent on acetylene and 90 percent on HCN.

$$C_2H_2 + HCN \rightarrow NH{-}CH{-}CH_2$$

The Knapsack-Griesheim process starts with acetaldehyde and HCN

$$CH_3CHO + HCN \rightarrow CH_3CHOHCN \xrightarrow{-H_2O} CH_2{=}CH{-}CN$$

It has never been used commercially in the U.S. since the Sohio process was developed more or less at the same time as the Wacker acetaldehyde route which, it was thought, would allow the Knapsack-Griesheim process to become competitive. The process formerly used by Union Carbide involved the reaction between HCN and ethylene oxide:

$$\underset{\diagdown\!O\!\diagup}{CH_2{-}CH_2} + HCN \rightarrow HOCH_2CH_2CN \xrightarrow{-H_2O} CH_2{=}CH{-}CN$$

DuPont employs a unique process at Beaumont, Texas. In this process, propylene and nitric oxide are the raw materials. The reaction is carried out at high temperatures (1300° F) in the presence of a silver catalyst. This process is essentially equivalent to the Sohio route, except that air and ammonia are combined before the reaction with propylene; in other words, it is the equivalent of a two-stage Sohio process. It is significant that the company adopted Sohio technology at its latest plant.

The investment cost for a 60.0 million pounds per year acrylonitrile unit using the Sohio process is $12.0 million. Production costs are as follows:

	Cost, ¢/lb acrylonitrile	
Propylene (2.3¢/lb)	2.6	
Ammonia (4¢/lb)	2.9	
Sulfuric acid	1.1	
Catalyst and chemicals	1.6	
		8.2
HCN by-product (12¢/lb)	-1.3	
Ammonium sulfate by-product	-1.7	
Net raw materials		5.2
Utilities	1.6	
Labor and overhead	0.7	
Capital charges	6.6	
Total		14.1

The raw material costs alone using alternative routes — via acetylene, ethylene oxide or the Knapsack-Griesheim process — required to make one pound of acrylonitrile would cost 9.0-10.0 cents per pound, 85 percent of the entire cost of production using the propylene-ammonia route. This latter process, or similar ones developed recently by Distillers and SNAM Progetti (the latter using a fixed-bed reactor), is therefore unlikely to be replaced — in fact, in the U.S. some closing down of acetylene-HCN plants has already begun and Sohio units are being erected in their place.

The U.S. producers as of mid-1966, their respective capacities and routes, are indicated below.

Producer	Location	Capacity, MM lbs/yr	Route
American Cyanamid	Fortier, La.	110.0	Acetylene-HCN
	Fortier, La.	120.0	Sohio
DuPont	Memphis, Tenn.	100.0	Sohio
	Beaumont, Texas	125.0	Propylene-NO
Goodrich	Calvert City, Ky.	45.0	Sohio
Monsanto	Texas City, Texas	140.0	Acetylene-HCN
	Chocolate Bayou, Texas	280.0	Acetylene-HCN
Sohio	Lima, Ohio	280.0	Sohio
Total		1200.0	

Source: Based on *Hydrogen Processing*, January 1967.

The domestic demand for acrylonitrile in the U.S. is growing at a rate of 19 percent per year, largely as a result of the growing demand for acrylic fibers and ABS resins. Until the mid-1960's, the U.S. was a large exporter of acrylonitrile, but the advent of the Sohio process encouraged foreign consumers to produce their own which considerably diminished the export material.

The domestic market distribution for acrylonitrile in the U.S. during 1965 was the following:

End-use	MM lbs, 1965
Acrylic and Modacrylic Fibers	360.0
ABS and SAN Resins	105.0
Nitrile Rubber	52.0
Export	165.0
Other	68.0
Total	750.0

Source: *Oil, Paint & Drug Reporter*, September 27, 1965.

4.1 Acrylic Fibers

Acrylonitrile, alone or as a copolymer, can be solvent-spun into a fiber which competes mainly with wool and silk, and to a lesser extent, with polyester fibers and texturized nylon. Their main characteristic is their resistance to weathering; they can also be made crease-proof with relative ease. "Acrylics," which make up the bulk of the market, contain more than 85 percent acrylonitriles. "Modacrylics" are those fibers in which acrylonitrile represents less than 85 percent by weight. Acrylic fiber is about 25 percent cheaper than wool and thus has captured most of the nonapparel market for wool, as well as a considerable portion of certain apparel applications.

Due to the lack of a polar group, pure polyacrylonitrile is not easy either to dye or to dissolve in commercially available solvents. To improve dyeability, acrylic fiber resins contain comonomers with polar groups, usually methyl methacrylate and other vinyl-type compounds.

Producer	Location	1964 Capacity MM lbs/yr	Fiber
American Cyanamid	Pensacola, Fla.	55.0	"Creslan"
Monsanto	Decatur, Ala.	200.0	"Acrilan"
Dow	Williamsburg, Va.	25.0	"Zefran"
DuPont	Camden, S.C. Waynesboro, Va.	210.0	"Orlon"
Tennessee Eastman	Kingsport, Tenn.	25.0	"Verel"
Union Carbide	S. Charleston, W.Va.	15.0	"Dynel"
Total		530.0	

Source: *Chemical and Engineering News,* October 31, 1966.

About 355.0 million pounds of acrylic fibers were used in 1965, requiring an approximately equal amount of monomer. Almost 100 percent of the fiber is produced as staple. The end-use distribution for acrylics is shown below.

End-use	MM lbs, 1965
Sweaters	71.0
Pile fabrics	35.0
Carpets and rugs	144.0
Blankets (woven)	17.0
Other knit	50.0
Broad wovens	25.0
Miscellaneous	13.0
Total	355.0

Source: *Chemical and Engineering News*, October 31, 1966.

The main producers tend to specialize in certain principal outlets. Thus, Acrilan is used mostly for carpets and blankets, whereas Orlon has its largest outlets in apparel applications. The world-wide demand for wool is expected to remain static at around 400.0 million pounds per year, the increment being taken up by acrylic fibers.

4.2 Acrylamide

Polyacrylamide is used as a flocculatant in the concentration of uranium and other ores, as well as in water, waste and sewage treatment. It is also a constituent of certain adhesives. Acrylamide polymers are also used in hair sprays, as cosmetic thickeners, and as soil stabilizers, despite their relatively high cost. A copolymer of acrylamide and acrylic acid is being widely used in paper manufacture to improve strength and wet-end behavior. Acrylamide makes up about 15 percent of certain thermosetting acrylic resins in the form of its methylol derivative.

The total consumption of acrylamide for all these uses amounted to around 14.0 million pounds in 1965. It is prepared by controlled sulfuric acid hydrolysis of acrylonitrile:

$$CH_2{=}CH{-}CN \xrightarrow[\text{H}_2\text{SO}_4]{\text{H}_2\text{O}} CH_2{=}CH{-}\overset{\displaystyle O}{\overset{\|}{C}}{-}NH_2$$

4.3 Other Acrylonitrile Derivatives

Some 35.0-40.0 million pounds per year of monosodium gluta-mate (MSG) made in the U.S. are produced by fermentation tech-nology. In Japan, however, the Ajinomoto synthetic process is now being used in a 10.0 million pounds per year plant. The route is an unusual application of the oxo reaction:

$$CH_2\!\!=\!\!CH\!\!-\!\!CN \xrightarrow{\;CO + H_2\;} CHO\!\!-\!\!CH_2\!\!-\!\!CH_2\!\!-\!\!CN \xrightarrow[NH_3]{HCN;}$$

$$
\underset{\textstyle NC\!\!-\!\!\overset{\displaystyle \overset{NH_2}{|}}{CH}\!\!-\!\!CH_2CH_2\!\!-\!\!CN}{} \xrightarrow[H_2O;\,NaOH]{H^+;} NaOOC\!\!-\!\!\overset{\displaystyle \overset{NH_2}{|}}{CH}\!\!-\!\!CH_2CH_2COOH
$$

The only use for MSG is in taste-enhancing products in food process-ing or at the consumer level. The main U.S. producers are Inter-national Minerals and Chemicals, Commercial Solvents and Merck (but by fermentation).

The heat and mildew resistance of cotton can be improved con-siderably by modification with acrylonitrile. Since one-third of the final product by weight is acrylonitrile, the process seemed un-attractive until the introduction of the Sohio process lowered the price of merchant acrylonitrile to 14.5 cents per pound. Cyano-ethylation is now once again being considered as a potentially im-portant application of acrylonitrile. Some acrylonitrile is already being used by General Electric for mildew-proofing kraft paper used in transformer insulation, and by American Cyanamid for their "Cyanocel" products.

The potential acrylonitrile market for cotton cyanoethylation is considered to be 500.0 million pounds per year, of which 10 per-cent is for heat-resistant applications and the rest for uses in which mildew-resistance is required.

Acrylonitrile is one of the raw materials for the manufacture of the well-known coccidiostat "Amprol," a quaternary ammonium derivative of pyridine.

In the field of surface-active agents, acrylonitrile is used to make fatty diamines of the type:

$$RNH_2 + CH_2\!\!-\!\!CH\!\!-\!\!CN \longrightarrow RNHCH_2CH_2CN \xrightarrow{H_2} RNHCH_2CH_2CH_2NH_2$$

These are used principally as asphalt additives.

4.4 Acetonitrile

About 0.1 pound of acetonitrile is produced per pound of acrylo-nitrile when employing the propylene-ammonia route.

Eastman and Union Carbide make acetonitrile from acetic acid and ammonia:

$$CH_3COOH + NH_3 \rightarrow CH_3CN + 2H_2O$$

With the enormous amounts of acetonitrile that are becoming available from acrylonitrile plants, this route will probably be abandoned.

The demand for acetonitrile in the U.S. is about 4.0 million pounds per year, most of which is used in feed-preparation for butadiene plants (Shell, Humble and Polymer Corp. use this method) to separate butylenes from butanes.

Acetonitrile is also an intermediate for making vitamin B_1, and a reaction medium for making steroid drugs. It is the spinning solvent for "Darvan," a fiber made from vinylidene cyanide and vinyl acetate, and has been used as a solvent for certain specialty coatings.

Acetoguanamine, made from acetonitrile and dicyandiamide, is a possible comonomer for melamine formaldehyde molding powders.

2-Methyl imidazole is a dyeing auxiliary for cationic dyeing of acrylic fibers. It has also been proposed as an epoxy resin hardener and a urethane catalyst.

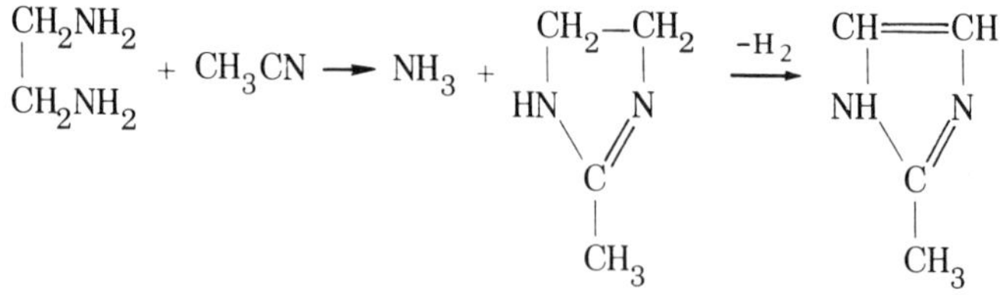

5 ALLYL CHLORIDE

High temperature chlorination of propylene leaves the double bond intact and gives mainly allyl chloride:

$$CH_3CH{=}CH_2 + Cl_2 \rightarrow ClCH_2CH{=}CH_2 + HCl$$

Fresh propylene is mixed with unreacted recycle in an approximately 1:3 ratio, dried, and preheated to 650° F in a fired heater. The combined feed is then mixed with chlorine and introduced into the reactor, a simple steel tube operating adiabatically. The exothermic reaction raises the temperature to 950° F. Chlorine-to-propylene ratio in the reactor feed is 1:4; propylene conversion is 25 percent per pass; chlorine conversion stoichiometric. The overall allyl chloride yield is 80-85 percent. The reactor effluent is cooled to 120° F and prefractionated to remove unreacted propylene and byproduct HCl, the latter being recovered as the usual 33 percent solution. The product stream is washed with NaOH to remove the last traces of acid before the product is fractionated into light ends (small amounts), allyl chloride and about 0.25 pound of dichloropropanes and propanes and other heavier chlorinated compounds per pound of product. A flow diagram is shown in Fig. 6-3.

The investment for a 30.0 million pounds per year allyl chloride plant is $2.2 million. Production costs are estimated below. All U.S. producers have captive sources of chlorine, so this is taken at 2.5 cents per pound; the feed must be pure propylene, so the 3.0 cents per pound figure applies.

Cost, ¢/lb allyl chloride

Net raw materials		3.4
Propylene	2.1	
Chlorine	2.8	
Less: HCl	(-0.9)	
Less: dichloropropane	(-0.6)	
(2.3 ¢/lb)		
Utilities		0.7
Labor and overhead		0.8
Capital charges		2.4
Total		7.3

The most important outlet for allyl chloride is the production of epichlorohydrin, starting point for certain synthetic glycerine routes and for epoxy resin manufacture. Other derivatives are allyl alcohol and diallyl amine, as well as pharmaceuticals such as "Seconal" and other barbiturates. Based on the output of related derivatives, the demand for allyl chloride can be estimated at 177.0 million pounds per year for 1965, distributed as follows:

End-use	MM lbs, 1965
Epichlorohydrin	153.0
Allyl alcohol	14.0
Others	10.0
Total	177.0

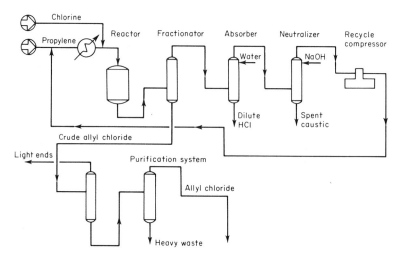

Fig. 6-3 Allyl Chloride.

There are two allyl chloride producers in the U.S.

Producer	Location	Capacity, MM lbs/yr
Dow Chemical	Freeport, Texas	120.0
Shell	Houston, Texas	180.0
Total		300.0

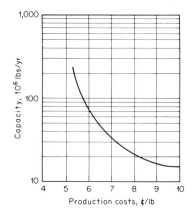

Fig. 6-4 Allyl Chloride.

5.1 Epichlorohydrin

The main outlet for allyl chloride is the manufacture of epichlorohydrin, which in turn is one starting point for synthetic glycerine production and for making epoxy resins.

The manufacture of epichlorohydrin resembles that of propylene oxide.   Allyl chloride reacts with hypochlorous acid, formed by passing chlorine countercurrent to water in a tower. The reactor effluent is separated into an aqueous phase, which returns to the chlorine tower, and an organic phase which is sent to a second reactor where it meets a lime slurry. The effluent from this reactor is stripped of the crude product, leaving behind a calcium chloride slurry. Crude epichlorohydrin, consisting of an azeotrope with water, is fractionated to yield 89 percent epichlorohydrin.   A further purification step is necessary if epoxy resin grade epichlorohydrin is required. Yields are around 80 percent; a flow diagram appears in Fig. 6-5.

$$ClCH_2CH-CH_2 + HOCl \rightarrow ClCH_2CHOHCH_2Cl \xrightarrow{Ca(OH)_2} ClCH_2CH\underset{O}{\overset{}{-}}CH_2$$

The cost of producing epichlorohydrin from captive allyl chloride is shown below.   Not all U.S. producers have their own source of allyl chloride; some epoxy resin manufacturers make their own epichlorohydrin from merchant allyl chloride. The investment for a 40.0 million pound epichlorohydrin plant is $1.7 million; chlorine is again taken at 2.5 cents per pound.

Raw materials	Cost, ¢/lb epichlorohydrin
Allyl chloride	7.6
Chlorine	2.4
Lime	0.2
Utilities	0.8
Labor and overhead	0.7
Capital charges	1.4
Total	13.1

Except for the manufacture of glycerine and epoxy resins, epichlorohydrin has few uses.   One is the manufacture of epichlorohydrin-polyamide resins, used in increasing amounts to impart wet-strength to neutral or alkaline paper stocks.

In 1964, the total demand for epichlorohydrin in the U.S. was 146.0 million pounds per year, distributed as follows:

End-use	MM lbs, 1964
Glycerine	89.0
Epoxy resins	54.0
Other	3.0
Total	146.0

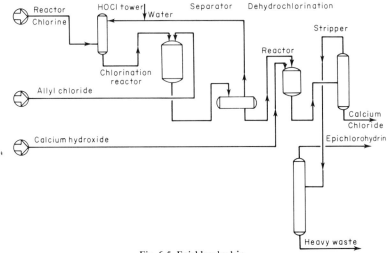

Fig. 6-5 Epichlorohydrin.

Apart from the two allyl chloride manufacturers mentioned above, epichlorohydrin is produced by two epoxy resin manufacturers:

Producer	Location	Capacity, MM lbs/yr
Ciba	Toms River, N.J.	20.0
Union Carbide	S. Charleston, W.Va.	20.0
Total		40.0

These two companies produce only high-purity epichlorohydrin. The other two convert part of their output to epoxy-grade, most of it being used to make glycerine.

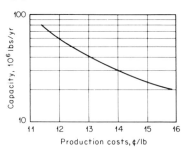

Fig. 6-6 Epichlorohydrin (from captive allyl chloride).

Glycerine

In 1965, about half of the glycerine made in the U.S. was still "natural," that is, a byproduct of soap manufacture and fat splitting for fatty acid production. The growth of detergent consumption has kept the production of soap and other natural glycerine sources fairly static over the years, and the demand for glycerine has been rising steadily. Thus synthetic glycerine has greatly increased its share of the total glycerine consumed in the U.S. In the rest of the world detergents have not yet acquired the relative importance they enjoy in the U.S. Even so, the same trend is at work. The income elasticity of the demand for the various glycerine derivatives is higher than that for soap, so every economy can be considered to be in danger of running into a glycerine shortage as it becomes more prosperous.   There are already two plants operating in Europe (in Holland and France), and one in the U.A.R. Construction is under way for a synthetic glycerine plant in the U.S.S.R.

Synthetic glycerine is made via several routes. Shell and Dow first developed the classical process via hydrolysis of epichlorohydrin with dilute NaOH.   Conversion is almost complete; the effluent is dilute glycerine containing sodium chloride, which is removed by multiple effect evaporation and crystallization. Vacuum distillation yields first 90 percent and then 99 percent glycerine. A flowsheet is shown in Fig. 6-7.

$$ClCH_2CH\underset{\underset{O}{\diagdown\diagup}}{—}CH_2 + NaOH + H_2O \rightarrow HOCH_2CHOHCH_2OH + NaCl$$

Next, Shell developed its process starting from acrolein (Fig. 6-8). The reactions are:

$$CH_2{=}CHCHO + CH_3CHOHCH_3 \rightarrow CH_2{=}CHCH_2OH + CH_3COCH_3$$

$$CH_3CHOHCH_3 + O_2 \rightarrow CH_3COCH_3 + H_2O_2$$

$$CH_2{=}CHCH_2OH + H_2O_2 \rightarrow HOCH_2CHOHCH_2OH$$

Allyl alcohol produced by isomerization of propylene oxide can be chlorinated and dehydrochlorinated with caustic soda to give epichlorohydrin, which is converted to glycerine as above. This route is used in the Olin plant, at present, not operating.

$$CH_2{=}CH—CH_2OH + Cl_2 \rightarrow ClCH_2CHClCH_2OH \xrightarrow{\ NaOH\ } ClCH_2CH\underset{\underset{O}{\diagdown\diagup}}{—}CH_2$$

Finally, Atlas makes synthetic glycerine by a unique method involving catalytic hydrogenation of sugar.

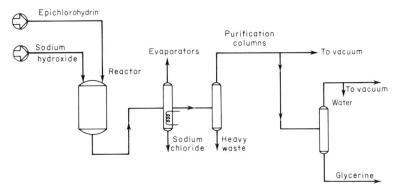

Fig. 6-7 Glycerine.

The economics of the first and third routes are compared below. The yield of epichlorohydrin conversion to glycerine is 95 percent, that of allyl alcohol conversion via chlorine between 75 and 80 percent. The capacity in either case has been taken at 40.0 million pounds per year, which is compatible with those assumed earlier for the various intermediates. Investment requirements are $1.3 million and $1.7 million, respectively.

Via epichlorohydrin	Cost, ¢/lb glycerine	Via allyl alcohol	Cost, ¢/lb glycerine
Epichlorohydrin	13.5	Allyl alcohol	11.1
Sodium hydroxide	2.1	Chlorine	2.4
Utilities	0.4	Caustic soda	1.1
Labor	0.4	Utilities	1.0
Capital charges	1.1	Labor	0.8
Total	17.5	Capital charges	1.4
		Total	17.8

Within the limits of the precision that can be expected from this kind of appreciation, therefore, the two routes appear to be equivalent. One advantage of the propylene oxide route is that it does not

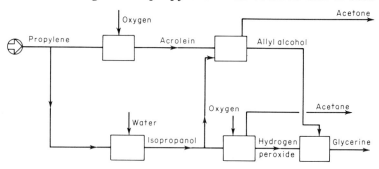

Fig. 6-8 Glycerine (Shell Process).

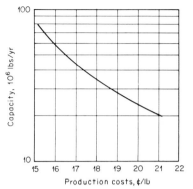

Fig. 6-9   Glycerine from Captive
Allyl Chloride (via Epichlorohydrin).

require pure propylene as a feedstock, whereas allyl chloride
manufacture does.   Nevertheless, the plant using the propylene
oxide route is not in operation.

The total demand for glycerine in the U.S. is increasing slowly
from around 300.0 million pounds per year.   Sales of natural
glycerine are expected to remain static while all increases will be
handled by synthetic glycerine.   Current distribution is 60 percent
for synthetic glycerine and 40 percent for natural glycerine.   The
rather static nature of the soap industry was compensated for by
the increasing demand for fat-splitting derivatives — fatty acids
for a variety of uses, fatty alcohols for detergents.   However, these
sources are decreasing or expected to decrease.   Furthermore,
synthetic glycerine is excluded from the food, drug, and other
markets.   The demand for glycerine can be distributed as follows:

End-uses	MM lbs, 1965
Alkyd resins	69.0
Cellophane	42.0
Tobacco	45.0
Foods and beverages	31.0
Drugs and cosmetics	59.0
Explosives	17.0
Polyurethanes	14.0
Exports	52.0
All other	17.0
Total	346.0

Source: *Oil, Paint & Drug Reporter*,
    May 23, 1966.

In the coatings industry, glycerine is used in formulating short and
medium oil alkyd resins.   Use in drugs and cosmetics is varied:

syrups, suppositories, and proprietary formulations, in general, are typical drug applications, and cosmetic uses include toothpaste, creams and lotions of all kinds. Glycerine in cosmetics is used as such and as the monoglyceride of various fatty acids. In the food industry, the chief outlets for glycerine are in margarine and shortening. Glycerine is also a humectant for cork products and for tobacco in competition with other polyfunctional alcohols such as propylene glycol and diethylene glycol.

At present, nitroglycerine is made by continuous nitration of high-purity glycerine. It is the main ingredient in dynamite; peacetime consumption in the U.S. is about 14.0 million pounds per year.

$$HOCH_2CHOHCH_2OH + 3HNO_3 \xrightarrow{H_2SO_4} \begin{matrix} CH_2-CH-CH_2 \\ | \quad | \quad | \\ ONO_2 \; ONO_2 \; ONO_2 \end{matrix} + 3H_2O$$

Triacetin, the triacetic ester of glycerine, is one of the preferred plasticizers for cellulose acetate tow, used to make cigarette filters. It is not a very effective plasticizer, but has the advantage of being nontoxic. About 9.0 million pounds per year are consumed in the U.S.

The following U.S. companies can produce synthetic glycerine. The list of natural glycerine producers is extensive, including all the major soap manufacturers and fat-splitting plants; Procter & Gamble alone can produce around 60.0 million pounds per year.

Producer	Location	Capacity MM lbs/yr	Process
Atlas	Atlas Point, Del.	25.0	Carbohydrates hydrogenation
Dow	Freeport, Texas	80.0	Allyl chloride
Olin	Brandenburg, Ky.	40.0	Propylene oxide*
Shell	Houston, Texas	110.0	Allyl chloride
	Norco, La.	35.0	Acrolein
Total		290.0	

*Not operating.

Source: *Oil, Paint & Drug Reporter*, May 23, 1966.

Epoxy Resins

The term "epoxidation" refers to the introduction of $\overset{O}{\overset{/\backslash}{C-C}}$ at some point in the molecule. This can be accomplished in two completely different manners. First, a component containing a double bond can react with hydrogen peroxide or a peracid to form the epoxy linkage. However, the term "epoxy resins" is usually reserved for the second manner. These resins are compounds made by reacting epichlorohydrin with a molecule containing two hydroxyl groups; polymerization occurs because both the chlorine atom and the epoxy linkage can react with the -OH group. Most epoxy resins are the product of reaction between epichlorohydrin and bisphenol-A; an example of this type of polymer is shown below.

$$\left[-O-\bigcirc-\underset{\underset{CH_3}{|}}{\overset{\overset{CH_3}{|}}{C}}-\bigcirc-OCH_2-\underset{\underset{OH}{|}}{CH}-CH_2- \right]$$

There are numerous varieties of epoxy resins and applications for them; they are considered to involve technical service costs over twice those for the chemical industry in general. Molecular weights can range from around 380, corresponding to the molar reaction between epichlorohydrin and bisphenol-A, to over 10,000 for the resins used in automobile coating primers.

Coatings represent the most important application for epoxy resins. They are extremely resistant to heat and corrosion, but have a tendency to chalk, that is, to fail in sunlight. Therefore, they are used either as primers or in applications where sunlight is no problem, such as inside gas-transmission lines. The excellent adherence of epoxy resins to metals accounts for their use as primers in the automobile and household appliances industries and as maintenance paints.

Structural and electrical laminates are next in importance, with television-tube laminates being largest outlet. Applications where heat and chemical resistance are especially important are usually made from epichlorohydrin and a conventional phenol-formaldehyde, Novolak, as the polyhydroxyl component. Potentially a very important application is as a protective agent for roads and especially for public works such as tunnels and bridges, to keep them from cracking under thermal shock. Although these constructions are often expensive, they are also of strategic importance. Despite the high cost involved this application may become very large. Epoxy resins are also used in adhesives, of which the best-known examples are the two-component household adhesives consisting of the resin itself and a hardening agent, which are mixed just before using. Tool-handles, especially in the aircraft industry, and applications in the electronic industry for potting and encapsulating heat-sensitive components and reinforced plastics round out the main end-uses.   The coal tar-epoxy mixtures are recent developments.   They have the advantage of reducing the cost of an epoxy coating by 50 percent and are used widely wherever the black color of the resulting mixture is no objection.   Applications such as tank linings, filament-wound vessels and other process-equipment applications have also been included under coatings.

The total market for epichlorohydrin-derived epoxies — including resins made from bisphenol-A, brominated bisphenol-A, cycloaliphatic epoxies, etc. — was 109.0 million pounds in 1965. Coatings, still the most important application, are used mainly in maintenance paints and by the automobile industry.   They have been losing ground in some applications, for example, to polyurethanes in heavy-duty wood finishes and to polybutadiene resins in can-linings.   In construction, epoxy resins are used in flooring and in road and bridge deck coating.   Epoxy resins represent about 5 percent of the total demand for reinforced plastics, main outlets being filament-wound pipe and vessels both in commercial and military applications, tooling and laminates for tubing, printed circuits, skis, etc.   The market distribution in 1965 was as follows:

		1965 Consumption, MM lbs/yr
Coatings		48.0
Plant maintenance	14.5	
Automotive primers	9.5	
Can and drum coatings	6.5	
Appliance finishes	5.0	
Trade paints and other	6.5	
Pipe coatings	6.0	
Bonding and adhesives		14.0
Road and bridge coating	2.0	
Flooring	4.5	
Adhesives	1.0	
Military uses and trade sales	6.5	
Reinforced plastics		17.0
Filament winding	4.0	
Tooling	3.0	
Laminates	10.0	
Exports		11.6
Other		18.4
Total		109.0

Source: *Modern Plastics*, January 1966.

Dow was the original manufacturer of epoxy resins, but today Shell is the major company in the field.   U.S. producers and capacities are given on the next page.

Producer	Location	Capacity, MM lbs, 1964
Celanese	Louisville, Ky.	12.0
Ciba	Toms River, N.J.	18.0
Dow	Freeport, Texas	35.0
Reichhold	Ballardvale, Mass.	10.0
Shell	Houston, Texas	55.0
Union Carbide	Marietta, Ohio	18.0
Total		148.0

Source: *Oil, Paint & Drug Reporter*, June 1,1964.

The demand for epoxy resins has been growing at a rate of over 10 percent per year.

Other Epichlorohydrin Derivatives

Copolymers of epichlorohydrin and bisphenol A having molecular weights around 30,000-40,000, known as phenoxy resins, are being used in the automotive industry to make molded parts such as instrument panels, arm-rests and wheel-house covers. Their best prospects, however, are coatings and especially structural adhesives to bind ornamental trim to the automobile body, avoiding galvanic corrosion.

The difference between phenoxy and epoxy resins is that the former contain no epoxy end-groups. The demand in 1965 was almost 1.0 million pounds, and the only producer in the U.S. is Union Carbide.

Epichlorohydrin can be homopolymerized or copolymerized with ethylene oxide to give elastomers having excellent fabrication properties, chemical stability and resistance to tear and abrasion.

$$\left[-CH_2-\underset{\underset{CH_2Cl}{|}}{CH}-O-CH_2-CH_2-O- \right]$$

These products are expected to compete with neoprene and nitrile rubber. Their limitations are poor low-temperature properties and, in the case of the copolymer, low flame-retardancy. Vulcanization is by means of difunctional amino-compounds which react with the chloromethyl group.

5.2 Allyl Alcohol

There are three routes for making allyl alcohol. The first is the hydrolysis of allyl chloride with sodium hydroxide.

$$ClCH_2CH{=}CH_2 + NaOH \longrightarrow HOCH_2CH{=}CH_2 + NaCl$$

Propylene oxide can be isomerized to allyl alcohol; this reaction was employed in the Olin synthetic glycerine process:

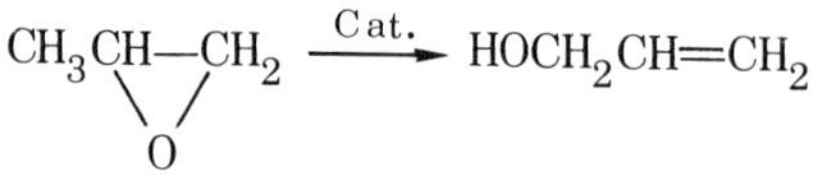

$$CH_3CH{-}CH_2 \xrightarrow{\;Cat.\;} HOCH_2CH{=}CH_2$$
$$\diagdown O \diagup$$

Finally, Shell at its Norco, La. plant uses a route to allyl alcohol involving the reaction between isopropanol and acrolein. This unit is part of a complex producing mainly glycerine and acetone.

Apart from its use by Olin as an intermediate in glycerine manufacture, allyl alcohol is used in a number of applications where its double functionality is of interest. The total demand for allyl alcohol is around 5.0 million pounds per year, in addition to its use as an intermediate in the Olin and Shell glycerine routes.

Allyl Alcohol Derivatives

The allyl ester of phthalic anhydride polymerizes to give resins used particularly in low-cost decorative laminates for furniture. The annual demand is 2.0 million pounds. Triallylcyanurate is made by reacting cyanuric chloride with allyl alcohol.

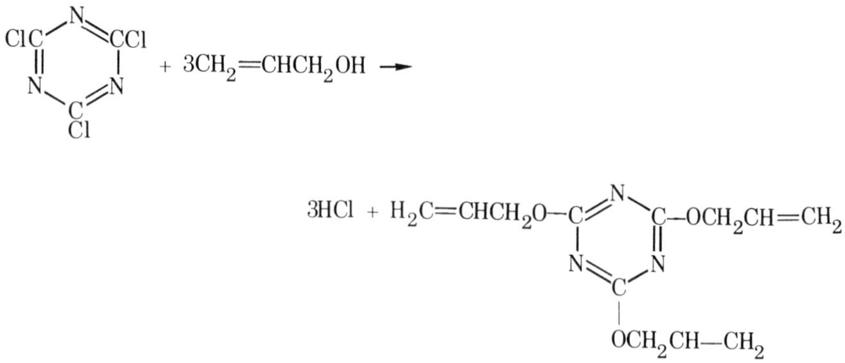

$$+ 3CH_2{=}CHCH_2OH \longrightarrow$$

$$3HCl + H_2C{=}CHCH_2O{-}C \cdots C{-}OCH_2CH{=}CH_2$$

It is used as a substitute for styrene in the formulation of high-temperature-resistance polyesters and in certain applications of methyl methacrylate.

Monsanto has built a plant to make allyl alcohol-styrene co-polymers, to be used in water-soluble paints.

Mono and diallyl ethers of trimethylopropane are recommended as crosslinking agents for baked acrylic coatings, adhesives, textile resins, polyester coatings and other applications.

5.3 Diallyl Amine

Allyl chloride reacts with ammonia to give diallyl amine, one of the raw materials for "Randox," an important herbicide.

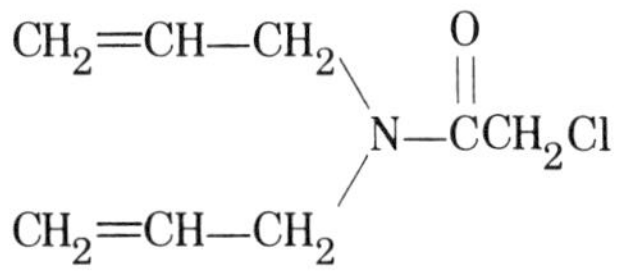

$$CH_2=CH-CH_2 \diagdown \atop CH_2=CH-CH_2 \diagup N-\overset{O}{\overset{\|}{C}}CH_2Cl$$

5.4   1,2-Dibromo-3-Chloropropane

This compound is used as a nematocide, especially on perennial plants such as citrus trees. It has the advantage of having to be aerated out of the soil just before planting time, as is the case with the more commonly used D-D. Consumption in the U.S. was around 3.5 million pounds per year.

$$CH_2=CHCH_2Cl + Br_2 \longrightarrow BrCH_2CHBrCH_2Cl$$

5.5 Dichloropropanes, Dichloropropenes

Dichloropropanes and dichloropropene are byproducts of the high-temperature chlorination of propylene to allyl chloride; di-chloropropanes are also obtained in the manufacture of propylene glycol.   About 32.0 million pounds per year of a mixture of 1,2-dichloropropane and 1,3-dichloropropene are marketed under the trade-names "Dowfume" and "D-D" for use as a soil fumigant, especially for tobacco field and seed-bed treatment. Consumption of these products represents about two-thirds of the total U.S. market for nematocides.   These byproducts can also be fed to chlorinated solvent plants together with other raw materials; this is done by producers both of allyl chloride and propylene oxide, which yields mainly 1,3-dichloropropane as a byproduct.

6 ACROLEIN

There are two routes to acrolein. The first, used by Union Carbide, involves the reaction between formaldehyde and acetaldehyde, followed by dehydration:

$$CH_3CHO + CH_2O \longrightarrow CH_3CHOHCHO \xrightarrow{-H_2O} CH_2{=}CH{-}CHO$$

Shell, the largest acrolein producer, uses direct oxidation of propylene with oxygen. Conversion per pass is low, so the process requires very pure feedstocks in order to diminish buildup of undesirable compounds in the recycle stream. Steam is used as a diluent.

$$CH_2{=}CHCH_3 + O_2 \longrightarrow CH_2{=}CH{-}CHO + H_2O$$

Acrolein is used mainly as the starting point of the Shell glycerine process. Union Carbide, however, makes a number of other derivatives from acrolein, the most important of which are 1,2,6-hexanetriol and glutaraldehyde. Acrolein is also a raw material for methionine.

An idea of the yield for this process can be gained from an oxidation route developed by Montecatini, which employs no diluent. The total aldehyde yield is 75 percent, of which 91 percent is acrolein and the rest mainly propionaldehyde, acetaldehyde, and formaldehyde. Raw material costs for this process are given below.

	Cost, ¢/lb acrolein
Propylene	3.3
Oxygen	0.3
Total	3.6

Credits for various byproducts are not counted. The capacities of the two U.S. producers are as follows:

Producer	Location	Capacity, MM lbs/yr
Shell	Norco, La.	40.0
Union Carbide	S. Charleston, W.Va.	10.0

Aside from the amounts used to make glycerine, the U.S. demand for acrolein was around 7.0 million pounds per year in 1965.

6.1 Methionine

In addition to the small amounts used in pharmaceuticals to treat liver disorders, the entire output of methionine is used as an animal feed supplement.   The total U.S. market at present is around 12.0 million pounds per year, but an increase in the price of natural amino acids such as fish-meal could have a profound impact on the demand.   Most of the methionine is made by the following route:

$$CH_2{=}CH{-}CHO + CH_3SH \longrightarrow CH_3{-}S{-}CH_2{-}CH_2{-}CHO$$

$$CH_3{-}S{-}CH_2{-}CH_2{-}CHO + HCN + (NH_4)_2CO_3 \longrightarrow$$

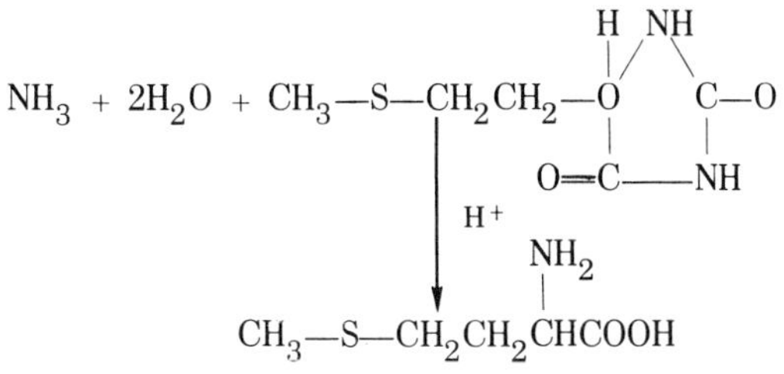

$$NH_3 + 2H_2O + CH_3{-}S{-}CH_2CH_2{-}\underset{\underset{\underset{CH_3-S-CH_2CH_2CHCOOH}{NH_2}}{\overset{H^+}{\downarrow}}}{\overset{\overset{H\ \ NH}{\underset{O=C\underline{\quad}NH}{\overset{|\diagup\diagdown}{C-O}}}}{O}}$$

The main producer is Dow which has announced the commercial use of a new, undisclosed, process in its latest plant expansion. In 1967 Monsanto announced its intention to build a 10.0 million pounds per year plant.

6.2 1,2,6-Hexane Triol

This polyfunctional alcohol is made by dimerizing acrolein, followed by hydrolysis and hydrogenation:

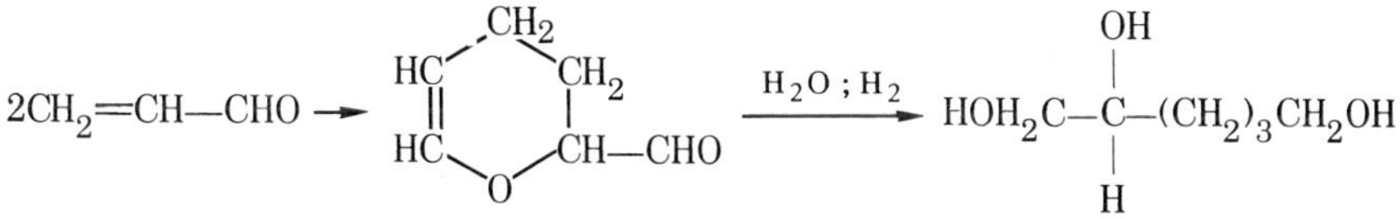

$$2CH_2{=}CH{-}CHO \longrightarrow \begin{array}{c} CH_2 \\ HC \diagup \quad \diagdown CH_2 \\ || \qquad | \\ HC \diagdown \quad \diagup CH{-}CHO \\ O \end{array} \xrightarrow{H_2O\,;\,H_2} HOH_2C{-}\underset{H}{\overset{OH}{\underset{|}{\overset{|}{C}}}}{-}(CH_2)_3CH_2OH$$

Its main use is in alkyd resins and polyester manufacture.

6.3 Glutaraldehyde

Acrolein and ethyl vinyl ether react to give 2-ethyl-3,4-dihydro-1,2-pyran, which can be hydrolyzed to glutaraldehyde and ethanol:

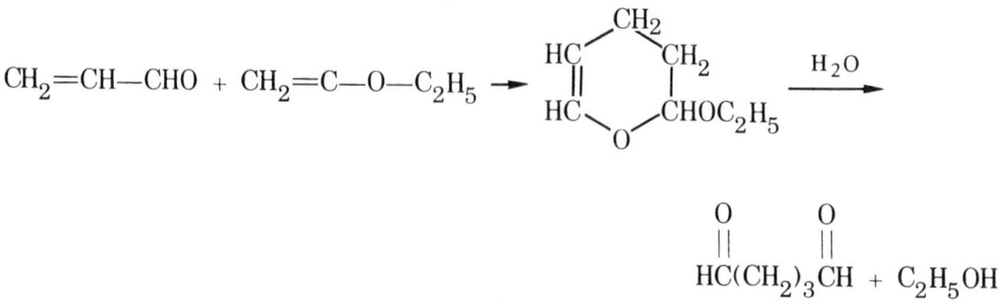

$$CH_2=CH-CHO + CH_2=C-O-C_2H_5 \longrightarrow$$

(structure)

$$\xrightarrow{H_2O}$$

$$HC(CH_2)_3CH + C_2H_5OH$$

Glutaraldehyde is used in such applications as textile finishings, glue insolubilization, and leather tanning, in competition with other aldehydes, mainly formaldehyde and glyoxal.

7 PROPYLENE OXIDE

The manufacture of propylene oxide from propylene is similar to that of ethylene oxide by the chlorohydrin route. In fact, propylene oxide operations in the U.S. are converted ethylene oxide plants, which had become obsolete by the advent of direct oxidation. Another source of propylene oxide is the Celanese LPG oxidation plant at Bishop, Texas, which recovers some 10.0 million pounds per year in addition to the many other oxygenated products.

Considerable research has been carried out for a viable process for making propylene oxide dispensing with the use of chlorine. In 1966, Scientific Design announced a route, to be used first commercially in the U.S. and in Spain, involving epoxidation of propylene by means of a hydroperoxide. One raw material suitable for oxidizing to a hydroperoxide is ethyl benzene, and styrene is obtained as a coproduct. Mild conditions are said to prevail throughout. Toyo Soda has claimed good yields by direct oxidation of propylene at around 350°F and 50 atmos.; Monsanto has developed liquid-phase, molecular oxygen direct oxidation. Finally, Celanese has announced a peracetic acid oxidation route:

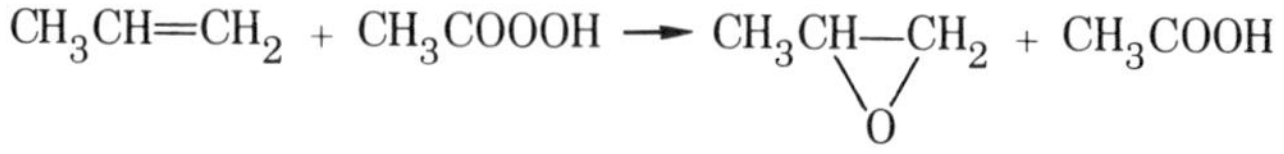

$$CH_3CH=CH_2 + CH_3COOOH \longrightarrow CH_3CH-CH_2 + CH_3COOH$$

The U.S. manufacturers of propylene oxide and their capacities for 1964 are as follows:

Producer	Location	1964 Capacity, MM lbs/yr
Celanese	Bishop, Texas	10.0
Dow	Freeport, Texas	125.0
	Plaquemine, La.	50.0
	Midland, Mich.	25.0
Olin	Brandenburg, Ky.	50.0
Jefferson	Port Neches, Texas	140.0
Wyandotte	Wyandotte, Mich.	140.0
Union Carbide	S. Charleston, W.Va.	200.0
		740.0

Source: *Oil, Paint & Drug Reporter*, April 20, 1964.

Propylene glycol is still the most important outlet for the oxide, although polyethers used in the manufacture of urethane foams and elastomers are rapidly closing the gap.   Total use of propylene oxide in 1964 was 500 million pounds distributed as follows:

End-use	MM lbs, 1964
Propylene glycol	170.0
Adducts for urethane foams	150.0
Higher propylene glycols	70.0
Surfactants	30.0
Miscellaneous	80.0
Total	500.0

Source: *Oil, Paint & Drug Reporter*,
April 20, 1964.

Some producers of unsaturated polyesters use propylene oxide instead of propylene glycol.

Production costs for 50.0 million pounds per year plant are estimated below.   Investment costs are $2.6 million.   Feed to the process can be low-assay propylene.   Chlorine is assumed captive at 2.5 cents per pound, and propylene dichloride credited at 2.3 cents per pound.

End-use		Cost, ¢/lb propylene oxide
Net raw materials		6.4
Propylene	2.1	
Chlorine	3.8	
Lime	0.9	
Less: propylene dichloride	(-0.4)	
Utilities		1.1
Labor and overhead		0.5
Capital charges		1.7
Total		9.7

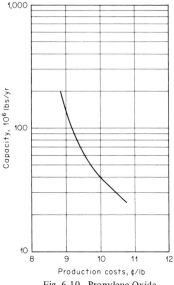

Fig. 6-10   Propylene Oxide.

Propylene dichloride is usually sold as feed to a chlorinated solvent plant.

In 1967, it was reported that the investment for a 50 million pounds per year propylene oxide plant, producing simultaneously 120,000 pounds per year of styrene, using the SD process, was estimated at $15.0 million.

7.1 Propylene Glycol

Hydration of propylene oxide yields propylene glycol; the process is analogous to the manufacture of ethylene glycol.

$$CH_3CH—CH_2 + H_2O \rightarrow CH_3CHOHCH_2OH$$
$$\diagdown O \diagup$$

All propylene oxide producers also make glycols; capacities approximate the following:

Producer	Capacity, MM lbs/yr
Atlas	5.0
Celanese	8.0
Dow	110.0
DuPont	2.0
Jefferson	40.0
Olin	35.0
Union Carbide	80.0

Wyandotte	30.0
Total	310.0

Source: *Oil, Paint & Drug Reporter,*
February 22, 1965.

The U.S. market consumed 257.0 million pounds in 1965, distributed as follows:

End-use	MM lbs, 1965
Unsaturated polyester	103.0
Cellophane	51.0
Hydraulic fluids	26.0
Tobacco	20.0
Polymeric plasticizers	13.0
Export	26.0
Miscellaneous	18.0
Total	257.0

Source: *Oil, Paint & Drug Reporter,*
February 22, 1965.

Propylene glycol is by far the most widely used difunctional alcohol for unsaturated polyester manufacture although diethylene glycol can be used.

The use of propylene glycol as a cellophane plasticizer is the result of regulations regarding toxicity; in the U.S. cellophane coming into contact with food can no longer be plasticized with ethylene glycol. Toxicity regulations also determine the use of propylene glycol in tobacco. Polymeric plasticizers are used mostly in vinyl polymers, especially film, sheeting and wire coating applications; they have the advantage of lower volatility and better resistance to migration and extraction than the usual monomerics. Among the miscellaneous outlets, one may mention that as a binder for cork bottle caps; as a humectant in cosmetics; and as an intermediate for propylene carbonate.

7.2 Polyethers

The term "polyethers," or "polyols," has come to designate a family of compounds made primarily from propylene oxide, used almost exclusively for the manufacture of polyurethane foams. The principle of polyurethane production is a condensation reaction between a multifunctional isocyanate and a resin containing between two and six -OH groups per molecule. This explains the "polyol" appellation for all compounds employed in urethane manufacture, including those of the polyester type.

When polyurethanes were first introduced, polyols were predominantly polyesters made from condensation of a difunctional acid (usually adipic) and ethylene glycol.

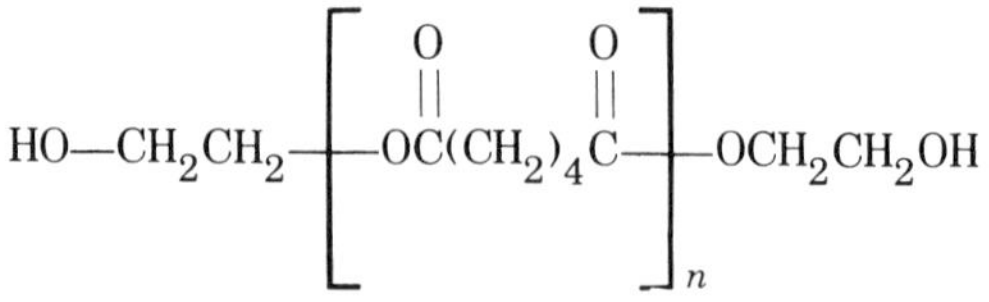

$$HO-CH_2CH_2-\left[-OC(CH_2)_4C-\right]_n-OCH_2CH_2OH$$

However, since the polyol represents a substantial portion of the raw material cost (40 percent for flexible foams, 35 percent for rigid), there was considerable incentive for developing a cheaper class of polyols, since adipic acid is an expensive intermediate. Accordingly, since 1960, a radical shift has taken place, with the result that by 1965 almost 90 percent of the polyols used in flexible foams and 85 percent of those for the rigid applications had been replaced by polyethers.

Polyethers are the product of reaction at high temperature and pressure between a mol of a di- or -poly-functional alcohol or amine with several mols of propylene oxide, using sodium hydroxide as a catalyst. Molecular weights are in the 400-4000 range, the higher values being used for flexible foams and the lower ones for rigid products. The polyethers are known as "diols," "tetrols," etc., according to the number of reactive groups on the parent compound. Diols are mostly propylene oxide adducts of ethylene, or propylene glycols; triols start from trimethylol propane or glycerine; tetrols from pentaerythritol or ethylene diamine; pentols from diethylene triamine; and hexols from sorbitol. The -OH groups on polyethers are secondary and those on polyesters are primary.

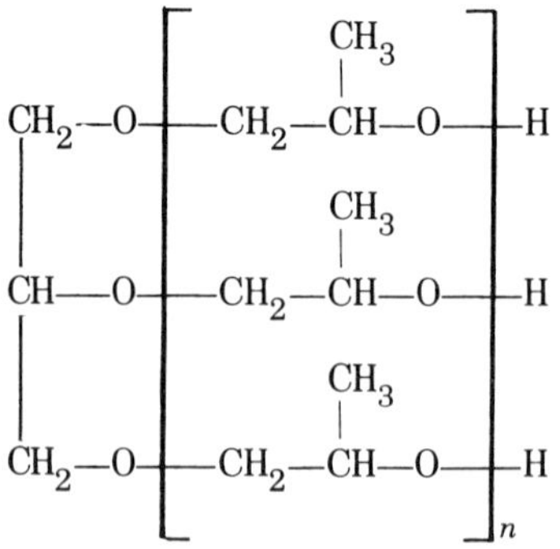

(a triol)

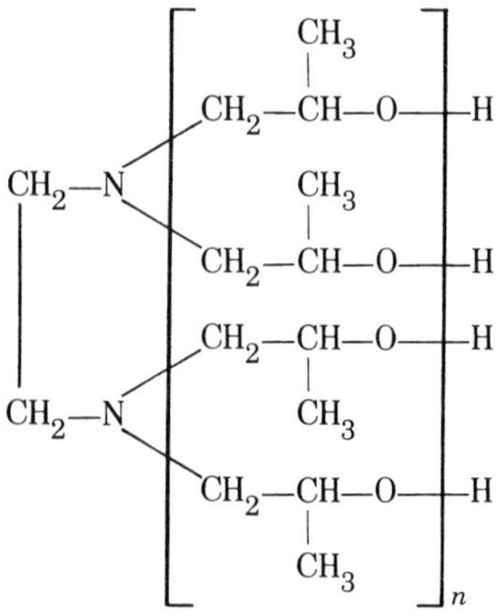

(a tetrol)

The properties of urethanes made from polyesters and polyethers present important differences. Polyethers are more stable to hydrolysis, but less stable to oxidation. Foams made from polyesters have a higher load-bearing capacity, but those using polyethers are more flexible at low temperatures. Polyesters are more stable than polyethers at elevated temperatures. At first, almost all urethane foams were flexible, thus no particular problems were involved in switching from polyesters to polyethers; catalyst solved the problem of reduced reactivity of the secondary compared with primary hydroxyl groups, and since polyethers give superior cushioning ability, and flexible foams go mostly into bedding and furniture, the new polyethers caught on rapidly. Acceptance was not as complete in rigid foams, where load-bearing capacity is important and the use of polyesters thus preferred, but the cost advantage of polyethers eventually captured most of this market as well. The type of application also influences choice of functionality. Triols are by far the most popular polyethers, and with the more rapid growth in rigid — that is, more highly cross-linked — foams, the demand for polyols with three or more groups has been rising. Reactivity, since it influences reaction times, also affects economics, which offsets a part of the cost advantage presented by polyethers. On the other hand, multi-functional isocyanates like polymethyl polyphenyl isocyanate (PAPI), derived from aniline, are cheaper as well as more reactive than tolylene diisocyanate (TDI). Their development is partly responsible for the acceptance of polyethers in rigid foam manufacture. The ratio of primary to secondary hydroxyls can be varied by the blending of, for example, the reaction products of propylene oxide with multifunctional alcohols or amines, which contain secondary groups, with others containing mostly primary hydroxyls.

Flexible urethane foams are still the largest outlet for polyehters, both because they still represent more than twice the market for rigid foams and because in flexible foams the polyol is a larger portion of the recipe. The total demand for polyethers in 1965 was 172.0 million pounds per year, distributed among the various types of polyurethane materials as follows:

End-use	MM lbs, 1965
Rigid foams	43.0
Flexible foams	103.0
Export	17.0
Other	9.0
Total	172.0

Source: *Oil, Paint & Drug Reporter*,
January 4, 1965.

Other uses for polyols include phthalic esters, used in water resistant floor polish formulations. In practice, the only area where polyesters still predominate over polyethers is in the manufacture of textile-foam laminates, since lamination is usually accomplished at temperatures unsuitable for polyether use. The following U.S. companies make polyethers based on propylene oxide:

Producer	Location	Capacity, MM lbs/yr
Allied	Baton Rouge, La.	20.0
Atlas	Wilmington, Del.	25.0
Dow	Freeport, Texas	60.0
Jefferson	Conroe, Texas	20.0
Olin	Brandenburg, Ky.	25.0
Union Carbide	S. Charleston, W.Va.	120.0
Witco	Chicago, Ill.	20.0
Wyandotte	Wyandotte, Mich.	45.0
	Washington, N.J.	20.0
Total		355.0

Source: *Oil, Paint & Drug Reporter*, January 4, 1965.

The capacity is not a very precise concept in the case of polyether production. The above table presents the relative importance of the various manufacturers.

7.3 Dipropylene Glycol

The most important use of dipropylene glycol is as a nontoxic plasticizer for cellophane. Other uses are similar to propylene glycol, and small amounts are employed in combination with diethylene glycol in the "Udex" aromatics extraction process.

7.4 Higher Propylene Glycols

Made by reacting several mols of propylene oxide with propylene glycol, polypropylene glycols have a better viscosity-temperature relationship than castor oil and hence are used widely in brake fluids. Low-molecular-weight polymers are used as lubricants in emulsion cutting oils, and those in the 2000 range as demulsifiers in API separators. Their fatty acid esters are nonionic detergents, as are the adducts of propylene oxide and alkyl-phenols or fatty alcohols. Polypropylene glycols are also included in ink and lacquer formulations, and as the polyhydroxyl component in epoxy resin manufacture in place of bisphenol A. Nonionic detergents, especially petroleum demulsifiers, account for almost 50 percent of the demand for higher propylene glycols.

7.5 Isopropanolamines

These compounds are made by the reaction between propylene oxide and ammonia, and, generally, are similar to ethanolamines. They are used, among other things, in detergents, in certain CO_2 removal processes (e.g., "Sulfinol"), and to make 2,4-D salts.

7.6 Propylene Carbonate

Propylene carbonate is made by reacting propylene oxide and CO_2

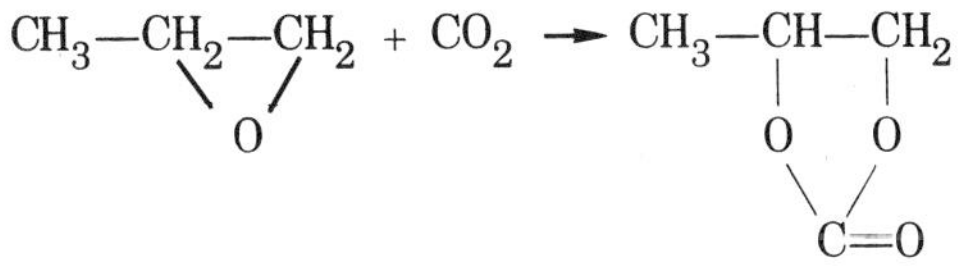

$$CH_3-CH_2-CH_2 + CO_2 \rightarrow CH_3-CH-CH_2$$

It is the solvent required in the Fluor process for removing acid constituents from natural gas.

7.7 1,3-Propylene Diamine

Reaction of 1,3-dichloropropane with ammonia gives 1,3-propylene diamine.

$$ClCH_2CH_2CH_2Cl + 2NH_3 \rightarrow H_2NCH_2CH_2CH_2NH_2 + 2HCl$$

Its main outlet is in the manufacture of wash-and-wear cotton finishes; it is also used to make other types of chemicals previously made primarily from ethylene diamine (paper wet-strength resins, surface active agents). The growth in demand for this amine has been such that one company, Jefferson Chemical, now makes 1,3-dichloropropane by direct synthesis instead of only converting that obtained from its propylene glycol operation.

7.8 Polypropylene Oxide Elastomers

Elastomers having qualities similar to natural rubber plus outstanding low-temperature properties and resistance to ozonolysis are being made by homopolymerizing propylene oxide or copolymerizing it with butadiene monoxide, allyl glycidyl ether, glycidyl acrylate and other unsaturated epoxides.

8 ISOPROPANOL

Most isopropanol is made from propylene in a manner similar to the production of ethanol from ethylene. In the future, however, the Wacker process for making acetone directly from propylene, without going through isopropanol, could account for much of the added acetone production capacity of the world.   Currently, by-product acetone from phenol production supplies a large percentage of U.S. output.   Therefore it is to be expected that production of acetone from isopropanol will not be used in any new projects.

The conventional isopropanol process (see Fig. 6-11) begins with the counterflow absorption of the liquefied feed stream, e.g., depropanizer overhead containing 65 percent propylene, in 75 percent sulfuric acid.   For ethanol, three vigorously agitated reactors are used in series.   The reaction takes place at 400 psig and 140°F.   The reactor effluent is hydrolyzed and the products removed from the acid by steam-stripping. The liquid stream is neutralized with 20 percent caustic soda and sent to purification, and the dilute acid sent to reconcentration.   After unreacted propylene and other light material is flashed off and sent to recompression, crude isopropanol goes to a column where an isopropyl ether-water-alcohol ternary azeotrope is taken overhead and returned to the absorber-reactors, leaving wet isopropanol at the bottom.   This wet isopropanol goes to a column where water and heavy polymers are removed and an 87 percent isopropanol — 13 percent water—azeotrope obtained as overhead. This overhead can be purified by azeotropic distillation if desired.   Yield is 93 percent on propylene feed. Investment costs for a 50.0 million pounds per year isopropanol plant are $3.0 million.   Since the feed need not be free of propane, it will be taken at 2.3 cents per pound propylene.   Production costs are:

Cost, ¢/lb isopropanol

Propylene	1.7
Utilities	0.3
Labor and overhead	0.4
Capital charges	1.9
Total	4.3

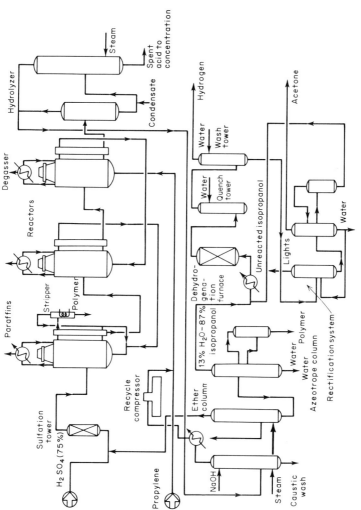

Fig. 6-11 Isopropanol-Acetone.

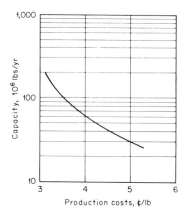

Fig. 6-12 Isopropanol.

Isopropanol producers in the U.S. are given below.

Producer	Location	Mid-1966 capacity, MM lbs/yr
Enjay	Baton Rouge, La.	480.0
Shell	Houston, Texas	330.0
	Dominguez, Cal.	180.0
	Martinez, Cal.	50.0
Union Carbide	S. Charleston, W.Va.	130.0
	Texas City, Texas	250.0
	Whiting, Ind.	280.0
Total		1700.0

Source: *Hydrocarbon Processing*, January 1967.

Consumption of isopropanol in 1965 was 1580 million pounds, distributed as follows:

End-use	MM lbs, 1965	
Chemical uses		
Acetone	860.0	
Glycerine	40.0	
Isopropyl acetate	30.0	
Amines	30.0	
Hydrogen peroxide	10.0	
Other esters and miscellaneous	120.0	
Total chemical uses		1090.0
Solvent uses		
Coatings	120.0	
Rubbing alcohol	50.0	
Drug and cosmetic	25.0	
Government use	10.0	
General de-icing	40.0	
Export	50.0	

All other	195.0
Total solvent uses	490.0
Total	1580.0

Source: *Oil, Paint & Drug Reporter*, November 17, 1966.

As a solvent, isopropyl alcohol is used mainly for gums, shellac and synthetic resins, competing on a price-performance basis with ethanol. It is also used widely as a rubbing alcohol.

8.1 Acetone

While the most important route to acetone both in the U.S. and other countries is the catalytic dehydrogenation of isopropanol, several other sources exist. Celanese, at its LPG-oxidation plant at Bishop, Texas, recovers acetone in addition to numerous other oxygenated compounds. Part of the acetone produced by the Shell plant, at Norco, La., is obtained by reacting acrolein and isopropanol. Although by 1965 no plants were announced in the U.S. (several were built elsewhere, mainly in Japan), the Wacker process for direct oxidation of propylene to acetone using a palladium catalyst promises to obsolete all competing processes except, of course, the byproduct acetone from the phenol-via-cumene process. In Rumania, some acetone is made by steam hydrolysis of acetylene over a zinc oxide catalyst. Figure 6-11 shows a flowsheet for one version of the process for making acetone from isopropanol. Isopropanol-water azeotrope is vaporized and fed to the dehydrogenation furnace, the effluent from which is quenched and washed with water. Two columns remove the water contained in the feed and the unreacted isopropanol, which is recycled. Overall yield is 95 percent.

A successful liquid-phase dehydrogenation process has been developed by the Institut Francais du Pétrole. Yield is 99.5 percent and utility requirements are reduced because reaction temperatures are lower. This process can be used with slight variations for making methyl ethyl ketone (MEK) or cyclohexanone from s-butyl alcohol and cyclohexanol, respectively. The 1124 million pounds per year of acetone used in the U.S. in 1965 can be distributed by source as follows:

	MM lbs, 1965
Isopropanol (inc. via acrolein)	897.0
Cumene	146.0
All other	81.0
Total	1124.0

Source: *Synthetic Organic Chemicals*, U.S. Tariff Commission, 1965.

Manufacturers of acetone and their capacities are:

Producer	Location	1965 capacity, MM lbs/yr	Process
Allied	Frankford, Pa.	140.0	Cumene
Chevron Chemical	Richmond, Cal.	35.0	Cumene
Celanese	Bishop, Texas	35.0	LPG oxidation
Clark Oil	Blue Island, Ill.	30.0	Cumene
Enjay	Linden, N.J.	110.0	Isopropanol
Hercules	Gibbstown, N.J.	15.0	Cumene
Monsanto	Alvin, Texas	75.0	Cumene
Shell	Dominguez, Cal.	150.0	Isopropanol
	Houston, Texas	180.0	Isopropanol
	Norco, La.	100.0	Isopropanol
Skelly	Eldorado, Kan.	30.0	Cumene
Tennessee Eastman	Kingsport, Tenn.	90.0	Isopropanol
Union Carbide	Institute, W.Va.	120.0	Isopropanol
	Texas City, Texas	130.0	Isopropanol
	Whiting, Ind.	120.0	Isopropanol
Total		1360.0	

Source: Based on *Oil, Paint & Drug Reporter*, November 1, 1965; November 7, 1966.

With only one exception (Tennessee Eastman), all producers of acetone from isopropanol have a captive raw material source.

Economics of the acetone market are conditioned by the fact that the phenol-via-cumene process yields a byproduct that can be credited at a very low value. The price at which it would be necessary to charge out byproduct acetone in order to equate the cost of phenol from cumene with that produced by the next-cheapest alternative is less than 3.0 cents per pound. This is lower than the out-of-pocket costs for making acetone from propylene. The investment for a 50.0 million pounds per year acetone plant from isopropanol using the Institut Français du Pétrole process is $500,000. Taking the feed at the previously established cost of 4.3 cents per pound, one arrives at the following cost for acetone:

	Cost, ¢/lb acetone	
Isopropanol	4.4	
Utilities	0.2	
Labor and overhead	0.3	
Capital charges	0.3	
		5.2
Hydrogen	-0.2	
Total		5.0

Acetone via vapor-phase dehydrogenation requires an investment of $600,000 for the same capacity, and production costs are:

	Cost, ¢/lb acetone	
Isopropanol	4.7	
Utilities	0.9	
Labor and overhead	0.3	
Capital charges	0.4	6.3
Hydrogen	− 0.2	
Total		6.1

The U.S. demand for acetone in 1965 can be broken down as follows:

End-use	MM lbs, 1965	
Chemical uses		
Methyl isobutyl ketone	286.0	
Methyl isobutyl carbinol		
Methyl methacrylate	198.0	
Bisphenol-A	33.0	
Other chemical uses	242.0	
Total chemical uses		759.0
Solvent uses		
Cellulose acetate	44.0	
Paint, varnish and lacquer	110.0	
Other solvent uses	176.0	
Total solvent uses		330.0
Exports		11.0
Total		1100.0

Source: *Oil, Paint & Drug Reporter*, November 1, 1965.

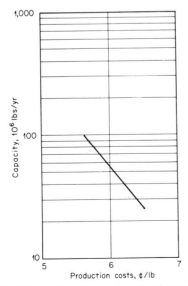

Fig. 6-13   Acetone [vapor-phase dehydrogenation].

Apart from these applications, a certain amount of acetone is at times cracked to ketene for conversion to acetic anhydride whenever there is an imbalance of the acetone market brought about by excessive production from phenol-via-cumene plants.

Diacetone Alcohol (DAA)

DAA is made by condensing two mols acetone. Around 5.0 million pounds per year are consumed mainly as a coating solvent and for pesticide emulsions. Once it was used widely in brake fluids, but has been replaced by hexylene glycol, which is more stable to alkaline hydrolysis:

$$2CH_3COCH_3 \rightarrow CH_3\underset{\underset{\displaystyle CH_3}{|}}{C}OHCH_2COCH_3$$

Methyl Isobutyl Ketone (MIBK)

The only important use of MIBK is as a solvent for surface coatings, in competition with butyl acetate. MIBK allows the preparation of high solids, low viscosity solutions with a lower volatility than MEK or ethyl acetate. It is used especially in spray formulations. About 160 million pounds per year are used in the U.S.

MIBK is made by dehydrating diacetone alcohol to mesityl oxide and hydrogenating the double bond; methyl isobutyl carbinol is obtained as a byproduct.

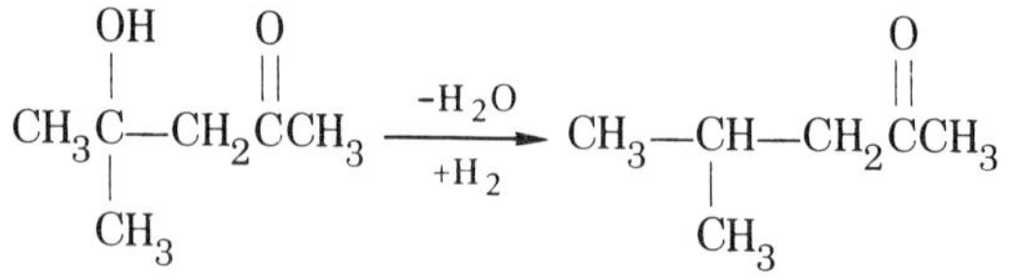

$$\underset{\underset{\displaystyle CH_3}{|}}{\overset{\overset{\displaystyle OH}{|}}{CH_3C}}-CH_2\overset{\overset{\displaystyle O}{\|}}{C}CH_3 \xrightarrow[+H_2]{-H_2O} CH_3-\underset{\underset{\displaystyle CH_3}{|}}{CH}-CH_2\overset{\overset{\displaystyle O}{\|}}{C}CH_3$$

U.S. producers and capacities are as follows:

Producers	Location	Capacity, MM lbs/yr
Enjay	Linden, N.J.	25.0
Shell	Dominguez, Cal.	25.0
	Houston, Texas	80.0

Eastman	Kingsport, Tenn.	20.0
Union Carbide	S. Charleston, W.Va.	30.0
	Texas City, Texas	50.0
Total		230.0

The demand in 1965 was:

End-use	MM lbs, 1965
Nitrocellulose lacquers	90.0
Vinyl coatings	40.0
Solvent extraction, etc.	30.0
Total	160.0

The investment for a 40.0 million pounds per year MIBK plant is $1.2 million.

Hexylene Glycol

The most important use for hexylene glycol is as a lubricant and viscosity regulator in heavy-duty brake fluids, of which some 10 million gallons are used annually. It is made by hydrogenation of diacetone alcohol.

$$\underset{\underset{CH_3}{|}}{CH_3COHCH_2COCH_3} + H_2 \longrightarrow \underset{\underset{CH_3}{|}}{CH_3COHCH_2CHOHCH_3}$$

Methyl Isobutyl Carbinol (MIBC)

Hydrogenation of mesityl oxide under more severe conditions than for making MIBK yields methyl isobutyl carbinol:

$$\underset{\underset{CH_3}{|}}{CH_3C=CHCOCH_3} \xrightarrow{H_2} \underset{\underset{CH_3}{|}}{CH_3-CH-CH_2CHOHCH_3}$$

Actually, most MIBC is obtained as a byproduct from MIBK manufacture. The main applications of this solvent are as a vehicle for lube-oil additives, and as a flotation chemical. Consumption in 1965 was 45.0 million pounds.

Isophorone

This very slow-boiling solvent is made from acetone by the following reaction:

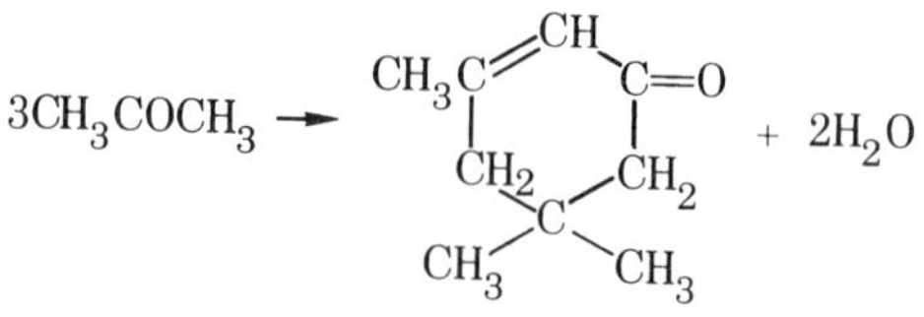

$$3CH_3COCH_3 \longrightarrow \text{(isophorone structure)} + 2H_2O$$

It is used in printing-ink formulations, and especially for vinyl coatings used in continuous metal coating. Metal precoating is taking hold rapidly in the U.S., since it allows considerable savings both in labor and in the form of reduced material losses. Vinyl coatings are used on about 40 percent of all precoated metal, the rest being mainly alkyd resins (50 percent) and acrylics.

8.2 Isopropylamines

Isopropyl alcohol reacts catalytically with ammonia in the vapor phase to form a mixture of mono-, di- and tri-isopropylamines. Recycling the unwanted homologs gives the desired product mix. The process is analogous to the manufacture of ethylamines, and in fact many alkylamine plants convert ethyl, isopropyl, butyl and cyclohexyl alcohols to amines in blocked-out operation.

The isopropylamine salt of dodecylbenzene sulfonic acid is used as a detergent normally added to the perchloroethylene used in coin-operated dry-cleaning machines; its function is to speed up the cleaning cycle. In the field of agricultural chemicals, isopropylamines are intermediates for "Atrazine," "Propazine," and "Avadex," all successful herbicides. The first two, developed by Geigy, are used for preemergency control of sugar and corn weeds, and the third to combat wild oats. Some rubber chemicals analogous to those derived from ethylamines are also made from isopropylamines.

The total demand for all three isopropylamines in 1965 was 18.0 million pounds per year, broken down as follows:

End-use	MM lbs, 1965
Triazine herbicides	6.0
Carbamate herbicides	4.0
Detergents	3.0
Rubber chemicals	2.0
Others	3.0
Total	18.0

The growth of the isopropylamine demand has been rapid. The surfactant mentioned above is being used not only in coin-operated but in existing dry-cleaning plants as a means of increasing capacity without further investment. The triazine and carbamate herbicides have become widely accepted all over the world. It has been predicted that the demand for isopropylamine will grow at 15 percent per year, although the dry-cleaning solvent outlet may soon approach saturation.

9 ISOPRENE

The great interest shown in isoprene results from the resemblance of stereospecific polyisoprene to natural rubber. Polyisoprene can now reproduce the structure of natural rubber so closely that it can be used in practically all applications once the exclusive domain of the natural product. Polyisoprene can be manufactured in the same plant as polybutadiene. The catalyst can be either an alkyl lithium, which yields a polymer of somewhat higher molecular weight than natural rubber, or of the Ziegler type which, as in the case of polybutadiene, gives a product having a higher cis linkage content as well as a lower molecular weight.

The recent introduction of radial-ply automobile tires may have a profound influence on the demand for isoprene. This tire-building technique requires natural rubber or polyisoprene and cannot be employed with SBR. Radial-ply tires are said to possess a much longer useful life than conventional types. Polyisoprene has also been shown to be superior to polybutadiene in heavy-duty and aircraft tires.

There are several routes to the isoprene monomer. The SNAM process starts with acetylene and acetone:

$$CH_3COCH_3 + C_2H_2 \longrightarrow HC{\equiv}C-\underset{\underset{OH}{|}}{\overset{\overset{CH_3}{|}}{C}}-CH_3 \xrightarrow{H_2} H_2C{=}CH-\underset{\underset{OH}{|}}{\overset{\overset{CH_3}{|}}{C}}-CH_3 \xrightarrow{-H_2O}$$

$$CH_2{=}CH-C(CH_3){=}CH_2$$

The process has received little attention due to the high price of acetylene, but in the future may become attractive if one of the processes said to be able to make acetylene for 3 to 5 cents per pound is proved commercially.

The Institut Français du Pétrole (IFP) process uses iso-butylene and formaldehyde as raw materials. It has the advantage of not requiring pure isobutylene, since under the mild conditions used for the reaction only the tertiary olefin reacts. The reactions are:

$$CH_3-\underset{\underset{CH_3}{|}}{C}=CH_2 + 2CH_2O \longrightarrow (CH_3)_2-C\underset{H_2C-CH_2}{\overset{O-CH_2}{\diagdown}}O \overset{\Delta}{\longrightarrow}$$

$$CH_2=CH-C(CH_3)=CH_2 + CH_2O + H_2O$$

The second mol of formaldehyde is recovered and recycled. The final cracking reaction is accomplished in a fluid bed catalytic reactor, followed by fractionation to remove and recycle dimethyl n-dioxane, and finally by isoprene purification. Two 90.0 million pounds per year plants in the USSR use a similar process, and the same route has been developed in Japan and Czechoslovakia.

Of the processes that synthesize isoprene from shorter molecules, the most successful in the U.S. has been the Goodyear-Scientific Design route from propylene. The first step is dimerization of propylene to 2-methyl-1-pentene, which is then isomerized to 2-methyl-2-pentene. The final step consists of vapor-phase cracking, in the presence of a hydrobromic acid catalyst, to isoprene and methane.

$$2CH_3CH=CH_2 \longrightarrow CH_2=\underset{\underset{CH_3}{|}}{C}-CH_2-CH_2-CH_3 \longrightarrow CH_3-\underset{\underset{CH_3}{|}}{C}=CH-CH_2-CH_3 \longrightarrow$$

$$CH_2=CH-C(CH_3)=CH_2 + CH_4$$

The only plant in the U.S. is Goodyear's at Beaumont, Tex. It has a capacity of 100.0 million pounds per year. Yields are around 60 percent of theoretical.

An improvement on this approach has been developed in Japan; the process consists of reacting one mol each of ethylene and propylene, and dehydrogenating the resulting C_5 olefin to isoprene. The process employs methyl aluminum as an intermediate. In addition to the synthetic routes to isoprene, two refinery operators — Humble at Baton Rouge and Shell at Torrance, Calif. — make isoprene monomer from olefin plant and refinery streams, respectively. Shell extracts isoamylene from a C_5 stream with sulfuric acid, in a manner similar to the production of pure iso-butylene. The isoamylene is then dehydrogenated by a process analogous to that for making butadiene from butylenes. Humble, on the other hand, uses a route in which not only isoamylene, but also isohexane (by cracking) can be converted to isoprene. These two processes are said to yield cheaper isoprene than any of the synthesis routes. The Humble process, however, yields isoprene that can be used to make butyl rubber but not polyisoprene.

The economics of the IFP and the Goodyear-SD processes are compared below. It can be seen that, especially if the producer also has a captive source of methanol, the IFP process is more attractive. The limitation of this route, however, is the low iso-butylene content of the usual C_4 cut, so that only a 200,000 barrels per day refinery or a 1.0 million pounds per year ethylene plant by cracking naphtha could supply the raw material for a 100.0 million pounds per year isoprene plant. The propylene required for a plant of the same size using the Goodyear-SD process can be obtained from a 400.0 million pounds per year ethylene plant cracking naphtha, or from a 100,000 barrels per day or slightly larger refinery that operates a catalytic cracker at fairly high severity. This second set of conditions seems much easier to fulfill, especially since the trend to larger petrochemical plants has not been accompanied by a similar trend in refinery sizes. The abandonment of the Rhône-Alpes isoprene project in 1967 was due mainly to the unavailability of sufficient isobutylene from a single

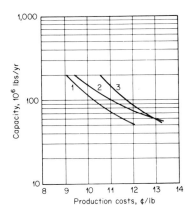

Fig. 6-14 Isoprene: (1) IFP Process (captive methanol); (2) Goodyear-SD process; (3) IFP Process (merchant methanol).

source.   The following assumptions regarding raw materials have been made:

Raw material	Cost, ¢/lb
Methanol (captive)	3.0
Methanol (merchant)	4.5
Isobutylene (contained in C_4 stream)	1.5
Propylene	2.3

The investment for a 100.0 million pounds per year plant using the IFP process is $700,000; a unit of the same size based on propylene dimerization requires an outlay of $14.8 million. Production costs are:

		Cost, ¢/lb isoprene	
Process:	Goodyear-SD	IFP (merchant methanol)	IFP (captive methanol)
Propylene	4.7	—	—
Methanol	—	4.5	3.0
Isobutylene	—	1.7	1.7
Utilities, catalyst and chemicals	0.5	1.6	1.6
Labor and overhead	1.0	0.8	0.8
Capital charges	4.9	3.2	3.2
Total	11.1	11.8	10.3

The potential cost of isoprene is thus around that of butadiene, and polyisoprene may well be one of the major elastomers of the future, as it already seems to be in the U.S.S.R.

The demand for polyisoprene is expected to rise to 180.0 million pounds per year by 1968. Most of the growth in demand will be for tires. Shell is the largest polyisoprene latex producer in the U.S., with plants at Torrance, Cal. (40.0 million pounds per year) and Marietta, O. (80.0 million pounds per year).

The demand for polyisoprene latex, as such, was around 15.0 million pounds (100 percent solids basis) in 1964. It is used to make dipped and foam products, adhesives and carpet backing. Goodyear produces only dry rubber; combined production of dry polyisoprene by Shell and Goodyear in 1964 was estimated at 80.0 million pounds.

Polymer Corp. at Sarnia, Ont. has a plant to produce 12,000 tons per year of balata, or trans-polyisoprene. Finally, butyl rubber contains an average of 3 percent isoprene, accounting for another 3.0 million pounds per year.

7

C$_4$ HYDROCARBONS

1 INTRODUCTION

Although propane and propylene can still be separated by distillation, the various four-carbon paraffins and olefins have boiling points so close that their separation requires quite different techniques.

The C$_4$ fraction (also known as "B-B" out) from a refinery vapor recovery unit consists of butane, isobutane, butylenes (1- and 2-) and isobutylene. Although it has several specific uses of its own, isobutylene is not suited as a feedstock for making butadiene. It is about 300 times more reactive than n-butylenes, and hence, can be removed by absorption in dilute (as low as 50 percent) sulfuric acid. If n-butylenes are to be produced from a mixed olefinic stream for feedstock to a butadiene unit, the raffinate from the isobutylene removal step can be extracted with furfural or acetonitrile for removal of the n-butylenes, leaving a stream containing only butane and isobutane. On the other hand, if pure butane is required as feed to a Houdry dehydrogenation unit, or pure 1-butylene for making polyolefins, the isobutylene-free C$_4$ stream can be fractionated into an overhead containing isobutane and 1-butylene, and bottoms consisting mainly of butane and 2-butylene. The olefins can then be solvent-extracted from each of these two streams.

Figure 7-1 shows a block flow diagram illustrating the distribution of the C$_4$ production.

2 BUTADIENE

Now that the wartime alcohol process (last used in 1955) has been definitely abandoned in the U.S., there are three sources of

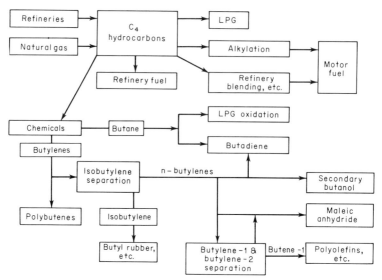

Fig. 7-1   Block flow diagram of distribution of C_4 production.

butadiene. These are:

Source	% of total, 1965
Butane dehydrogenation	40%
Butylene dehydrogenation	43%
Olefin plant recovery	17%

Butadiene can be recovered from the C_4^+ residue of an olefin plant by distilling off a fraction containing most of the butadiene, catalytically hydrogenating the higher acetylenes to olefins and separating the product from other olefins and isobutene by extraction. The butadiene content of the C_4 from an olefin plant is as high as 30 percent when cracking naphtha at high severity.

When the raw material is butane, the Houdry one-step catalytic dehydrogenation method is preferred:

$$CH_3CH_2CH_2CH_3 \rightarrow CH_2{=}CH{-}CH{=}CH_2 + 2H_2$$

All U.S. plants starting from butane, except the Phillips unit at Borger, Texas, employ this route. The process consists mainly of a battery of fixed-bed catalytic dehydrogenation reactors which operate on a cycle, during which the reactor is sequentially heated by burning off the coke deposit, cooled by surrendering heat to the butane feed, and purged of hydrocarbon vapors so the cycle can be repeated. To remain close to optimum conditions during the cycle, temperature variations must be kept within close limits. Thus the cycle must be fairly short. The usual procedure is a 21-minute cycle: 9 minutes on stream, 9 minutes regenerating the catalyst and the rest for purging and switching valves. It is expected that eventually butane dehydrogenation will be carried out continuously as better catalysts are developed.

The reactor effluent is compressed and sent to the vapor recovery unit. From there a stream containing around 11 percent butadiene goes to the butylene splitter. Unreacted butane and 2-butylene are recycled to the dehydrogenation reactors, and butadiene is separated from the splitter overhead by extractive distillation with furfural or cuprous ammonium acetate (CAA) extraction. The raffinate from this extraction is 1-butylene, which can also be recycled or used for other purposes. Overall yields on butane are around 56 percent; a flowsheet is shown in Fig. 7-2.

The Phillips process dehydrogenates butane to butylenes, which, after separation from light and heavy byproducts, are mixed with superheated steam and further dehydrogenated to butadiene. Unreacted butane is recycled from the butylene recovery unit, and unreacted butylenes from the butadiene purification train, which employs furfural extraction.

The main source of butadiene in the U.S., however, is still the catalytic dehydrogenation of butylenes. This is due partly to the large role played by catalytic cracking in the country's overall refining scheme. Thus a far greater proportion of olefins is obtained than

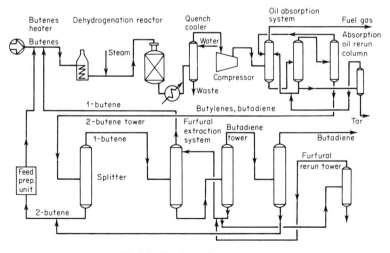

Fig. 7-2 Butadiene (from butenes).

would, for example, be obtained by the topping-reforming scheme often employed in Europe. Figure 7-2 describes the feed preparation unit for a butadiene from butylenes plant, and the butadiene unit itself.

The so-called Esso Research steam dilution process is used by U.S. plants starting from butylenes in conjunction with catalyst developed by Dow or by Shell. The reactor is adiabatic just as in the Houdry process, but the heat of reaction is supplied by superheated steam instead of by heating the catalyst bed. The feed, after the removal of isobutylene and saturates, contains 80-85 percent butylenes and is preheated to 1100° F, mixed with 8 mols/mol superheated steam and passed continuously through the fixed-bed reactors. Although the steam reduces the butylene partial pressure, some coke always forms on the catalyst bed and has to be burned off periodically. The effluent is quenched and then further cooled in a waste-heat boiler before entering the vapor-recovery unit. The off-gas from the unit, being rich in hydrogen, can be used not only for fuel but also as feed to a synthesis-gas unit or in hydrogen treating. From the vapor-recovery unit a 20 percent butadiene stream is sent to a butadiene recovery unit where first 2-butylene is separated from the lighter olefins, and then butadiene is recovered with furfural, CAA, or other solvent. A final rerun, which separates butadiene from any remaining butylenes, yields a 98 percent product suitable for synthetic rubber manufacture. Overall yields are 72 percent, including losses on recovery. Shell has a butylene dehydrogenation process using iodine as catalyst, which was used for the first time at Berre, France.

The conventional extraction solvents, CAA and furfural, produce butadiene that must be catalytically hydrogenated to eliminate higher acetylenes if the product is to be fed to a stereo rubber unit; the significance of the recent N-methylpyrrolidone extraction process is that it produces stereo-rubber-grade butadiene without further purification. Another extractant, dimethyl formamide (DMF), is used in the B. F. Goodrich process; it is reported that a two-stage extraction of the butadiene with DMF contained in a naphtha-pyrolysis C_4 stream eliminates the need to hydrogenate the higher acetylenes. In Europe, feedstocks from cat-cracking are relatively less important than in the U.S. On the other hand, olefin plants yield a higher proportion of butadiene because most of them crack naphtha instead of the lighter feedstocks used in the U.S. Houdry plants exist at Marl, W. Germany (Huels) and Ravenna, Italy (ANIC). Phillips uses its two-step process from butane at a 60.0 million pounds per year plant at Bordeaux, France. ANIC also makes butadiene from acetylene via acetaldehyde. During World

War II, butadiene was made in Germany by hydrogenation and subsequent dehydration of butynediol, an acetylene derivative. During the same period, in the U.S. enormous amounts of butadiene were made from ethyl alcohol:

$$C_2H_5OH \longrightarrow CH_3CHO + H_2$$

$$C_2H_5OH + CH_3CHO \longrightarrow CH_2{=}CH{-}CH{=}CH_2$$

However, after the war, C_4 hydrocarbon feeds became plentiful and techniques for dehydrogenating them were developed which made the ethanol process obsolete. Except for the U.S.S.R., where the process actually originated, butadiene is still being made from fermentation alcohol in the sugar-producing regions Pernambuco, Brazil, and Uttar Pradesh, India; the process is also used by ANIC at Ravenna, Italy, starting from petrochemical acetylene which is hydrated to acetaldehyde, one half being hydrogenated to ethanol to provide the other raw material. U.S. producers, their capacities and raw material appear below. It should be pointed out that some plants, for example, Petro-Tex, have both Houdry and steam-dilution units, so that one part of the feed to the latter consists of fresh butylenes and the other of partially dehydrogenated butane from the Houdry unit.

Butadiene producers Company	Location	1965 Capacity, MM lbs/yr
From butane		
El Paso	Odessa, Tex.	130.0
Firestone	Orange, Tex.	220.0
Phillips	Borger, Tex.	224.0
Petro-Tex	Houston, Tex.	220.0
Shell	Torrance, Calif.	140.0
Sinclair	Channelview, Tex.	242.0
Sub total		1,176.0
From butylenes		
Copolymer	Baton Rouge, La.	120.0
Goodrich-Gulf	Port Neches, Tex.	320.0
Enjay	Baytown, Tex.	66.0
PCI (Cities Service)	Lake Charles, La.	160.0
Texas-U.S.	Port Neches, Tex.	320.0
Petro-Tex	Houston, Tex.	280.0
Sub total		1,266.0
Olefin plant C_4		
Chevron Chem.	El Segundo, Calif.	32.0
Dow	Freeport, Tex.	64.0
Enjay	Baton Rouge, La.	110.0

Butadiene producers Company	Location	1965 Capacity, MM lbs/yr
Mobil	Beaumont, Tex.	50.0
Monsanto	Alvin, Tex.	100.0
Union Carbide	Seadrift, Tex., etc.	140.0
Tidewater	Delaware City, Del.	14.0
Sub total		510.0
Grand total		2,952.0

Source: *Oil, Paint & Drug Reporter*, October 24, 1966.

In 1965, butadiene consumption in the U.S. reached 2,710 million pounds distributed as follows:

End-use	MM lbs, 1965
SBR	1,630.0
Polybutadiene rubber	380.0
Nitrile rubber	100.0
SB resins and ABS plastics	410.0
Adiponitrile	190.0
Total	2,710.0

Source: *Oil, Paint & Drug Reporter*,
October 24, 1966.

The investment for a 100.0 million pounds per year Houdry-process butadiene plant is $17.0 million. Production costs are given below and in Fig. 7-3.

	Cost, ¢/lb butadiene
Butane (at 1.4 ¢/lb)	2.6
Utilities	0.7
Catalyst, chemicals and royalties	1.2
Labor and overhead	0.8
Capital charges	5.8
	11.1
Hydrogen credit	-0.4
Total	10.7

The investment for an extraction plant from an olefin plant residue containing 8.0 million pounds per year of butadiene is $1.8 million. Thus, the value of butadiene in a byproduct stream depends on the amounts involved. That contained in the residue from a 200.0 million pounds per year ethylene plant fed on refinery off-gas is worth slightly more than its value as fuel. On the other hand, the byproduct value of butadiene from a large naphtha-cracking plant can be as high as 5.0 to 6.0 cents per pound. The cost of extracting butadiene from large naphtha-cracking olefin plant residue is given in the following tabulation.

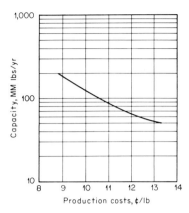

Fig. 7-3 Butadiene (Houdry Process).

	Cost, ¢/lb butadiene
Utilities	1.2
Chemicals	0.3
Labor and overhead	0.9
Capital charges	2.9
Total	5.3

Although butylenes still account for the major part of all butadiene produced in the U.S., the Houdry process has been used for almost the entire capacity added since 1955, except for that recovered from olefin plants.   One advantage of the Houdry process is that if suitable hydrocarbon raw materials are available, the same plant can produce either butadiene or isoprene; it is therefore a convenient process for providing feedstock to a stereo-rubber plant.

2.1 Styrene-Butadiene Rubber (SBR)

Synthetic rubber generally represents over 60 percent of the total rubber used in the free world, and 80 percent of that consumed in the U.S.   The 1057 thousand tons of SBR consumed annually in the U.S. are over 70 percent of the synthetic rubber used.   A breakdown of the U.S. rubber market is given below.

Tires and tire products		62.9%
Non-tire products		37.1%
Mechanical goods	14.5	
Footwear (heels and soles)	5.8	
Latex foam products	3.5	
Wire and cable covering	1.1	
Others	12.2	
Total		100.0%

Source: *Rubber Age*, January 1966.

Until the appearance of stereo rubbers, 70 percent seemed to be the maximum percentage of total rubber demand that could be met by synthetic rubbers. This was due mainly to the properties required for truck tires and conveyor belts: for example, low heat build-up, resistance to flexing, low abrasions, etc. These properties were not to be found in any of the earlier synthetics. With the advent of polybutadiene and polyisoprene, the share of natural rubber should eventually settle at no more than 15 percent of total demand.

SBR is still the most important type of synthetic rubber, despite the development of elastomers having better properties. It is a copolymer of butadiene and styrene where the proportion of the two monomers is around 70:30 by weight. Polymerization is carried out at 41° F, forming a suspension of fine rubber particles. After polymerization reaches the desired point, the suspension is flashed to remove unreacted butadiene and then steam-stripped to recover styrene. The latex is run into a blend tank, and from there sent to tanks where it is coagulated in the presence of acid inorganic salts. The rubber slurry is filtered and washed free of inorganic material, and finally dried and baled.

Phillips and Firestone have announced solution-polymerization techniques for making SBR using alkyl lithium catalysts similar to those used in certain stereo-rubber processes. The product has a higher percentage of cis-linkages, which are supposed to account for the superior abrasion and heat build-up properties of the product. Non-rubber content of the product is also lower than in conventional SBR. It is predicted that the proportion of all SBR made by solution polymerization will show rapid growth.

SBR was made originally by polymerization at 122° F, the result being a copolymer in which short chains acted as plasticizers that made the rubber workable. Later, it was discovered that one could polymerize at low temperatures ("cold rubber"), which gave only long chains of more uniform length; plasticizing could then be accomplished with oil, a procedure known as oil-extension. This had an important effect on the economics of synthetic rubber production since the petroleum fraction used as extender represents 20-25 percent of the final product and is much cheaper than either styrene or butadiene.

Except for the two comonomers, other chemicals make up the polymerization recipe. These include: antioxidants; a peroxide which acts as the free-radical initiator; emulsifiers; reaction stoppers, such as hydroquinone; metal-ion sequestrants; *t*-dodecylmercaptan which is used to regulate the chain length distribution;

and others.   Producers of SBR, their capacities and locations are given in the table below.

Producer	Location	1965 Capacity, 1000 long tons/yr
American Synthetic Rubber Corp.	Louisville, Ky.	125.0
Copolymer Rubber & Chem.	Baton Rouge, La.	116.0
Dewey & Almy	Cambridge, Mass.	3.0
	Owensboro, Ky.	2.0
Firestone Tire & Rubber	Akron, Ohio	43.0
	Lake Charles, La.	222.0
General Tire & Rubber	Odessa, Tex.	65.0
Goodrich-Gulf Chem.	Port Neches, Tex.	155.0
	Institute, W.Va.	103.0
B.F. Goodrich Industrial Products	Derby, Conn.	13.0
Goodyear Tire & Rubber Co.	Houston, Tex.	260.0
	Akron, Ohio	32.0
International Latex	Cheswold, Del.	8.0
Phillips Petroleum	Borger, Tex.	72.0
Shell Chemical Co.	Torrance, Calif.	98.0
Southeast Polymers, Inc.	Chattanooga, Tenn.	0.5
Texas-U.S. Chemical Co.	Port Neches, Tex.	140.0
United Rubber & Chem.	Baytown, Tex.	60.0
U.S. Rubber Co.	Naugatuck, Conn.	32.0
Wica Chem.	Charlotte, N.C.	0.5
Total		1,550.0

Source: *Rubber World*, June 1966.

It can be seen that most SBR producers are tire manufacturers. American Synthetic and Copolymer Rubber are joint ventures by several rubber consumers, the former led by American Cyanamid and the latter by Armstrong Rubber. Shell and Phillips are the only producers not in the tire or rubber goods business; their strength lies in their raw material positions, and in the fact that gasoline stations are important outlets for replacement tires, which represent almost two-thirds of the total tire market.

These capacities represent total solid rubber output, which includes not only the oil used for extension, but also that part of total carbon black used for master-batching. In this technique part of the carbon black required for use of SBR in tires is added to the polymer suspension. Instead of milling the total quantity into the rubber on the Banbury mixer it not only imparts better qualities to the final product but lowers processing time and power consumption, both important cost components. The use pattern for SBR is the following (1000 long tons per year):

End-use	1000 long tons, 1964
Tires and tire-related items	826.0
Mechanical goods	87.0
Shoes	87.0
Foams	50.0
Miscellaneous	200.0
Total	1,250.0

Source: *Oil, Paint & Drug Reporter*, December 14, 1964.

The influence of stereospecific rubbers is expected to reduce the SBR demand to 100 tons per year by 1970, although the total demand for rubber will continue growing at 3 percent per year.

2.2 SBR Latex

Rubber latex foam, the largest outlet for SBR in latex form, has been receiving increasing competition from plastic foams such as vinyl and urethane.   Hence the demand has been fairly constant at around 240.0 million pounds per year.   Some of the markets for SBR latex still expanding are carpet underlay (as a felt substitute) and paper saturation.   For carpet-back sizing, most of it is now made of carboxylated SBR resins. An approximate breakdown of the market as of 1965 is given below.

End-use	MM lbs, 1965
Foam rubber	100.0
Paper saturation	25.0
Carpet underlay	18.0
Water-based adhesives	14.0
Textile coating	15.0
Dipped tire-cord	14.0
Carpet-back sizing	32.0
Other uses	20.0
Total	240.0

Carboxylated SBR latexes, made by copolymerizing butadiene and styrene with a small amount of acrylic or methacrylic acid, are thermosetting and self-curing.   Their particular application is in carpet-backing, in competition with other similar resins such as natural rubber and polyisoprene latex.

The total demand for synthetic rubber latex in 1965 was 315.0 million pounds distributed as follows among the various types of polymers:

Latex	MM lbs, 1965
SBR	240.0
Nitrile rubber	30.0
Neoprene	30.0
Polyisoprene	15.0
Total	315.0

2.3 Stereospecific Rubbers

Butadiene and isoprene are very easy to polymerize. Polybutadiene and polyisoprene elastomers only became commercially important after it was discovered, around 1958, how to control the orientation of the individual molecules in the polymer chain. These elastomers have therefore come to be known as stereospecific rubbers.

Shell and Firestone use alkyl-lithium catalysts, Goodrich-Gulf has developed a cobalt catalyst and other polybutadiene processes are based on Ziegler-type catalysts consisting of alkyl aluminum compounds and titanium halides. Whereas the Goodrich-Gulf and Ziegler catalysts produce polymers consisting of around 95 percent cis-linkages, the s-butyl lithium catalyst yields polybutadiene in which trans-linkages (55-65 percent) predominate. Phillips has developed a process ("Solprene") in which polybutadiene is made directly from the C$_4$ cut, obtained by cracking naphtha to olefins. The expense of extracting the naphtha is thus avoided. Stereo rubbers created a sensation when they first appeared because they do not compete mainly with SBR, but with natural rubber. Up to that time, the latter, having better abrasion resistance and lower heat build-up than SBR, was indispensable in truck and passenger car tire treads. The new types of rubber enabled the advanced countries to reduce their dependence on the rubber supplies of the Far East. There has been a great deal of discussion as to the merits of these two types of product — truly stereospecific cis-polybutadiene, and that produced by lithium catalysts. The volume of data published to date is difficult to summarize, but it seems that cis-polybutadiene does indeed have better properties. Whereas, for example, Goodrich-Gulf has announced the use of pure polybutadiene plus processing aids in passenger tire treads, for which twice the life of natural rubber treads is claimed. The lower cis-content rubbers have until now been used only as 50-50 blends. On the other hand, the tire treads made from either type of stereo rubber have proved satisfactory and superior to natural rubber.

Treads containing polybutadiene have better abrasion resistance, traction and lower hysteresis loss than those made of natural rubber.   Low heat build-up also favors its use in blends with SBR for truck tire carcasses.   An important advantage over EPT rubbers is that it can be blended.   Choosing a process is a complex matter. It depends in part on how the particular type of rubber fits in with the resource structure of a country or producer.   Brazil, for example, chose a lithium catalyst process because, still being a natural rubber producer, it had no desire to eliminate its use entirely.   Countries having no natural rubber production or access to plantations in underdeveloped countries would tend to prefer the processes yielding high cis-content rubber. Although several plants in the U.S. have been converted from SBR to polybutadiene production, the two processes differ in several respects.   Whereas in SBR production polymerization is accomplished in batch reactors, stereo rubbers are made continuously in a number of reactors connected in series.   SBR is made by emulsion polymerization, but butadiene and isoprene are polymerized in a pure paraffinic solvent such as hexane.   After polymerization, the solution is flashed to remove unreacted monomer and solvent, which are separated, purified and stored for reuse. Subsequent filtering, drying and baling operations are similar to those for SBR.   The processes using Ziegler catalysts require an additional step to wash the catalyst from the polymer solution. This additional step increases investment costs over the alkyl-lithium process, where the catalyst is in such low concentration that it is allowed to remain in the final product.   Two advantages of SBR over polybutadiene have been overcome.   Stereo rubbers can now be both oil-extended and master-batched with carbon black. Oil-extension has the obvious advantages of cost reduction.   Master-batching, which consists of mixing part of the carbon black requirements into the SBR suspension, reduces the power required if all carbon black was blended on Banbury mills.   For SBR this blending is a simple operation, but special techniques had to be developed for polybutadiene because of the difference in polymerization methods.   Columbian Carbon developed a process called "Hydrodispersion Solution Master-batching," in which a highly turbulent carbon black-water slurry is blended with this viscous polymer solution.   Solvent is then steam-stripped from this slurry, and the remaining polymer crumbs dried.   This technique permits economic use of fine carbon black grades such as SAF, HAF, and ISAF, which impart better properties but are more expensive since they require more power and, above all, more processing time to achieve a given dispersion rating.

Stereo rubber consumption in the U.S. was 180,000 long tons in 1965, of which 90 percent was used in tires. Producers of both stereo rubbers are listed below:

Producer	Location	1965 Capacity, MM lbs/yr	Product
Shell	Marietta, Ohio	60.0	Polyisoprene
	Torrance, Calif.	30.0	Polyisoprene
Firestone	Orange, Tex.	110.0	Polybutadiene
Goodrich-Gulf	Institute, W.Va.	30.0	Polybutadiene
Phillips	Borger, Tex.	65.0	Polybutadiene
Goodyear	Beaumont, Tex.	65.0	Polybutadiene
	Beaumont, Tex.	60.0	Polyisoprene
American Rubber	Louisville, Ky.	100.0	Polybutadiene
Total		520.0	

Source: *Oil, Paint & Drug Reporter*, May 3, 1965.

Apart from its use in tires, some polybutadiene is being used in place of SBR to make high-impact polystyrene because of its excellent color properties. Except for raw material costs, there is little overall difference between production economics of SBR and stereospecific elastomers. SBR production involves somewhat higher investment costs, but stereospecific elastomers require higher catalyst and chemical inputs. Utility costs are also somewhat greater for SBR. The investment for a 100.0 million pounds per year synthetic rubber plant is around $18.0 million, and average production costs, aside from the value of the monomer, are given in the table below.

	Cost, ¢/lb synthetic rubber
Utilities	1.6
Labor and overhead	2.0
Catalyst and chemicals	2.6
Capital charges	5.9
Total	12.1

Yields on raw materials are taken at 96 percent.

2.4 Polybutadiene Resins

The most important use of unvulcanized liquid polybutadiene resins is in can linings because of their chemical stability and

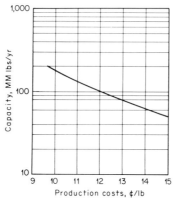

Fig. 7-4 Synthetic rubber (production costs only).

freedom from odor and taste. Most tin-plate beer cans are lined with a polybutadiene base, which adheres well to metals, followed by a vinyl topcoat. Polybutadiene linings are also used in many fruit and vegetable cans. The U.S. manufacturers of polybutadiene can linings are DuPont and Enjay.

Food Machinery makes a family of epoxidized polybutadiene resins ("Oxirons") employed in surface coatings, reinforced plastics and other typical epoxy-resin applications. Polybutadiene resins are also used to make fast-drying surface coatings.

2.5 Nitrile Rubbers

Nitrile Rubbers, also known as Buna N, are 70:30 to 50:50 emulsion copolymers of butadiene and acrylonitrile. The outstanding properties of these materials are their resistance to lube-oils and gasoline, abrasion and heat-aging. They cannot be used in the presence of aromatic or chlorinated solvents. Oil resistance increases with acrylonitrile content, but on the other hand, processability decreases. Their uses are in automotive parts where resistance to lubricants is critical, and in fuel hoses, fuel cell liners, printing rolls and other mechanical goods. U.S. producers and their capacities are shown below.

Producer	Location	Capacity, MM lbs/yr
Firestone	Akron, Ohio	25.0
Goodrich	Louisville, Ky.	80.0
Goodyear	Akron, Ohio	35.0
U.S. Rubber	Baton Rouge, La.	40.0
International Latex	Dover, Del.	20.0
Total		200.0

In 1965, U.S. consumption was around 102.0 million pounds of which 30 percent was in the form of latex for making oil-resistant paper, and for leather and textile coating. Nitrile latexes are also used as binders for nonwoven fabrics. Nitrile rubber is often used in blends with SBR to improve the oil-resistance of the latter. Nitrile rubbers cost about twice as much as SBR, since their inherent dryness makes them difficult to process. Blends with SBR, phenolics and other materials are therefore used as cost-property compromises. One of the best prospects for nitrile rubbers are blends with polyvinyl chloride to give high-impact PVC molding resins.

A 10.0 million pounds per year nitrile rubber plant was reported to have cost $2.4 million.

2.6 "Sulfolane"

Tetramethylene sulfone, or "Sulfolane," is a solvent used increasingly for CO_2 removal ("Sulfinol" process) and aromatics extraction. Apart from its selectivity, sulfolane has the advantage over glycols and other extraction solvents in that it is chemically inert, thus reducing losses through degradation. Shell, the licensor of the "Sulfinol" process (the solvent of which is a mixture of sulfolane and di-isopropanolamine), recommends it for use with its partial oxidation synthesis gas process. By 1965 there were about 10 plants in the world using sulfolane extraction, and the process seems to be gaining in popularity.

"Sulfolane" is made by Shell in Britain, and, more recently, by Phillips in the U.S., by the reactions:

$$SO_2 + CH_2{=}CH{-}CH{=}CH_2 \rightarrow \underset{\substack{\displaystyle | \\ O \diagdown \! \diagup O \\ \displaystyle S}}{\overset{HC{=\!=}CH}{H_2C \diagdown \diagup CH_2}} \xrightarrow{H_2} \underset{\substack{\displaystyle | \\ O \diagdown \! \diagup O \\ \displaystyle S}}{\overset{H_2C{-\!-}CH_2}{H_2C \diagdown \diagup CH_2}}$$

2.7 Other Copolymers

Vulcanizable copolymers of butadiene and ethylene have been developed by Phillips. They will probably be used in polyethylene pipe, and wherever else crosslinkable polyethylene can be employed.

2.8 Cyclodienes

A 25.0 million pounds per year plant to make mainly 1,3- and 1,5-cyclooctadiene from butadiene has been built by Cities Service at Lake Charles, La.

$$2C_4H_6 \longrightarrow$$

$$
\begin{array}{c}
\text{H} \qquad\qquad \text{H} \\
\diagdown\qquad\quad\diagup \\
\text{C}=\!=\text{C} \\
\diagup\qquad\quad\diagdown \\
\text{H}_2\text{C}\qquad\qquad\text{CH}_2 \\
|\qquad\qquad\qquad| \\
\text{H}_2\text{C}\qquad\qquad\text{CH}_2 \\
\diagdown\qquad\quad\diagup \\
\text{C}=\!=\text{C} \\
\diagup\qquad\quad\diagdown \\
\text{H}\qquad\qquad \text{H}
\end{array}
$$

The main uses for the 1,5-isomer is as the third comonomer in ethylene-propylene rubber. Cyclooctadienes can also be epoxidized, or hydrogenated and oxidized to suberic acid which can be used to make polyamides or synthetic lubricants, typical uses for difunctional acids.

In Germany, Huels has built a plant to make nylon 12 (a polymer of lauryllactam) from 1,5,9-cyclododecatriene, a cyclic trimer of butadiene. It has lower water-retention and much better dimensional stability than other polyamides, which makes it especially suitable for molding powders. The process is analogous to the manufacture of caprolactam from cyclohexane.

2.9 1,4-Hexadiene

A process for making 1,4-hexadiene from butadiene has been developed in Japan:

$$C_4H_6 + C_2H_4 \longrightarrow H_2C\!=\!CH\!-\!CH_2\!-\!CH\!=\!CH\!-\!CH_3$$

This unconjugated diene is potentially important as the third comonomer in EPT rubber, in competition with cyclic butadiene dimers and dicyclopentadiene.

3 ISOBUTYLENE

Isobutylene of 99 percent + purity is required for such applications as butyl rubber and aromatics alkylation; in other processes,

such as the production of polybutylenes or t-butyl alcohol, isobutylene can react in the presence of other C_4's, both saturated and
unsaturated.

In the U.S. the usual process for producing pure isobutylene is
extraction with 65 percent sulfuric acid. This has the disadvantage
of causing polymer formation problems when butadiene is present in
the feed. When starting from a refinery C_4 cut, for example, from
a cat-cracker vapor recovery unit, this is not much of an inconvenience. But when it is desired to recover isobutylene from
olefin plant residue, this factor becomes serious due to the butadiene
concentration, which can reach 30 percent of the feed.

The Compagnie Française de Raffinage-Badger process over-
comes this difficulty by using 50 percent sulfuric acid as the
extractant. Side advantages of this process are higher product
purity, reduction of sulfuric acid losses and of coke formation (both
are favored by higher acid concentrations), better yields on isobutylene contained in the feed and a clean separation between isobutylene and n-butylenes which can be advantageous when it is
desired to produce polymer-grade 1-butene in addition to pure
isobutylene. Since ethylene plants in Europe are relatively more
important as a source of C_4 feedstocks than they are in the U.S., the
process is expected to be more successful there. The first unit
was built at Grangemouth, U.K.

The process consists of a three-stage countercurrent extraction
system. The effluent from the final stage is flashed and goes to a
regenerator where heat reverses the reaction and regenerates both
acid and isobutylene. Temperature is kept below the point where
polymerization would become a problem. Sulfuric acid is removed
from the bottom of the regenerator and recycled after reconcentration. The regenerator off-gas, containing isobutylene and some
light polymer plus t-butyl alcohol, is fractionated to produce 99
percent + isobutylene, and the bottoms can be further fractionated
into t-butyl alcohol and di-isobutylene. A flowsheet is shown in
Fig. 7-5. Investment for a 30.0 million pounds per year isobutylene extraction plant is $2.5 million. Assuming the raw-material
is worth 1.5 cents per pound production costs are as follows:

Raw material	1.5
Utilities	0.2
Labor and overhead	0.5
Capital charges	2.8
Total	5.0

In 1964, 600 million pounds of isobutylene were consumed in the
U.S. as petrochemical feedstock. The distribution was as follows:

End-use	MM lbs, 1964
Butyl rubber	302.0
Polybutylenes	102.0
Polyisobutylene	28.0
Alkylated aromatics	25.0
Oxo alcohols	48.0
Di-isobutylene, tri-isobutylene	47.0
t-Butyl alcohol, acetate, peroxide, etc.	27.0
t-Butylamine	12.0
Other uses	19.0
Total	610.0

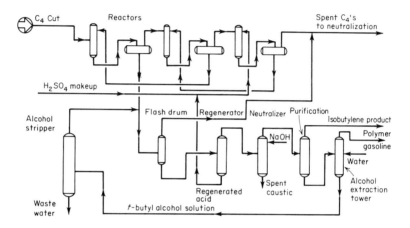

Fig. 7-5   Isobutylene (Badger Process).

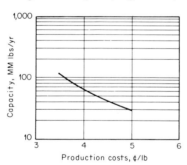

Fig. 7-6   Isobutylene.

3.1 Butyl Rubber

In contrast to natural rubber or SBR, which even after vulcanization retains a certain number of double bonds, butyl rubber is almost completely saturated. This absence of reactive double bonds gives it exceptional resistance to heat, oxidation, ozonolysis and chemical degradation.

Butyl rubber is a copolymer of isobutylene and 1-3 percent isoprene, the function of the latter being to supply the double bonds needed for vulcanization. It is made by copolymerization in the presence of AlCl$_3$, in methyl chloride solution, and at very low temperatures (-130°F) because isobutylene is extremely reactive.

Chemical stability and impermeability to gases account for its use in inner tubes. Such automotive parts as weather stripping are often made of butyl rubber because of its resistance to ozonolysis. Butyl rubber tires give a softer ride and make less noise on corners than other rubbers. However, they have poor tread wear, and a long curing cycle which increases production costs. Butyl rubber, developed by Esso Research is used in cable covering, especially in urban underground distribution networks where there is danger of overloading. All butyl rubber plants built so far are either operated or licensed by Enjay. U.S. capacities are:

Producer	Location	1965 Capacity, MM lbs/yr
Enjay	Baton Rouge, La.	136.0
	Baytown, Tex.	169.0
Columbian Carbon	Lake Charles, La.	38.0
Total		343.0

Source: *Oil, Paint & Drug Reporter*, February 8, 1965.

The consumption of butyl rubber in the U.S. for 1965 is broken down below.

End-use	MM lbs, 1965
Tires and tire products	100.0
Wire and cable	7.0
Other uses	45.0
Export	73.0
Total	225.0

Source: *Oil, Paint & Drug Reporter*, February 8, 1965.

The market for butyl tires may grow as a result of the master-batching technique developed by Columbian Carbon. This allows finer carbon blacks to be used economically in butyl rubber, thereby reducing the excessive tread wear which has been the main drawback of butyl tires. Inner tubes will at best remain a static market, as tubeless tires become more and more accepted. Mechanical goods include conveyor belts, steam hose, gaskets and seals.   In automotive use such as weather stripping and cooling system parts, butyl rubber has been losing ground to ethylene-propylene terpolymers.

Halogenated butyl rubbers are superior to straight butyl in a number of properties: for example, heat resistance, adhesion to other rubbers and fabrics, tensile strength, and curing speed. Chlorinated butyl rubber is made by Enjay and brominated butyl by B.F. Goodrich. They compete with straight butyl in applications of mechanical goods.

3.2   Isobutylene Polymers

There are two main types of polymerized isobutylene. The first, and more important commercially, consists of viscous liquid co-polymers of isobutylene and n-butylenes having molecular weights below 1500; products of this type are referred to as "polybutylenes." Isobutylene can also be polymerized at low temperatures in a hydro-carbon solvent. The resulting gums or solids are known as "poly-isobutylenes"; their properties resemble those of uncured butyl rubber, but they cannot be vulcanized since no extra double bonds are present.

In the Cosden Petroleum polybutylenes process, a mixed C_4 feed stream is dried to avoid hydrolysis of the acid catalyst and intro-duced into a cooled reactor, where it is mixed with a recycle stream of light polymers and catalyst.   Temperature is the important variable in determining molecular weight, and, thus, viscosity. Low temperatures favor high viscosity polymers.   The reactor effluent is allowed to settle, and a catalyst slurry is recycled.   The product stream is flashed to remove unreacted C_4's, and at this point saturated hydrocarbons and most of the n-butylenes are re-moved from the system. The polymer is stripped of light material, from which butylene dimer and trimer can be isolated. The sepa-ration between iso- and n-butylenes is thus accomplished by poly-merization of the former.   Figure 7-7 shows a flowsheet of the operation.   Investment for a 25 million pounds per year plant for polybutylenes is reported to be $3.4 million.

Polybutylenes are stable liquids that do not harden or dry since there are no extra double bonds in the molecule, nor do they de-compose by depolymerization as is the case with isobutylene dimer. U.S. producers of polybutylenes are:

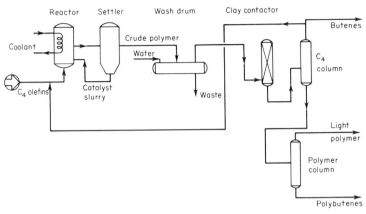

Fig. 7-7 Polybutenes (Cosden process).

Producer	Location	Capacity, MM lbs/yr
American Oil	Wood River, Ill.	80.0
	Yorktown, Va.	20.0
	El Dorado, Ark.	16.0
Cosden Petroleum	Big Spring, Tex.	20.0
California Chem.	Richmond, Calif.	50.0
Total		186.0

The consumption of polybutylenes is fairly static at 110.0 million pounds per year. The main use of these compounds is as viscosity improvers in lubricating oils, a market which they share mainly with the higher polymethacrylates. Of the 160.0 million pounds per year total market for viscosity improvers, polybutylenes account for 55.0 million pounds. The market for automotive lubricants is growing slowly, if at all. The number of grease fittings and average crankcase content per automobile is decreasing while the change interval for crankcase and automatic transmission oils is increasing. The next most important use is in caulking and sealing compounds, where the main competitors are polysulfides. The market for mastics is being invaded not only by polysulfides but, more recently, by polyurethane and silicone rubbers. Best known among the polyisobutylene compounds are the "Vistanex" series of lube-oil additives marketed by Enjay. About 28.0 million pounds per year of polyisobutylenes are used annually in the U.S., almost entirely as tackiness agents for nonsplashing bearing oils, viscosity index improvers for oils, and in functional fluids subject to large temperature variations. An important class of heavy-duty additives is made by reacting polyisobutylenes with phosphorus pentasulfide. Other lube-oil additives are copolymers of isobutylene and styrene.

3.3 Heptene

Isobutylene and propylene react to give highly branched C_7 olefins known as "heptene." This is the feedstock for making the C_8 oxo alcohol known as iso-octanol. The reaction is carried out in the presence of aluminum chloride or phosphoric acid. Around 43.0 million pounds of isobutylene went into heptene manufacture in 1964.

3.4 Di-isobutylene

Most of the di-isobutylene consumed in the U.S. is used to make octylphenol, a raw material for nonionic detergents. The rest is fed to oxo units to make nonyl alcohol, which competes with other alcohols in the same chain-length range (C_8 to C_{10}).

Di-isobutylene is a byproduct of isobutylene extraction with sulfuric acid, and of polybutylene manufacture.

3.5 Other Isobutylene Derivatives

t-Butyl alcohol is formed by hydrolysis of t-butyl hydrogen sulfate, the result of reaction between isobutylene and sulfuric acid during extraction of the former from a C_4 stream. It is used as a solvent and as a medium for certain reactions.

t-Butyl acetate is formed by reaction of isobutylene with acetic acid. It has been suggested as an additive for antiknock fluids, especially those going into premium gasolines.

t-Butyl peroxide, formed by reaction between t-butyl hydrogen sulfate and hydrogen peroxide, is stable up to 165° F and used as a polymerization catalyst.

Numerous aromatic compounds are made by alkylation with isobutylene. The most important is 2,6 di-t-butyl-p-cresol, an antioxidant with a large number of applications.

4 s-BUTYL ALCOHOL

s-Butyl alcohol is made from mixed butylenes. The process is similar to the manufacture of isopropanol from propylene; in fact, at least one U.S. plant has been designed to produce either one.

$$CH_3CH{=}CH{-}CH_3 + H_2SO_4 \rightarrow CH_3CH_2{-}\underset{\underset{\displaystyle HSO_4}{|}}{C}HCH_3 \xrightarrow{H_2O} CH_3CH_2CHOHCH_3$$

The consumption of *sec-*butyl alcohol in the U.S. was 260.0 million pounds in 1965, of which all but 10 percent, used mainly as a solvent or to make s -butyl acetate also a solvent, went into manufacture of methyl ethyl ketone.

Apart from dehydrogenation to butadiene, s -butyl alcohol is the main outlet for butylenes. The increasing availability of n -butenes from naphtha crackers has led Bayer, in Germany, to develop a process for the conversion of n-butenes to acetic acid.

A flowsheet for the production of MEK from s-butanol by liquid-phase dehydrogenation is shown in Fig. 7-8. The alcohol is passed over a dehydrogenation catalyst. After separation of the unreacted alcohol, the product stream is flashed and the vapors sent to a purification section where hydrogen is separated from the product. Yields are around 95 percent.

Celanese and Union Carbide produce MEK by direct oxidation of butane, along with numerous other oxygenated compounds. The Wacker process for direct oxidation of olefins to carbonyl compounds has been adapted to the manufacture of MEK. It is expected to reduce production costs considerably. It is reported that the Sinclair plant, the latest to be built in the U.S., uses this process.

All but 15 percent of the MEK produced in the U.S. is used as a fast-boiling solvent. As such, it competes mainly with ethyl acetate; in less developed economies, the cheaper acetate is the more widely used of the two, but in the U.S. more than twice as much MEK as ethyl acetate is consumed. MEK has the advantage of forming lower viscosity solutions, for a given solids content, than ethyl acetate. Apart from its use in lacquers, adhesives and coatings, MEK is the preferred solvent for dewaxing of lube-oil stock. An outlet that will become increasingly important is that as a reaction medium in the Mobil Chemical terephthalic acid process. MEK requirements are considered to be 0.5 pound per pound of product. A list of MEK producers in the U.S. is given below.

Producer	Location	1965 Capacity, MM lbs/yr	Route
Shell	Dominguez, Calif.	50.0	Butylenes
	Houston, Tex.	100.0	Butylenes
	Norco, La.	50.0	Butylenes
Enjay	Linden, N.J.	145.0	Butylenes
Union Carbide	Brownsville, Tex.	30.0	LPG oxidation
Celanese	Pampa, Tex.	25.0	LPG oxidation
Sinclair	Channelview, Tex.	40.0	Butylenes
Total		440.0	

Source: *Oil, Paint & Drug Reporter*, March 8, 1965.

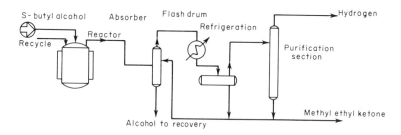

Fig. 7-8   Methyl Ethyl Ketone.

In 1964, MEK consumption in the U.S. was 250 million pounds distributed as follows:

End-use	MM lbs, 1964
Coatings (lacquers, rubber and plastic)	175.0
Lube-oil dewaxing	25.0
Miscellaneous (adhesive, chemical, other)	42.5
Export	7.5
Total	250.0

Source: *Oil, Paint & Drug Reporter*, November 2, 1964.

Methyl Ethyl Ketoxime

The specific application of methyl ethyl ketoxime is as an anti-skinning agent for paints, especially those sold in aerosol packaging where the formation of a skin would render the product unusable. It is the result of the reaction between MEK and a salt of hydroxylamine:

$$CH_3CH_2CCH_3 + NH_2OH \xrightarrow{H^+} CH_3CH_2CCH_3$$

About 3.0 million pounds of this antiskinning agent are consumed in the U.S.

Methyl Ethyl Ketone Peroxide

Methyl ethyl ketone peroxide is used as a polymerization catalyst, especially by small producers of acrylic and polyester polymers since it is one of the safest peroxides to handle. Total consumption in the U.S. was some 1.7 million pounds per year in 1965. It is made by the following reaction:

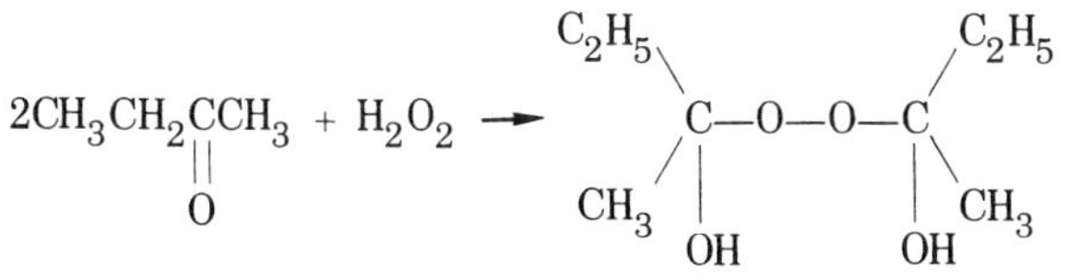

$$2CH_3CH_2\underset{\underset{O}{\|}}{C}CH_3 + H_2O_2 \longrightarrow \underset{\underset{OH}{|}}{\underset{CH_3}{|}}\overset{C_2H_5}{C}-O-O-\underset{\underset{OH}{|}}{\underset{CH_3}{|}}\overset{C_2H_5}{C}$$

Actually, a mixture of several peroxides and hydroperoxides is formed; the product is sold as a solution in dimethyl phthalate.

Methyl Pentynol

Methyl pentynol is made by the high-pressure reaction between acetylene and MEK. Consumption is about 1.0 million pounds per year, mainly as a corrosion inhibitor, but also as an intermediate for certain perfumery products.

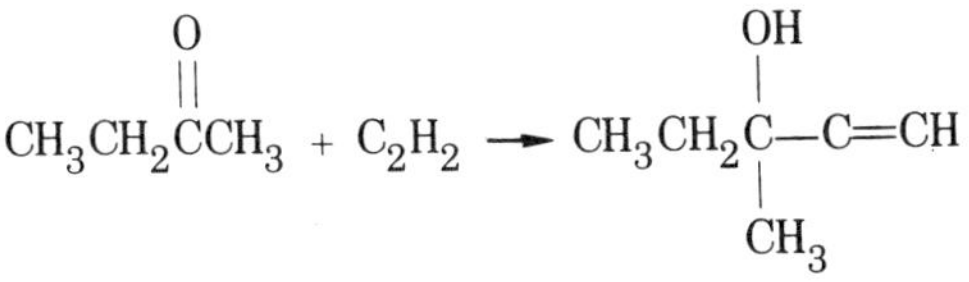

$$CH_3CH_2\overset{\overset{O}{\|}}{C}CH_3 + C_2H_2 \longrightarrow CH_3CH_2\underset{\underset{CH_3}{|}}{\overset{\overset{OH}{|}}{C}}-C\equiv CH$$

4.1 *s*-Butyl Lithium

s-Butyl alcohol reacts with hydrochloric acid to form *s*-butyl chloride, which can be combined with lithium metal to form *s*-butyl lithium:

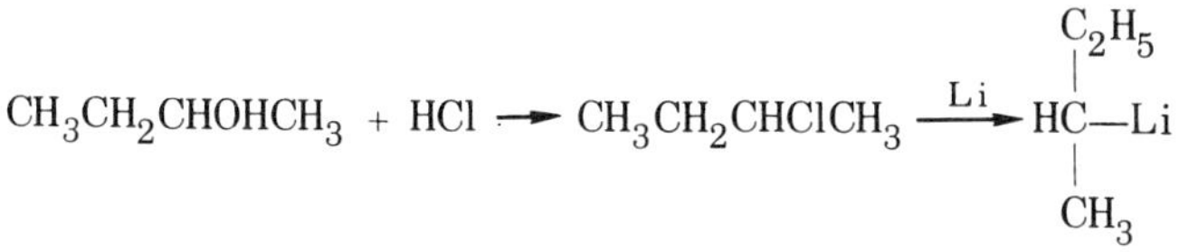

$$CH_3CH_2CHOHCH_3 + HCl \longrightarrow CH_3CH_2CHClCH_3 \xrightarrow{\text{Li}} \underset{\underset{CH_3}{|}}{\overset{\overset{C_2H_5}{|}}{H}C}-Li$$

The main use for the organometallic compound is as a catalyst in stereospecific rubber manufacture. It is used by producers such as Firestone and Shell not licensed under the Ziegler patents.

When used to make polyisoprene, *s*-butyl lithium produces a rubber having cis-content of 93 percent and a molecular weight above that of natural rubber, which increases processing costs. Ziegler catalysts yield a product very similar to natural rubber. When used to make polybutadiene, lithium catalyst gives a product much lower in cis-content than that obtained from Ziegler catalysts.

Consumption of butyl lithium in the U.S. is around 0.25 million pounds per year, of which the overwhelming majority is used to make stereo rubber.   Concentrations are around 0.1 percent, low enough that, contrary to the Ziegler processes, the catalyst may remain in the product.

The total U.S. capacity is around 0.35 million pounds per year; producers   are   Lithium   Corp.,   Foote   Minerals,   and   American Potash.

5 1-BUTYLENE

Pure 1-butylene is used on an increasing scale in copolymers with other olefins obtained by Ziegler-type techniques.   High-density copolymers with polyethylene, for example, are used to make blow-molded detergent bottles.   PetroTex developed pure polybutylene resins, also using Ziegler catalysts, which have the property of being carbon-black-extensible.   The product is tougher than the lower polyolefins, and because of its properties and low cost is expected to make inroads into the plastic pipe market.

An increasingly important application of 1-butylene is in the manufacture of butylene oxide and 1,2-butylene glycol, made by the same process as their propylene homologs.   Butylene glycol is used in the manufacture of polymeric plasticizers and is a stabilizer for chlorinated solvents.   The total demand for pure 1-butylene was 25.0 million pounds per year in 1964.

8

AROMATIC HYDROCARBONS

1 INTRODUCTION

The original method for obtaining aromatic raw materials —
benzene, toluene, xylenes, naphthalenes — was coal-tar distillation.
Most coal-tar is a byproduct of the steel industry, but a certain
amount is also produced by merchant coke-oven operators who
sell coke to nonintegrated users. However, two factors have been
reported as contributing to the insufficient amounts of aromatics
derived from these two sources. First, technological improvements have steadily reduced the amount of coke required to produce a ton of steel, therefore, the incremental demand for steel
did not bring about a corresponding increase in the availability of
coal-tar derivatives. Second, many of the principal derivatives of
benzene, toluene, and xylene have shown an income elasticity of
demand far in excess of that for steel. These two factors are responsible for the fact that by 1965, over 80 percent of all benzene
and even larger proportions of toluene and xylene were derived
from petroleum.

Although crude oil contains a certain percentage of aromatics,
the principal source of petroleum-derived aromatics is catalytic
reforming. The purpose of reforming is to upgrade straight-run
naphtha, which has a very low octane number and has consequently
become less and less suitable for the U.S. automotive fuel market.
Reforming consists of dehydrogenating virgin naphtha over a
platinum catalyst, the main effect of which is to convert cycloaliphatic hydrocarbons to benzene, toluene, xylenes and higher
aromatics.

A certain amount of isomerization and cyclization also take place during catalytic reforming; for example, both n-heptane and ethyl-cyclopentane are transformed into toluene. Since gasoline-range aromatics have high octane numbers, this is the cheapest way of obtaining good quality motor fuel and it has become the most important contributor (about 35 percent) to the U.S. gasoline pool.

Pure aromatics for petrochemical use are recovered from catalytic reformate by solvent extraction. The most common method in the U.S. is extraction with a mixture of water and diethylene glycol, a process known as "Udex" (Universal Oil Products). Recently, such processes as "Sulfinol" (Shell uses sulfolane) and that of the Institut Français du Pétrole using dimethyl sulfoxide, have also become widely used, primarily outside the U.S. The "Udex" process consists of two steps. In the first, aromatics are selectively dissolved in the extractant, and distilled from the solvent which is recycled. In the second step, the aromatics stream is fractionated to separate benzene and toluene from the heavier aromatics, which are either fed to xylene isomer production units or incorporated into the gasoline pool.

Another source of aromatics is the residue of olefin plants, especially those which use naphtha as a feedstock. This particular source is therefore relatively more important in Europe and Japan, and becoming more so since plants are designed for higher severity cracking, thus producing pyrolysis gasoline containing aromatics in higher concentrations. Recovery is accomplished by the same techniques as from catalytic reformate but only after the highly olefinic naphtha has been hydrogen-treated. In such plants a source of hydrogen may be required in addition to that obtained from the pyrolysis step.

The predominant aromatic present in coal-tar is benzene. However, the opposite is true of catalytic reformate. Hence, reforming produces much more toluene and xylenes than required for petrochemical feedstock, and a number of processes have been developed for dealkylating toluene to benzene. Since hydrogen is necessary to accomplish this reaction, the key to these processes is to dealkylate without hydrogenating the aromatic ring.

This operation can be accomplished catalytically or thermally. In the U.S. the catalytic route has found more popularity, having

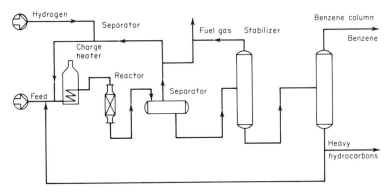

Fig. 8-1   Benzene from toluene by dealkylation.

been selected for six of the first nine units built. Among these six plants, four used the "Hydeal" process promoted by Universal Oil Products.   The most important thermal process is Houdry's "Detol." Catalytic dealkylation has the advantage of lower hydrogen consumption, higher yields and conversion per pass. The main advantage of thermal dealkylation is greater flexibility in regard to feedstock.   This means that a larger proportion of xylenes can be fed to the unit.

In the usual versions of dealkylation (see Fig. 8-1), hydrogen from the reforming unit (which is a net producer of hydrogen, even after requirements for feed desulfurization are considered) is compressed and mixed with the toluene-xylene feed and sent first through a feed-effluent exchanger, then through the feed heater and finally to the reactor.   After appropriate heat recovery, the effluent goes to a separator where hydrogen is given off and recycled after removal of a slip-stream. The flashed product stream is stabilized and sent to the benzene fractionation tower. The bottoms are fractionated once more to remove heavy aromatics before being recycled to the feed-preheating section. Toluene is the easiest aromatic to dealkylate. Xylene has a tendency to form polymers and coke. The process is rather delicate, since high temperatures and pressures are required to achieve satisfactory yields and avoid hydrogenation of the aromatic ring.   Yields are reported to be around 97 percent of theoretical.

Coke-oven operators have also begun to upgrade their aromatics by solvent extraction and dealkylation, although in the latter case an independent source of hydrogen must be found. This is generally coke-oven gas, which contains 48-53 percent hydrogen and can be upgraded by low temperature fractionation. Another

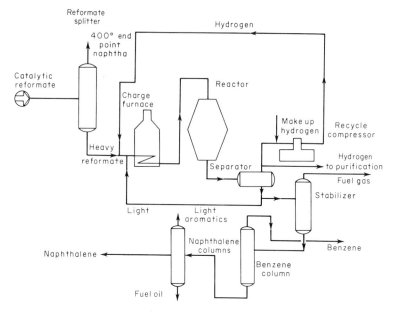

Fig. 8-2 Naphthalene (from heavy reformate).

practice that is becoming common is for refineries to process light coal-tar oils in their aromatics extraction units.

The feedstock to a catalytic reformer producing both high-octane gasoline and aromatics has an end-point of around 370° F, since the reformate end-point is somewhat higher than that of the feed. However, aromatics in the C_6-C_8 range are obtained by reforming the lighter part of a straight-run naphtha stream. Aromatics can be produced in two ways: 1) by splitting the feed and running the reformer in blocked operation to maximize aromatics formation from the light fraction; 2) by reforming the entire feed at once, and fractionating the reformate into two portions, the lighter to the Udex unit. The first method is generally preferred because of the higher yields, despite higher investment and operating costs.

The history of naphthalene supply sources parallels that of other aromatics. Whereas the demand for naphthalene as a starting point for phthalic anhydride has grown at about the same pace as vinyl chloride and alkyd resins the supply from coal-tar has, in fact, diminished over the years. Not only has the amount of coke required per ton of steel declined steadily, the naphthalene yield per ton of coal has also been falling. Thus, due to chronic over-capacity in the steel industry, coking is often carried out at lower temperatures (to avoid shutting down plants by lengthening the coking cycle) thereby reducing the naphthalene output per ton of coal input.

Thus, around 1960, petroleum refiners decided to produce naphthalene by dealkylation of a heavy aromatics stream such as is found in heavy reformate or in cat-cracker cycle stock. This resulted in a product having better purity characteristics than coal-tar naphthalene, particularly with respect to sulfur content, since the charge to a reforming unit is invariably hydrogen-treated to remove sulfur-products which would otherwise poison the catalyst.

To produce naphthalene, however, it becomes necessary to reform a 400° F end-point fraction in order to convert the naphthalene precursors to alkylated naphthalenes, which can then be dealkylated. The gasoline rerun bottoms from reforming contain about 10 percent naphthalene and 55 percent methylnaphthalenes; however, only a very large refinery can produce such a fraction in sufficiently large amounts to justify a naphthalene plant. A typical flow-sheet for making naphthalene from petroleum is shown in Fig. 8-2. Reformate is split into 400° F end-point gasoline and bottoms; the latter, after being mixed with hydrogen and passing through a heat pick-up train, is sent to the dealkylation reactor. Effluent is sent to a separator, where hydrogen is given off and recycled after part of the stream has gone through a clean-up circuit. The product stream goes to a stabilizer, and the stabilized dealkylate is clay-treated first and then fed to a benzene splitter, since a certain amount of benzene is formed from the toluene remaining in the feed. Two further fractionation towers produce a pure naphthalene heart-cut in addition to heavy and light fractions that can be incorporated into other refinery streams.

Petroleum naphthalene has a purity of 99 percent, whereas the 78° F product commonly obtained from coal-tar is only 95 percent pure. However, these impurities are primarily methylnaphthalenes, which also are converted to phthalic anhydride. Thus, the market was divided into two separate sectors. Petroleum naphthalene, selling at 2-2.5 cents per pound above coal-tar naphthalene, began to be used by phthalic anhydride operators who found that the higher yields and intrinsically cheaper fluid-bed operation were worth the difference in raw material costs. Fixed-bed operators, on the other hand, continued to use coal-tar raw materials. However, petro-naphthalene had just become established when orthoxylene began to drop in price until numerous phthalic plants operating on naphthalene actually were forced to shut down, and others had to convert to an o-xylene feed. Finally, in 1967 the price of o-xylene rose

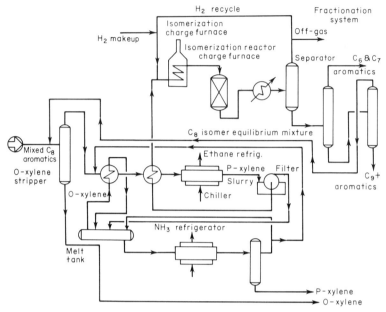

Fig. 8-3 *p*-Xylene.

sharply to the point where the processes based on the various raw materials were again competitive. However, *o*-xylene is expected to be the most desirable of the three raw materials.

Interest in pure xylene isomers has been growing. This is due to the rapidly increasing demand for the aromatic difunctional acids which can be produced from them. The ortho- and meta-isomers boil about 9°F apart and can be separated by superfractionation. However, the para-isomer cannot be removed by this technique since it has almost the same boiling point as *m*-xylene. Fortunately, *p*-xylene melts at 56°F, about 69°F above the nearest isomer (*o*-xylene) and therefore can be removed by fractional crystallization. A flowsheet combining crystallization of *p*-xylene and isomerization of the unwanted isomers is shown in Fig. 8-3. Feed and recycle are cooled against other process streams and finally against ethane in a scraped chiller. The resulting slurry is filtered under vacuum, the mother-liquor being sent to other units of the refinery or, for example, in this case, to isomerization. The filter-cake, consisting of mixed xylenes, is molten and sent to another scraped chiller where this time it is cooled to 0°F against propylene, ammonia, or a fluorocarbon so that essentially only *p*-xylene settles out.   Then the slurry is filtered again, the mother-liquor being recycled to the previous chilling stage. The product is 99.5 percent p-xylene. The crystals from the two distilling stages can also be removed by centrifuging and using crystallizers instead of scraped chillers.   In 1967, Institut Francais du Pétrole announced a version of this process in which pure p-xylene can be obtained in one stage from a continuous crystallization column.

Cosden Petroleum, after removing ethylbenzene from the incoming C_8 stream, separates p-xylene by tieing it up in a Werner-type complex, known as a clathrate, which is filtered off. The clathrate is heated and then pure p-xylene is separated from the clathrating agents (λ-picoline and nickel isothiocyanate).

The usual C_8 reformate extract cut consists of roughly 45 percent m-xylene, 20 percent each of o- and p-xylene and 15 percent ethylbenzene.   It so happens, however, that there is much less demand for m-xylene than for either of the other two isomers. Several companies are therefore using a process called "Octafining," developed by Atlantic Refining.   In this process, m-xylene is isomerized over a platinum catalyst into an equilibrium mixture of 45 percent m-xylene and roughly equal amounts of the other isomers mentioned above, which is recycled to the crystallization step.   Hence, about 70 percent of the m-xylene originally present can be converted to its o- and p-isomers.   "Octafining" is suited for the type of xylene plant described above, where the effluent from the isomerization reactor is reintroduced into the crystallization-superfractionation unit together with the fresh feed. With the rise in demand for p-xylene, isomerization units are frequently built onto existing xylene-isomer separation facilities.

To complete the commercially important aromatic raw materials, there has been increasing interest in durene, mesitylene and pseudocumene.   The derivatives of these aromatics, as is the case of the xylene isomers, are the respective acids and anhydrides obtained by oxidation.   Pseudocumene is removed from a C_9 aromatics cut by superfractionation.   Mesitylene can be formed from pseudocumene, which is the main constituent of the C_9 fraction, by isomerizing pseudocumene in a manner similar to "Octafining."

Durene, on the other hand, has a high melting point and is therefore obtained by fractional crystallization from a C_{10} aromatics stream; the availability of durene can also be increased by methylating pseudocumene:

It is difficult to discuss the economics of producing aromatics from petroleum production because so much depends on the cyclic content of the reformate.   The sum of benzene and toluene in reformates obtained from the naphthas commonly refined in Europe and the U.S. can vary between 19 percent for Kuwait to 27 percent for certain Venezuelan oils.   This will determine the throughput required for the reforming unit, "Udex" extraction and dealkylation unit, and, in turn, influence investment requirements. This has a great effect on the potential cost of aromatics once a certain benzene output has been specified.

The most difficult cost component in the economics of producing aromatics is the value of the aromatics themselves. They have very high octane numbers and add value to the gasoline as premium fuel.   However, sometimes their removal from the gasoline pool still permits the refiner to achieve the balance between regular and premium motor fuels required by the market in which he operates. Thus, removal of the aromatic hydrocarbons does not down-grade the value of the fuel.

As an illustration, the cost of producing benzene has only been estimated for a particular set of conditions.   The system consists of a catalytic reformer, an aromatics extraction unit and a dealkylation unit which converts to benzene all toluene and those C_8 aromatics dealkylated with reformer hydrogen.

The material balance shows that the raw material value of benzene under the assumptions made is 1.4 cent per pound.

	lbs/lb feed		Value, ¢/lb	¢/lb feed	
				+	−
Feed	1.00		1.32	1.320	
Product					
Fuel gas		0.08	0.7		0.056
LPG		0.21	1.4		0.294
Raffinate		0.34	1.2		0.408
High octane gasoline		0.11	1.8		0.198
Benzene		0.26	1.4		0.364
	1.00	1.00	6.5	1.320	1.320

Taking as a specified basis the production of 250.0 million pounds per year of benzene, the investment costs will be as shown below.

Unit	$MM
Desulfurizer and catalytic reformer	3.4
Aromatics extraction and fractionation	3.7
Dealkylation and hydrogen purification	3.0
Total	10.1

The production costs will then be as follows:

	Cost, ¢/lb benzene
Feed	1.4
Utilities	0.5
Labor and overhead	0.1
Catalyst and chemicals	0.1
Capital charges	1.3
Total	3.4

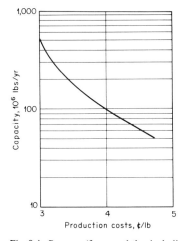

Fig. 8-4 Benzene (from naphtha, including reforming, alkylation and dealkylation).

A 100.0 million pounds of benzene per year catalytic dealkylation unit operating on pure toluene would cost $1.8 million. Assuming toluene at 2.2 cents per pound and hydrogen at 30 cents per thousand cubic feet, the cost of benzene from such a plant would be as shown below.

	Cost, ¢/lb benzene
Toluene	2.7
Hydrogen	0.2
	2.9
Less: net off gas	(−0.4)
Net raw materials	2.5
Labor, overhead, utilities	0.4
Capital charges	0.6
Total	3.5

In general, it is not considered very profitable to dealkylate toluene because of its importance to the refinery's octane pool.

At best, these figures can be considered representative. For the purposes of this discussion benzene will be taken at 25 cents per gallon, or 3.4 cents per pound.

Investment for a 75.0 million pounds per year naphthalene plant from heavy reformate is $2.9 million, and the production costs are as follows:

	Cost, ¢/lb naphthalene
Feed	2.0
Utilities	0.5
Labor and overhead	0.1
Catalyst and chemicals	0.2
Capital charges	1.3
Total	4.1

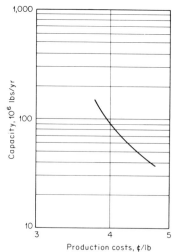

Fig. 8-5 Naphthalene (from heavy reformate).

Sun Oil makes naphthalene from a 450-550° F catalytic cycle oil, containing approximately 50 percent methylnaphthalenes. Feed preparation for such a unit is cumbersome since methylnaphthalenes must be concentrated by solvent extraction to reduce hydrogen requirements. This process is advantageous in that it upgrades the extract; however, the reverse happens when anything heavier than gasoline is reformed. Production costs, exclusive of feed stock, are 1.6 times those from heavy reformate, and investment costs about double.

Again, it must be emphasized that only a detailed study can determine the real economics of a given aromatics project, after all the local factors, for example, fuel markets, transportation costs, and refinery economics as a whole have been taken into account.

The investment for a 50.0 million pounds per year p-xylene plant, including isomerization, is $6.0 million. The production costs are given below.

	Cost, ¢/lb p-xylene
Raw material	2.7
Utilities	0.7
Labor and overhead	0.6
Capital charges	4.0
Total	8.0

The relatively high price assumed for p-xylene contained in the feed is a reflection of its octane value.

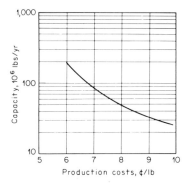

Fig. 8-6 p-Xylene.

The superfractionation unit for making 60.0 million pounds per year o-xylene costs around $1.2 million, and costs exclusive of raw material are as follows:

	Cost, ¢/lb o-xylene
Utilities	0.5
Labor and overhead	0.2
Capital charges	0.7
Total	1.4

Despite these much lower production costs, by 1965 o-xylene prices had fallen to the point where the refiner would reap a profit by converting it to p-xylene by isomerization.

The demand for benzene in the U.S. is growing at the rate of 5.5 percent per annum. Consumption in 1964 was 4,640 million pounds, distributed as follows:

End-use	MM lbs, 1964
Styrene	1,982.0
Phenol	1,016.0
Cyclohexane	623.0
Synthetic detergents	272.0
Aniline and nitrobenzene	176.0
Maleic anhydride	176.0
DDT	88.0
Mono- and dichlorobenzene	88.0
Cumene (other than phenol)	51.0
Miscellaneous	168.0
Total	4,640.0

Source: *Chemical Week*, March 6, 1965.

This estimate refers only to benzene actually extracted. The amount of cyclohexane produced is larger than indicated above, since large amounts are extracted from natural gasoline or made directly from toluene.

The potential sources of benzene in the U.S. as of 1965 can be broken down as shown in the table below.

Source		% of total
Petroleum		84.7
Reformate extraction –	59.5	
Olefin plants	3.2	
Dealkylation	22.0	
Coal		15.3
Tar distillers	1.8	
Coke-ovens	13.5	
Total		100.0

Source: *Oil, Paint & Drug Reporter*, June 14, 1965.

During 1965, 3,996.0 million pounds of toluene were produced in the U.S. with 95 percent coming from petroleum and 5 percent from coke-ovens. The main applications of this output were the following:

Destination	MM lbs, 1965
Motor fuel	956.0
Aviation gasoline	253.0
Solvents	362.0
Dealkylation to benzene	1,665.0
Chemicals	543.0
Exports	217.0
Total	3,996.0

Source: *Hydrocarbon Processing,* February 1966.

The operations of making benzene and cyclohexane from toluene have been reported elsewhere. The demand for toluene as a starting point for chemical synthesis can be broken down further as follows:

End-use	MM lbs, 1965
Toluene diisocyanate	87.0
Sulfates	43.0
Phenol	51.0
TNT	72.0
Miscellaneous chemicals	290.0
Total	543.0

Source: *Hydrocarbon Processing,* February 1966.

The U.S. producers of aromatics, benzene and toluene from petroleum are shown in the table below.

Producer	Location	Benzene and toluene capacity, 1965 MM lbs/yr		
		Benzene from petroleum		Toluene
		Extraction	Dealkylation	
Allied	Winnie, Tex.	30.0		
Amoco	Texas City, Tex.	110.0		145.0
Ashland	Buffalo, N.Y.	75.0		60.0
	Catlettsburg, Ky.	100.0	45.0	80.0
Atlas Processing	Shreveport, La.	75.0		
Conoco	Lake Charles, La.	45.0		
	Ponca City, Okla.	45.0		
Cosden	Big Spring, Tex.	65.0	110.0	110.0
Crown Central	Houston, Tex.	45.0	95.0	70.0
Dow	Bay City, Mich.		145.0	125.0
	Freeport, Tex.		220.0	
Enjay	Baton Rouge, La.	175.0		110.0
	Baytown, Tex.	180.0	220.0	360.0
Gulf	Philadelphia, Pa.	110.0	90.0	110.0
	Port Arthur, Tex.	230.0		35.0

Continued

Producer	Location	Benzene and toluene capacity, 1965 MM lbs/yr		
		Benzene from petroleum		Toluene
		Extraction	Dealkylation	
Hess	Corpus Christi, Tex.	220.0		130.0
Leonard	Mount Pleasant, Mich.		20.0	25.0
Marathon	Detroit, Mich.	55.0*		120.0*
	Texas City, Tex.	45.0		85.0
Monsanto	Alvin, Tex.	250.0	185.0	230.0
Phillips	Sweeney, Tex.	160.0		
Pontiac	Corpus Christi, Tex.	65.0		95.0
Richfield	Wilmington, Calif.	130.0		175.0
Shell	Houston, Tex.	220.0		220.0
	Odessa, Tex.	35.0	110.0	70.0
	Wilmington, Calif.	110.0		70.0
	Wood River, Calif.	220.0		110.0
Signal	Houston, Tex.	20.0	140.0	110.0
Sinclair	Houston, Tex.			145.0
	Marcus Hook, Pa.			45.0
Socony-Mobil	Beaumont, Tex.	220.0		180.0
South Hampton	Silsbee, Tex.		45.0	45.0
Standard (Calif.)	El Segundo, Calif.	180.0		175.0
	Richmond, Calif.	70.0		60.0
Sun	Marcus Hook, Pa.	110.0		180.0
Sunray-DX	Tulsa, Okla.	80.0	90.0	30.0
Suntide	Corpus Christi, Tex.	70.0	110.0	95.0
Tenneco	Chalmette, La.	110.0		60.0
Texaco	Port Arthur, Tex.	220.0		145.0
Union-Atlantic	Nederland, Tex.	130.0		145.0
Union Carbide	S. Charleston, W.Va.	75.0		70.0
Union Oil	Lemont, Ill.	160.0		70.0
Vickers	Potwin, Kans.	20.0		35.0
Subtotals		4,340.0	1,625.0	4,125.0
Total from petroleum		5,965.0		4,125.0
Total from coal		130.0		50.0
Grand Total		6,095.0		4,175.0

Source: Based on *Oil, Paint & Drug Reporter*, June 14, 1965; *Hydrocarbon Processing*, February 1966.

*Toluene and benzene shipped as a blend to Dow at Bay City, Mich., and finally processed there.

The following chemical companies have their own aromatics source from coal-tar distillation:

Producer	Location	Capacity, MM lbs/yr	
		Benzene	Toluene
Allied Chem.	Syracuse, N.Y.	75.0	10.0
American Cyanamid	Bound Brook, N.J.	45.0	6.5
Koppers	Kobuta, Pa.	4.0	4.0

Source: Benzene: *Oil, Paint & Drug Reporter*, June 14, 1965;
Toluene: *Oil, Paint & Drug Reporter*, February 25, 1963.

By 1965, 97 percent of the xylene produced in the U.S. came from petroleum. Of total production, which amounted to 3450.0 million pounds, 64 percent went to fuel and another 10 percent to solvents for alkyd coatings. Of the remainder, only 94.0 million pounds were used without separation of the three isomers; the rest of the demand was for pure o-, m- and p-xylene, all three of which have gained enormously in importance as raw materials and are likely to continue doing so. The total domestic demand for xylene as a chemical raw material can be broken down as follows:

Xylene isomers	*MM lbs, 1965*
Mixed xylenes	94.0
p-Xylene	400.0
o-Xylene	350.0
m-Xylene	45.0
Total	889.0

The U.S. was the first country to produce xylene isomers on a commercial scale, and is therefore a large net exporter. This explains the difference between domestic demand and published production figures.

Almost the entire output of o-xylene is used to make phthalic anhydride. Domestic U.S. demand was 100.0 million pounds in 1965, but this cannot be taken as a reference point because several phthalic plants can switch back and forth between naphthalene and o-xylene without difficulty. However, this depends only on raw material prices compared with the inconvenience of shutting down to change catalyst.

The U.S. producers of o-xylene are listed below. Since o-xylene production involves only conventional fractionation, capacities are flexible and reflect rather the potential o-isomer of the plant's C_8 aromatic stream.

Producer	Location	Capacity, MM lbs/yr
Cities Service	Lake Charles, La.	120.0
Cosden	Big Spring, Tex.	12.0
Crown Central	Houston, Tex.	20.0
Hess	Corpus Christi, Tex.	75.0
Pontiac	Corpus Christi, Tex.	10.0
Enjay	Baytown, Tex.	175.0
Chevron	Richmond, Calif.	100.0
	El Segundo, Calif.	30.0
Sinclair	Houston, Tex.	75.0
Suntide	Corpus Christi, Tex.	32.0
Tenneco	Chalmette, La.	22.0
Total		671.0

Source: *Hydrocarbon Processing,* April 1966.

The usual phthalic anhydride plant feed is 95 percent *o*-xylene; upgrading to 99 percent is possible at an additional cost of 0.5 cent per pound. The only important use for pure *m*-xylene is production of isophthalic acid. Only a fraction of what is produced is isolated; the rest either goes to motor fuel or to an isomerization process that converts it to equilibrium mixture of the four C_8 isomers. The total demand for *m*-xylene in the U.S. was 45.0 million pounds in 1965. Producers of *p*-xylene in the U.S. are given below.

Producer	Location	Capacity,* MM lbs/yr	
		1965	1967
Amoco	Texas City, Tex.	—	200
Chevron	El Segundo, Calif.	70	90
	Pascagoula, Miss.	—	300
	Richmond, Calif.	60	110
Cities Service	Lake Charles, La.	—	36
Cosden	Big Spring, Tex.	6	6
Enjay	Baytown, Tex.	105	210
Hercor	Guayanilla, Puerto Rico	—	100
Shell	Houston, Tex.	—	100
Signal	Houston, Tex.	15	15
Sinclair	Houston, Tex.	120	200
	Marcus Hook, Pa.	15	15
Suntide	Corpus Christi, Tex.	65	170
Total		456	1,552

*Some capacities may be higher due to debottlenecking.
Source: *Oil, Paint & Drug Reporter,* June 6, 1966.

Practically all *p*-xylene is converted to polyester fiber precursors.

In 1965, the demand for naphthalene was estimated at 680.0 million pounds. Some 75 to 80 percent of all phthalic anhydride produced that year was based on naphthalene. About 250.0 million pounds came from petroleum, some 60 percent of the installed petro-naphthalene capacity. The demand for naphthalene in 1965 can be broken down as follows:

End-use	MM lbs, 1965
Phthalic anhydride	535.0
Mothballs	15.0
β-Naphthol	25.0
1-Naphthyl methylcarbamate	70.0
Surfactants	5.0
Synthetic tanning agents	20.0
Others	10.0
Total	680.0

Source: *Oil, Paint & Drug Reporter,*
March 14, 1966.

The producers of naphthalene are listed below. Notice that the two main coal-tar naphthalene producers are chemical producers, and not steel companies.

Producer of naphthalene	Location	Capacity, MM lbs, 1965
From coal tar		
Allied	Chicago, Ill.	10.0
	Detroit, Mich.	10.0
	Frankford, Pa.	80.0
	Ironton, Ohio	35.0
Coopers Creek	West Conshohocken, Pa.	1.0
Granite City Steel	Granite City, Ill.	5.0
Koppers	Chicago, Ill.	30.0
	Follansbee, W.Va.	150.0
	Woodward, Ala.	2.0
Productol	Santa Fe Springs, Cal.	12.0
Reilly Tar	Cleveland, Ohio	7.0
	Granite City, Ill.	5.0
Republic	Chicago, Ill.	1.2
	Cleveland, Ohio	4.9
	Gadsden, Ala.	0.7
	Thomas, Ala.	0.3
	Warren, Ohio	0.3
	Youngstown, Ohio	2.2
U.S. Steel	Clairton, Pa.	80.0
	Fairfield, Ala.	10.0
	Gary, Ind.	30.0
	Morrisville, Pa.	9.0
Woodward Iron	Woodward, Ala.	1.2
Total		486.8

From petroleum

Ashland	Catlettsburg, Ky.	90.0
Collier-Tidewater	Delaware City, Del.	100.0
Monsanto	Alvin, Tex.	100.0
Sun	Toledo, Ohio	110.0
Total		400.0

Source: *Oil, Paint & Drug Reporter*, March 14, 1966.

2 STYRENE

Benzene can be alkylated with ethylene using aluminum chloride or, as in the UOP process, with phosphoric acid as the catalyst.

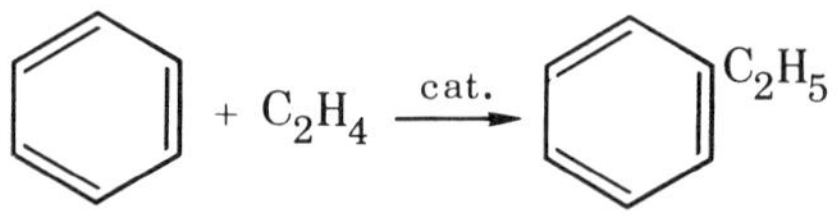

$$\bigcirc + C_2H_4 \xrightarrow{\text{cat.}} \bigcirc C_2H_5$$

The reaction is carried out in the presence of ethyl chloride. Ethylbenzene can be dehydrogenated to form styrene:

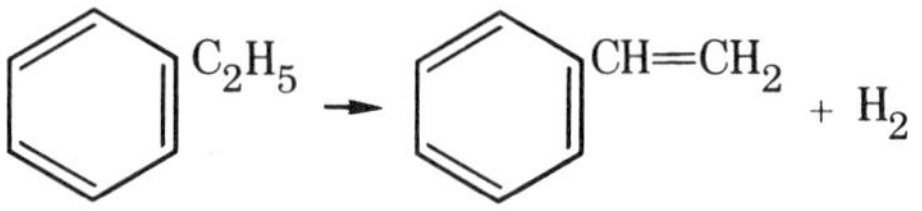

$$\bigcirc C_2H_5 \rightarrow \bigcirc CH{=}CH_2 + H_2$$

The ethylene stream need not be pure to effect the alkylation. In fact, a process known as "Alkar" (Universal Oil Products) has been developed to accomplish this alkylation in the presence of BF_3 from a stream containing as little as 10 percent ethylene. This has the advantage of using a cheaper raw material, and producing no polyalkylated benzene. Also, several plants have been built in the U.S. to make styrene from the "natural" ethylbenzene present in a C_8 aromatic extract cut. This can be separated by a super-fractionation technique requiring three 200-ft distillation columns in series, all operating at high reflux rates, first employed by Cosden Petroleum Corp.

Styrene itself is usually produced by vapor-phase dehydrogenation of ethylbenzene over a ferric oxide catalyst, with steam used as the diluent. After the reactor effluent has been condensed and the water removed, the crude styrene is sent to a column where benzene and toluene byproducts are removed, then sent to an ethyl-benzene tower. From this tower unreacted feed is returned to the reactor. A finishing column removes the heavy ends and any tar that may have formed by polymerization. The fractionation system operates under high vacuum to increase the relative volatility of the components and to avoid high temperatures which could bring about polymerization. The yield on ethylbenzene feed is 87 percent (see Fig. 8-7). While most styrene processes limit themselves to conversions of 35-40 percent per pass, to avoid degradation of ethylbenzene to toluene and benzene, some licensors (Badger, Scientific Design) have developed catalysts allowing over 55 percent conversion. Since utilities represent about 15 percent of production costs, the effect of higher conversion at equal yields can be quite significant; thus, improving conversion per pass from 40 to 60 percent lowers costs by around 5 percent.

The economics of styrene production are shown below for a production of 100.0 million pounds per year, starting from ethylene and benzene. Superfractionation is said to yield ethylbenzene at 1-1.5 cent per pound less than alkylation.

The investment for the benzene alkylation unit is $1.4 million. With ethylene at 4.0 cents per pound and benzene at 3.4 cents per pound, production costs are as follows:

	Cost, ¢/lb ethylbenzene
Ethylene	0.8
Benzene	2.5
Utilities	0.5
Labor and overhead	0.2
Capital charges	0.5
Total	4.5

The styrene unit requires an investment of $3.0 million. and production costs are as shown in the following and Fig. 8-8, p. 419.

		Cost, ¢/lb styrene
Ethylbenzene		5.4
Less: hydrogen	-0.1	
benzene-toluene	-0.1	-0.2
Net raw materials		5.2
Utilities		1.1
Labor and overhead		0.3
Capital charges		1.0
Total		7.6

The investment for a 70.0 million pounds per year styrene plant starting from catalytic reformate aromatics extract is $6.5 million.

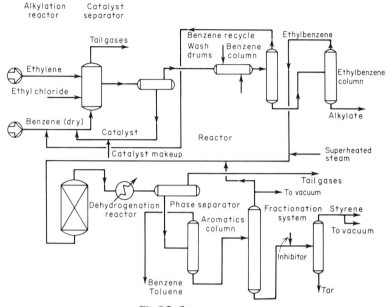

Fig. 8-7 Styrene recovery.

The demand for styrene in the U.S. has been growing at a rate of 6 percent per year, which promises to prevail until the end of the decade. Domestic demand for styrene in 1965 was 2,900.0 million pounds, distributed as shown in the table below.

End-use	MM lbs, 1965
Polystyrene, general-purpose grades	725.0
Polystyrene, high-impact grades	610.0
All other styrene resins	
(incl. ABS and SAN)	230.0
SBR elastomer	635.0
SBR copolymer	175.0
Export	380.0
Miscellaneous	145.0
Total	2,900.0

Source: *Oil, Paint & Drug Reporter*, April 18,1966.

Below is a list of styrene monomer producers in the U.S. in 1965.

Producer	Location	Styrene capacity, MM lbs, 1965
Amoco	Texas City, Tex.	250.0
Cosden	Big Spring, Tex.	100.0
Dow	Freeport, Tex.	500.0
	Midland, Mich.	300.0

El Paso	Odessa, Tex.	86.0
Foster Grant	Baton Rouge, La.	200.0
Marbon	Baytown, Tex.	125.0
Monsanto	Texas City, Tex.	650.0
Shell	Torrance, Calif.	210.0
Sinclair-Koppers	Houston, Tex.	70.0
	Kobuta, Pa.	200.0
Suntide	Corpus Christi, Tex.	60.0
Union Carbide	Institute, W. Va.	110.0
	Seadrift, Tex.	300.0
Total		3,161.0

Source: *Oil, Paint & Drug Reporter*, April 18, 1966.

The following companies produce ethylbenzene by superfractionation.

Producer	Location	Capacity, MM lbs/yr
Cosden	Big Spring, Tex.	22.0
Enjay	Baytown, Tex.	150.0
Monsanto	Alvin, Tex.	40.0
Signal Oil	Houston, Tex.	35.0
Sinclair-Koppers	Houston, Tex.	75.0
Suntide	Corpus Christi, Tex.	30.0
Tenneco	Chalmette, La.	22.0
Total		374.0

This is about 12.0 percent of total styrene capacity.

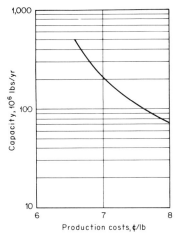

Fig. 8-8 Styrene monomer.

Integration is found to be the rule in styrene production. Of the twelve monomer producers, seven have their own sources of aromatics, of which six are from petroleum and one (Koppers) from coal-tar. By 1966 only three companies producing ethylbenzene were not also styrene manufacturers themselves.

2.1   Polystyrene

Styrene polymerization for the most part is carried out thermally (Dow process) or in suspension, in the presence of a catalyst, normally benzoyl peroxide. The main properties of polystyrene are its relatively low density (1.03), freedom from odor, taste and toxicity, good optical properties, and, above all, ease of molding, dimensional stability and colorability. Its disadvantages are poor impact and chemical resistance and that it is degraded by UV and therefore cannot be used in outdoor applications.

Polystyrene resins, both "general-purpose" or mixed with styrene-butadiene copolymers or polybutadiene to increase impact resistance, are the most important plastic materials for molding applications. Use of the more expensive high-impact grades will tend to surpass general-purpose resins in the U.S.; their use is growing faster than that of polystyrene as a whole and by 1965 almost 40 percent of all polystyrene was high-impact. High-impact grades were formerly mechanical mixtures of styrene polymer and SBR. At present, most grades are made by graft-polymerization of styrene onto an SBR or polybutadiene matrix.

The investment for a 40.0 million pounds per year styrene polymerization plant is $3.1 million, and polymerization costs are as follows, assuming captive monomer costs 7.5 cents per pound.

	Cost, ¢/lb polystyrene
Monomer	7.8
Utilities, catalyst and chemicals	0.3
Labor and overhead	1.0
Capital charges	2.6
Total	11.7

A comparison between the table of polystyrene producers and list of styrene manufacturers shows that almost 90 percent of the polymer capacity belongs to monomer producers.

Producer	Location	Capacity, MM lbs/yr
Brand Plastics	Willow Springs, Ill.	35
	Torrance, Calif.	12
	Medina, Ohio	12

Cosden Oil & Chemical	Big Spring, Tex.	120
Dow	Allyns Point, Conn.	
	Midland, Mich.	
	Torrance, Calif.	500
	Riverside, Mo.	
Foster-Grant	Leominster, Mass.	80
	Peru, Ill.	40
Monsanto	Addyston, Ohio	
	Long Beach, Calif.	300
	Springfield, Mass.	
Rexall Chemical	Holyoke, Mass.	
	Ludlow, Mass.	46
	Santa Ana, Calif.	14
Richardson	West Haven, Conn.	30
Shell	Wallingford, Conn.	40
Sinclair-Koppers	Kobuta, Pa.	100
Solar Chemical	Leominster, Mass.	40
Union Carbide	Bound Brook, N.J.	
	Marietta, Ohio	150
Others		38
Total		1,557

Source: *Chemical & Engineering News*, June 18, 1966.

The polystyrene market in 1965 can be broken down by type of fabrication as follows. This excludes SBR latex, as well as co-polymer resins.

Type of fabrication	% of total
Molding	66
Extrusion	19
Foam	11
Others	4
Total	100

Molding and extrusion applications were as shown below.

End-use	MM lbs, 1965
Packaging	265.0
Refrigeration	68.0
Major appliances	23.0
Small appliances	15.0
Radio and television	45.0
Housewares	125.0
Lighting and signs	36.0
Toys, premiums, novelties	145.0
Combs, brushes, eyeglasses	24.0
Pipe	20.0
Monofilament brushes	4.0
Foamed sheet (from crystal)	4.0
Miscellaneous (auto, telephone, business machines, tile, export, etc.)	289.0
Total	1,063.0

Source: *Modern Plastics*, January 1966.

Polystyrene foam is manufactured by polymerizing at low temperatures in the presence of a volatile compound (e.g., methylene chloride or propylene) so that the resulting polymer beads expand on heating.   Most styrene foam, with the exception of Dow's "Styrofoam," is made by this process.   The demand in 1965 was 125.0 million pounds, making it the most important plastic foam after flexible urethanes.   Styrene foams are rigid, and their most important application is in the manufacture of insulation board. A breakdown of the market is shown below:

End-use	MM lbs, 1965
Expanded styrene board	35.0
Expandable styrene beads	
Insulating cups	15.0
Packaging	22.0
Building	24.0
Molding, export, etc.	23.0
Miscellaneous	6.0
Total	125.0

Source: *Modern Plastics*, January 1966.

Polystyrene accounts for about 45 percent of all molded plastics made in the U.S., 10 percent of extruded plastics and 27 percent of plastic foams, not including foam rubber.   The trend towards higher-specification resins will also result in a partial substitution of polystyrene by ABS resins.   However, although they are more expensive, their low-temperature and strength properties and chemical resistance are better than high-impact polystyrene. In terms of relative growth, the best prospects are for foam, especially formed and extruded products, and film. Polystyrene film is the cheapest of all shrink materials; in addition, it is clear, crisp and printable, and has excellent aging characteristics.

In wartime as much as 300.0 million pounds per year of polystyrene has been used to make "napalm."

2.2   ABS and SAN Resins

ABS resins are the most important of the recently developed and increasingly important family of engineering plastics, that is, those plastics with such physical properties as higher softening points, tensile strength, toughness, and better aging properties. These properties permit them to compete for markets from which PVC, polystyrene and the polyolefins are excluded. There are three ways to make ABS resins: 1) by copolymerizing styrene with a copolymer of acrylonitrile and butadiene; 2) by blending acrylonitrile-butadiene and acrylonitrile-styrene copolymers; 3) by grafting styrene and acrylonitrile onto a preformed polybutadiene matrix. Overall, ABS resins are copolymers of 50 percent styrene and 25 percent by weight butadiene and acrylonitrile; however, they are usually tailored to suit each particular end-use. This accounts for the very large number of grades on the market. Most of the markets taken over by ABS resins from other plastics formerly belonged to high-impact polystyrene; impact strength is imparted by butadiene, whereas high styrene and acrylonitrile contents are used to produce tougher and more heat-resistant resins. Higher styrene-acrylonitrile resins have better properties, but are more difficult to process and fabricate; the present trend is towards a lower overall butadiene content.

In automobiles, the most important use of ABS resins has been as a housing for instrument cluster boards, but the biggest potential application is as a substitute for chrome-plated metal parts since a technique for plating ABS has been successfully developed. The appearance of polyurethane insulation for refrigerators on the market has resulted in increasing use of ABS for food cabinet liners, freezer sections and door liners. This is due mainly to the fact that impact polystyrene is subject to attack by the fluorocarbon blowing agents in urethane foams and ABS is not. Its superior strength enables refrigerators to be made with thinner walls, and increases the load-bearing capacity of doors (for bottles, etc.). Other major appliances are also consumers of ABS, as are such minor appliances as electric can openers, blenders and others. Plastic pipe of ABS can be used in general-purpose plumbing applications, which is not the case with, for example, PVC; it is widely employed for drain and sewage pipes.

Other applications include textile equipment accessories such as bobbins, as well as power tool housing and a variety of less important applications. Sheet tends to become the most important form of ABS (the predicted share of the market for 1970 is 43 percent) followed by injection molding (34 percent), extrusion (15 percent) and blow-molding (8 percent). The market breakdown for 1965 was as follows:

End-use	MM lbs, 1965
Automotive	32.0
Major appliances	25.0
Small appliances	16.0
Luggage	10.0
Business machines, telephones	16.0
Shoe heels	15.0
Pipe fittings	26.0
Miscellaneous, incl. export	35.0
Total	175.0

Source: *Modern Plastics*, January 1966.

United States manufacturers of ABS resins are the following:

Producer	Location	Capacity, MM lbs/yr
Borg-Warner	Washington, W.Va.	140.0
Dow	Midland, Mich.	30.0
B.F. Goodrich	Louisville, Ky.	20.0
Monsanto	Addyston, Ohio	90.0
Rexall	Joliet, Ill.	33.0
U.S. Rubber	Baton Rouge, La. Scotts Bluff, La.	80.0
Union Carbide	Bound Brook, N.J.	40.0
Total		433.0

Source: *Oil, Paint & Drug Reporter*, May 24, 1965.

SAN, a copolymer of styrene (2/3) and acrylonitrile (1/3), has also obtained most of its markets from impact polystyrene. About 30.0 million pounds were being used in 1965, of which some 50 percent was for dinnerware. The major growth recently has been for a glass-reinforced thermoplastic in automobiles. The major manufacturer of these resins is Dow (under the trade name of "Tyril").

2.3 SB Resins

Contrary to SBR and rubber latices, SB resins contain about 2/3 styrene. They are used in water-based paints, where due to poor resistance to UV they are restricted to indoor applications.

Whereas smaller coating manufacturers have tended to prefer polyvinyl acetate emulsion paints, large paint makers still employ the cheaper SB formulations for indoor applications. In 1965, rubber-based paints consumed 42.0 million pounds of SB resins, representing 20 percent of the total market for latex coatings. Another 142.0 million pounds per year of SB resins were used for textile and paper treatment. Here the trend is towards incorporation of an unsaturated carboxylic acid (methacrylic or acrylic) as a third comonomer, which makes the resins thermosetting and self-curing by reaction of the -COOH groups. Carboxylated resins have better filming properties, higher elasticity and are more resistant to solvents. On the other hand, their laundering resistance is affected. The total use of SB resins in 1965 was 184.0 million pounds.

2.4 Chlorinated Styrene

Dow has built a 1.0 million pounds per year chlorinated styrene unit at Midland, Mich.

Unsaturated polyester resins formulated with chlorinated styrene cure much faster than those containing only styrene.

2.5 Divinylbenzene

Divinylbenzene is obtained by dehydrogenating diethylbenzene.

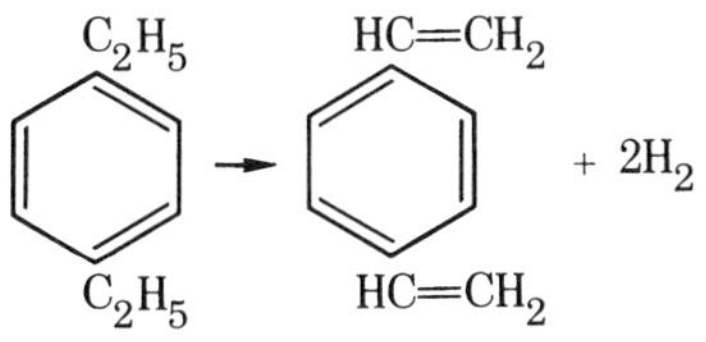

$$
\begin{array}{c} C_2H_5 \\ \bigcirc \\ C_2H_5 \end{array}
\longrightarrow
\begin{array}{c} HC{=}CH_2 \\ \bigcirc \\ HC{=}CH_2 \end{array}
+ \ 2H_2
$$

It is used as a crosslinking agent in ion exchange resins, SBR and stereo rubber.

2.6 Ion Exchange Resins

About 35.0 million pounds per year of ion exchange resins are used in the U.S. The greater part of these resins are sulfonated copolymers of styrene and divinylbenzene, used for domestic water softening. These resins make up 50 percent of the total, and industrial applications are likewise mainly for cation exchange. Anion exchange resins are also based on styrene polymers, the active groups being quaternary ammonium ions:

Cation Exchange

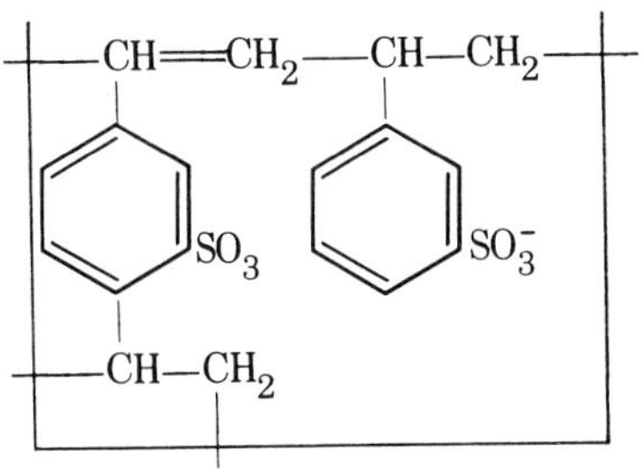

$$
\begin{array}{c}
\vert\!\!-\!CH{=}CH_2\!-\!CH\!-\!CH_2\!-\!\vert \\
\bigcirc\!-\!SO_3 \quad \bigcirc\!-\!SO_3^- \\
\vert\!\!-\!CH\!-\!CH_2 \\
\vert
\end{array}
$$

Anion Exchange

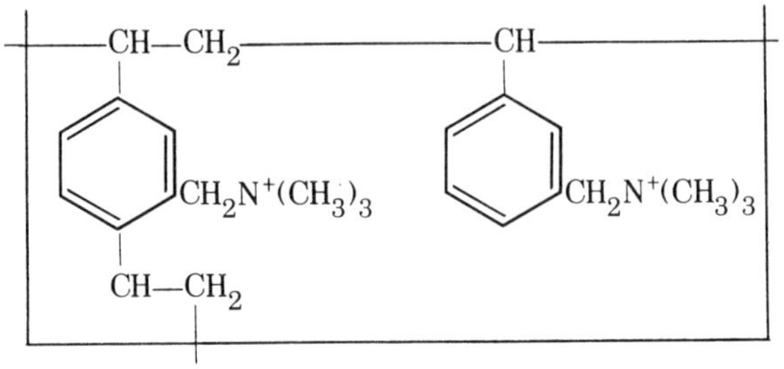

3 CYCLOHEXANE

Catalytic hydrogenation of benzene yields cyclohexane:

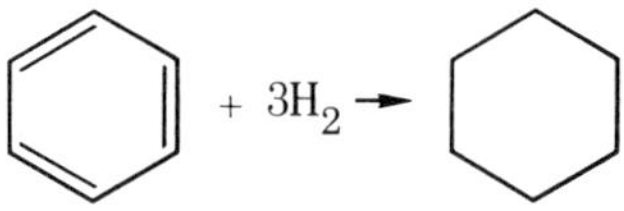

$$+ 3H_2 \rightarrow$$

The reaction is usually carried out in the liquid phase, although vapor-phase hydrogenation is used in the U.S. by DuPont; yields are close to stoichiometric in either case. The advantage of vapor-phase operation is that there is no need to separate suspended nickel catalyst from the product stream, a rather complicated operation; on the other hand, vapor-phase technology requires larger equipment, in particular, the reactor with its catalyst inventory. Liquid-phase hydrogenation must be carried out at around 300 psig, whereas only 10 psig are required for vapor-phase operation; this, however, is seldom an advantage, as the overwhelming majority of cyclohexane made in the world uses reformer off-gas hydrogen which is already available at 250–300 psig.

Processes differ from one another mainly in the method employed for removing the heat of reaction. In the UOP process, since the liquid-phase hydrogenation reaction is practically zero-order with respect to benzene, this is accomplished by circulating large amounts of benzene without affecting the reaction rate. The IFP process employs two reactors in series, the first operating in the liquid-phase and the second in the vapor-phase. Any benzene not converted in the first reactor is hydrogenated in the trim reactor. Other processes employ cooling tubes, involving a more complicated reactor but a lower catalyst inventory. Raney nickel is the usual catalyst.

There are two other important sources of cyclohexane. The first is an 85 percent grade of cyclohexane on a large scale by Phillips from Venezuelan naphthenic crude virgin naphtha and from natural gasoline. This product is used as a solvent for medium-pressure polyethylene production, but can also be fed to a cyclohexanol-cyclohexanone plant. Second, one U.S. producer makes cyclohexane from toluene by simultaneous dealkylation and double-bond hydrogenation.

These other sources account for some 35 percent of U.S. production; the remaining 65 percent are obtained by hydrogenating benzene, but this has tended to rise.

The U.S. produces most of the cyclohexane consumed in the world. Because of the enormous amounts of benzene derived from catalytic reformate (in contrast to the still predominant position of coal-tar aromatics in Europe), benzene and hydrogen are available at internal costs that enable cyclohexane to be sold for slightly more than benzene. The investment for a 100.0 million pounds per year liquid-phase hydrogenation plant employing cyclohexane recirculation is around $500,000, not including facilities for cleaning up reformer hydrogen, which still contains hydrocarbons as heavy as C_6. The production costs are given below, assuming the internal value of hydrogen at 25 cents per thousand cubic feet.

	Cost, ¢/lb cyclohexane
Benzene (at 3.4 ¢/lb)	3.15
Hydrogen	0.38
Labor and overhead	0.08
Utilities, catalyst	0.03
Capital charges	0.16
Total	3.80

This amounts to 25 cents per gallon, the same price at which benzene was taken. Note that raw materials account for over 90 percent of the total cost, so that economics are extremely sensitive to the internal values of benzene and hydrogen.

The IFP process, which provides for recovery of reaction heat in the form of steam, requires a 20 percent higher investment but shows a slight credit for utilities consumption so that costs are the

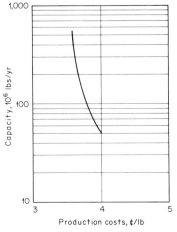

Fig. 8-9 Cyclohexane.

same.   However, at higher capacities, the IFP process shows a slight advantage.

Cyclohexane producers, as of 1965, with their respective capacities and processes are given below.

Producer	Location	1965 capacity, MM lbs/yr	Process
Ashland	Catlettsburg, Ky.	130.0	Liquid-phase hydrogenation
Atlantic-Richfield	Watson, Calif.	65.0	Liquid-phase hydrogenation
Continental	Ponca City, Okla.	260.0	Liquid-phase hydrogenation
DuPont	Belle, W.Va.	100.0	Vapor-phase hydrogenation
	Orange, Tex.	100.0	Vapor-phase hydrogenation
Gulf	Port Arthur, Tex.	195.0	Liquid-phase hydrogenation
Phillips	Borger, Tex.	195.0	Fractionation from petroleum
	Sweeny, Tex.	350.0	Fractionation from petroleum
	Sweeny, Tex.	160.0	Liquid-phase hydrogenation
Pontiac	Corpus Christi, Tex.	80.0	Liquid-phase hydrogenation
Signal	Houston, Tex.	80.0	Liquid-phase hydrogenation
South Hampton	Silsbee, Tex.	20.0	Liquid-phase hydrogenation
Union Oil	Smiths Bluff, Tex.	195.0	Liquid-phase hydrogenation
Sub total		1,930.0	
On stream in 1966:			
Cosden	Big Spring, Tex.	50.0	Liquid-phase hydrogenation
Enjay	Baytown, Tex.	120.0	Liquid-phase hydrogenation
Texaco	Port Arthur, Tex.	195.0	Liquid-phase hydrogenation
Texas-Eastman	Longview, Tex.	20.0	Liquid-phase hydrogenation
Sub total		385.0	
Grand total		2,315.0	

Source: *Hydrocarbon Processing,* October 1965.

Cyclohexane consumption is growing at a rate of 10 percent per year. In 1965, the use breakdown was as follows:

End-use	MM lbs, 1965
Nylon 6/6	670.0
Nylon 6	120.0
Adipic acid (non-nylon)	90.0
Other non-nylon	20.0
Export	720.0
Total	1,620.0

Source: *Hydrocarbon Processing*, October 1965.

3.1   Cyclohexanol-Cyclohexanone

In   addition   to   its   small   use   as   a   medium   for   the   medium-pressure solution polymerization of ethylene, most of the cyclohexane used is first oxidized to a mixture of cyclohexanone and cyclohexanol (also known as "K-A oil," for ketone-aldehyde). This mixture is the starting point for making adipic acid and caprolactam.

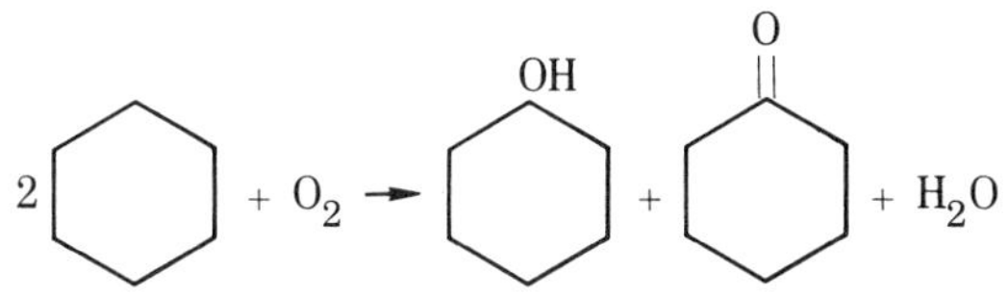

$$2 \; \bigcirc + O_2 \rightarrow \bigcirc^{OH} + \bigcirc^{O} + H_2O$$

A flowsheet for a typical cyclohexane oxidation unit is shown in Fig. 8-10.   Liquid-phase oxidation takes place with air at 330° F and around 200 psig.   The gaseous effluent is condensed to separate unreacted cyclohexane from what is essentially nitrogen.   The cyclohexane is recycled, and the liquid-effluent is also flashed. Then the product stream can be treated in several ways.   In the BASF process, caustic hydrolysis first frees the cyclohexanol that may have been esterified by any of the diacids formed during the oxidation step — adipic, succinic, glutaric.   A vacuum distillation train then separates the unreacted cyclohexane which goes back to the reactor.   Then the product fraction is freed from the heavy oxidation products.   Another technique used by Stamicarbon employs steam distillation of the reactor effluent to reduce the partial pressure of the desired products and break down the cyclohexyl esters.

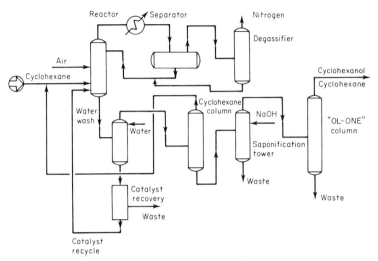

Fig. 8-10 Cyclohexane oxidation.

The yield on cyclohexane from any of these conventional oxidation processes is around 70 percent. Two air oxidation processes,
one developed by the Institut Français du Pétrole, the other by
Scientific Design, now claim yields of almost 90 percent, which may
have considerable effect on the economics of nylon production all
over the world.   Since low yields are due mainly to the fact that
oxidation does not stop at cyclohexanol or cyclohexane, but goes
on to a number of oxidation products such as the above-mentioned
dicarboxylic acids, the basic innovation of these processes was to

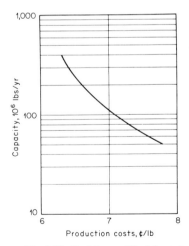

Fig. 8-11 Cyclohexanol-Cyclohexane
("K-A Oil").

carry out the reaction in the presence of boric acid which ties up the cyclohexanol as the respective ester as soon as it forms, thereby preventing its further oxidation.   The ester is then hydrolyzed to cyclohexanol and the acid regenerated. In practice, however, the consumption of boric acid more or less balances the increase in yield.

Investment for a 100.0 million pounds per year conventional cyclohexane oxidation plant is \$4.0 million. Production costs, assuming a 70 percent yield on cyclohexane, can be estimated as follows:

	Cost, ¢/lb "K-A oil"
Cyclohexane (at 4.0¢/lb)	4.8
Utilities, catalyst and chemicals	0.7
Labor and overhead	0.3
Capital charges	1.3
Total	7.1

The original raw material for nylon 6/6 was phenol. In use it is first hydrogenated to cyclohexanol and this product oxidized to cyclo-hexanone, the precursor of adipic acid, the principal reactant for nylon 6/6.

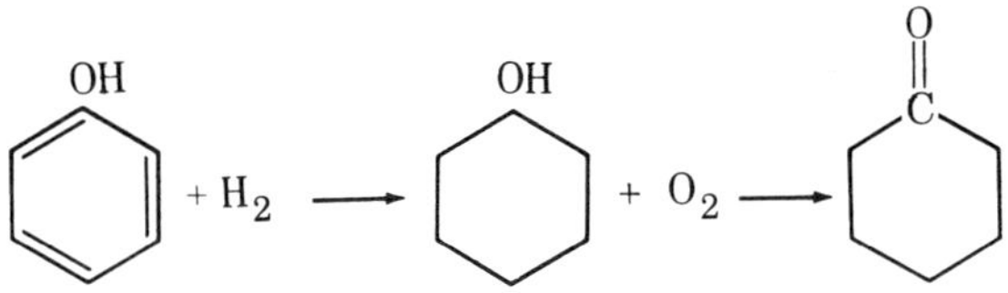

Cyclohexane oxidation has been substituted since it makes the desired product at a lower cost.

Cyclohexylamine

Cyclohexylamine was once made by hydrogenating aniline.

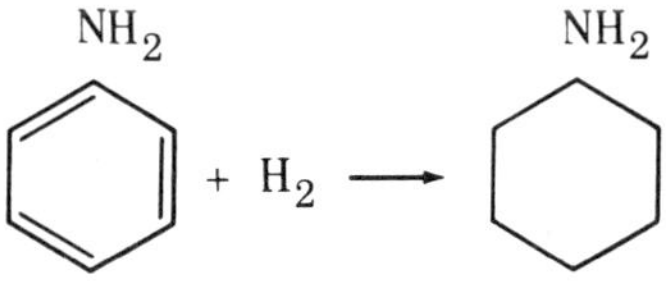

Today most producers of aliphatic amines operate flexible plants that can produce many different amines by the interaction of ammonia and the respective alcohols. In this manner, cyclohexanol is used as the starting point.

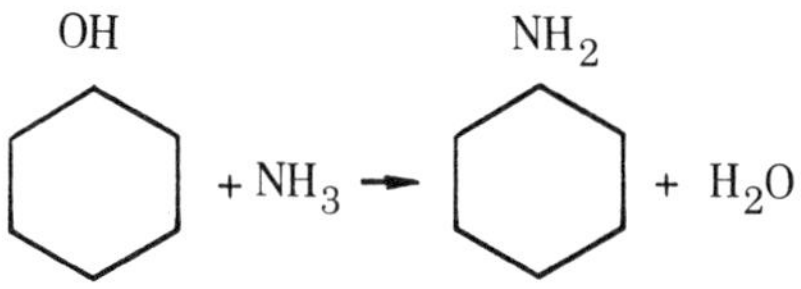

$$\text{OH} \quad + NH_3 \longrightarrow \quad \text{NH}_2 \quad + H_2O$$

Yields are better than from aniline, which tends to over-hydrogenate to ammonia and cyclohexane, and for companies having captive cyclohexanol facilities there is also a raw material cost advantage.   The only producer of cyclohexylamine from aniline is Abbott, which is also the only one not producing ethyl, isopropyl and butylamines.

Besides the production of cyclamates, the most important derivative of cyclohexylamine is the respective sulfenamide, a delayed-action rubber accelerator. Corrosion inhibitors are made from cyclohexylamine by forming salts with fatty acids. Other applications are textile chemicals, plasticizers and adhesives.

The total demand for cyclohexylamine in 1965 was 11.5 million pounds distributed as follows:

End-use	MM lbs, 1965
Cyclamates	6.5
Rubber accelerator	3.0
Corrosion inhibitors	1.0
Other uses	1.0
Total	11.5

The U.S. producers of cyclohexylamine are given below.

Producer	Location	Capacity, MM lbs/yr	Route
Abbott.	Wichita, Kans.	10.0	Aniline
DuPont	Beaumont, Tex.	3.0	Nitrocyclohexane
Pennsalt	Wyandotte, Mich.	8.0	Cyclohexanol
Jefferson	Port Neches, Tex.	5.0	Cyclohexanol
Monsanto	Monsanto, Ill.	2.0	Cyclohexanol
Virginia Chem.	W. Norfolk, Va.	8.0	Cyclohexanol
Total		36.0	

Note: All capacities are very flexible.
Source: *Oil, Paint & Drug Reporter*, December 28, 1964.

The rapid growth in demand for cyclamates is due to several factors.   First, the urban communities of the more advanced countries are becoming more and more conscious of the health hazards caused by obesity, brought about by higher standards of living and shorter working hours.   Second, on a sweetness-equivalent basis, cyclamates are as cheap as sugar at its lowest quotations, and in addition, are not subject to the kind of price fluctuations that sometimes raised the cost of sugar to many times its average quotation.   Hence, cyclamates (and saccharin) have begun to be widely used not only by individuals worried about their weight, but also by large-scale food processors.   The sugar equivalent of synthetic sweeteners used in the U.S. in 1965 reached 700,000 metric tons, an impressive figure if it is considered that the entire Cuban crop amounts to 6 million tons.

The process for cyclamate manufacture operates according to the equations below.   Sulfamic acid can be used instead of chloro-sulfonic; it is more expensive on an equivalent basis, but has the advantage of giving off ammonia instead of hydrochloric acid, thus reducing investment requirements by permitting the use of carbon steel equipment.

$$2\ \underset{NH_2}{\bigcirc} +\ ClSO_3H \longrightarrow\ \underset{NHSO_3}{\bigcirc} +\ \underset{NH_3}{\bigcirc} \xrightarrow{Ca(OH)_2}\ \left[\underset{NHSO_3^-}{\bigcirc}\right]_2 Ca^{++}$$

Abbott was the first to commercialize cyclamates, and is still the largest producer.   Since the process is simple, however, numerous companies have entered the field — some because they are food processors, like Pillsbury and Norse, others such as Pfizer, on account of their position in the drug field.   The list of producers, as of 1965, is given below.

Producer	Location	Capacity, MM lbs/yr
Abbott	N. Chicago, Ill.	8.0
Monsanto	St. Louis, Mo.	4.0
Pfizer	Groton, Conn.	4.0
Union Starch	Edinburg, Ind.	3.0
Cyclamate Corp.	Newark, N.J.	4.0
Pillsbury	Painesville, Ohio	3.0
Drew Chem.	Boonton, N.J.	1.5
Norse	Cudahy, Wis.	2.5
Total		30.0

Source: *Oil, Paint & Drug Reporter*, February 22, 1965.

The demand for cyclamates is growing at 20 percent per year, and stood at 10.0 million pounds in 1965. By far the major part of this was consumed by the diet beverage industry, mostly for bottled drinks, but also for dry beverage bases. A breakdown is given below:

End-use		MM lbs, 1965
Diet beverages:	bottled drinks	5.3
	beverage bases	1.7
Dietetic foods		1.3
Retail		1.2
Miscellaneous		0.5
Total		10.0

Source: *Oil, Paint & Drug Reporter*, February 22, 1965.

3.2 Cyclohexanone

In addition to its use as solvent and in the manufacture of poly-ketone resins used in certain printing ink formulations, the only use for pure cyclohexanone is as the starting point for caprolactam, a raw material for nylon 6. Whereas the cyclohexane oxidation products, a mixture of cyclohexanol and cyclohexanone, can be fed to an adipic acid plant directly, in order to make caprolactam the mixture must be further dehydrogenated to convert the entire feed to cyclohexanone.

A flowsheet of the IFP liquid-phase dehydrogenation process is shown in Fig. 8-12.

Vapor-phase dehydrogenation can also be used. This route is employed by Allied Chemical, the largest caprolactam producer in the U.S.

Caprolactam

The first step in the classical route to caprolactam from cyclo-hexanone is oximation, in which cyclohexanone reacts with an excess of hydroxylamine sulfate, which in turn is prepared in a six-step process from ammonia and sulfur dioxide.

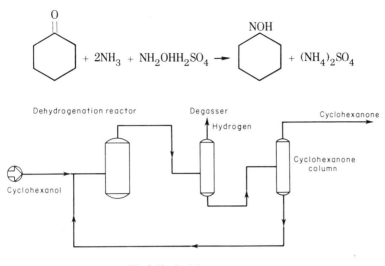

Fig. 8-12   Cyclohexanone.

The organic phase is separated by decantation, and the inorganic phase extracted with cyclohexanone before being sent to a concentration unit producing fertilizer-grade ammonium sulfate.

Under the action of oleum, the oxime undergoes the Beckmann rearrangement to caprolactam:

The reaction is quantitative and two mols of acid are required per mol of oxime.   The excess sulfuric acid is neutralized with ammonia, and the product recovered by decantation or solvent extraction with an aromatic such as benzene.   About 4.5 pounds of ammonium sulfate are produced per pound caprolactam.   A caprolactam unit using this route is thus necessarily tied to captive ammonia and sulfuric acid plants, and the producer thus finds himself turning out large quantities of a fertilizer that is declining in relative importance.   Hence a search has been made for processes that bypass the Beckmann rearrangement, especially by potential producers not already in the fertilizer business.

Not all processes starting from cyclohexane begin with an oxidation step. DuPont, for example, uses nitric acid to produce nitrocyclohexane, which is partially hydrogenated to cyclohexanone oxime in the presence of a Zn–Cr catalyst.

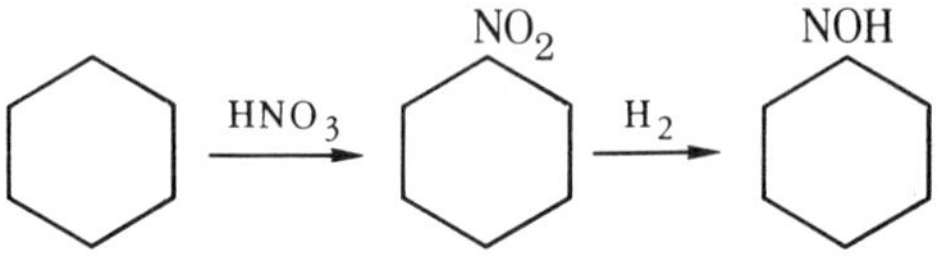

The disadvantage of this process is that it is impossible to avoid formation of large amounts of cyclohexylamine.

The Snia Viscosa process starts from toluene, which is first converted to benzoic acid and then hydrogenated to hexahydrobenzoic acid. This reacts with nitrosylsulfuric acid to form caprolactam. Nitrosylsulfuric acid in turn is made by reacting N_2O_3, obtained by ammonia oxidation, with sulfuric acid. This process has been licensed to Allied Chemical in the U.S.

$$C_6H_5CH_3 \xrightarrow{(O)} C_6H_5COOH \xrightarrow{H_2} C_6H_9COOH$$

$$C_6H_9COOH + NOHSO_4 \longrightarrow C_6H_8CONH + H_2SO_4 + CO_2$$

This route has the advantage of starting from a cheaper hydrocarbon, but also produces around four pounds of ammonium sulfate per pound product. The preparation of the inorganic reactant is simpler than in the oximation route.

The Toyo Rayon photonitrosation process avoids the use of hydroxylamine sulfate. The oxime is formed from cyclohexane and nitrosyl chloride in the presence of a mercury lamp placed inside the reaction mass.

$$NOHSO_4 + HCl \longrightarrow NOCl + H_2SO_4$$

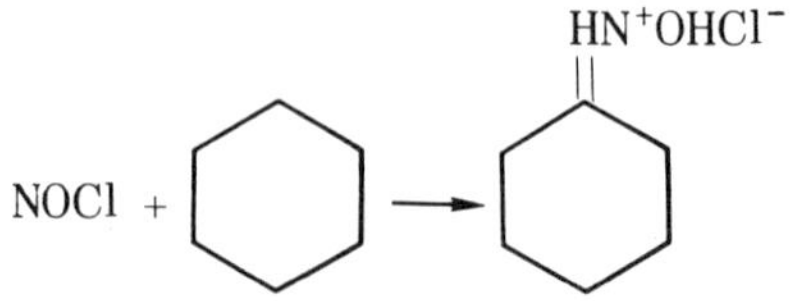

Sulfuric acid can be returned to the oleum tower. The rest of the process is similar to the cyclohexanone route, but because there is no oxidation step, roughly half as much (2.5 pounds per pound) ammonium sulfate is produced.

Union Carbide has developed a process which avoids the production of ammonium sulfate entirely. Starting from cyclohexanone, the process employs peracetic acid oxidation to caprolactone.

$$
\underset{\text{cyclohexanone}}{\bigcirc\!\!=\!\!O} + CH_3COOOH \rightarrow
\begin{array}{c}
H_2C\!-\!CH_2 \\
| \quad\quad | \\
H_2C \quad\quad | \\
| \quad\quad O \\
H_2C \quad\quad | \\
| \quad\quad | \\
H_2C\!-\!C\!\!=\!\!O
\end{array}
$$

This is converted to caprolactam by ammoniation at temperatures around 750° F.

$$
\begin{array}{c}
H_2C\!-\!CH_2 \\
| \quad\quad | \\
H_2C \quad\quad | \\
| \quad\quad O \\
H_2C \quad\quad | \\
| \quad\quad | \\
H_2C\!-\!C\!\!=\!\!O
\end{array}
+ NH_3 \rightarrow
\begin{array}{c}
H_2C\!-\!CH_2 \\
| \quad\quad | \\
H_2C \quad\quad | \\
| \quad\quad NH \\
H_2C \quad\quad | \\
| \quad\quad | \\
H_2C\!-\!C\!\!=\!\!O
\end{array}
+ H_2O
$$

In addition to acetic acid, the process yields adipic acid as a byproduct. Caprolactam can be made for 18.5 cents per pound in a 50.0 million pounds per year plant using this route, provided the producer is already active in the field of acetyl chemicals as is Union Carbide.

Dow-Badische uses a process based on reacting cyclohexane, activated by irradiation, with nitrosyl chloride. Sulfuric acid converts the reaction product to caprolactam.

Conversion of caprolactam to a polyamide is accomplished by formation of the ω-amino-acid, and subsequent condensation to the polyamide.

$$
\begin{array}{c}
H_2C\!-\!CH_2 \\
| \quad\quad | \\
H_2C \quad\quad | \\
| \quad\quad NH \\
H_2C \quad\quad | \\
| \quad\quad | \\
H_2C\!-\!C\!\!=\!\!O
\end{array}
\xrightarrow{H_2O} H_2N(CH_2)_5COOH \xrightarrow{-H_2O}
\left[-NH(CH_2)_5\overset{\displaystyle O}{\overset{\displaystyle \|}{C}} - \right]_n
$$

The producers of caprolactam in the U.S. follow:

Producer	Location	1965 Capacity, MM lbs/yr	Route
Allied Chem.	Hopewell, Va.	250.0	Phenol
Dow-Badische	Freeport, Tex.	88.0	Cyclohexane
DuPont	Beaumont, Tex.	50.0	Cyclohexane-nitric acid
Sub total		388.0	
On stream in 1966:			
Columbia Nitrogen (PPG & DSM)	Augusta, Ga.	44.0	Cyclohexane
Grand total		432.0	

Source: *Oil, Paint & Drug Reporter*, March 21, 1966.

In 1965, the consumption in the U.S. was 310.0 million pounds of which all but 13 percent went into fibers. The remainder was used to make molding resins and for export.

Production costs of caprolactam by the oxime-Beckmann rearrangement route are estimated below, with ammonium sulfate being credited at its raw material cost starting from sulfuric acid and captive ammonia (1.6 cents per pound). Investment for a 80.0 million pounds per year plant, including cyclohexane oxidation and hydroxylamine production facilities, is \$20.5 million. Cyclohexane is taken at 4.0 cents per pound and captive ammonia and sulfuric acid sources are assumed.

	Cost, ¢/lb caprolactam
Cyclohexane	4.9
Ammonia	4.1
Sulfur dioxide and oleum	3.4
Other chemicals	1.5
Utilities	2.3
Labor and overhead	2.4
Capital charges	8.5
	27.1
Less: ammonium sulfate	(-6.9)
Total	20.2

Depending on the process adopted, caprolactam polymer is therefore 15-30 percent cheaper than nylon 6/6 polymer. Aside from this differential in raw material costs, the main advantage of nylon 6 is that merchant caprolactam is available to the independent spinner, and technology both for monomer production and spinning has for some time been readily obtainable from a number of licensors. Nylon 6/6, however, until recently was the exclusive secret of a small number of producers — DuPont, Monsanto, and Celanese in the U.S.; BASF, Rhône-Poulenc and ICI in Europe.

The basic disadvantage of caprolactam as a raw material is that all processes involve a large output of byproducts. A caprolactam unit is necessarily tied to a fertilizer complex, or, in the case of Union Carbide's process, to an acetyl derivatives complex. However, a nylon 6/6 plant, provided ammonia can be acquired, is essentially independent of any byproduct.

Lysine

This amino-acid is used as a feed supplement. The U.S. market in 1965 was around 500,000 pounds per year, much less than that for cheaper amino-acids. It is produced by two U.S. companies via fermentation.

Dutch State Mines have built a plant using a process starting from caprolactam that is expected to reduce the price of lysine to the point where it will become one of the more popular feed supplements.

$$
\begin{array}{c}
H_2C-CH_2 \\
| \quad\quad | \\
H_2C \quad\quad | \\
| \quad\quad NH \quad + \; COCl_2 \;\rightarrow \\
H_2C \quad\quad | \\
| \quad\quad | \\
H_2C-C{=}O
\end{array}
\quad
\begin{array}{c}
H_2C-CH_2 \\
| \quad\quad | \quad\; O \\
H_2C \quad\quad | \;\; || \\
| \quad\quad N-CCl \;\xrightarrow[H_2]{HNO_3} \\
H_2C \quad\quad | \\
| \quad\quad | \\
HC{=}C-Cl
\end{array}
\quad
\begin{array}{c}
H_2C-CH_2 \\
| \quad\quad | \\
H_2C \quad\quad | \\
| \quad\quad NH \;\xrightarrow{H_2O} \\
H_2C \quad\quad | \\
| \quad\quad | \\
HC-C{=}O \\
| \\
NH_2
\end{array}
$$

$$
\begin{array}{c}
NH_2 \\
| \\
H_2N(CH_2)_4-CH-COOH
\end{array}
$$

3.3   Adipic Acid

Oxidation of the cyclohexanol-cyclohexanone mixture obtained by oxidizing cyclohexane gives adipic acid in yields of over 90 percent:

OH

$$
\bigcirc \xrightarrow{(O)} HOOC(CH_2)_4COOH
$$

Oxidation is normally accomplished with nitric acid; although one plant has been built using air oxidation, the nitric acid process is still used by all nylon manufacturers. Air oxidation yields a large number of degradation products which make commercially viable yields rather doubtful. There are great incentives for developing a workable air-oxidation process, since it takes about 1.5 cent of nitric acid to make one pound of adipic acid. This corresponds to a differential in yield of around 35 percent. In addition to the problem of yields, however, there is another concerning impurities. Nylon fiber manufacture is extremely sensitive to the purity of the raw materials, and it appears that the established producers of nylon 6/6 are reluctant to abandon a technology on which are based not only the production of adipic acid, but also all subsequent steps through manufacture of the fiber itself. It is noteworthy that the only producer of adipic acid by air oxidation in the U.S. is also the only one not in the nylon business, the entire output being used to make various esters and polymers.

The nitric acid process consists of three operations: oxidation; crystallization; and nitrogen oxides recovery. The feed is oxidized in a series of agitated reactors, the nitrogen oxides leaving overhead and being sent to the recovery section. The effluent is transferred to a crystallizer, where adipic acid settles out and is separated by centrifuging; this process is repeated to meet fiber-grade specifications. The mother-liquid from the first centrifuging operation, consisting of dilute nitric acid, is sent to an acid reconcentration tower. Nitrogen oxides are recovered in an oxidation-absorption unit resembling the absorption section of a nitric acid plant. Nitric acid losses occur because part of it is reduced to nitrogen or to nitrogen oxides that cannot be recovered. Cyclohexanol losses are due to degradation to glutaric and succinic acids. A flowsheet appears in Fig. 8-13.

The investment for a 130.0 million pounds per year adipic acid plant is $3.0 million. Production costs for this capacity are as follows:

	Cost, ¢/lb adipic acid
Cyclohexanol-cyclohexanone	5.4
Nitric acid	1.5
Utilities	0.3
Labor and overhead	0.8
Capital charges	0.8
Total	8.8

The main uses of adipic acid during 1965 are listed below. It takes about 0.68 pound of adipic acid as acid to make 1.0 pound of

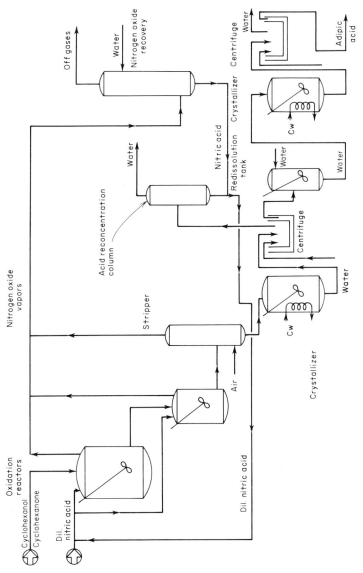

Fig. 8-13 Adipic acid.

nylon 66 fiber.   To this must be added another 1.05 pounds per pound required to make the corresponding amount of hexamethylene diamine. Since in the U.S., however, the largest producer, DuPoint, uses butadiene as the raw material for a large part of their diamine needs, the ratio between nylon consumed and adipic acid produced is lower than would be expected.

The second largest nylon 66 producer in the U.S., Chemstrand, now produces hexamethylene diamine from acrylonitrile.   Thus there appears to be a general trend away from adipic acid as a source of the diamine.

The demand for adipic acid in 1965 was 830.0 million pounds, broken down as follows:

End-use	MM lbs, 1965
Nylon 6/6	750.0
Esters for plasticizers and synthetic lubes	40.0
Polyurethane resins	25.0
All other	15.0
Total	830.0

Source: *Oil, Paint & Drug Reporter*, January 10, 1966.

Producers of adipic acid in the U.S. together with their source of cyclohexanol are listed below.   All, with the exception of Rohm & Haas, use nitric acid oxidation. Costs are charted in Fig. 8-14.

Producer	Location	1965 Capacity, MM lbs/yr	Source
Allied Chem.	Hopewell, Va.	20.0	Phenol
DuPont	Belle, W.Va.	210.0	Cyclohexane
	Orange, Tex.	210.0	Cyclohexane
Monsanto	Luling, La.	30.0	Phenol
	Pensacola, Fla.	320.0	Cyclohexane
Rohm & Haas	Louisville, Ky.	20.0	Cyclohexane
Nylon Industries (Celanese)	Bay City, Tex.	35.0	Cyclohexane
El Paso-Beaunit*	Odessa, Tex.	40.0	Cyclohexane
Total		885.0	

*On stream in 1966.
Source: *Oil, Paint & Drug Reporter*, January 10, 1966.

Nylon 6/6

Nylon 6/6 can be made from adipic acid in a series of steps, beginning with the conversion of adipic acid to adiponitrile by reaction with ammonia:

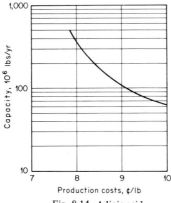

Fig. 8-14 Adipic acid.

$$HOOC(CH_2)_4COOH + 2NH_3 \longrightarrow NC(CH_2)_4CN + 4H_2O$$

The reaction can be carried out in the liquid or the vapor phase; it takes place in the presence of a large excess of ammonia, which must be recovered and recirculated.

Next, adiponitrile is hydrogenated catalytically at around 300 psig to give hexamethylene diamine (HMD). Reactor effluent is stripped of the solvent in which the reaction is carried out and then sent to a purification unit where a large number of trays and very high reflux ratios are required to remove the last traces of impurities which might have adverse effects on the spinning process.

$$NC(CH_2)_4CN + 4H_2 \longrightarrow H_2N(CH_2)_6NH_2$$

Other routes to hexamethylene diamine are employed also. DuPont uses butadiene as a starting point, via the following reactions:

$$CH_2{=}CH{-}CH{=}CH_2 + Cl_2 \longrightarrow \begin{array}{c}\text{mixture of isomeric}\\ \text{dichlorobutenes}\end{array} \xrightarrow{\text{NaCN}}$$

$$NCCH_2CH{=}CHCH_2CN \xrightarrow{H_2} H_2N(CH_2)_6NH_2$$

In the future, direct reaction between butadiene and HCN may replace this sequence.

Below, a comparison is made between the raw material economics of the two routes.

Butadiene

	¢/lb	¢/lb HMD
Butadiene	11.0	6.5
HCN	8.5	4.5
Caustic + chlorine	2.5	3.7
		14.7

Adipic Acid

	¢/lb	¢/lb HMD
Adipic acid	8.8	13.6
Ammonia	2.5	1.0
		14.6

The process is suited to a company having access to cheap butadiene, and captive caustic-chlorine, ammonia and HCN sources.

Monsanto has developed an acrylonitrile hydrodimerization route in which the first step to adiponitrile is carried out electrolytically. Conventional hydrogenation then yields the diamine. Considering that captive acrylonitrile from a unit using the Sohio process can be made available at around 12.0 cents per pound, and that yields are close to stoichiometric, this process may well become of great interest to all horizontally integrated synthetic fiber producers.

$$2CH_2{=}CH{-}CN + H_2O \longrightarrow NC{-}CH_2{-}CH_2{-}CH_2{-}CH_2{-}CN + \tfrac{1}{2}O_2 \xrightarrow{\ H_2\ }$$

$$NH_2(CH_2)_6NH_2$$

Finally, it has been reported that Nylon Industries converts cyclohexanone to caprolactone, which can be converted to hexamethylene diamine by reductive amination. In this manner, much less nitric acid per pound nylon 66 polymer is required than if both raw materials are made from adipic acid.

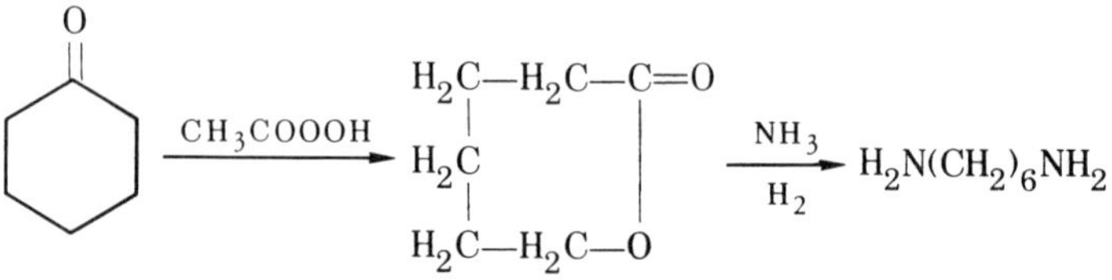

The production of the nylon 66 resin required for melt spinning into a fiber takes place in two steps. First, adipic acid and HMD react to give a salt:

$$NH_2(CH_2)_6NH_2 + HOOC(CH_2)_4COOH \longrightarrow NH_2(CH_2)_6N^+H_3^-OOC(CH_2)_4COOH$$

Next, polymerization takes place in the presence of a catalyst and some acetic acid, which acts as a chain-length regulator:

$$\left[\begin{array}{c} \quad\quad O \quad\quad O \\ \quad\quad \| \quad\quad \| \\ -NH(CH_2)_6NHC(CH_2)_4C- \end{array}\right]$$

The investment for the unit required to convert 130.0 million pounds of adipic acid to approximately 80.0 million pounds per year nylon 6/6 resin is $6.3 million, and the production costs are as follows:

	Cost, ¢/lb nylon 6/6 resin
Adipic acid	14.2
Ammonia	0.8
Hydrogen	0.2
Catalyst and chemicals	1.5
Labor and overhead	2.4
Utilities	1.8
Capital charges	2.6
Total	23.5

The total capacities of U.S. producers of nylon 6/6 are given below.

Producer	Location	1965 Capacity, MM lbs/yr
DuPont	Camden, S.C. Chattanooga, Tenn. Richmond, Va. Martinsville, Va. Seaford, Del.	450.0
Monsanto	Pensacola, Fla. Greenwood, S.C.	200.0
Nylon Industries	Greenville, S.C.	40.0
Total		690.0

Source: *Chemical Week,* November 12, 1966.

Polyamide Fibers

The success of nylon fibers has been due mainly to their elasticity, high tensile strength and resistance to abrasion. Their resistance to salt and hydrolysis has gained them entry into such areas of production as fishing nets and sails. Because of their excellent resistance to oils, polyamides have been used to make filter cloths. Their exceptional durability permits their use in mechanical goods such as hose and conveyors.

Polyamide fibers are made by melt-spinning through an orifice, followed by individual drawing of the air-cooled filaments to produce the desired degree of orientation. Other operations such as twisting and texturizing allow a wide range of visual and mechanical characteristics to be obtained. Nylon 6 is usually polymerized and spun continuously. Nylon 6/6 polymer is usually allowed to solidify first, and then is cut into chips before being sent to the charge hopper where these chips are remelted, filtered and sent on to the spinnerets. A continuous process has recently been developed by ICI.

The orientation phase adds greatly to the investment and operating costs of synthetic fiber production. These costs, in turn, vary considerably with the gauge of the fiber; thinner gauges are considerably more costly to produce. Since polyamides are incompatible with natural fibers, they must be spun into filament which is inherently more expensive than polyester or acrylic staple. Furthermore, melt-spinning is also a more costly operation than spinning from a solvent.

The table below compares the cost of producing nylon filament (an average distribution among the numerous gauges was assumed), polyester melt-spun staple, and acrylic staple from a dimethyl formamide solution. Production in all cases was assumed to be 20.0 million pounds per year, but economics are relatively insensitive to scale of operation.

	Nylon filament	Acrylic staple	Polyester staple
Investment, $ MM	15.8	9.0	11.2
Production costs, ¢/lb			
Raw materials	29.0	16.6	22.1
Utilities, chemicals	4.0	5.7	3.9
Labor and overhead	45.5	21.7	20.5
Capital charges	26.1	14.9	18.5
Total	105.5	57.9	65.0

The total U.S. polyamide fiber production in 1965 was 850.0 million pounds, of which 220.0 million pounds was nylon 6, therefore accounting for 26 percent of the U.S. nylon fiber market. In most applications, either type of nylon can be employed, but nylon 6 has better resistance to abrasion and is therefore preferred for automotive upholstery. Nylon 6 is also used to the exclusion of nylon 66 in safety belts. The most important outlet for polyamide fibers is tire cord, especially in replacement tires which constitute almost 70 percent of the automobile tire business. Tire cord was originally made of cotton which was subsequently replaced by rayon; rayon, in turn, was largely displaced by nylon. Rayon is used only in original-equipment tires, since the conventional nylons have a tendency to flat-spot. Even so, nylon is used in original-equipment tires for nonautomotive applications — earth-moving equipment, industrial vehicles and trucks, where aesthetic considerations are of no importance. New types of polyamide fiber are expected to capture the original-equipment market for nylon and thus practically the entire tire-cord demand. Allied Chemical, for example, employs a process whereby nylon is crosslinked with isocyanate vapors, thereby solving the problem. Running counter to this trend, however, are recent significant improvements in rayon technology, and the introduction of polyester tire cord by Goodyear in 1965.

Almost half the nylon used for apparel is textured yarn, also known as "nylon stretch"; nonstretch applications are mainly in knit-wear, but nylon is also used in woven goods. It has captured a large portion of the carpeting market which, until recently, was held by rayon; nylon is also used in carpet-backing, where the main competitor is jute. Industrial applications include filter media, conveyors, seat-belts (increasingly used as a standard automotive accessory), tarpaulins and rope. The overall breakdown of the U.S. polyamide fiber output of 1965 is shown below.

End-use	1965, MM lbs/yr		
	Nylon 6/6	Nylon 6	Total
Tire cord	180.0	65.0	245.0
Apparel	145.0	70.0	215.0
Home furnishings	160.0	60.0	220.0
Industrial	117.0	25.0	142.0
Total	602.0	220.0	822.0

Source: Based on *Chemical Week*, November 12, 1966; *Oil, Paint & Drug Reporter*, July 18, 1965.

The total demand for tire cord and fabrics was 495.0 million pounds, 240.0 million pounds being nylon and about 250.0 million

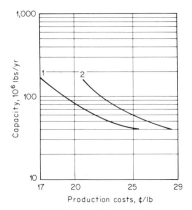

Fig. 8-15 Polyamide fiber raw
materials. 1-Caprolactam (via
oxime); 2-Nylon 6/6 polymer.

pounds being rayon. Nylon is used in 70-75 percent of replacement tires, for both passenger vehicles and trucks, and in 80 percent of the original equipment for trucks. By 1965, hardly any nylon tires were used for passenger car original equipment.

In 1965, DuPont began producing a polyamide fiber called N-44, a mixture of 72 percent nylon 6/6 and 28 percent of the condensation product of adipic acid and m-xylylene diamine. This material reduces flat-spotting, and permits the use of polyamide tire cord for original equipment.

Polyamide Plastics

Both nylon 6/6 and 6 have molding and extrusion applications. In addition, a polyamide made from HMD and sebacic acid (the C_{10} dicarboxylic acid) is used to make monofilament, which finds its most important applications in fishing lines, brushes and bristles.

Sebacic acid is made from ricinoleic acid (castor oil) by caustic hydrolysis:

$$CH_3(CH_2)_5CHOHCH_2CH{=}CH(CH_2)_7COOH \xrightarrow{\text{NaOH}} NaOOC(CH_2)_8COONa +$$

$$CH_3(CH_2)_5CHOHCH_3$$

Nylon is an engineering plastic for special applications where it is mainly a substitute for metals. Among the numerous outlets for molded nylon are the following: noiseless gears in household appliances; washing machine valves; furniture coasters; cams; door latches; electric shaver housings; and certain textile-machine parts. Nylon is also beginning to replace phenolic resins in such electrical applications as switches and connectors. Its main advantages over ordinary thermoplastics are excellent dimensional stability, high tensile and impact strength, resistance to abrasion, and self-lubricating properties. Extruded applications, aside from monofilament, include gasoline lines and speedometer cable housings for automobiles. Nylon extruded film is finding increasing use in the packaging of processed meat products and in surgical instruments that can be sterilized while still in their package since nylon is permeable to water vapor.

Domestic nylon plastic applications amounted to 60.0 million pounds in 1965, which can be broken down by type of polyamide.

Type of polyamide	MM lbs, 1965
Nylon 6/6	30.0
Nylon 6	15.0
Nylon 6/10	15.0
Total	60.0

Below is the breakdown by application.

End-use	MM lbs, 1965
Extrusion	21.0
Machinery parts	5.8
Consumer goods	7.0
Electrical parts	6.2
Appliance parts	4.2
Automotive	15.0
Miscellaneous	1.6
Total	60.8

Source: *Modern Plastics*, January 1966.

The table below lists companies that produce polyamide resins.

Producer	Location	1965 Capacity, MM lbs
Allied Chem.	Chesterfield, Va.	10.0
DuPont Chem.*	Parkersburg, W.Va.	50.0
Foster Grant	Manchester, N.H.	5.0
Firestone	Pottstown, Pa.	3.0
Gulf	Henderson, Ky.	6.0
Monsanto	Pensacola, Fla.	10.0
Total		84.0

*Estimate.
Source: *Oil, Paint & Drug Reporter*, July 18, 1967.

Non-Nylon Uses of Adipic Acid

Most of the other uses of adipic acid resemble those of phthalic anhydride and other long-chain diacids. Monomeric plasticizers using adipic instead of phthalic impart better low-temperature flexibility to vinyl compounds; however, they have the disadvantage of being more volatile than the corresponding phthalates. Most plasticizers containing adipic acid (about 70 percent of the total) are polymeric. Monomeric plasticizers are the diesters of octyl and decyl oxo alcohols, or of 2-ethylhexanol, which are also used as lubricants. Polymeric plasticizers contain adipic as well as other dicarboxylic acids, for example, fatty acid dimer, obtained by dimerization of an unsaturated fatty acid along one of its double bonds. The usual alcohols are propylene, butylene and neopentyl glycols, or 2-ethyl 1,3 hexanediol. They are used mainly in vinyl film, sheeting and wire-coating, in applications where the plasticizer is highly subject to extraction. Their advantages over monomeric plasticizers in general are good resistance to migration and extraction. However, they are also considerably more expensive, have no stabilizing effect, and are generally formulated with 4 percent of an epoxy plasticizer in addition to the usual soaps.

Monomeric adipates could probably replace phthalates, at least in part, in a number of applications, and had not phthalic anhydride become available at low cost long before adipic acid, the plasticizer market would presently be consuming a much higher proportion of adipates. As it is, the use of phthalates in PVC compounds has become so firmly established that the consumption of monomeric adipates has grown slowly. On the other hand, the use of polymerics is increasing at a rate of 12 percent per year and in 1965 stood at 50.0 million pounds. The main producers of polymeric plasticizers are given below.

Producer	Capacity, MM lbs/yr
Monsanto	15.0
Rohm & Haas	30.0
Eastman	5.0
Archer-Daniels-Midland	10.0
Emery	10.0
Harchem	5.0
Reichhold	5.0
Others	15.0
Total	95.0

The market for polymerics can be broken down as follows:

End-use	MM lbs/yr
Vinyl plasticizers	44.0
Cellulosics, rubber, etc.	6.0
Total	50.0

Resins employed to react with diisocyanates to form poly-urethanes were originally polyesters. However, the development of polyethers almost completely replaced the more expensive poly-esters. They are still used in several specific applications. In foams, for example, polyesters are employed to make textile-foam laminates of which 20.0 million pounds per year are used generally as lining for clothing, shoes, rainwear, etc. The majority of these products are flame-laminated, and the temperatures encountered are too high for the more volatile polyethers. Diethylene glycol is the usual alcohol for urethane polyester resins.

Wet-strength resins of the polyamidamine family are used when the paper stock is neutral or alkaline, in contrast to the more common amino resins, which are acid-curing. They are made by first reacting adipic acid with diethylene triamine, and then adding epi-chlorohydrin, which reacts with the secondary amine hydrogen. This side-chain acts as the crosslinking agent.

$$\left[\begin{array}{c} \overset{O}{\overset{\|}{C}}-(CH_2)_4\overset{O}{\overset{\|}{C}}NH-CH_2CH_2-\underset{\underset{\underset{\underset{O\diagup\underset{\diagdown}{|}}{CH}}{|}}{\underset{|}{CH_2}}}{N}-CH_2CH_2NH- \\ \\ \\ \\ CH_2 \end{array} \right]_n$$

Adipic acid esters are of some importance in the manufacture of monomeric synthetic lubricant bases. Acids from natural sources such as sebacic, azelaic, and pelargonic are being re-placed by adipic, but, in general, the use of monomeric esters as lubricants is declining; the trend is towards the use of higher molecular-weight lubricants usable above 400°F. Total use of monomeric adipates as lubricants was 10.5 million pounds in 1965.

As a food acidulant, adipic acid is used only in making certain gelatine desserts, although it has the advantage of not being hygroscopic and therefore noncaking. Most of this market is expected to be taken over by cheaper fumaric acid.

4 PHENOL

Several processes are used to make phenol and all except one start from benzene. The exception is a route starting with toluene developed by Dow. From an economic standpoint, processes can be divided into two categories: 1) those that depend on the commercialization of byproducts; and 2) those that yield no major saleable products apart from phenol.

Among the processes yielding byproducts the most important are the sulfonation route employed by Reichhold Chemical and Monsanto, the Dow benzene chlorination process, and the widely used cumene hydroperoxidation process. Sulfonation process byproducts are sodium sulfite and sodium sulfate, which are used in the paper industry. The Dow process yields a number of byproducts, for example, diphenyl oxide, dichlorobenzenes, phenylphenols, and others.   Since the market for these byproducts is limited, the process is not likely to be used in any new installations. The cumene process yields mainly acetone.

The sulfonation process begins by liquid-phase sulfonation of benzene to benzene sulfonic acid. Excess acid is neutralized with caustic soda, producing sodium sulfate.

Benzene-sulfonic acid is neutralized with sodium sulfite (produced in the next stage) to form sodium benzenesulfonate and SO_2. The sodium salt is then fused with caustic soda to form sodium phenate and sodium sulfite.

Finally, SO_2 produced during the neutralization step is used to acidify sodium phenate. This operation is known as "springing," yielding phenol and more sodium sulfite.

The overall reaction can be written as follows:

When the producer has captive sulfuric acid but no caustic soda, limestone can be substituted to neutralize the excess sulfuric acid. Gypsum is then the second byproduct instead of sodium sulfate and CO_2 is used instead of SO_2 in the neutralization step.

The Dow chlorination process is used in what was the largest single phenol unit in the world until Union Carbide's plant in Bound Brook, New Jersey was completed. The steps involved are chlorination of benzene to chlorobenzene and HCl, sodium hydroxide fusion of chlorobenzene to sodium phenate and more HCl, and the use of part of the HCl to neutralize the sodium phenate to phenol. Since the process is based on a caustic-chlorine complex, sodium chloride is returned to the electrolysis plant. The main byproducts from the process are di- and polyphenyls, diphenyl ether, dichlorobenzene and phenylphenols.

Most important among the byproduct processes is the cumene route. Cumene is made by alkylating benzene with propylene, using aluminum chloride or phosphoric acid as the alkylation catalyst.

Cumene is oxidized with air or enriched air, in an alkaline medium, often in the presence of a fatty amine which acts as a promoter. The resulting cumene hydroperoxide is decomposed to phenol and acetone at 140° F by dilute sulfuric acid hydrolysis in an agitated vessel. A large recycle of either phenol or acetone keeps peroxide concentration down to around 1 percent, thereby limiting the danger of explosion. The reactor effluent is decanted into a water phase which returns it to the cleavage reactor, and an organic phase which is washed with water to remove the last acid traces and sent to final purification. First, acetone is removed and sent to a tower where it is separated from mesityl oxide. Next, two vacuum towers remove first cumene, which is recycled, then α-methylstyrene, which, in the version known as the "long process," is recycled via a small hydrogenation unit which converts it back to cumene. Phenol is obtained as overhead from a final vacuum tower and can be purified further by crystallization. The bottoms are tar, containing some acetophenone which can also be recovered if desired. A flowsheet appears in Fig. 8-16.

The main byproduct is acetone, of which 0.60 pound per pound phenol is recovered.

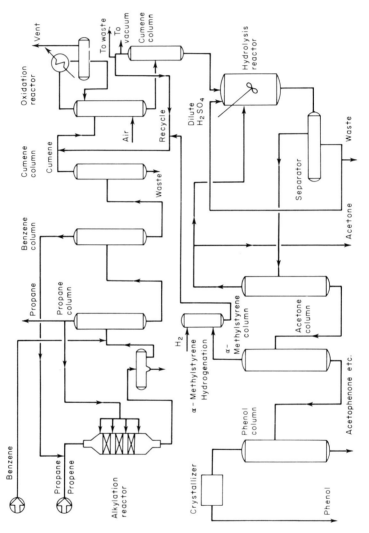

Fig. 8-16 Phenol (via cumene).

The modified Raschig process, developed by Hooker, starts by oxychlorinating benzene to monochlorobenzene; dichlorobenzenes are also formed.

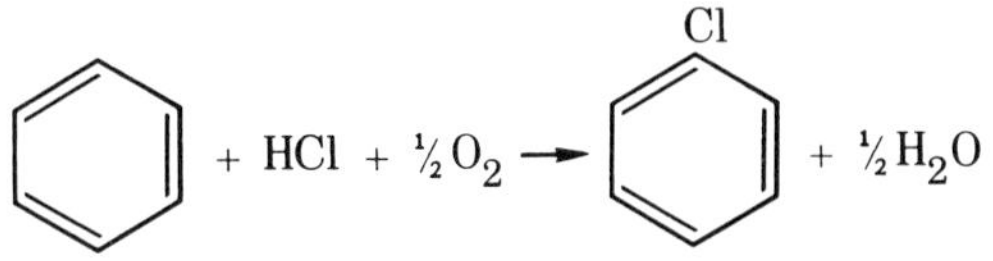

Effluent from the reactor goes to a fractionation unit where un-reacted benzene is taken off and recycled.

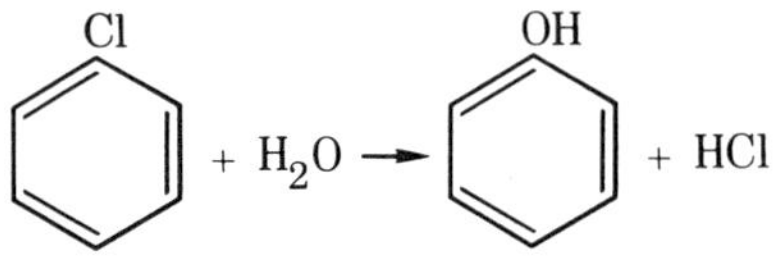

The chlorobenzene stream is mixed with the effluent from a second in which hydrolysis takes place, and sent to a fractionation system where three main streams are produced: the phenol product; unre-acted monochlorobenzene; and a phenol-dichlorobenzene azeotrope which is also recycled to the second reactor. The formation of byproducts is avoided because the hydrolysis catalyst also promotes the following reaction:

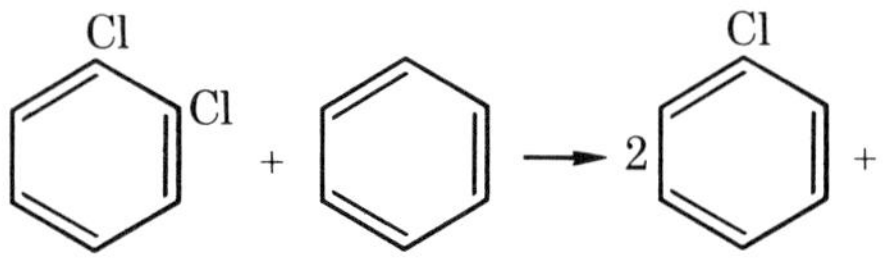

Therefore in the end no dichlorobenzenes are left over. A block flow diagram is shown in Fig. 8-17.

Dow has a process which starts from toluene. The first step is conventional air oxidation to benzoic acid over a cobalt catalyst at around 30 psig, with yields around 95 percent. Next, benzoic acid is mixed with steam and air and enters a catalytic decarboxylation reactor, where it passes through a salicylic acid stage before being converted to phenol and carbon dioxide.

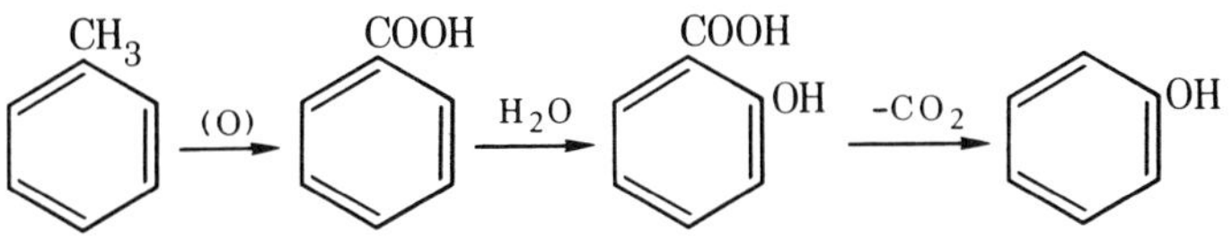

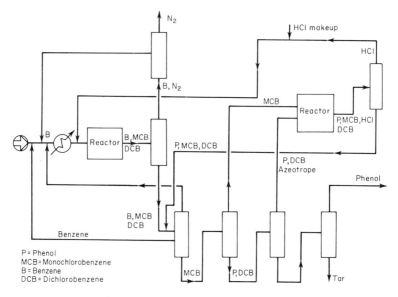

Fig. 8-17 Phenol (modified Raschig process).

There have been reports of a direct benzene oxidation process used by one U.S. plant but detailed data are unavailable.

The economics of the two most successful processes (modified Raschig and cumene) are compared below.   The yields from the cumene route are based on use of the "long process." This process requires a higher investment, some hydrogen consumption and more labor and utilities.   These are more than compensated for by a better yield referred to cumene. Acetone is credited at 3.0 cents per pound.   Capacities in both cases are 100.0 million pounds per year.

	Cost, ¢/lb phenol	
	Modified Raschig process	Cumene process
Investment, $MM	8.3	8.1
Raw materials		
Benzene	3.2	3.3
Hydrochloric acid	0.1	—
Propylene	—	1.1
Utilities, chemicals	1.2	1.7
Labor and overhead	0.5	0.8
Capital charges	2.7	2.7
	7.7	9.6
Less: actone credit	—	-1.8
Total	7.7	7.8

Source: *Hydrocarbon Processing*, January 1966.

Thus, at this capacity the two routes are equal at the low credit taken for acetone.   This is why the cumene route is used in seven out of the 15 U.S. plants, including the most recently built one. The total breakdown of synthetic phenol capacity (about 45.0 million pounds per year are recovered from coal-tar and caustic-wash refining operations) by route is given below.

Route	Capacity, MM lbs/yr	% of total	No. of plants
Cumene	550.0	43	7
Sulfonation	205.0	16	2
Raschig (modified)	240.0	19	3
Chlorination	230.0	18	1
Toluene	40.0	3	1
Benzene oxidation	20.0	1	1
Total	1,285.0	100	15

Source: *Hydrocarbon Processing*, January 1966.

The capacities and routes of U.S. plants are shown below.

Producer	Location	1965 Capacity, MM lbs/yr	Route
Allied Chem.	Frankford, Pa.	200.0	Cumene
Dow Chem.	Kalama, Wash.	40.0	Toluene
	Midland, Mich.	230.0	Chlorination
Monsanto	Monsanto, Ill.	115.0	Sulfonation
	Alvin, Tex.	140.0	Cumene
Clark Oil	Chicago, Ill.	30.0	Cumene
Chevron Chem.	Richmond, Calif.	50.0	Cumene
Hercules	Gibbstown, N.J.	30.0	Cumene
Hooker	Tonawanda, N.Y.	65.0	Modified Raschig
	South Shore, Ky.	65.0	Modified Raschig
Reichhold	Tuscaloosa, Ala.	90.0	Sulfonation
Shell	Houston, Tex.	50.0	Cumene
Skelly Oil	El Dorado, Kans.	50.0	Cumene
Union Carbide	Marietta, Ohio	110.0	Modified Raschig
Schenectady	Rotterdam Junction, N.Y.	20.0	Benzene oxidation
Total		1,285.0	

Source: *Hydrocarbon Processing*, January 1966.

Cumene is entirely petroleum-based, and there are three merchant producers in the U.S., all petroleum companies, who operate no phenol facilities themselves. Among the phenol-from-cumene producers, only Allied Chemical had no captive raw material source by 1965. The merchant cumene plants are shown below.

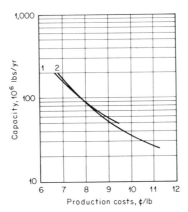

Fig. 8-18 Phenol. 1-Cumene.
2-Modified Raschig.

Producer	Location	Capacity, MM lbs/yr
Amoco	Texas City, Tex.	50.0
Gulf	Philadelphia, Pa.	180.0
Texaco	Westville, N.J.	140.0
Total		370.0

In addition, Dow has a 10.0 million pounds per year plant at Midland, Mich., whose output is dehydrogenated to α-methylstyrene. The cost of producing cumene at a refinery is estimated below. A 160.0 million pounds per year unit requires an investment of $1.7 million.

	Cost, ¢/lb cumene
Benzene	2.4
Propylene	0.8
Utilities	0.2
Labor and overhead	0.2
Capital charges	0.4
Total	4.0

It takes 1.38 pounds of cumene to make 1.0 pound of phenol; although the quoted price for merchant cumene is around 11 cents per pound, it can be seen that a non-integrated phenol producer could not possibly pay over 5-5.5 cents per pound and remain competitive. The U.S. consumed 1,225.0 million pounds phenol in 1965, distributed as follows:

End-use	MM lbs, 1965
Phenolic resins	615.0
Bisphenol A	120.0
Caprolactam	230.0
Adipic acid	50.0
Lube-oil refining	60.0
Surfactants	50.0
Others	100.0
Total	1,225.0

Source: *Oil, Paint & Drug Reporter*,
April 11, 1966.

4.1   Phenolic Resins

The main outlet for phenol is the production of phenolformalde-
hyde resins.   Phenol and formaldehyde can be condensed in acid
medium using a mol-ratio slightly less than 1:1.   This gives linear
polymers that can be crosslinked by the action of hexamethylene
tetramine.   These resins are known as "novolaks."   A mol-ratio
of 1.8:1 (formaldehyde:phenol), on the other hand, yields cross-
linked resins known as "resols"; they are usually prepared in
alkaline medium and crosslinking is accomplished by heating.
Molding powders and foundry resins are novolaks; most bonding
and laminating resins are resols, with crosslinking taking place
during the hot-pressing phase.

Condensation takes place in a steam-jacketed stainless steel or
clad kettle; after about 12 hours the reaction is arrested by neutralizing the alkaline catalyst; yields are around 90 percent.   A continuous process, developed by Booty Research, reports yields of
99 percent after a residence time of 20 minutes.   This process
should place the production of phenolic resins within the reach of
many consumers who previously bought them from large integrated
manufacturers.

Phenolic moldings are the cheapest of all molding materials,
since they contain about 50 percent inert filler, usually sawdust.
Their main properties are heat resistance, excellent dielectric
qualities and ease of transformation.   Their disadvantages are
poor impact resistance and inability to make colored objects from
phenolic powders, thereby restricting their use to objects that
are functional but not attractive. With growing prosperity, the advanced economies are beginning to require beauty where functionality once sufficed.   For example, telephone sets, which were
first made exclusively from phenolics, are now made in color
from more expensive materials.

Phenolic laminates are well known in such applications as institutional furniture, wall-covering and restaurant furnishings. They also have innumerable industrial applications, mainly in the electrical field; the market distribution between decorative and industrial laminates is about 70:30.   In decorative laminates, the top layer is impregnated with a melamine formaldehyde resin, which is tougher and can be colored.

Shell-molding resins are another important outlet for phenolics. The process is used by 20-30 percent of all U.S. foundries to cast precision ferrous and brass parts. Phenolic resins are also used to make hot-box cores.

Most abrasive wheels are bonded with phenolic resins; so are mineral and especially glass wool used as thermal and acoustical insulation for construction and refrigerators.   Many adhesives contain phenolics as modifiers; phenolic resins are also the bonding agent in brake linings and in organic fibers for use in furniture or automobile upholstery.

Phenolic resins also have an important role in plywood manufacture. The market is divided fairly sharply between exterior and marine grades, on the one hand, and interior grades, on the other. The former require water-resistant plywood and therefore resins stable to hydrolysis; this market belongs entirely to phenolic resins. Starch and urea-formaldehyde resins, being cheaper, are used for interior grades, but the declining price of phenol has enabled phenolics to gain entry into this field as well. However, not all plywood is manufactured with synthetic resins; starch and more recently dialdehyde starch are also used, especially on Douglas fir. The total output of plywood in the U.S. is 10.7 billion square feet, of which two-thirds is exterior-grade.

Phenolic resins are used as binders to make particle board out of chips, sawdust or wood shavings.   However, since most of its applications are interior, urea-formaldehyde resins hold over 80 percent of this market.   Although more expensive, the use of particle board is growing faster than that of plywood because it allows more efficient production of furniture. But it still represents less than 10 percent of its volume at present.

The breakdown for phenolic resin use is given below. It should be pointed out that these statistics consolidate the information for resins made not only from phenol, but also alkylphenols (mainly *t*-butyl phenol), cresols, resorcinol, and others. Alkylphenols are used mostly in surface-coating applications, and to increase compatibility with neoprene in rubber adhesives.   Resorcinol resins are used as adhesives, particularly in the tire industry. *o*-Cresol resins are used in laminates, molding powders and such special applications as can linings and coating.

The total of 810.0 million pounds of phenolic resins reported for 1964 can be broken down by weight as shown below.

Raw material	MM lbs, 1964
Phenol	650.0
Cresylics	80.0
Alkylphenols	25.0
Resorcinol	10.0
Formaldehyde (less water condensation)	80.0
Total	810.0

The table below shows the breakdown of the demand in 1965 which was 895.0 million pounds.

End-use	MM lbs, 1965
Molding materials	255.0
Decorative laminates	126.0
Abrasive bonding	22.0
Brake linings, clutch facings, etc.	32.0
Particle-board binder	24.0
Plywood binder	115.0
Foundry resins	71.0
Coatings	30.0
Insulation bonding	119.0
Other bonding applications	39.0
All other	62.0
Total	895.0

Source: *Modern Plastics*, January 1966.

Below are some of the most important applications for phenolic molding powders.

End-use	MM lbs, 1965
Appliance parts	33.0
Utensil handles	23.0
Electrical controls	25.0
Electrical switchgear	46.0
Wiring devices	28.0
Washing machines	22.0
Automotive	26.0
Telephone and intercom	16.0
Closures	15.0
Business machines	5.0
Machine parts, etc.	10.0
Miscellaneous	6.0
Total	255.0

Source: *Modern Plastics*, January 1966.

The principal producers of phenolic resins in the U.S. in 1964 were:

Producer	Capacity, MM lbs/yr
American Cyanamid	20.0
Borden	40.0
Catalin	25.0
General Electric	50.0
Hooker	180.0
Monsanto	90.0
Owens Corning	60.0
Plastics Engineering	40.0
Reichhold	90.0
Simpson Timber	20.0
Union Carbide	150.0
Total	765.0

4.2   Chlorinated Phenols

The most important chlorinated derivatives of phenol are p-chlorophenol, pentachlorophenol and 2,4-dichlorophenol. 2,4,5-trichlorophenol is made from tetrachlorobenzene by caustic fusion. p-Chlorophenol is an intermediate for the production of quinizarin, a raw material for anthraquinone dyes:

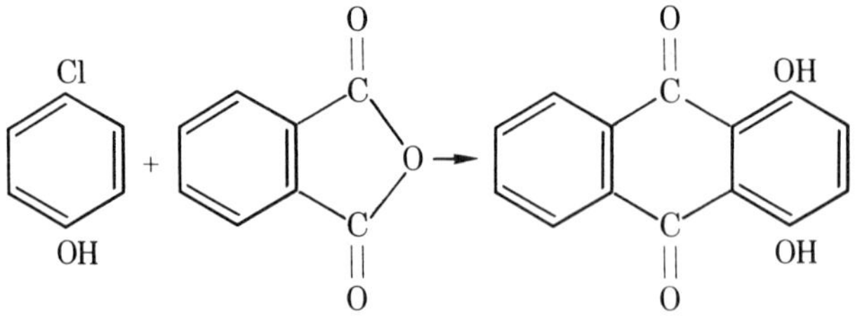

p-Chlorophenol is also one of the raw materials for the well-known wool fungicide Mitin FF.

2,4-Dichlorophenol is an intermediate for the preparation of 2,4-D (2,4-dichlorphenoxyacetic acid), made by reacting the sodium phenate with monochloroacetic acid:

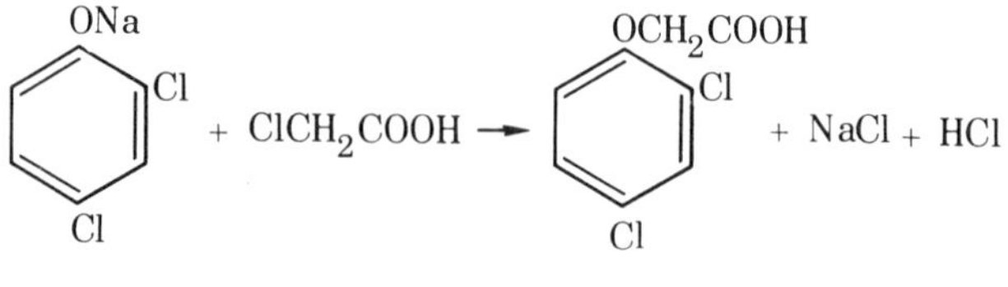

The reaction is carried out in the presence of excess 2,4-dichlorophenol, which is recovered for reuse by distillation.

2,4-D, as well as its esters and soluble salts, are the most important type of herbicides developed so far, despite the fact that they are also the oldest organic compounds used for this purpose. They act by upsetting plant hormone balance, and are used almost exclusively to fight broad-leaf plants, in grainfields — rice, corn, etc. In the U.S., there has been a trend towards the use of pre-emergents which has cut into the demand for 2,4-D and its derivatives; furthermore, their use is restricted in areas where broad-leaf crops, such as cotton, are grown extensively. Nevertheless, the equivalent of about 13.0 million pounds per year of the acid are produced in the U.S.; esters and salts are used to the extent of 63.0 million pounds per year of which esters such as isopropyl or isooctyl represent the major portion. The most important salt is that of dimethylamine.

The investment for 10.0 million pounds per year 2,4-D plant, starting from phenol and monochloroacetic acid, is $1.1 million. Assuming captive phenol (at 7.5 cents per pound), chlorine and monochloroacetic acid sources (the usual case), production costs are as follows:

	Cost, ¢/lb 2,4-D acid
Phenol	3.8
Chlorine	2.0
Caustic soda	1.8
Monochloroacetic acid	8.0
Utilities	1.3
Labor and overhead	4.8
Capital charges	3.6
Total	25.3

2,5-Dichlorophenol is the starting point for making "Banvel," an important herbicide for sugar-cane weed control.

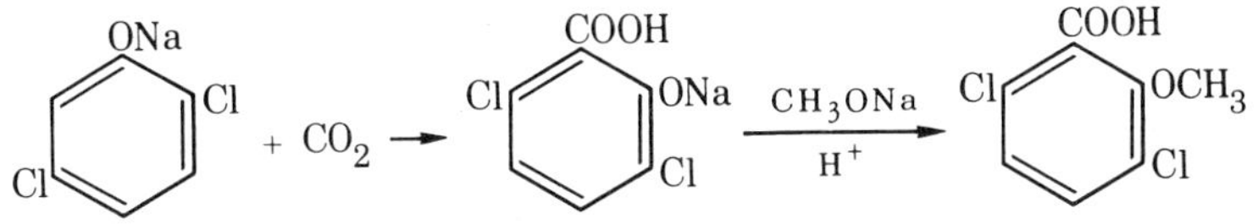

The derivatives of 2,4,5-trichlorophenoxyacetic acid (2,4,5-T), made in a similar way, have somewhat different applications. They are used to fight shrubs and bushes on grazing land; the output is around 12 million pounds per year. "Ronnel," a phosphate in-

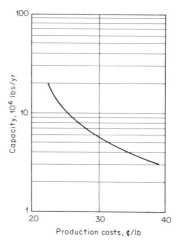

Fig. 8-19   2,4,D (acid).

secticide having low mammalian toxicity, is also derived from 2,4, 5-trichlorophenol. Over 10.0 million pounds per year are made in the U.S. by Dow.

Actually, 2,4,5-trichlorophenol is not a phenol derivative. Isolation or concentration of the γ-isomer of benzenehexachloride ("Lindane") leaves over substantial amounts of worthless isomers, which can be cracked to 1,2,4-trichlorobenzene and HCl. Further chlorination and caustic fusion produce 2,4,5-trichlorophenol.

2,4,5-Trichlorophenol is the starting point for making hexa-chlorophene, a well-known product in the cosmetic field as the active ingredient in antiperspirants and deodorant soaps. Generally, it is used in combination with other products that do not contain a phenol group, which makes it a skin irritant. The total demand for germicides by the soap and cosmetic industry is 4.5 million pounds per year, of which hexachlorophene represents 2.5 million pounds. Givaudan is the largest manufacturer of hexachlorophene in the U.S. with a $2.0 million plant at Clifton, N.J.

$$2 \; \text{(2,4,6-trichlorophenol)} + CH_2O \longrightarrow \text{(methylene-bis-trichlorophenol)} + H_2O$$

2,4,5-Trichlorophenol is used increasingly as an industrial fungicide. The demand for all these applications is around 18.0 million pounds per year of trichlorophenol.

Chlorinated phenols are also used to make other herbicides such as "Falone," "Erbon," "Silvex" and others of less importance.

Finally, pentachlorophenol and the respective sodium phenate are used widely as wood preservatives, especially in such applications as residential construction where creosote would be undesirable. It is also used as a fungicide in paint, adhesives and in paper mills. The demand for pentachlorophenol in the U.S. is 40.0 million pounds per year of which about 15.0 million pounds are used in mixtures with creosote. It can also be made by caustic fusion of hexachlorobenzene, which is a cheaper route but gives a product contaminated with NaCl.

Although pentachlorophenol is made by batch chlorination in most countries, large manufacturers in the U.S. use continuous processes. Chlorination is accomplished in two reactors in series, with fresh phenol being fed to the first and incoming chlorine to the second. Unreacted chlorine off-gas from the second reactor is absorbed in the phenol feed and the two raw materials sent to the first. The effluent from the first reactor is trichlorophenol, which goes to the second chlorination step.

Pentachlorophenol is applied from hydrocarbon solutions, whereas sodium phenate is water-soluble. Application from oil has the inconvenience that the solvent tends to migrate to the surface carrying along some pentachlorophenol, which makes finishing difficult. In the Koppers "Cellon" process, used primarily to make architectural woods, the preservative is introduced from an LPG vehicle by a pressure-vacuum sequence. Creosote is the main competing product. Most pentachlorophenol is now consumed as prills, the preferred form for wood preservation.

The investment for a 5.0 million pounds per year pentachlorophenol plant is $1.5 million. Assuming captive phenol at 7.5 cents per pound, as well as a captive chlorine source, production costs are as follows.

	Cost, ¢/lb pentachlorophenol
Phenol	2.8
Chlorine (net)	3.0
Utilities	0.4
Labor and overhead	4.8
Capital charges	3.2
Total	14.2

It is assumed that chlorine recovered as HCl is worth 50 percent of fresh chlorine.

The main U.S. producers of pentachlorophenol are as follows:

Producer	Location	1965 Capacity, MM lbs	Captive phenol
J.H. Baxter	Long Beach, Calif.	2.0	—
Dow	Midland, Mich.	15.0	*
Frontier	Wichita, Kans.	7.0	—
Monsanto	E. St. Louis, Ill.	15.0	*
Reichhold	Tacoma, Wash.	7.0	*
	Tuscaloosa, Ala.	3.0	*
Sanford Chem.	Port Neches, Tex.	10.0	—
Total		59.0	

*Some capacities are flexible since the same equipment can be used to make other products.

Source: *Oil, Paint & Drug Reporter*, February 12, 1966.

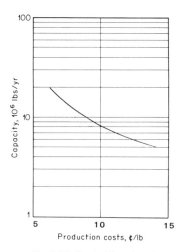

Fig. 8-20 Pentachlorophenol.

4.3   Bisphenol A

Phenol and acetone react under acid catalysis to form di-phenylol propane (DPP), also known as bisphenol A.

$$2 \; C_6H_4OH + CH_3COCH_3 \longrightarrow CH_3-\underset{\underset{}{}}{C}-CH_3 + H_2O$$

The production of bisphenol A is a delicate process, since a number of isomers and tri- or mono-hydroxy byproducts are formed. While a product containing these impurities is suitable for the manufacture of epoxy resins, it cannot be used to make polycarbonates.

Early processes eliminate these impurities by crystallization, which greatly increased the cost of the purified product. The Hooker process, however, employs distillation and extractive crystallization under pressure to purify crude bisphenol A continuously instead of batchwise, which has greatly improved economics. Distillation of bisphenol A is especially difficult because numerous elements, some in minute concentrations, decompose the product or promote the formation of polymers and tar.

A flowsheet is shown in Fig. 8-21. Acetone and excess phenol are mixed, saturated with HCl and sent to the reaction vessel. There, conversion takes place in the presence of promoters. After reaction, the crude product is stripped of HCl and water of reaction. Then the overhead is decanted into an organic phase, consisting mainly of phenol, which is recycled, and an aqueous phase. The latter goes on to an HCl recovery unit, and water is sent to disposal.

Bottoms from the HCl still are sent to the purification section. Excess phenol is taken overhead and recycled to the reactor. Crude bisphenol A then goes through two stills; the first removes low-boiling byproducts and isomers which are recycled, and the second eliminates tarry material. Distillate from the second still goes to the extraction operation which produces a continuous slurry of pure crystals. Mother liquor is stripped to recover the solvent,

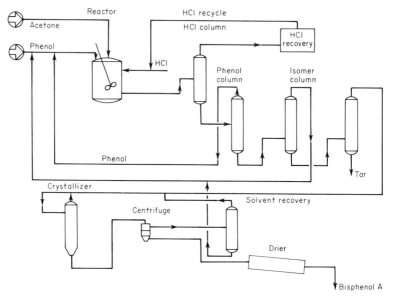

Fig. 8-21 Bisphenol A (Hooker process).

and the impurities are recycled to the reactor where they are
ultimately converted to bisphenol A (see Fig. 8-21). Producers of
bisphenol A usually manufacture epoxy resins also; the only exception is Monsanto, a merchant producer and manufacturer of
polycarbonates. The U.S. capacities are given below.

Producer	Location	Capacity, MM lbs/yr	Main captive use
Dow	Midland, Mich.	40.0	Epoxy resins
Monsanto	St. Louis, Mo.	30.0*	Polycarbonates
Shell	Houston, Tex.	45.0	Epoxy resins
Union Carbide	Marietta, Ohio	25.0	Epoxy resins Polysulfones
General Electric	Mt. Vernon, Ohio	20.0*	Polycarbonates
Total		160.0	

*Two plants came on stream in 1966: 15.0 million lbs/yr for Monsanto at
St. Louis, Mo., and the General Electric plant.
   Source: *Oil, Paint & Drug Reporter*, January 17, 1966.

The economics of bisphenol A production are evaluated below.
Acetone is assumed captive at 5.0 cents per pound. Actually,
acetone is available to a phenol-from-cumene producer for as
little as 3.0 cents per pound. Investment for a 20.0 million pounds

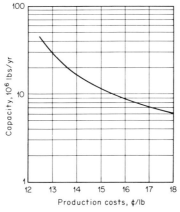

Fig. 8-22   Bisphenol A.

per year plant is $1.9 million and production costs can be estimated as shown in the table below.

	Cost, ¢/lb bisphenol A
Phenol	7.2
Acetone	1.4
Catalyst and chemicals	0.1
Utilities	1.0
Labor and overhead	0.9
Capital charges	3.1
Total	13.7

In 1965, the demand for bisphenol A in the U.S. was 100.0 million pounds, distributed as follows:

End-use	MM lbs, 1965
Epoxy resins	85.0
Polycarbonates and other	15.0
Total	100.0

"Epoxy resins" include the amount used to make tetrabromo bisphenol A, for fire-resistant epoxies and polycarbonates, and hydro bisphenol A, made by hydrogenation of the benzene rings, which goes into construction resins.

Other miscellaneous uses are in the manufacture of UV absorbers and in brake fluids. There are also some alkylated bisphenol A rubber antioxidants on the market.

Polysulfones

This class of engineering plastics based on bisphenol A and monochlorobenzene was developed by Union Carbide. Their main properties are flame-resistance, dimensional stability, a useful temperature range from -150 to 300°F, oxidation resistance and flexibility.

The main potential applications are pipe and sheet. The only commercial plant was built in 1965 by Union Carbide at Marietta, Ohio, and has a capacity of 10.0 million pounds per year.

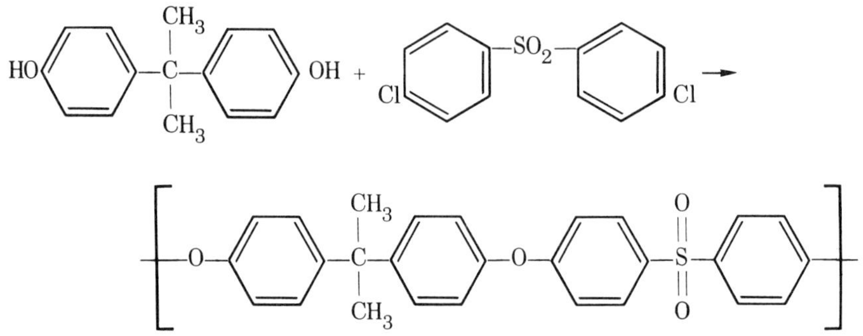

4.4   Salicylic Acid

The production of salicylic acid still follows the original Kolbe synthesis for making carboxylic acids from aryl hydroxy compounds:

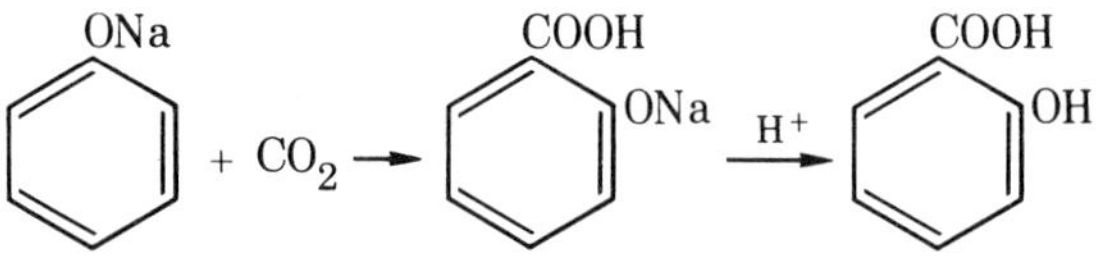

The main manufacturers of salicylic acid in the U.S. are given below.

Producer	Location	Capacity	Captive phenol
Dow	Midland, Mich.	14.0	x
Monsanto	St. Louis, Mo.	15.0	x
Rexall	Kearny, N.J.	1.0	–
Sterling Drug	Cincinnati, Ohio	4.0	–
Tenneco	Garfield, N.J.	8.0	–
Total		42.0	

Source: *Oil, Paint & Drug Reporter*, June 6, 1965.

The main outlet for salicylic acid is in the manufacture of aspirin; other uses, however, are of considerable importance. A breakdown of the 40.0 million pounds per year consumed in the U.S. in 1964 appears below.

End-use	MM lbs/yr
Aspirin	26.0
Salicylate esters and salts	6.0
Rubber retarder	2.0
Azo dye intermediate	2.0
Miscellaneous	4.0
Total	40.0

Source: *Oil, Paint & Drug Reporter*, June 7, 1965.

Methyl salicylate, the methyl ester of salicylic acid, is also known as "oil of wintergreen" and used as a flavoring compound in mouthwash and certain food products. Other salicylates are used as antioxidants and UV protectors; the most important are *t*-butyl and phenyl salicylate ("salol"). The latter, which is also employed as a denaturant for ethyl alcohol, is made by Friedel-Crafts reactions between phenol and salicylic acid.

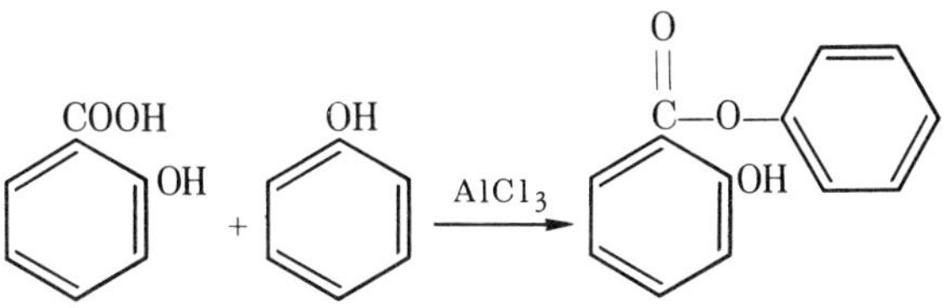

Salicylates are the cheapest UV stabilizers on the market, but are being replaced by more expensive products that have less of a tendency to break down with time. Their outlets are chiefly in cosmetics and sun-tan lotions, which consume about 700,000 pounds per year. Retarders are used in rubber formulations to minimize the risk of scorching during vulcanization. As a dye intermediate, salicylic acid is used mainly to make premetallized red, brown and orange diazo products for wool and nylon dyeing.

As a dyeing aid for polyester and cellulose acetate fiber, salicylic acid or its esters act as swelling agents, making it easier for disperse dyes to penetrate the fiber. It is also an intermediate for salicylamide, used in numerous analgesic-antipyretic formulations, and for salicylanilide, an important fungicide for textile fibers.

Aspirin

Acetylsalicylic acid (aspirin) is the most widely used of the mild analgesics. It is made by acetylating salicylic acid with acetic anhydride, acetic acid being recovered as a byproduct.

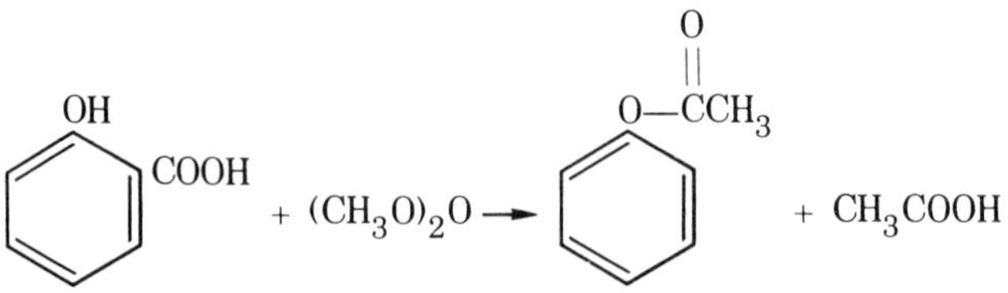

$$\underset{\text{OH}}{\bigcirc}\text{COOH} + (CH_3O)_2O \longrightarrow \underset{O-CCH_3}{\overset{O}{\bigcirc}} + CH_3COOH$$

Of the 30.0 million pounds per year consumed in the U.S., roughly 15.0 million pounds are sold pure; the rest is used in mixtures with other pain relievers, or antipyretics, such as phenacetin, APAP and salicylamide.

Salicylic acid manufacturers are the main producers of aspirin. Only two companies make aspirin from purchased salicylic.

Producer	Location	Capacity, MM lbs/yr	Captive salicylic
Dow	Midland, Mich.	10.0	x
Monsanto	St. Louis, Mo.	12.0	x
Sterling	Trenton, N.J.	10.0	x
Rexall	Kearney, N.J.	1.0	x
Miles Lab.	Zeeland, Mich.	2.0	–
Norwich	Norwich, Conn.	2.0	–
Total		37.0	

Source: *Oil, Paint & Drug Reporter*, March 1, 1965; January 17, 1966.

Salicylanilide

This mildew-proofing agent is made from salicylic acid and aniline:

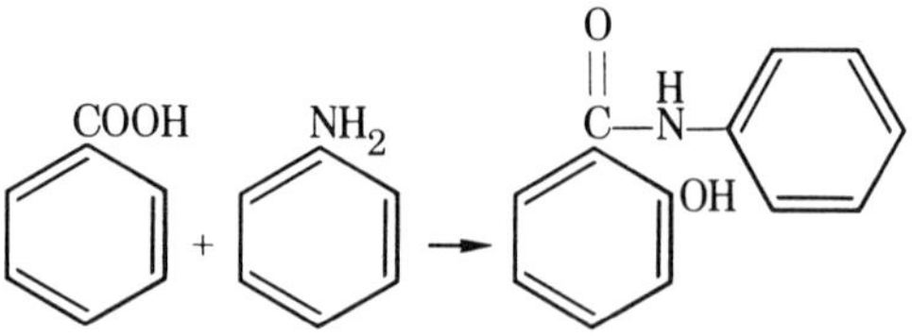

$$\underset{\text{OH}}{\bigcirc}\text{COOH} + \underset{}{\bigcirc}\text{NH}_2 \longrightarrow \underset{\text{OH}}{\bigcirc}\overset{O}{C}-\underset{H}{N}\bigcirc$$

It is used mainly on cotton goods, but as its sodium salt, it can be used for wool.

Brominated salicylanilides are germicides used in the soap and cosmetics industry.

4.5   p-Hydroxybenzoic Acid

If instead of sodium phenate, potassium is subjected to the Kolbe synthesis, p-hydroxybenzoic acid is formed.

Its methyl and *n*-propyl esters are used as preservatives in the cosmetic and pharmaceutical fields, and to a minor extent, in the food industry; they are known commercially as "Parabens." Consumption in 1965 was 0.5 million pounds.

4.6   Alkylphenols

Phenol can be alkylated via conventional alkylation means — aluminum chloride or phosphoric acid catalysis. There are two important ranges of alkylphenols. Alkylation with isobutylene or amylene yields products used to make phenolic resins and antioxidants; isobutylene dimer or propylene trimer gives higher alkylphenols which can be ethoxylated to form nonionic surface-active agents and lube-oil additives, as in the case of dodecylbenzene production. The preferred feedstocks for lube-oil additives are heavy alkylate fractions.

The total demand for alkylphenols in the U.S. was around 140 million pounds in 1965, but the future of nonionic detergents based on these products is doubtful. Although a considerable part of nonionic detergents end up in industrial compounds, they are also used in household products. However, in this application, they become vulnerable to the "biodegradability" issue which has greatly modified the entire picture for detergent raw materials since 1964. The demand can be broken down as shown below.

End-use	MM lbs, 1965
Surfactants	60–68
Lube-oil additives	32–35
Plastic and chemicals	35
Rubber chemicals	5–8
Total	132–146

Source: *Chemical Week*, January 30, 1964.

The U.S. manufacturers of alkylphenols are given below.

Producer	Location	Capacity	Captive phenol	Captive ethylene oxide
Chevron Chem.	Richmond, Calif.	10.0	x	–
Enjay	Linden, N.J.	3.0	–	–
General Aniline	Linden, N.J.	10.0	–	x
	Calvert City, Ky.	25.0	–	x
Jefferson	Port Neches, Tex.	15.0	–	x

Monsanto	Kearny, N.J.	25.0-28.0	x	-
Productol	Santa Fe Springs, Calif.	5.0	x*	-
Rohm & Haas	Philadelphia, Pa.	20.0-25.0	-	-
Stepan	Millsdale, Ill.	5.0-7.0	-	-
Union Carbide	Marietta, Ohio	20.0	x	x
UOP	E. Rutherford, N.J.	4.0	-	-
Other		6.0-12.0		
Total		148.0-164.0		

*"Natural" phenol.
Source: *Chemical Week*, May 30, 1964.

The investment cost for an alkylphenol plant is similar to that for alkylation plants in general; a 22.0 million pounds per year plant was reported to have cost $1.4 million.

4.7 Salicylaldehyde

The reaction between phenol and chloroform and subsequent hydrolysis gives salicylaldehyde:

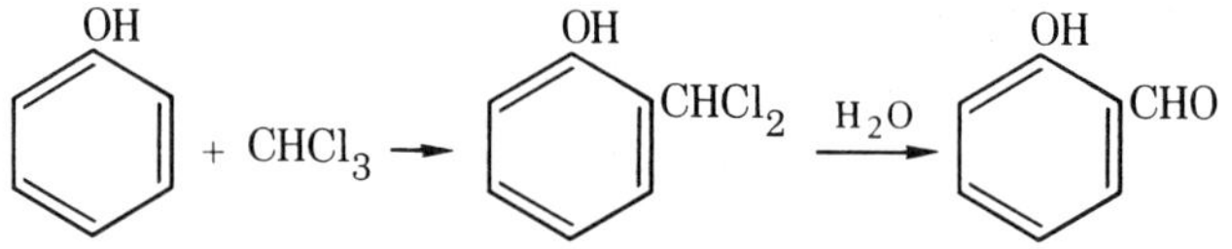

It is an intermediate for the production of coumarin, used widely as a synthetic substitute of Tonka bean flavor, of which about 1.3 million pounds per year were consumed in the U.S. during 1965.

The most important outlet for salicylaldehyde is the manufacture of N,N disalicylidene ethylene (or propylene) diamine of which 2.0 million pounds per year go into motor fuel as metal deactivators. The propylene diamine derivative is preferred because of its better solubility.   The compounds act by tieing up as complexes traces of Fe and Cu, which could catalyze undesirable reactions.

Total production for salicylaldehyde is about 2.5 million pounds per year.

4.8 Rubber Antioxidants

In addition to the numerous alkylated phenols, mention may be made of "Wingstay S," a styrene-phenol condensation product used by many SBR producers as the antioxidant that is added to the latex blending tank before coagulation.

4.9   Phosphate Esters

Although the most important member of this group of chemicals is tricresyl phosphate, several phosphate esters contain phenol. Triphenyl phosphate is used mainly as a plasticizer for cellulose acetate molding powders and photographic film. It is produced by Stauffer, Celanese, FMC, Frontier, Montrose, and Monsanto, and the demand is steady at 10.0 million pounds per year. Octyldiphenyl phosphate has become widely established as a fire-retardant plasticizer for vinyl film and sheet; it imparts better low-temperature flexibility than tricresylphosphate, and has been approved for use in food packaging. Demand is around 12.0 million pounds per year; it is made by Monsanto under the name "Santicizer 141." Cresyldiphenylphosphate originally replaced tricresylphosphate as a gasoline additive for controlling the formation of preignition coke deposits, but in turn has been partly replaced by methyl diphenylphosphate, which contains more active ingredient per pound. The total demand for these two additives in 1965 was around 18.0 million pounds per year, of which 16.0 million was cresyldiphenylphosphate.

Also during 1965, about 25.0 million pounds of phenol went into phosphate ester production.

4.10   Byproducts from Phenol Production

Diphenyl Oxide (Chlorination Process)

The eutectic mixture of diphenyl oxide and diphenyl (75:25) is well known as a heat-transfer medium under the trade-name of "Dowtherm." It is stable up to 800° F, where most liquids decompose. "Dowtherm" systems can operate in the vapor or the liquid phase. Although vapor-phase systems involve lower investment requirements because heat-transfer coefficients are higher, the trend is towards liquid-phase systems operating under pressure because losses of the heat-transfer medium are lower. The chlorination process yields about 0.1 pound diphenyl oxide per pound of phenol.

With the decline in the importance of the chlorination process, diphenyl oxide is made synthetically from phenol and monochlorobenzene (for example, by ICI) to meet the steady rise in demand for heat-transfer media.

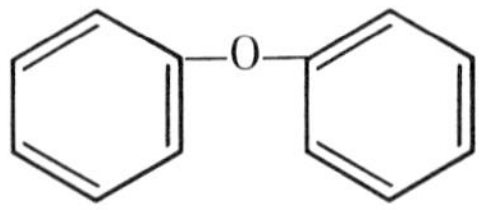

Westinghouse has developed a family of thermosetting polymers, trade-named "Doryl," derived from diphenyl oxide. These resisns, polymers of methylated diphenyl oxide, can be used as a varnish or impregnating resin, and retain their properties up to 480° F. They are expected to compete with epoxy resins. Chloromethylated diphenyl oxide, obtained by reacting DPO with formaldehyde and HCl, can be polymerized by a Friedel-Crafts catalyst to thermoplastic resins used as adhesives of bonding agents for glass, mica, etc. Certain chloromethylated DPO resins can react with phenol to produce Novolak-type products having properties superior to phenolics.

DPO is also an intermediate for making polyphenyl ethers (via chlorination), which have found some use as high-temperature lubricants. Other applications of DPO are in perfumery, and in the manufacture of surface-active agents by alkylation and sulfonation.

Diphenyl and Polyphenyls (Chlorination Process)

The demand for diphenyl has grown faster than its availability as a phenol byproduct. Progil, Bayer, Monsanto, and ICI all make diphenyl from benzene by thermal decomposition.

$$2C_6H_6 \xrightarrow{\Delta} C_{12}H_{10} + H_2$$

Besides being a constituent of "Dowtherm," diphenyl is used in treating citrus trees against certain diseases. Chlorinated diphenyls find application as plasticizers for chlorinated rubber and vinyls, and are used widely as heat-transfer agents. Under the name "Aroclor," chlorinated polyphenyls are recommended for use in fire-retardant epoxy resins, nonflammable hydraulic fluids, and lubricants for certain severe conditions.

Dichlorobenzenes (Chlorination Process)

The most important use for p-dichlorobenzene is as a fumigant for mothproofing woolen goods, both raw and processed, a market that once belonged to refined naphthalene. It is a dye intermediate, via 2,5-dichloroaniline, which, in turn, is an intermediate for making "Mitin," a mothproofing agent for wool.

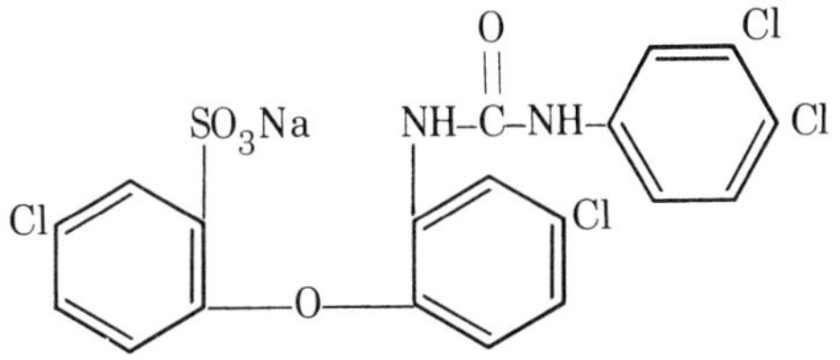

Consumption of p-dichlorobenzene in 1965 amounted to 67.0 million pounds.

o-Dichlorobenzene has been growing in demand over the last few years mainly as a reaction medium for phosgenation. Since for every pound of o-DCB 3.0 pounds of the p-isomer are obtained, the rising demand for polyurethanes caused a shortage of the former. At least two plants have been built to produce the o-isomer to the exclusion of p-DCB. They are presented below.

Producer	Location	Capacity, MM lbs/yr
Monsanto	Monsanto, Ill.	10.0
Dover	Dover, Ohio	3.0
Total		13.0

In addition, about 20.0 million pounds per year are produced as byproducts in the combined monochlorobenzene plants of the country.

o-DCB is also employed as a heat transfer medium, as a reaction medium in the production of dye intermediates, and as an intermediate for anthraquinone dyes. In addition, it can be nitrated and then reduced to 3,4-dichloroaniline, which is used in the manufacture of herbicides. "Stam" and "Dicryl" are, respectively, amides of propionic and metacrylic acid. "Diuron," of which about 10.0 million pounds per year are produced, is a very successful preemergence herbicide used mainly for extermination of cotton weed.

Caustic fusion of o-dichlorobenzene is the preferred route for making catechol; diorthotolylguanidine dicatechol borate, known as "Permalux," is a fast accelerator for neoprene and chlorinated butyl rubber.

The total o-DCB consumption in the U.S. was 60.0 million pounds in 1965; the growth in demand parallels that of urethane foams.

α-Methylstyrene (Cumene process)

Copolymers of various materials with α-methylstyrene have excellent heat and UV-degradation resistance. The most important application of this property is heat-resistant polystyrene. Polyesters, alkyd resins and polymethylmethacrylate can also be improved with respect to heat-resistance and UV-degradation by addition of α-methylstyrene. Its main disadvantage is its slow rate of reaction. The total demand for α-methylstyrene is around 12.0 million pounds per year, most of it for heat-resistant polystyrene used in washing machine parts, etc. Producers of α-methylstyrene are listed below; only Dow is not a producer of phenol via cumene.

Producer	Location	Capacity, MM lbs/yr
Dow	Midland, Mich.	10.0
Allied	Frankford, Pa.	10.0
Hercules	Gibbstown, N.J.	3.0
Total		23.0

Acetophenone (Cumene process)

The market for acetophenone is around 4.0 million pounds per year, some 25 percent of which is recovered from the combined phenol-from-cumene plants. It is used in perfumery, but its most important outlet is as starting point for the synthesis of chloramphenicol, the only entirely synthetic antibiotic.

Cumene Hydroperoxide (Cumene process)

The particular application of this peroxide is as a redox cross-linking agent for polyethylene and ethylenepropylene rubber. Peroxide vulcanization has the disadvantage of yielding a malodorous product. Other products are under development (e.g., trichloromelamine). In 1964, 600,000 pounds per year of cumene hydroperoxide were consumed.

5   MONOCHLOROBENZENE

Besides its use as an intermediate in the manufacture of phenol, aniline and DDT, monochlorobenzene has a number of important uses as a solvent and intermediate.

Both monochlorobenzene and *o*-dichlorobenzene are used widely as the solvents in which phosgenation reactions take place. Use of monochlorobenzene alone for this purpose is around 25.0 million pounds per year.

As an intermediate, chlorobenzene has a number of important applications, which required another 55.0 million pounds in 1965. These can be broken down as follows:

	MM lbs, 1965
"Trithion"	4.0
Nitrochlorobenzenes	45.0
Chloroanthraquinone	2.0
Others	4.0
Total	55.0

Below is a list of plants producing monochlorobenzene in the U.S. as well as the main captive applications for each.

Producer	Location	1965 Capacity, MM lbs/yr	Captive end-use
Allied Chem.	Syracuse, N.Y.	25.0	DDT, solvent
Dover Chem.	Dover, Ohio	1.0	—
Dow	Midland, Mich.	350.0	Phenol, aniline
DuPont	Deepwater, N.J.	10.0	Nitrochlorobenzenes
Geigy	McIntosh, Ala.	20.0	DDT
Hooker	Niagara Falls, N.Y.	15.0	—
	N. Tonawanda, N.Y.	65.0	Phenol
	S. Shore, Ky.	65.0	Phenol
Monsanto	E. St. Louis, Ill.	15.0	Nitrochlorobenzenes
Montrose	Henderson, Nev.	70.0	DDT
Olin	Huntsville, Ala.	30.0	DDT
Pittsburgh Plate Glass	New Martinsville, W.Va.	50.0	Solvent
Total		716.0	

Source: *Oil, Paint & Drug Reporter*, April 5, 1965.

5.1 Nitrochlorobenzenes

Chlorobenzene can be nitrated in a mixture of nitric and sulfuric acid, giving 30 percent ortho- and 70 percent para-nitrochlorobenzene. The latter is of greater commercial importance, but, nevertheless, there is an excess of *o*-nitrochlorobenzene on the market. The two isomers are separated by crystallization.

The demand for nitrochlorobenzenes in 1965 was 55.0 million pounds, broken down as follows:

End-use	MM lbs, 1965
Parathion	28.0
p-Nitroaniline	13.0
Sulfur dyes	9.0
Substituted benzidines	5.0
p-Phenetidine	6.0
Others	4.0
Total	65.0

p-Phenetidine

p-Nitrochlorobenzene is reacted with sodium ethylate and reduced to give p-phenetidine.

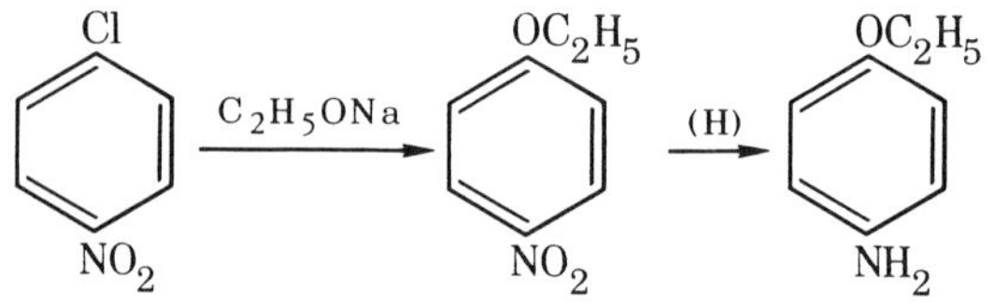

p-Phenetidine has some use as a dye intermediate, but its main applications are the manufacture of 6-ethoxy-1,2-dihydro-2,2,4-trimethylquinoline, a widely used antioxidant not only for rubber but also in forage and feed.

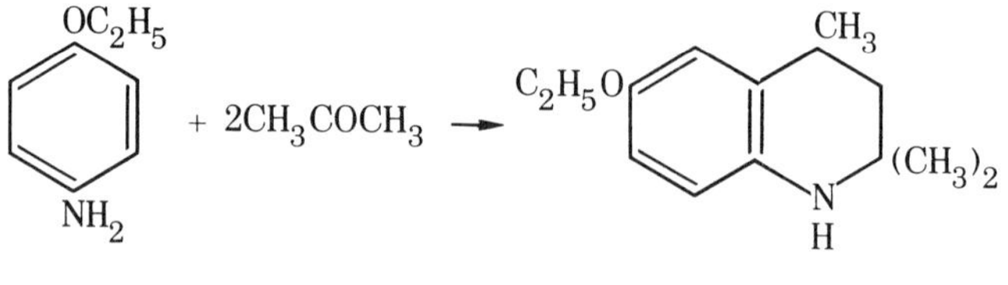

p-Anisidine

This compound is made in the same way as p-phenetidine, with sodium methylate used instead of the ethylate. Anisidine can be nitrated to 2-nitro-4-methoxy-aniline, the starting point for making 6-methoxy-8-aminoquinoline, which is the key intermediate in the manufacture of such antimalarial drugs as "Aralen" and "Primaquine."

("Primaquine")

NH

H_3C—CH—$(CH_2)_3$—NH_2

p-Anisidine is also a dye intermediate.

p-Chloroaniline

p-Nitrochlorobenzene can be reduced to p-chloroaniline, an intermediate in the production of such herbicides as "Neburon" and "Urox."

Diamino 4,4 diphenylether, a derivative of p-chloroaniline, is an intermediate for the production of polyimides, a class of promising high-temperature resins.

p-Nitroaniline

p-Nitrochlorobenzene reacts with ammonia to form p-nitroaniline, an important intermediate for dyes and pigments.

However, the main end-use of this product is as an intermediate for the preparation of numerous antioxidants and antiozonants of the N-substituted p-phenylenediamine type. These compounds are made by first reacting p-nitroaniline with an alcohol or ketone, and then sec-butyl-p-phenylenediamine, an antioxidant for gasoline of which some 2.0 million pounds per year are used in the U.S.

The compounds of this family are also very important as rubber antiozonants. The shift to smaller and lower-pressure tires, and the use of greater oil-extension ratios to cheapen the rubber stock, have enhanced the danger of cracking due to ozonolysis, and these compounds are therefore growing in importance. They produce a dark color in rubber, and thus find use in tires and other industrial goods; hindered phenols are preferred when nonstaining antioxidant action is required. In 1965, the rubber industry used 9.0 million pounds of these antiozonants. "Botran," a well-known fungicide, is a chlorinated derivative of p-nitroaniline.

Approximately 11.0 million pounds per year of p-nitroaniline are consumed in the U.S., of which 60 percent is for p-phenylenediamine derivatives and 20 percent as a dye intermediate.

p-Nitrophenol

Caustic fusion of p-nitrochlorobenzene yields sodium p-nitrophenate.

Its most important outlet is as an intermediate in the production of methyl and ethyl parathion, two insecticides that together are consumed to the extent of 30.0 million pounds per year.

$$(RO)_2 \overset{\overset{S}{\|}}{P}Cl + NaO\!\!\left\langle\!\!\bigcirc\!\!\right\rangle\!\!NO_2 \longrightarrow (RO)_2\overset{\overset{S}{\|}}{P}\!-\!O\!-\!\!\left\langle\!\!\bigcirc\!\!\right\rangle\!\!NO_2 + NaCl$$

By far the most important outlet for parathion is in cotton pest-control. Two-thirds of the total output is methyl parathion because of its specific action against the boll-weevil.

The major manufacturers are given below.

Producer	Location	1965 Capacity, MM lbs/yr
DuPont	Deepwater, N.J.	5.0
Monsanto	E. St. Louis, Ill.	15.0
Southern Dyestuffs	Sodyero, N.C.	1.0
UOP Chemical	Shreveport, La.	5.0
Total		26.0

Source: *Oil, Paint & Drug Reporter*, August 15, 1966.

A 10.0 million pounds per year parathion plant requires an investment of $2.0 million. By 1965, all plants used batch processes, but a continuous process has been developed and used for the first time in India.

Parathions account for 44 percent of all phosphorothioate poisons produced in the U.S. Although they have been on the market for some time, they are so highly toxic that immunities take a very long time to develop.

p-Nitrophenol is also an intermediate for making EPN, a phosphate insecticide developed by DuPont for use against rice stem borers. The reduction of p-nitrophenol gives p-aminophenol; acetylation gives N-acetyl-p-aminophenol, known as acetaminophen, a widely used antitussive and antipyretic.

Diazotization, reaction with sodium arsenite, and neutralization yield p-aminophenylarsonic (arsanilic acid), used as a growth stimulant in swine and poultry feed. Its potential market has been estimated at 4.0 million pounds per year.

2,4-Dinitrochlorobenzene

This product is obtained by nitration of *p*-chloronitrobenzene. It can be converted by caustic fusion to 2,4-dinitrophenol, a starting point for making Sulfur Black, the most important type of sulfur dye.

Sulfur dyes are well-suited to high-speed continuous application, and are among the cheapest types of color. On the other hand, their chlorine fastness is poor. The demand in 1965 was around 47.0 million pounds.

Partial reduction of 2,4-dinitrophenol gives 2-amino-4-nitrophenol, an important intermediate in the manufacture of premetallized dyes for wool and nylon. Dyeing with premetallized colors can be accomplished under almost neutral conditions, instead of the low pH needed when using acid azo dyes. This gives a softer and more uniformly dyed textile than can be obtained when dyeing from an acid bath.

Ammoniation gives 2,4-dinitroaniline, also a dye intermediate.

The total demand for 2,4-dinitrochlorobenzene as a dye intermediate is around 16.0 million pounds per year, most of which goes into the making of Sulfur Black. 2,4-Dinitrophenol can be alkylated to 4,6-dinitro-*o-sec*-butylphenol (DNBP). Around 5.0 million pounds per year are being used, mainly in the form of its alkanolamine salts, as a preemergence herbicide for potato, peanut and corn crops, and for blossom thinning of fruit trees.

o-Phenylenediamine

Ammoniation and reduction of *o*-nitrochlorobenzene gives *o*-phenylenediamine.

As a dye intermediate, *o*-phenylenediamine is used mainly to make optical bleaches. One of its main applications is as the starting point for Merck's livestock antihelminthic, 2-(4-thiazolyl)-benzimidazole, known as "Thiabendazole," which has taken over most of the market that once belonged to phenothiazine.

o-Phenylenediamine is the starting point for making benzotriazole UV stabilizers. These are more durable than the more common salicylates, but are also much more expensive. They are used mainly in PVC and polyolefin films.

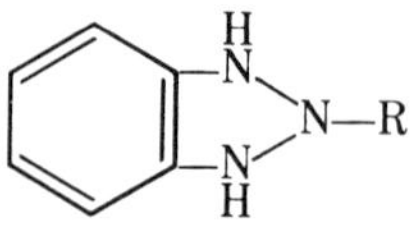

Another derivative of *o*-phenylenediamine is mercaptobenzimidazole, a rubber anti-aging agent.

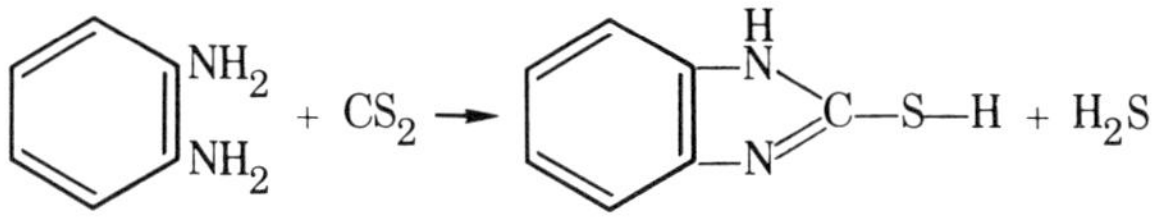

5.2 "Trithion"

Stauffer manufactures this important phosphate insecticide from two sources: diethylphosphorochloridothionate and *p*-chlorophenylthiomethylchloride. The latter is obtained via the following sequence:

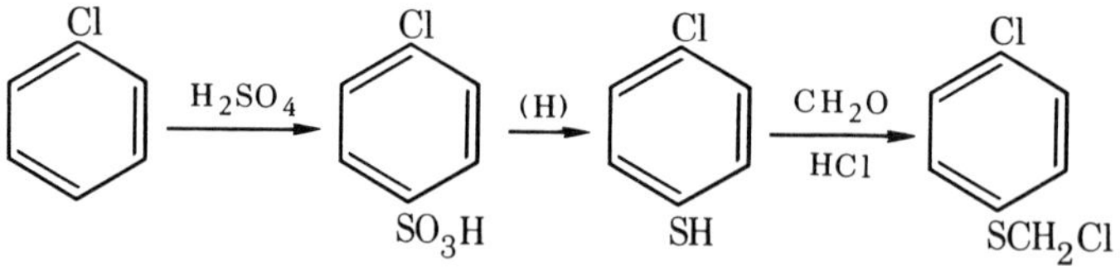

Annual consumption of this miticide and insecticide, which is used mainly on fruit trees and against onion maggots, is around 5.0 million pounds per year.

6 ALKYLBENZENE SULFONATES

Excellent detergent power and low production cost make sodium salts of alkyl benzyl sulfonic acids the most common raw materials for the formulation of household detergents. The original alkylation material was predominantly propylene tetramer, which has now been replaced almost completely by derivatives of normal paraffins.

$$\text{benzene} + \text{RCH}=\text{CH}_2 \longrightarrow \underset{\text{RCH}_2\text{CH}_2}{\text{benzene}} \xrightarrow[\text{NaOH}]{\text{H}_2\text{SO}_4;} \underset{\text{RCH}_2\text{CH}_2}{\text{benzene}-\text{SO}_3\text{Na}}$$

Alkylation of benzene takes place according to the flowsheet in Fig. 8-23.

Benzene and the alkylation material — an olefin or an n-paraffin chloride — react in the presence of anhydrous aluminum chloride. The reactor effluent is fractionated to remove unreacted benzene, which is recycled and then distilled under vacuum to remove light and heavy alkylate. The product then is sent to the sulfonation unit. Alkylbenzene is produced by petroleum companies and sold to detergent manufacturers who operate sulfonation and spray-drying facilities.

Light alkylate byproduct is used as a solvent. Heavy alkylate, resulting from the reaction of more than one olefin molecule with one of benzene, can also be sulfonated to give industrial surfactants used in cutting oils, pesticide emulsions and especially in lubricants. When the alkylation material is linear, this heavy fraction does not have the same properties as those obtained from propylene tetramer. This resulted in a shortage of raw materials for lube-oil additives which developed after alkylate makers switched to n-paraffins. Thus manufacturers had to turn to other aromatic petroleum fractions as sulfonation feedstock. Heavy alkylate makes up about 12-16 percent of the detergent-range material produced.

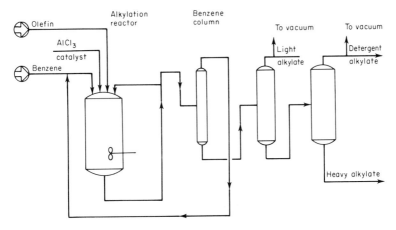

Fig. 8-23 Detergent alkylate.

The investment for a 50.0 million pounds per year alkylate plant is $1.5 million. The production costs shown below assume that light and heavy alkylates are worth the same as detergent alkylate. Propylene tetramer is taken at 4.0 cents per pound, assumed captive.

	Cost, ¢/lb detergent alkylate
Benzene	1.8
Propylene tetramer	4.2
Catalyst and chemicals	0.6
Utilities	0.2
Labor and overhead	0.5
Capital charges	1.0
Total	8.3

Approximately 1.25 pounds of saleable products are obtained. Assuming the byproducts are worth 4.0 cents per pound, the captive cost of dodecyl benzene is 7.3 cents per pound.

In the usual sulfonation process, oleum reacts with the feed-stock in an agitated kettle. The mass is then pumped from the kettle to a settler where spent acid is separated and sent to re-concentration. The sulfonic acid is neutralized with caustic soda and then sent to a crutcher, where builders such as sodium tripoly-phosphate, sodium chloride, sodium silicate and others are mixed with the sulfonate. The addition of these builders does not affect the detergency of the surface-active agent. The mixture is finally spray-dried to obtain the product in powder form; this operation represents one-third of the investment and about half of the op-erating costs in the manufacture of detergents by this process. The investment for a 100.0 million pounds per year household deter-gents plant is $1.6 million, and production costs, excluding raw materials, amount to 1.6 cent per pound.

Since some sulfuric acid is always left in the mixture even after settling, the neutralized product contains sodium sulfate. Although solid detergents contain sodium sulfate, this has the disadvantage of making the product unsuitable for formulating clear liquid detergents, which are growing faster in popularity than the overall detergent industry.

Several processes using SO_3 as the sulfonating agent instead of sulfuric acid have been developed to overcome this inconvenience. The three such processes which have achieved commercial success are: the Ballestra process, which operates at atmospheric pres-sure; Chemithon, which was a high-speed mixing technique at a pressure of 15 psig; and the "Sulfan" process, promoted by Allied

Chemical, which uses merchant sulfur trioxide that has been stabilized with boron compounds. The advantage of using the latter process is that the need for a sulfur-burning plant is eliminated; thus since this unit represents by far the major part of the total investment for an integrated SO_3 sulfonation unit, small producers find the process attractive despite the fact that the price of sulfur in "Sulfan" is necessarily higher than that of elemental sulfur. The Ballestra process has been the most successful worldwide, but Chemithon and "Sulfan" have found more application in the U.S. Economics favor the use of elemental sulfur; the investment requirements for the three alternatives — sulfuric acid, "Sulfan" and sulfur burning — appear in the table below for a 13.0 million pounds per year unit.

Price of sulfur assumed $35/ton

	Investment
Oleum sulfonation	$ 30,000
SO_3 (outside supply) sulfonation	90,000
SO_3 (captive source) sulfonation	340,000

Most detergent alkylates are produced either by oil companies or chemical companies having petroleum-refining subsidiaries, as is the case of Monsanto and Allied Chemical. The product of these plants is sold to such concerns as Procter & Gamble, Lever and Colgate, which do the sulfonation and finishing. As seen from the list of companies using sulfur trioxide sulfonation, the three major detergent producers still have not gone over to this technique. This is due partly to the fact that it presents fewer advantages for making solid detergents (which are dominated by the three largest companies) than for liquids, where the "independents" predominate.

Nearly all detergent alkylate producers have converted to soft alkylate, although in 1965 there was still a domestic demand for around 110.0 million pounds per year for such nonhousehold detergent uses as lube additives, emulsifiers and in glass slurry processes.

However, not even LAS is considered to be the final answer to the biodegradability issue, and the sodium sulfonates of benzene alkylates generally tend to be replaced by ethylene- or n-paraffin-derived, long-chain, alcohol ether sulfates.

The U.S. also remains a large exporter of hard alkylate. Below is a list of U.S. producers and capacities.

Producer	Location	Capacity, MM lbs/yr	
		Soft	Hard
Allied Chem.	Buffalo, N.Y.	20.0	–
Chevron Chem.	Richmond, Calif.	75.0	200.0
Monsanto	E. St. Louis, Ill.	150.0	–
Union Carbide	S. Charleston, W.Va.	150.0	–
Witfield	Watson, Calif.	30.0	–
Continental Oil	Baltimore, Md.	150.0	50.0
Atlantic-Richfield	Aireco, Tex.	50.0	–
Phillips	Pasadena, Tex.	30.0	–
Total		655.0	250.0

Source: *Oil, Paint & Drug Reporter*, March 21, 1966.

Some of the "independents" making sulfur trioxide derivatives are:

Producer	1964 Capacity, MM lbs/yr
Witco, Ultra Division	10.0
Retzloff	6.0
Tennessee Corporation	6.0–10.0
Richardson	20.0
Textilana	8.0
Texsize	6.0–10.0
Pilot Chem.	80.0–100.0
Continental Chem.	14.0
Continental Oil	20.0–30.0
Total	170.0–208.0

Source: *Chemical Week*, April 18, 1964.

The products from these plants are used mainly in "private label" liquid detergents, or specialty products such as agricultural emulsifiers, windshield washing formulations, textile wet-

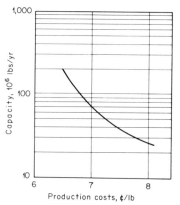

Fig. 8-24 Detergent alkylate.

processing aids, petroleum additives and polymerization emulsifiers. The demand for trioxide derivatives in 1965 was around 150.0 million pounds, although the major producers, accounting for 75 percent of household detergent production in the U.S., had not yet begun replacing oleum sulfonation derivatives even in their liquid formulations.

Alkyl benzene sulfonates are used mainly in heavy-duty household detergents, especially solid formulations. Since these are expected to show little growth in the future, the demand is expected at best to remain fairly static. An estimate of the demand for alkyl benzene sulfonate in 1965 and 1970 is given below.

End-use	Estimate MM lbs	
	1965	1970
Heavy duty, high-foaming solids	254.0	190.0
Heavy duty, low-foaming solids	54.0	60.0
Heavy duty, high-foam liquids	14.0	11.0
Heavy duty, low-foam liquids	2.0	0-3.0
Light duty, liquid	105.0	138.0
Light duty, solid	11.0	8.0
Industrial	70.0	60.0
Other (bubble baths, scouring cleansers, etc.)	32.0	28.0
Total	542.0	498.0

Source: *Chemical Week*, May 30, 1964.

The proportion of sulfonated detergent alkylate in household detergents has on the average been declining steadily. This is due to the shift towards low-foaming solids, which contain higher proportions of nonionics and correspondingly less LAS. The formerly more popular high-foaming formulations contain around 12 percent LAS, compared to 5.3 percent for the low-foaming formulations. By 1964, the average content for all household products was less than 9 percent, and the fact that LAS may be a more efficient detergent than ABS may drive this proportion down even further.

7 MALEIC ANHYDRIDE

Benzene can be oxidized in the vapor phase to maleic anhydride over a vanadium pentoxide catalyst.

Overall stoichiometric yields of around 65 percent, corresponding to 80-85 percent by weight, are obtained by the better processes. A flowsheet of the Scientific Design process is shown in Fig. 8-25. Air and benzene are mixed and sent to a fixed-bed catalytic reactor at essentially atmospheric pressure. The temperature is controlled by circulating a coolant. The hot exit gases are cooled in a waste-heat boiler and finally against cooling-water. Then they are absorbed in a wash-tower where all maleic anhydride is absorbed in water and thereby converted to maleic acid. This is then dehydrated back to maleic anhydride and sent to a discontinuous vacuum distillation system for final purification.

Maleic anhydride is usually sold in the form of tablets. Some manufacturers use a switch condenser to recover over 85 percent of the product directly as the anhydride, and convert the remainder to fumaric acid. By absorption of the remaining 15 percent in water, yields are thereby increased since the losses during the hydration-dehydration step are avoided. This option requires a lower air: benzene ratio, and thus implies somewhat lower yields in the oxidation step.

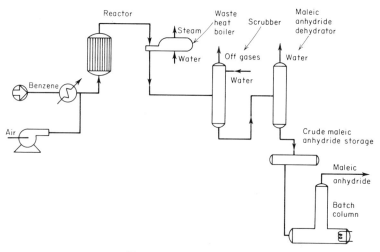

Fig. 8-25 Maleic anhydride.

Assuming a theoretical yield of 65 percent, economics for the production of maleic anhydride are estimated below. The investment for a 25.0 million pounds per year plant is $3.5 million.

	Cost, ¢/lb maleic anhydride
Benzene	5.2
Utilities	0.6
Labor and overhead	1.6
Capital charges	4.6
Total	12.0

Another process which arrives at maleic anhydride by oxidations of n-butylenes has been commercialized in the U.S.

$$CH_3CH{=}CHCH_3 \xrightarrow{O_2}$$

Yields are around 45 percent on a stoichiometric basis, or 80 percent by weight. However, despite the use of a cheaper raw material, production costs are about the same since investment requirements are 45 percent higher than for benzene oxidation. The process was used until 1966 by Petro Tex, a major producer of butadiene and therefore also of butylenes; however, the company is reported to have switched to benzene as a result of a decision to use butylenes as a raw material for neoprene. On a pilot-plant scale, good yields of maleic anhydride have been obtained by oxidizing n-butane at around 850° F in the presence of cobalt molybdate.

The plants producing maleic anhydride in the U.S. are given on the following page.

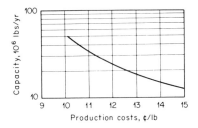

Fig. 8-26 Maleic anhydride.

Producer	Location	1965 Capacity, MM lbs/yr	Process
Allied Chem.	Moundsville, W.Va.	20.0	Allied
California Chem.**	Richmond, Calif.	20.0	Badger Reichhold
Tenneco	Fords, N.J.	24.0	n.a.
Koppers	Bridgeville, Pa.	20.0	S.D.
Monsanto	E. St. Louis, Mo.	50.0	S.D.
Petro Tex	Houston, Tex.	10.0	Butene oxidation*
USS Chem.	Neville Island, Pa.	20.0	S.D.
Reichhold	Elizabeth, N.J.	20.0	Reichhold
Total		184.0	

*Converted to benzene in 1966.
**Currently not running.
  Source: Capacity data *Oil, Paint & Drug Reporter*, June 27, 1966.

The demand for maleic anhydride in 1965 was 123.0 million pounds which includes 22.0 million pounds isomerized to fumaric acid. The market for maleic anhydride as such was therefore 101.0 million pounds, and is growing at a rate of 15 percent per year after having remained static for several years. The growing use of polyester resins is the main reason for this rapid increase in demand. The breakdown is given in the table below.

End-use	MM lbs, 1965
Polyester resins	57.0
Pesticides	14.0
Surface coatings	7.0
Plasticizers	5.0
Lubricants	7.0
Copolymers	5.0
Fumaric acid	22.0
Malic acid and others	6.0
Total	123.0

Source: *Oil, Paint & Drug Reporter*,
June 27, 1966.

7.1 POLYESTER RESINS

Polyesters can be made by the reaction between any dicarboxylic acid and a difunctional alcohol. When speaking of polyesters, however, a distinction is made between three kinds of polymers:

1. alkyd resins, made by reacting a saturated dicarboxylic acid, usually phthalic anhydride, with a trifunctional alcohol, most commonly glycerine. The extra -OH groups are esterified by a monofunctional acid;

2. polyester fibers and films, made from linear condensation polymers, terephthalic acid, or its methyl ester and ethylene glycol. The resulting polymer is thermoplastic and can be spun from a melt in the same way as nylon;

3. polyester resins are crosslinked polymers. This is accomplished by reacting both saturated (phthalic, isophthalic and others) and unsaturated (mainly maleic anhydride) dicarboxylic acids with a difunctional alcohol such as propylene glycol, and crosslinking the resulting polyester by vinyl-type polymerization of the double bond. This class will be discussed here.

Polyester resins are often modified by copolymerization with other monomers, chiefly styrene. Other modifiers are used, such as methacrylates or triallyl cyanurate for high-resistance polyesters. Since most polyester resins are sold for such structural applications as boats and process equipment, they are therefore reinforced, mostly with glass fibers; other inorganic fillers can also be added. The proportion of polyester resins going into the manufacture of reinforced plastics approaches 75 percent; reinforced polyesters in turn represent almost 95 percent of the total consumption of reinforced plastics, the best being mainly reinforced epoxy resins.

The total demand for polyesters is growing at a rate of 20 percent per year. However, although at present having captured two-thirds of the total market for pleasure-boat shells, this growth-rate is likely to decline. Nevertheless, the demand for polyester resins is expected to continue growing in the fields of automobile body construction, building construction and others.

End-use	MM lbs, 1965	
Reinforced plastics*		369.0
Construction	72.0	
Containers, trays, industrial housing	16.0	
Electrical	18.0	
Aircraft and missiles	37.0	
Boats	71.0	
Transportation	75.0	
Pipes, ducts, tanks	23.0	
Appliances	14.0	
Miscellaneous	16.0	
Consumer products	27.0	
Nonreinforced applications		91.0
Surface coatings	8.0	
All others	72.0	
Export	11.0	

*Data for reinforced plastics include resin and reinforcement while data for nonreinforced uses is for polymer only.
Source: *Modern Plastics*, January 1966.

Principal construction applications are flat and corrugated sheeting, which account for 45 percent of the total; other outlets are ventilator hoods and concrete forms. Structural shapes and garage doors are among the more recent applications. Polyesters are used in industrial equipment where corrosion resistance is required, in such institutional products as trays; consumer goods include packaged dinner trays. Electrical applications are chiefly in switchgear. Military outlets consist of filament-wound vessels and piping, also significant in the process equipment field. Other uses in this category are aircraft partitions, missile nose-cones and heat shields. Almost 65 percent of all pleasure-boat shells is made from reinforced polyester. There appears to be a trend towards longer boats, such as fishing boats and minesweeper shells. Of the amount used in transportation equipment, 37 percent is for structural applications, the best known of which is sport-car bodies. Initially, this had the serious inconvenience of not permitting the kind of finish obtainable on steel, but this difficulty has since been overcome, and the number of models incorporating a reinforced plastic body is expected to grow quickly. The remaining 63 percent consists of kick plates, window panel moldings, step wells for buses (for corrosion protection against salt in winter) and others. In the appliance field, polyester is used where heat-resistance is important, especially where cheaper and less decorative materials were formerly used.

Besides maleic anhydride, other dicarboxylic acids not containing a reactive double bond are used to make polyester resins. The estimated distribution of the most important acids used in unsaturated polyesters during 1965 is given below.

	MM lbs, 1965
Maleic anhydride	60.0
Phthalic anhydride	90.0
Isophthalic acid	25.0
Fumaric acid	10.5
Total	185.5

7.2   Pesticides

Maleic hydrazide is made by the reaction between maleic anhydride and hydrazine. It is used for the specific purpose of "suckering" tobacco plants, that is, removing shoots from the lower stem. This growth regulator is said to affect both taste and the filling capacity of cigarette machines, and hence tobacco companies have tried to discourage its use by farmers. Currently 40 percent of all tobacco acreage is treated with maleic hydrazide; this represents a consumption of 6.5 million pounds per year, requiring 5.5 million pounds per year maleic anhydride. It is also used for weed control along highways and rights of way.

Next to the parathions, the most important phosphate insecticide is malathion, made from the ethyl ester of maleic anhydride.

$$
\begin{array}{c}
\overset{O}{\overset{||}{HC-C-OC_2H_5}} \\
\underset{||}{\overset{|}{HC-C-OC_2H_5}} \\
\overset{}{O}
\end{array}
\; + \; (CH_3O)_2\overset{S}{\overset{||}{P}}SH
\;\longrightarrow\;
(CH_3O)_2\overset{S}{\overset{||}{P}}S-\underset{\underset{O}{\underset{||}{CH_2C-OC_2H_5}}}{\overset{O}{\overset{||}{CH-C-OC_2H_5}}}
$$

Malathion is considerably less hazardous and toxic to mammals than methyl parathion. It is used as a broad-spectrum insecticide particularly against household, dairy and livestock insects, and on fruits and vegetables. A recent technique of spraying pure malathion from planes in concentrations of 8 ounces per acre is expected to bring about a rapid increase in demand. In addition, this represents a fundamental change in the technique of applying insecticides in general, since this is the first instance of undiluted pesticide spraying. In 1965, the demand for malathion was 14.0 million pounds per year, requiring 5.0 million pounds maleic anhydride.

7.3   Tetrahydrophthalic Anhydride

The Diels-Alder reaction between maleic anhydride and butadiene gives tetrahydrophthalic anhydride.

It can be used to make alkyd resins in competition with other difunctional acids such as phthalic and isophthalic. Solventless coatings made with THPA have excellent adhesion, flexibility and light stability, cure rapidly and provide excellent gloss. They are being developed for use on furniture, appliances and automobiles, the latter in an attempt to win back the market lost by alkyds to acrylic coatings. THPA is also a curing agent for epoxy resins, but dianhydrides are used more often for this purpose.

THPA reacts with ammonia to form tetrahydrophthalimide, which combines with perchloromethylmercaptan to give "Captan," a fungicide used mainly on fruit to combat rot and for protection during the time between harvesting and distribution.

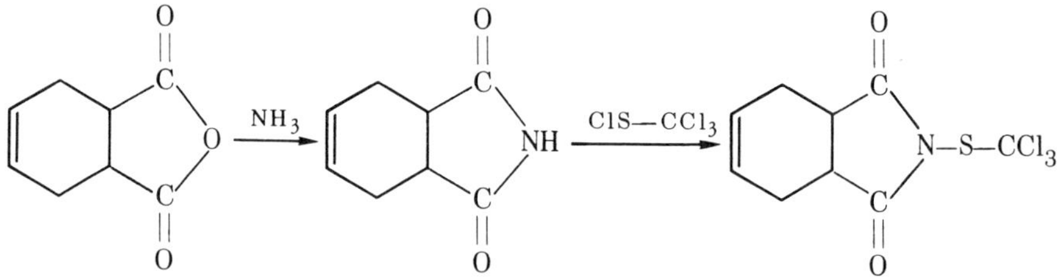

Consumption in the U.S. is 10.0 million pounds per year, mostly by the apple industry.

7.4   Succinic Anhydride

Maleic anhydride can be hydrogenated to succinic anhydride.

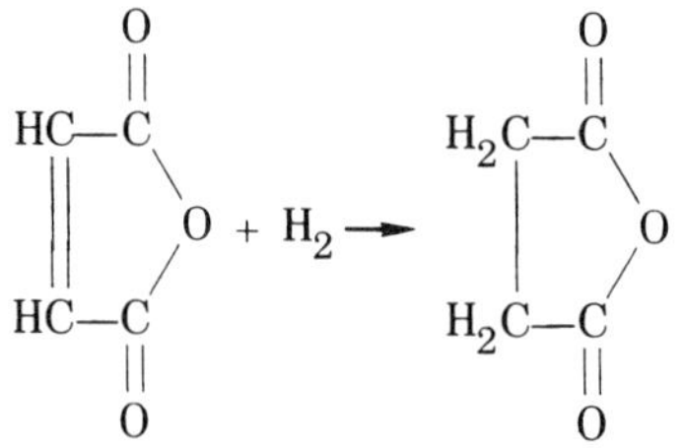

Allied Chemical employs this method. DuPont produces succinic acid by recovery of the byproduct diacids formed during the nitric acid oxidation of cyclohexanol to adipic acid. As a diacid, succinic acid (or anhydride) is an intermediate for modified alkyd resins, plasticizers and other typical application of esters, but it has specific applications. As a raw material for making nylon 4-10, a polyamide developed by DuPont for making toothbrush bristles, succinic acid is the precursor of tetramethylenediamine, which is reacted with sebacic acid.

Succinic anhydride is one of the raw materials used to make quinacridone pigments. It has been reported that these pigments are to red and maroon what phthalocyanine pigments have become to blue and green. They are exceptionally durable and light-fast, and find their most important outlets in automotive finishes. In the U.S. quinacridone pigments are made by DuPont under the trade-name "Monastral." Other uses of succinic anhydride are in the succinic ester of the active compound (e.g., tocophenyl succinate).

Finally, reaction with amines gives the respective succinimides, a family of compounds used as lube-oil additives.

7.5   Malic Acid

Catalytic hydration of maleic anhydride with steam gives malic acid:

$$
\begin{array}{l}
\text{HC—C} \\
\quad\quad \diagdown \\
\quad\quad\quad \text{O} + 2\text{H}_2\text{O} \longrightarrow \\
\quad\quad \diagup \\
\text{HC—C}
\end{array}
\quad\quad
\begin{array}{l}
\text{HOHC—COH} \\
\quad | \\
\text{H}_2\text{C—COH}
\end{array}
$$

Its only use is as a food acidulant. Although in 1964 only 1.5 million pounds were used, it is predicted by some that lower prices may allow it to capture some of the market held by citric acid and reach 10.0 million pounds per year by 1968. Allied Chemical is the sole U.S. producer, with a 20.0 million pounds-per-year plant.

7.6   Fumaric Acid

Maleic acid in solution can be isomerized catalytically to its trans-isomer, fumaric acid.

$$
\begin{array}{ccc}
\text{H} \quad\quad \text{H} & & \text{HOOC} \quad\quad \text{H} \\
\diagdown \quad \diagup & \overset{\text{cat.}}{\longrightarrow} & \diagdown \quad \diagup \\
\text{C}=\text{C} & & \text{C}=\text{C} \\
\diagup \quad \diagdown & & \diagup \quad \diagdown \\
\text{HOOC} \quad\quad \text{COOH} & & \text{H} \quad\quad \text{COOH}
\end{array}
$$

Thiourea (Allied) and ammonium salts (Scientific Design) among others, have been proposed as catalysts. The main advantage to maleic producers of making fumaric acid is that maleic acid recovered by absorption in water can be converted to a saleable product without dehydration to the anhydride, a step which involves considerable loss of yield. Practically all fumaric acid manufacturers are basically maleic anhydride producers. The exceptions in the U.S. are Pfizer and Stepan Chemical, both users of a fermentation technique that yields no maleic anhydride.

Besides its use as a food acidulant, applications of fumaric acid resemble those of maleic anhydride. Being a trans-isomer, fumaric is far less reactive. Since a given reaction takes much longer to complete, fumaric derivatives are more expensive and, therefore, are considered premium products. On the other hand, both physical and chemical properties are improved; polyesters made from fumaric have better heat and chemical resistance; alkyd resins are harder and also more resistant to heat, etc. In contrast to maleates, plasticizers made from fumaric are nontoxic, and thus are used in vinyl acetate emulsions that come into contact with food. As a food acidulant, fumaric was only recently approved by the FDA. Since then, its use in this field has been growing rapidly. It is cheaper per pound, and even cheaper, on an equivalent basis, than citric. Hence most of this growth has been at the expense of citric acid. The main limitation is its poor solubility in water, thereby limiting its use in beverages; these represent 20 percent of the total market for acidulants. Several ways of promoting fumaric acid solubility are being studied. It is predicted that eventually 15.0 million pounds per year will be used in dry beverage bases, lemon pie fillings and gelatin desserts, all of which are applications where poor solubility is not a disqualifying characteristic. A breakdown of the market for food acidulants in 1965 is given below.

	MM lbs, 1965
Citric	33.0
Fumaric	13.0
Maleic	1.5
Phosphoric	21.2
Adipic	2.0
Lactic	6.8
Tartaric	4.0
	81.5

Other uses are as an intermediate for optical bleaches. The demand for fumaric acid in the U.S. was around 36.0 million pounds in 1965, broken down as follows:

End-use	MM lbs, 1965
Polyester resins	4.0
Hard resins	9.0
Food acidulant	13.0
Paper sizing	4.0
Intermediate and miscellaneous	6.0
Total	36.0

Source: *Oil, Paint & Drug Reporter*, December 5, 1966.

Producers of fumaric acid are given below.

Producer	Location	Capacity, MM lbs/yr	Food grade
Allied Chem.	Moundsville, Pa.	15.0	x
Monsanto	St. Louis, Mo.	15.0	-
Petro Tex	Houston, Tex.	8.0	-
Pfizer	Terre Haute, Ind.	10.0	x
USS Chem.	Neville Island, Pa.	5.0	-
Stepan Chem.	Keyport, N.J.	3.0	-
Tenneco	Garfield, N.J.	10.0	x
Total		66.0	

Source: *Oil, Paint & Drug Reporter*, December 5, 1966.

7.7 Other Maleic Anhydride Derivatives

A certain amount of maleic anhydride is used as the diacid in alkyd resins to accelerate the reaction and improve resin color. It replaces 5 percent of phthalic anhydride on the average.

The most important plasticizer derived from maleic anhydride is dibutyl maleate. It is used as an internal plasticizer in vinyl acetate emulsions for latex paints. Not much growth is expected in this field. Other difunctional unsaturated acids used in the coating field are made from maleic anhydride by Diels-Alder reaction with cyclopentadiene or terpenes.

Several copolymers of maleic anhydride have become commercially important. Styrene-MA resins are used as latex paint modifiers and in floor-polish formulations; Monsanto's "Stymer" is recommended for cellulose acetate sizing. Paint thickeners and foundry core binders made from ethylene-MA copolymers have been developed by Monsanto. Under the trade-name "Gantrez," General Aniline & Film makes a family of copolymers of MA with methyl vinyl ether for use (in competition with polyacrylate salts and other water-soluble resins) in fields such as nylon and spandex sizing, adhesives and detergents.

The main types of surfactant made from maleic anhydride are the sulfosuccinates, obtained by reacting an ester of maleic anhydride (2-ethyl-hexyl, for example) with sodium sulfite; they are well known under the trade-name "Aerosols." They are used in the textile industry.

$$
\begin{array}{c}
\underset{\displaystyle \underset{O}{\|}}{HC-\overset{\displaystyle \overset{O}{\|}}{C}-OR} \\
HC-\underset{\displaystyle \underset{O}{\|}}{C}-OR
\end{array}
\; + \; NaHSO_3 \;\longrightarrow\;
\begin{array}{c}
HC-\overset{\displaystyle \overset{O}{\|}}{C}OR \\
NaSO_3-\underset{\displaystyle \underset{H}{|}}{C}-\overset{\displaystyle \overset{O}{\|}}{C}-OR
\end{array}
$$

Diels-Alder adducts of maleic anhydride with diolefins are employed as lubricants and jet-fuel additives.

The addition of maleic anhydride or fumaric acid to rosin (abietic acid) produces resins that find some use in paper-sizing. The soaps obtained by hydrolyzing these adducts impart better water resistance per pound than simple rosin soaps.

In the pharmaceutical field, antihistamines are often used in the form of their maleic acid salts (pyrilamine maleate, for example). Dibutyl tin maleate is used as a condensation reaction catalyst. Finally, the homopolymer of maleic anhydride has a potential market in competition with the polymers of acrylic acid.

8 NITROBENZENE

Benzene can be nitrated to nitrobenzene in a mixture of nitric and sulfuric acids:

$$
\underset{\text{benzene}}{\bighexagon} + HNO_3 \xrightarrow{\;H_2SO_4\;} \underset{\text{nitrobenzene}}{\bighexagon}^{NO_2} + H_2O
$$

To remove the water formed in the reaction, the reactor effluent is decanted into an aqueous phase that is extracted with incoming benzene and then sent to acid reconcentration, and an organic phase which is neutralized and stripped with steam. As is usually the

case, if nitrobenzene is to be sent on to an aniline unit, crude nitrobenzene needs no further treatment since reduction to aniline also
forms water. A further vacuum distillation step is necessary when
pure nitrobenzene is desired. The reaction can also be accomplished
with concentrated nitric acid slightly below the H_2O-HNO_3 azeotrope
which will suffice without the need for sulfuric. In this process,
spent nitric acid is continually withdrawn from a countercurrent reactor and sent to reconcentration; makeup to the system is 55 percent nitric acid.

By far the most important outlet for nitrobenzene is as the
starting point for aniline; only one aniline plant in the U.S. (Dow,
at Midland, Mich.) produces aniline via ammoniation of monochlorobenzene. Besides the amounts used to make aniline, 45.0 million
pounds per year of nitrobenzene are used for other purposes.
These include benzidine, m-pnenylenediamine and m-chloroaniline.

As a solvent, nitrobenzene is used mainly in shoe-polish formulations and for dye manufacture. Capacities of U.S. producers are
shown below; all except Monsanto are aniline manufacturers.

Producer	Location	1964 Capacity, MM lbs/yr
Allied Chem.	Buffalo, N.Y.	10.0
	Moundsville, W.Va.	50.0
American Cyanamid	Bound Brook, N.J.	60.0
	Willow Island, W.Va.	30.0
DuPont	Gibbstown, N.J.	110.0
Monsanto	Monsanto, Ill.	10.0
Rubicon Chem.	Geismar, La.	30.0
Total		300.0

Source: *Oil, Paint & Drug Reporter*, December 21, 1964.

8.1   Benzidine

Benzidine is made from nitrobenzene by reduction with zinc:

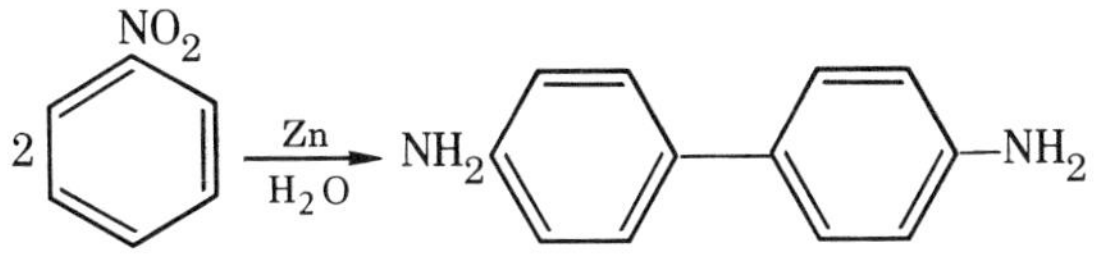

Benzidine is used in the manufacture of numerous dyes and pigments. The most important dye derived from benzidine is Direct

Black EW, which in 1965 accounted for 2.1 million pounds of this intermediate. Benzidine yellow is a widely used pigment of which 3.0 million pounds per year are used, requiring some 1.5 million pounds per year of benzidine. Other red, brown and orange disazo dyes, and pigments such as Ciba's "Chromophthal" series, make up the rest of the 4.5 million pounds per year demand for benzidine.

Diphenylhydrazine, obtained as an intermediate during benzidine production, is the starting point for making "phenylbutazone," a well-known analgesic and antirheumatic.

8.2   *m*-Chloroaniline

As in the case with other *m*-substituted aniline derivatives, *m*-chloroaniline is made from nitrobenzene.

The most important uses for *m*-chloroaniline are in the herbicide field. "Carbyne" is derived from *m*-chloroaniline and butynediol; it is recommended for use against wild oats. "CIPC," made by Pittsburgh Plate, Monsanto and United States Industries, is a more effective preemergence crab-grass control agent than 2,4,D; it is used in cotton and soybean weed repression.

8.3   *m*-Phenylenediamine

Nitrobenzene can be nitrated again and both nitro groups reduced, giving *m*-phenylenediamine.

m-Phenylenediamine is a coupling agent for several azo dyes, among them Direct Black EW. As an epoxy resin curing agent, *m*-phenylenediamine is staining and toxic, but is used in applications requiring medium-range, heat-distorting temperatures and for applications where these disadvantages can be tolerated.

In 1965 *m*-phenylenediamine began to be used on a large scale in the manufacture of polyamide and polyimide fibers. "Nomex," a polyamide made from *m*-PDA and isophthalic acid, has a softening point around 500° F; polyimides are based on pyromellitic dianhydride and aromatic diamines. Both are made in the U.S. by DuPont.

8.4   Aniline

Nitrobenzene can be converted to aniline by means of a convenient reducing agent. The traditional technique was liquid-phase reduction with iron filings. However, as plants began to grow, the average distance from which these filings had to be transported increased to the point where the process became unfeasible due to high freight costs. At present, aniline is made by liquid or vapor-phase catalytic hydrogenation.

$$\text{NO}_2 \text{ ring} + 3\text{H}_2 \longrightarrow \text{NH}_2 \text{ ring} + 2\text{H}_2\text{O}$$

In the American Cyanamid vapor-phase process, nitrobenzene is vaporized and fed to a fluidized bed reactor, containing a copper-silica hydrogenation catalyst. In the liquid-phase process, nitrobenzene is hydrogenated under pressure using a Raney nickel catalyst. Vapor-phase yields are 98 percent, and slightly lower for liquid-phase hydrogenation. In either case, following the unreacted hydrogen recycle the reactor effluent goes to a separator, where two phases form. The aqueous layer, formed by the water of reaction, contains some aniline which is recovered by extraction with incoming nitrobenzene. The organic phase contains water, and is fractionated in a two-tower system to remove heavy residue and water from the product. Figure 8-27 shows a flowsheet of an integrated nitrobenzene-aniline plant employing liquid-phase hydrogenation.

The investment for a 40.0 million pounds per year aniline plant, including benzene nitration facilities, is $3.2 million. Production costs are estimated assuming that nitric and sulfuric acids are available at $30 per ton.

	Cost, ¢/lb aniline
Benzene	3.1
Nitric acid	2.4
Hydrogen	0.8
Catalyst and chemicals	0.3
Utilities	0.4
Labor and overhead	0.6
Capital charges	2.6
Total	10.2

Strategically, aniline is one of the most important of all chemicals. From it are made numerous dyes, drugs, rubber chemicals, antioxidants that stabilize all kinds of products — petroleum derivatives, animal feeds, rubber, plastics — photographic chemicals, etc.   It is not an exaggeration to state that aniline was the basis of Germany's domination of the world's organic chemical business until well into the twentieth century.

The demand for aniline during 1964 is broken down below by field of application.

Field of application	MM lbs, 1964
Rubber chemicals	103.0
Dyes and dye intermediates	25.0
Pharmaceuticals	10.0
Hydroquinone	15.0
Plastics	11.0
Miscellaneous	7.0
Total	171.0

Source: *Oil, Paint & Drug Reporter*,
December 15, 1965.

United States aniline producers are listed below. The only user of the chlorobenzene-ammonia route is Dow; all others employ reduction of nitrobenzene.

Producer	Location	1965 Capacity, MM lbs/yr
Allied Chemical	Moundsville, W. Va.	40.0
American Cyanamid	Bound Brook, N. J.	60.0
	Willow Island, W. Va.	40.0
Dow	Midland, Mich.	25.0
DuPont	Gibbstown, N. J.	100.0
First Chemical	Pascagoula, Miss.	50.0
Mobay	New Martinsville, W. Va.	50.0
Rubicon	Geismar, La.	35.0
Total		400.0

Source: *Oil, Paint & Drug Reporter*, December 15, 1965.

Vulcanization accelerators

Sulfur-vulcanized rubbers are usually cured in the presence of accelerators, without which vulcanization would take place at an uneconomically slow rate.

For many years, mercaptobenzothiazole (MBT) was the most important of all general-purpose vulcanization accelerators. With the advent of furnace blacks, however, a fundamental change was introduced. These carbon blacks, being much finer than the older channel blacks, not only were needed to impart better properties to synthetic rubbers, but to permit the incorporation of extension oils in

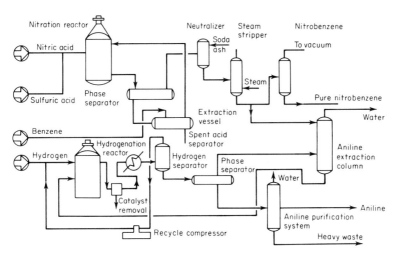

Fig. 8-27 Nitrobenzene-aniline.

higher proportions without impairing the properties of the finished products. Furnace blacks, however, have an alkaline surface pH, which activates the traditional accelerators to the point where they become too scorchy. While mercaptobenzothiazole is still the starting point for most of accelerators in use today, these are now mostly sulfenamides made by the reaction between MBT and an amine under various oxidizing conditions.

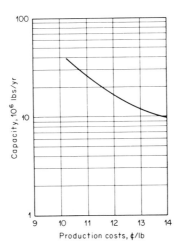

Fig. 8-28 Aniline.

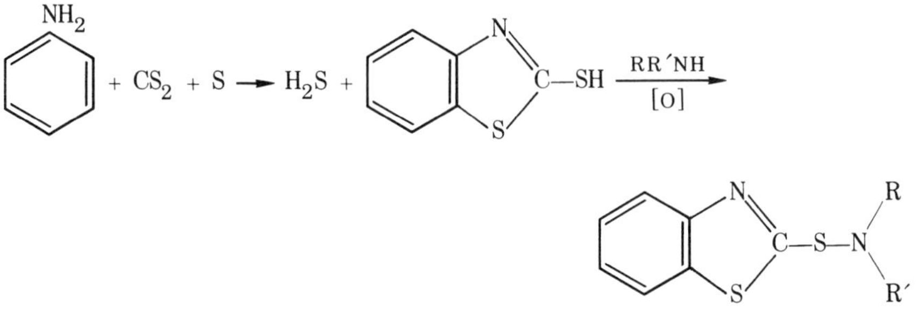

These accelerators are much safer than MBT itself. Another widely employed product is MBTS, made from two mols of MBT:

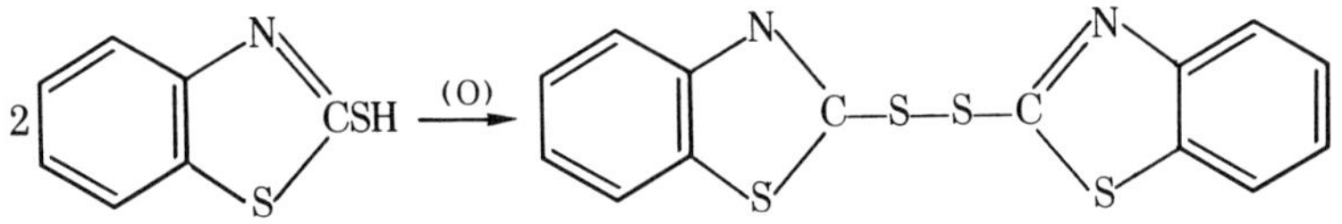

Guanidines, the most important example of which is diphenyl guanidine, are used as activators for the thiazole accelerators. They activate the vulcanization of thick articles and products requiring a high modulus, and in goods that are to be cured in hot air or live steam. DPG is made from aniline and cyanogen chloride

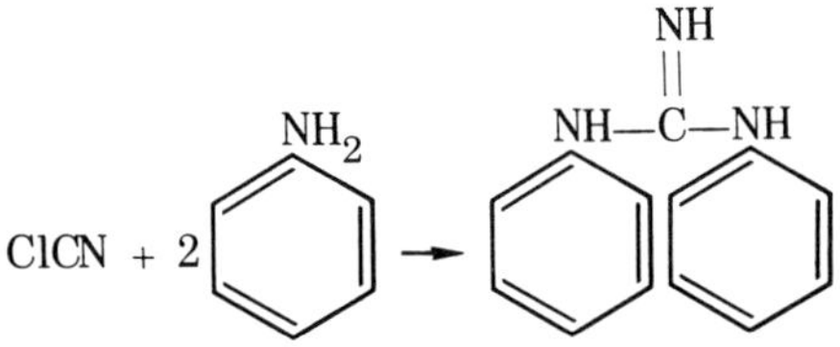

Dithiocarbamates are made from secondary aliphatic amines and CS_2

$$2R_2NH + 2CS_2 + ZnO \longrightarrow R_2N-\overset{\overset{\displaystyle S}{\|}}{C}-S-Zn-S-\overset{\overset{\displaystyle S}{\|}}{C}-NR_2$$

They are very active accelerators, and usually employed as their zinc salts. Sodium salts are employed when water-solubility is required, for example, in rubber latices.

Thiuram disulfides can be made by oxidizing sodium dithiocarbamates.

$$2R_2\overset{\overset{\displaystyle S}{||}}{N}C-SNa \xrightarrow{\ [o]\ } R_2\overset{\overset{\displaystyle S}{||}}{N}C-S-S-\overset{\overset{\displaystyle S}{||}}{C}-NR_2$$

They are used in combination with delayed-action accelerators, for example, the sulfenamides, to increase the curing speed while retaining most of the safety characteristics inherent in the use of thiazol derivatives. They are also used as vulcanization agents in the absence of sulfur to produce goods having excellent heat-aging characteristics.

Aldehyde-amine condensates are usually made from aniline with butyraldehyde or acetaldehyde as the aldehyde. However they are declining in importance.

The market for these accelerators in 1965 stood at around 107.0 million pounds, about 2.5 percent of the total rubber consumption.

	MM lbs, 1965
Thiazole derivatives	71.0
Guanidines	10.0
Thiocarbamates	13.5
Thiuram sulfides	7.5
Aldehyde-amine condensates	1.5
Others	3.5
Total	107.0

In general, the demand for accelerators grew rapidly during the period in which SBR, which requires 40 percent more accelerators than natural rubber, was gaining at the expense of the latter in tire application. Now that the SBR-to-natural rubber ratio has become fairly stable, the growth is reduced to about that of the population.

The demand in 1965 for mercaptobenzothiazole as such, and for its various derivatives, can be broken down as shown below.

	MM lbs, 1965
MBT	8.0
MBTS	20.0
MBTZn	3.0
2-(N-morphilinothio) benzothiazole	11.0
N-cyclohexyl-benzothiazole-2-sulfenamide	8.0
N-t-butyl-benzothiazole-2-sulfenamide and others	21.0
Total	71.0

The investment for a 5.0 million pounds per year MBT plant is $1.1 million, and production costs have been calculated for captive aniline at 10.0 cents per pound (Fig. 8-29).

	Cost, ¢/lb MBT
Aniline	6.8
Carbon disulfide	2.7
Other chemicals and solvent	4.1
Labor and overhead	1.6
Utilities	1.5
Capital charges	7.3
Total	24.0

Antioxidants

A number of important antioxidants for rubber and other materials are made from aniline.

Phenyl β-naphthylamine is made from aniline and β-naphthol.

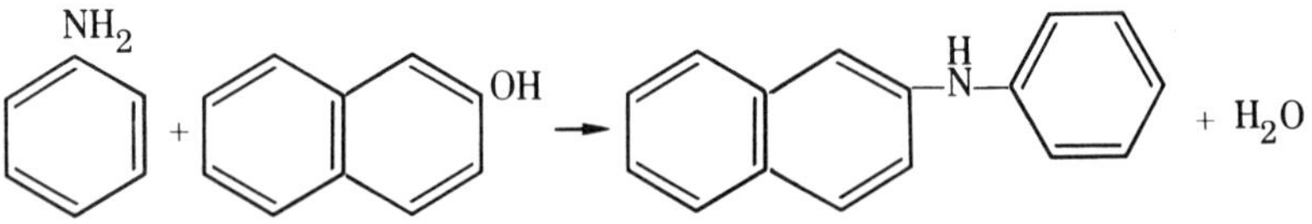

Annual consumption is around 30.0 million pounds, requiring some 15.0 million pounds per year of aniline. It is used mainly as an additive to stripped SBR latex before it goes on to coagulation and filtering.

Aniline reacts with α-naphthylamine in the presence of iodine to give phenyl α-naphthylamine, which has a lower melting point and better solubility than phenyl-β-naphthylamine.

Aniline is used to make a number of condensation products with aldehydes and ketones, for example, "Antox" (aniline-butyraldehyde) and "VGB" (aniline-acetaldehyde).

Amines are 75 percent of all rubber antioxidants, of which around 128.0 million pounds per year are consumed in the U.S. The production of antioxidants can be broken down as shown below.

Product		*MM lbs/yr, 1965*
All amines		98.0
Aldehyde- and acetone-amine	9.5	
Substituted *p*-phenylene diamines	41.5	
Other amines	47.0	
Phenolics and phosphites		30.0
Total		128.0

Source: "1965 Synthetic Organic Chemicals," U. S.
    Tariff Commission.

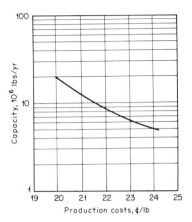

Fig. 8-29 Mercaptobenzothiazole
(captive aniline).

About 70 percent of all antioxidants are used in the manufacture of
tires.

Diphenylamine

Aniline hydrochloride, made by reacting aniline with hydrochloric acid, reacts with a second mol of aniline to form diphenylamine:

During wartime, large amounts are used to stabilize gunpowder,
but normally its main uses are in the rubber chemicals field.
N-Nitrosodiphenylamine is a retarder; diphenylamine-acetone condensation products are antioxidants known as BLE and "Aminox";
alkylated diphenylamines are the least staining of the amine antioxidants, and are therefore widely used as a compromise between
products of the diamine and hindered phenol types.

Diphenylamine can be fused with sulfur to produce phenothiazine.

This was formerly the most important of all livestock anti-helminthics; it has now been largely displaced by more active compounds. Diphenylamine itself is a constituent of many livestock smears used against screw-worms.

The total peacetime demand for diphenylamines was around 25.0 million pounds in 1964.

End-use	MM lbs, 1964
Antioxidants	15.0
Livestock products	6.5
Others	3.5
Total	25.0

Hydroquinone

Aniline sulfate can be oxidized to quinone, using manganese dioxide as the oxidant, and then hydrogenated to hydroquinone.

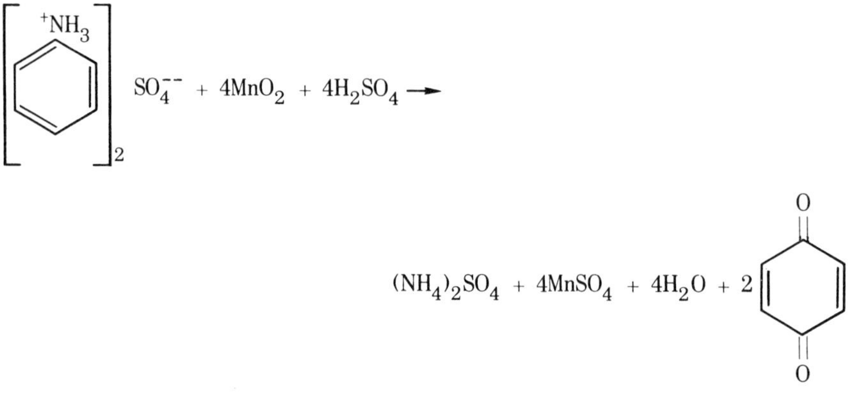

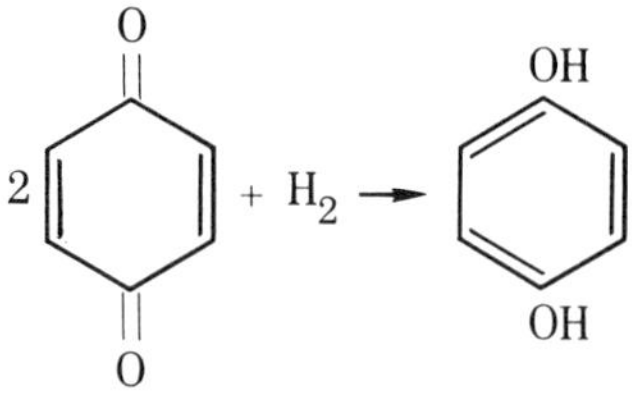

$$\left[\underset{}{C_6H_4 \cdot {}^+NH_3} \right]_2 SO_4^{--} + 4MnO_2 + 4H_2SO_4 \longrightarrow$$

$$(NH_4)_2SO_4 + 4MnSO_4 + 4H_2O + 2\, C_6H_4O_2$$

$$2\, C_6H_4O_2 + H_2 \longrightarrow C_6H_4(OH)_2$$

Although other routes to hydroquinone have been described, for example, via p-diisopropyl benzene in a manner similar to the phenol-from-cumene process, aniline oxidation continues to be the only process in use. A new plant using this method was built as late as 1962. The main problem with the hydroperoxide route is the high cost of purification; Signal Oil has developed an economically viable process via this route, but by 1967 it had not yet been used commercially.

The best-known use for hydroquinone is as a photographic developer; the main manufacturer in the U.S. is Eastman, which produces about 50 percent of the entire world demand for photography chemicals. Technical-grade hydroquinone is produced to the extent of some 1.5 times the photographic grades, including in this latter category the amount that goes into the manufacture of other developers.

Hydroquinone and its alkylated derivatives are widely used as polymerization inhibitors in monomers that must be stored and shipped. The usual concentration level is around 200 ppm, but less can be used if it is known that the monomer is not expected to have a long shelf-life; the rate of polymerization can thereby be improved considerably.

The monomethyl ether of hydroquinone, obtained by reaction with dimethyl sulfate, can be alkylated with isobutylene to t-butyl-p-hydroxyanisole, also known as BHA. This is a well-known food antioxidant, used especially for imparting oxidation resistance (at frying temperatures) to oils, potato chips and other materials. The demand in 1965 was around 3.0 million pounds.

An important use for hydroquinone is the manufacture of N,N' diphenyl p-phenylenediamine, an antioxidant and antiozonant for rubber and gasoline. It is obtained by reaction with two mols of aniline. The total production capacity for hydroquinone has been estimated to be about 20.0 million pounds per year with sales at about 14.0 million pounds per year.

End-use	1966, MM lbs
Antioxidants	6.0
Photographic chemicals	5.0
Miscellaneous	3.0
Total	14.0

Source: *Oil, Paint & Drug Reporter*,
April 17, 1967.

The above estimates probably are low, particularly for photographic chemicals. Until recently, the use of hydroquinone as a developer in photography was fairly static since its usuage was limited only to black-and-white photography. However, most of the incremental demand has been for color film. A much smaller quantity of hydroquinone is required for development of color film.

Antioxidants and antiozonants are the main growth areas for hydroquinone. U.S. producers are given below.

Producer	Location	Capacity, MM lbs/yr
Carcus Chem.	LaSalle, Ill.	4.5
Eastman	Kingsport, Tenn.	13.0
Manganese Chem.	Baltimore, Md.	2.5
Total		20.0

Source: Based on *Oil, Paint & Drug Reporter*, April 17, 1967.

Acetanilide

Acetanilide is made by the reaction between aniline and glacial acetic acid.

The p-isomer of the reaction product between acetanilide and chlorosulfonic acid is the starting point for the production of sulfa drugs.

The chlorine atom reacts with the amino-hydrogen of several amines, forming such drugs as sulfapyridine, sulfadiazine, and sulfathiazole; this is followed by hydrolysis of the acetyl group.

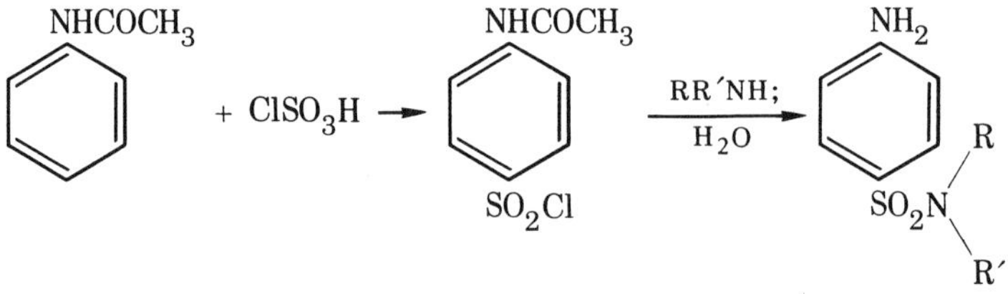

The total production of sulfa drugs in the U.S. was 5.0 million pounds in 1965, requiring around 4.0 million pounds of acetanilide. The major portion of these drugs goes into veterinary applications as feed supplements.

Aniline-formaldehyde resins

Amino resins based on aniline are reported to have electrical properties superior to those of any other. Roughly 2.5 million pounds per year of aniline go into these resins in the U.S.

Dicyclohexylamine

The production of cyclohexylamine by hydrogenation of aniline, as practiced in the U.S. by Abbott Laboratories, gives dicyclohexylamine as a byproduct:

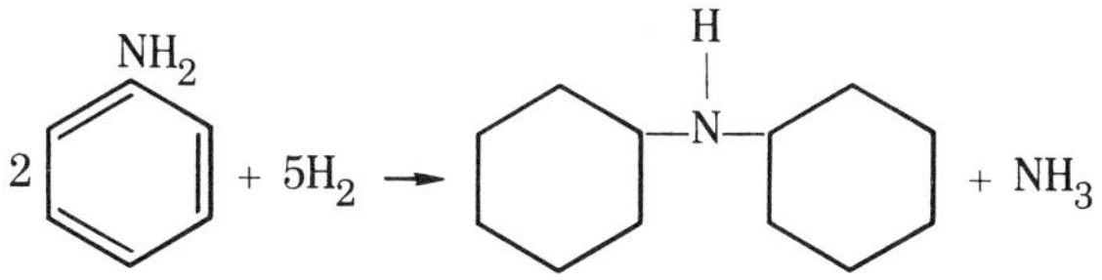

$$2 \;\; \text{(benzene ring with NH}_2\text{)} + 5H_2 \longrightarrow \text{(dicyclohexylamine)} + NH_3$$

The demand is around 3.0 million pounds per year, mainly for production of fatty acid salts used as vapor-phase corrosion inhibitors. They act by neutralizing the CO_2 present in steam condensate.

9. RESORCINOL

Resorcinol is obtained by sulfonating benzene to benzene disulfonic acid and hydrolyzing.

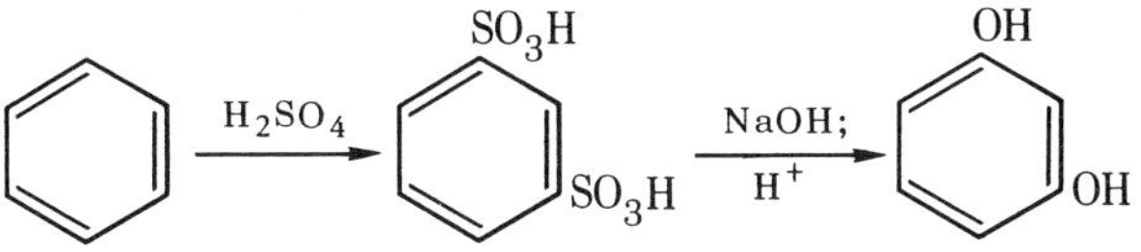

$$\text{(benzene)} \xrightarrow{H_2SO_4} \text{(benzene with SO}_3\text{H and SO}_3\text{H)} \xrightarrow[H^+]{NaOH;} \text{(benzene with OH and OH)}$$

Until 1965, Koppers was the only large-scale producer of resorcinol in the U.S., with a 15.0 million pounds per year plant at Petrolia, Penn. A second plant was built by U.S. Pipe and Foundry at Birmingham, Ala., with a capacity of around 5.0 million pounds per year. This company was previously a producer of benzene sulfonic acids for use as corrosion inhibitors in the steel industry.

Resorcinol, like phenol and cresols, forms condensation polymers with formaldehyde. The best-known application for these resins is in the manufacture of tire-cord adhesives for rayon and nylon, usually mixtures of a butadiene-styrene-vinylpyridine terpolymer and a resorcinol-formaldehyde copolymer. Resorcinol resins are also used for making wood laminates, especially when room temperature curing is required. The fact that softwood is beginning to be used as a plywood raw materials is certain to cause an increase in demand for these resins; they are also used as the bonding agents for sporting goods and structural beams. Outside the plywood field, resorcinol-formaldehyde adhesives are used to bond vinyls to metals. Since resorcinol is much more expensive than phenol, mixtures of phenol and resorcinol resins are used instead of pure resorcinol-formaldehyde whenever possible.

Resorcinol is also used in the manufacture of monoacetate (by reaction with acetic anhydride), an anti-dandruff agent in hair lotions; β-methyl umbelliferone, obtained from resorcinol and ethyl acetoacetate, an intermediate in the manufacture of optical bleaches used in soap; and of eosine and other lipstick dyes.

Several resorcinol derivatives have been developed as UV absorbers for vinyls and polyolefins. They are the 2,4-hydroxy (or alkoxy) benzophenones.

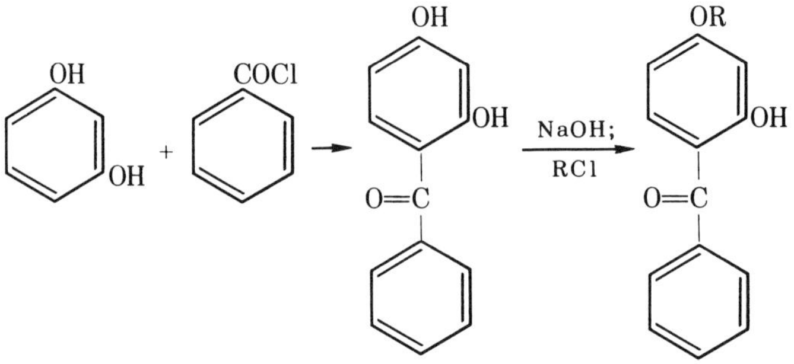

One example is 2-hydroxy-4-methoxy-benzophenone. The use for these absorbers is growing at the expense of the cheaper, but less permanent, aryl- or alkylaryl-salicylates. The demand for these substituted benzophenones in 1965 was around 0.8 million pounds, of which 30 percent were PVC, 25 percent, polyolefins and 20 percent, polyesters, and minor amounts in cellulosics, polystyrene and in such nonplastic applications as cosmetics and surface coatings. In the polyester resins, their function is to stabilize products made from chlorinated dicarboxylic acids such as chlorendic anhydride, the use of which is growing in the manufacture of flame-resistant polymers.

The estimated demand for resorcinol in 1966 was 10.0 million pounds, with the following breakdown:

Product	MM lbs/yr, 1966
Adhesives	3.5
Resins	3.0

Dyestuffs	1.5
Pharmaceuticals	1.0
Others	1.0
Total	10.0

Source: Based on *Chemical Week*, June 24,
1967; *Chemical and Engineering
News*, June 26, 1967.

10   BENZENE HEXACHLORIDE

Benzene can be chlorinated in the presence of actinic light to give BHC.

Of several possible stereo-isomers, only one, the so-called gamma-isomer, is active as an insecticide. The γ-isomer content of the crude product is about 14 percent, but this is usually concentrated to 40 percent. The active isomer can be isolated by fractional crystallization. Since one of the remaining isomers has a violently unpleasant odor, this separation is accomplished when BHC is to be used in household products or other products such as livestock smears and fumigants, where the odor cannot be tolerated.

Benzene hexachloride is used mainly in fighting cotton pests, particularly, the boll weevil. Despite its very low price, most of its market lately has been taken over by the parathions, as cotton pests have, over the years, developed immunities to chlorinated poisons. About 7.0 million pounds per year are used, on a gross basis, equivalent to 2.0 million pounds per year active isomer. The U.S. manufacturers are Hooker, Diamond, and Pittsburgh Plate Glass, all major caustic-chlorine producers.

The BHC remaining after γ-isomer extraction is of no use as an insecticide, and can be cracked to HCl and 1,2,4-trichlorobenzene, a solvent and intermediate for pentachloronitrobenzene, a soil fungicide, higher chlorobenzenes, and chlorophenols.

11   SIDE-CHAIN CHLORINATION DERIVATIVES OF TOLUENE

Chlorination of toluene in the presence of light causes substitution to take place on the side-chain.

The monosubstituted compound, benzyl chloride, is the most important of the three substituted compounds. Its main use is in the manufacture of butylbenzyl phthalate, of which nearly 50.0 million pounds per year are used as a vinyl chloride plasticizer. In 1964, consumption of benzyl chloride was 38.0 million pounds (see below for breakdown).

End-use	MM lbs, 1965
Butylbenzylphthalate	21.0
Miscellaneous captive	11.0
Open market	7.0
Total	39.0

Source: *Oil, Paint & Drug Reporter*, January 20, 1964.

Manufacturers of toluene side-chain chlorination products are listed below. Capacities are given in terms of benzyl chloride, but almost all producers make the three substitution products. The main captive outlets for benzyl chloride are indicated.

Producer	Location	1964 Capacity, MM lbs/yr	Main captive use
Benzol Products	Nixon, N.J.	8.0	Phenylacetic acid, benzyl alcohol
Grace (Hatco)	Fords, N.J.	10.0	Butylbenzyl phthalate
Tenneco	Fords, N.J.	5.0	—
Hooker	Niagara Falls, N.Y.	5.0	—
Monsanto	Bridgeport, N.J.	20.0	Butylbenzyl phthalate
	Monsanto, Ill.	5.0	—
Velsicol	Chattanooga, Tenn.	4.0	Benzyl alcohol
Total		57.0	

Source: *Oil, Paint & Drug Reporter*, January 20, 1964.

11.1 Butylbenzyl Phthalate

This plasticizer is made by mixed esterification of phthalic anhydride with butanol and benzyl chloride. It is used mainly for PVC, especially for flooring and plastisols. In flooring it has the advantage of making possible a larger asbestos-to-resin ratio than other phthalate plasticizers. These uses make up 90 percent of the total demand, which, in 1965, had been around 50.0 million pounds per year. The remaining 10 percent go into vinyl acetate coatings and adhesives, and nitrocellulose coatings.

11.2   Benzyl Alcohol

Benzyl chloride can be hydrolyzed to the respective alcohol.
Either as the alcohol itself, or in the form of its esters, about
4.0 million pounds per year of benzyl alcohol are consumed by the
perfume industry. It can be esterified to benzyl acetate by acetic
anhydride, which is used as a low-cost perfume in the soap industry;
annual consumption is 1.3 million pounds.

11.3   Phenylacetic Acid

Benzyl chloride can be reacted with sodium cyanide and subsequently hydrolyzed to phenylacetic acid:

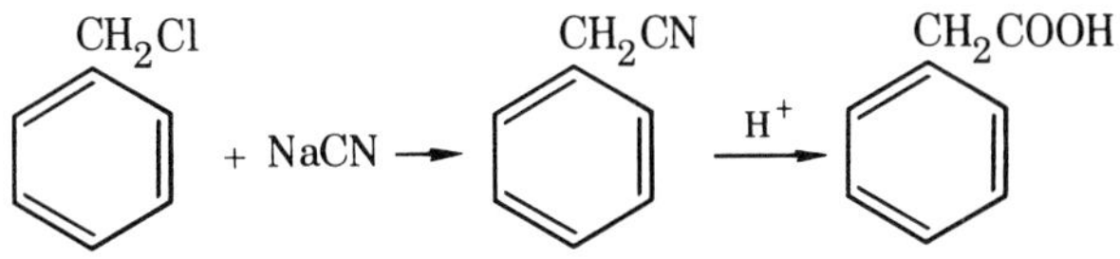

Among its many important uses in the pharmaceutical field,
phenylacetic acid is the precursor for penicillin G, which accounts
for 80 percent of the penicillin made in the U.S. For the 1.0
million billion units (BU) a year consumed, some 1.2 million pounds
per year of phenylacetic acid were required. Another application
is as a starting point for making amphetamine, a stimulant better
known as "benzedrine." Although several routes are possible, the
best-known is still that starting from phenylacetic and acetic acids.

Benzyl cyanide is also a starting point for making "Phenobarbital," a barbiturate of which 0.3 million pounds per year are
used in the U.S. Several routes exist, but all begin with benzyl
cyanide.

11.4   Benzaldehyde

The substitution of two side-chain hydrogen atoms by chlorine
gives benzalchloride.

Apart from some application as a dye intermediate, the principal use for this product is its hydrolysis to benzaldehyde.

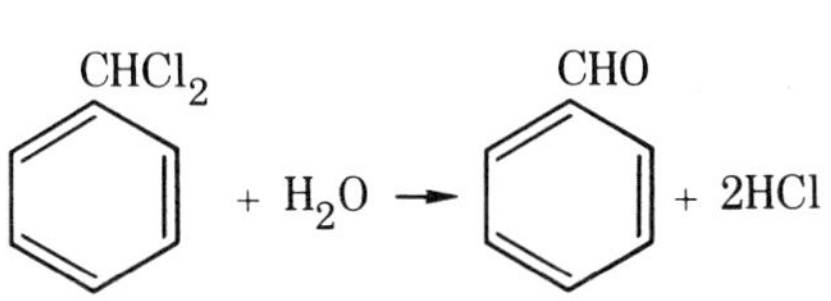

Benzaldehyde can also be made by direct oxidation of toluene, using molybdenum oxide as a catalyst.

This route has the advantage of giving a chlorine-free product.

In 1965, 4.0 million pounds of benzaldehyde were used in the U.S. Principal applications are as intermediates for certain perfumery chemicals, such as cinnamaldehyde, of which 900,000 pounds per year are used, and benzalacetophenone.

It is an intermediate for certain triarylmethane dyes, which consume another 800,000 pounds per year; and for making benzyl benzoate, which was a very important product during World War II as an insecticide against certain disease carriers encountered in the Pacific. This is still probably the most important outlet for benzaldehyde.

11.5   Benzotrichloride

The chief use of benzotrichloride is in the manufacture of benzoic acid

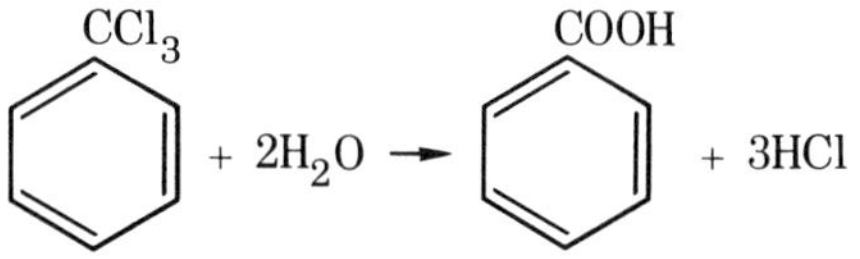

$$\text{C}_6\text{H}_5\text{CCl}_3 + 2\text{H}_2\text{O} \longrightarrow \text{C}_6\text{H}_5\text{COOH} + 3\text{HCl}$$

Benzoic acid reacts with benzotrichloride, giving benzoyl chloride:

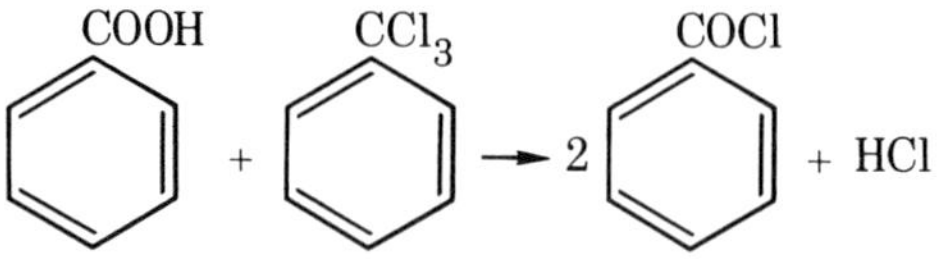

$$\text{C}_6\text{H}_5\text{COOH} + \text{C}_6\text{H}_5\text{CCl}_3 \longrightarrow 2\ \text{C}_6\text{H}_5\text{COCl} + \text{HCl}$$

It is a dye intermediate for several members of the anthraquinone and stilbene classes, but its most important outlet is in the manufacture of benzoyl peroxide

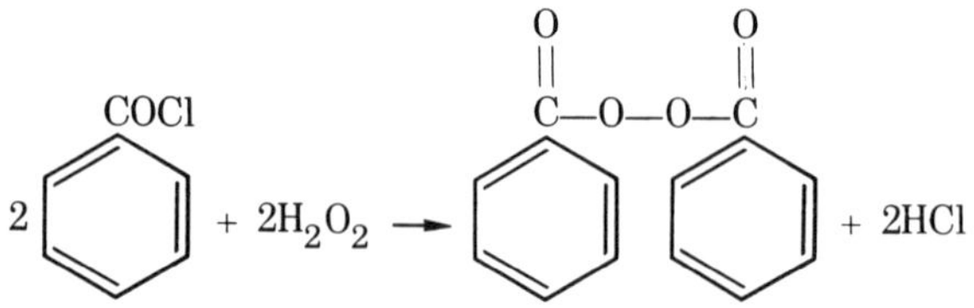

$$2\ \text{C}_6\text{H}_5\text{COCl} + 2\text{H}_2\text{O}_2 \longrightarrow \text{C}_6\text{H}_5\text{C(O)-O-O-C(O)C}_6\text{H}_5 + 2\text{HCl}$$

In addition to some use in flour bleaching and vegetable oil purification, benzoyl peroxide is the most widely used of the numerous peroxides employed to initiate polymerizations of the "vinyl" type, that is, those which take place along a double-bond. Consumption was close to 3.0 million pounds in 1965, chiefly for making polystyrene, polyesters and polyacrylates. Because benzoyl peroxide is extremely hazardous in pure form, it is sold in non-separating dispersions in various diluents such as tricresyl phosphate, or, more recently, in stable mixtures with water.

The demand can be broken down as shown in the table below.

End-use	MM lbs, 1965
Polyesters and acrylics	1.5
Polystyrene	1.0
Polyvinyl acetate and others	0.5
Total	3.0

Benzotrichloride is also an intermediate for some anthraquinone dyes.

12   BENZOIC ACID

There are several routes to benzoic acid, the cheapest being direct oxidation of toluene.

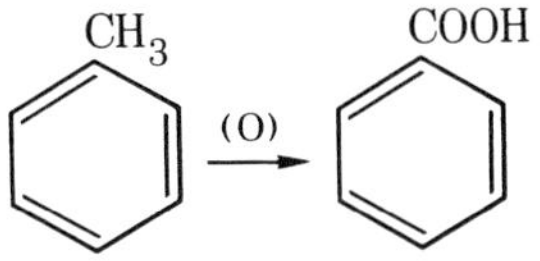

Some producers still employ the chlorination route, which involves hydrolysis of benzotrichloride. This is the approach used by companies that also make derivatives of the other side-chain chlorination products. This process has the disadvantage of giving a product unsuitable for use in food, the main outlet for benzoic acid and its salts.

Finally, a process exists for decarboxylating phthalic anhydride with steam. The final product remains contaminated with unconverted raw material, and purification is difficult; it is particularly suited for use in alkyd resins and similar fields, where phthalic anhydride is part of the formulation. The process has almost been abandoned in the U.S. since production costs via any of the other two routes are much lower.   Economics of the toluene oxidation route are evaluated below.   The investment for a 5.0 million pounds-per-year plant is $0.8 million, and production costs are shown below.

	Cost, ¢/lb benzoic acid
Toluene	2.4
Other raw materials	0.6
Labor and overhead	4.7
Utilities	0.5
Capital charges	5.3
Total	13.5

Producers of benzoic acid in the U.S. are listed below, with their respective routes.

Producer	Location	Capacity, MM lbs/yr	Routes
Monsanto	St. Louis, Mo.	10.0	Oxidation, phthalic anhydride decarboxylation
Velsicol	Chattanooga, Tenn.	10.0	Chlorination
Tenneco	Garfield, N.J.	8.0	Oxidation
Vulcan	Newark, N.J.	4.0	Oxidation
Pfizer	Terre Haute, Ind.	2.0	Oxidation
Total		34.0	

Source: *Oil, Paint & Drug Reporter*, June 12, 1967.

The principal use for benzoic acid is, in the form of its sodium salts, as a food preservative in canned goods and fruit beverages. Its main competitors are not only other chemicals such as sorbic acid, but also processing techniques such as pasteurization and vacuum packing.   As a modifier for alkyd resins, benzoic acid improves hardness and resistance to alkaline hydrolysis and reduces drying time; p-tert-butylbenzoic acid is also used in this application. Several vinyl plasticizers, such as glycol dibenzoates, are used to lower processing times and temperatures; another plasticizer is sucrose benzoate, used in extruded PVC to lower the viscosity without increasing softness, which cannot be tolerated in products such as pipe.

Benzoic acid can be converted to benzonitrile, which reacts with dicyandiamide to form benzoguanamine.

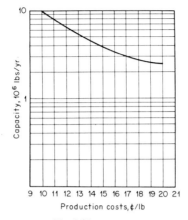

Fig. 8-30 Benzoic acid.

This is used as a comonomer with melamine and formaldehyde to make stain-resistant molding powders.   In Japan, benzonitrile is made by direct ammoxidation of toluene.

Yields of this type of reaction are as high as 95 percent.

Benzoic acid is a dye carrier for polyester fiber coloring and has some use as a dye intermediate.

The demand for benzoic acid in the U.S. which can be considered fairly stable was 18.0 million pounds in 1966, distributed as follows:

End-use	MM lbs, 1965
Benzoic acid for salts	8.0
Plasticizers	6.5
Alkyd resin modifier	1.7
Pharmaceuticals and flavoring	0.5
All others	1.3
Total	18.0

Source: *Oil, Paint & Drug Reporter*, June 12, 1967.

13   TOLUENE SULFONYL CHLORIDES

Toluene reacts with chlorosulfonic acid to form a mixture of *o*- and *p*-toluene sulfonyl chlorides, which can be separated by fractional crystallization

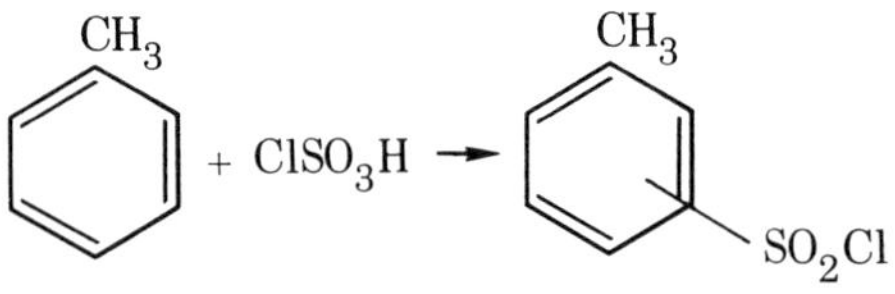

Both isomers can be ammoniated to the respective toluenesulfonamide. *o*-Toluenesulfonamide is one starting point for making saccharin.

Mixtures of both sulfonamides can be used as cheap plasticizers for cellulosic molding powders.

Tolbutamide, made from *p*-toluenesulfonamide, *n*-butylamine and phosgene, is an oral antidiabetic for adult patients.

Monsanto is the largest producer of toluenesulfonamides and their derivatives in the world and supplies *o*-toluenesulfonamide to other saccharin producers.

13.1 Saccharin

Monsanto, the major producer of this synthetic sweetener in the U.S., uses the classical route to saccharin via *o*-toluenesulfonamide.

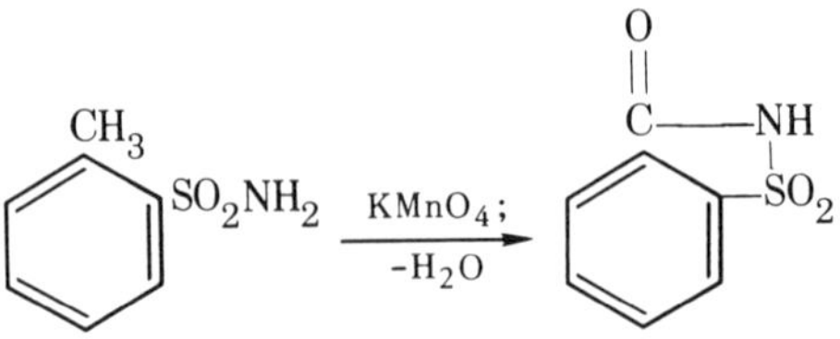

The second most important producer, Maumee Chemicals, starts from phthalic anhydride via anthranilic acid.

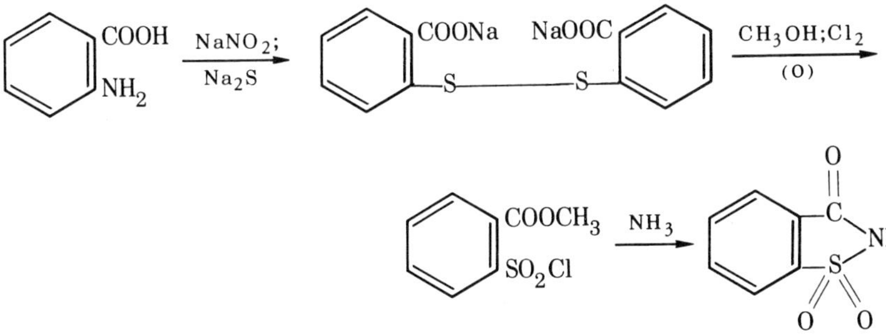

This route has the advantage of eliminating the costly separation of the two chlorosulfonation isomers. The yield from anthranilic acid is around 60 percent.

Saccharin, at around $1.50 per pound, is equivalent to sugar at 0.38 cents per pound. Since this is about one-seventh the lowest price sugar has reached since 1945, it is easy to see that synthetic sweeteners have a very great potential. The U.S. manufacturers of saccharin are given below.

Producer	Location	Capacity, MM lbs/yr
Monsanto	St. Louis, Mo.	3.0
Maumee	St. Bernard, Ohio	1.5
Norse	Cudahy, Wis.	0.5
Pillsbury	Painesville, Ohio	0.5
Total		5.5

Source: *Oil, Paint & Drug Reporter*, November 7, 1966.

Consumption in 1965 was 3.0 million pounds.

Unlike the cyclamates, which find most of their outlets in the food processing field, saccharin is sold mostly via retail outlets for individual sweetening requirements. It leaves a somewhat acrid aftertaste, especially that made via o-toluenesulfonamide; consequently, sweetening formulations are usually a compromise mixture of saccharin and cyclamates, which are more expensive per sugar equivalent but better tasting. About one-third of the saccharin output ends up in industrial applications such as tobacco flavoring and toothpaste.

14   NITROTOLUENES

Toluene can be nitrated in mixed acid in much the same way used to make nitrobenzene. The monosubstituted derivatives are predominantly the o- and p-isomers.

o-Nitrotoluene can be reduced to o-toluidine, an intermediate for thioindigoid and other dyes, and also for "Permalux," a rubber chemical. Under different conditions, reduction yields o-tolidine (a homolog of benzidine), an important intermediate for the production of orange and purple disazo dyes. 5-Nitro-o-toluene sulfonic acid is a dye intermediate of which some 8.5 million pounds per year are refined in the U.S.

p-Nitrotoluene can also be reduced to the respective toluidine, which can be nitrated again to m-nitro-p-toluidine, an intermediate for making such important pigments as Hansa Yellow and Toluidine Red, of which 1.0 and 2.5 million pounds per year, respectively, are consumed in the U.S. Other dye intermediates derived from p-toluidine are o-chloro-p-toluidine (also a starting point for making "Solan," a post-emergence herbicide used in tomato-beds); and the m-sulfonic acid of p-nitrotoluene, used to make a number of azo pigments.

Oxidation of p-nitrotoluene, followed by reduction of the nitro group, gives p-aminobenzoic acid (PABA). It is an intermediate for the production of benzocaine, novocaine and other well-known local anesthetics. p-Aminobenzoic acid can be converted to 4-carboxyphenylhydrazine, an intermediate for several optical bleaches used as brighteners in soap; PABA also has other uses as a dye intermediate, and is a starting point for making folic acid. The total demand for p-aminobenzoic acid is around 0.7 million pounds per year.

One of the most important derivatives of p-nitrotoluene is aminostilbene, a dye intermediate especially for optical bleaches. This class of dyes is being used increasingly as a substitute for chlorine bleaches. In 1965, about 19.0 million pounds of these dyes were used, of which 12.0 million pounds were in soap, 5.0 million pounds in paper and most of the remainder in textiles. Demand for stilbenesulfonic acid was around 4.0 million pouds per year.

$$2 \underset{CH_3}{\overset{NO_2}{\bigcirc}} \xrightarrow{(H)\,;\,(O)} H_2N \bigcirc CH{=}CH \bigcirc NH_2$$

14.1   Explosives

Toluene can be nitrated in the 2, 4 and 6 positions to trinitro-toluene, known as TNT, a well-known military explosive.

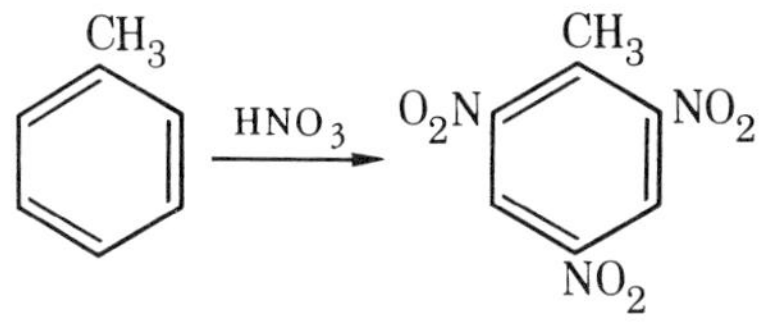

$$CH_3 \quad \xrightarrow{HNO_3} \quad O_2N-C_6H_2(CH_3)(NO_2)-NO_2$$

Nitration is carried out continuously, which greatly reduces the inventory of explosive material in the system compared with older batch techniques.

The output of TNT varies primarily with military conditions.

15   OTHER TOLUENE DERIVATIVES

p-Chlorotoluene, obtained by chlorination of toluene, can be reacted with sodium methylate to form anethole.

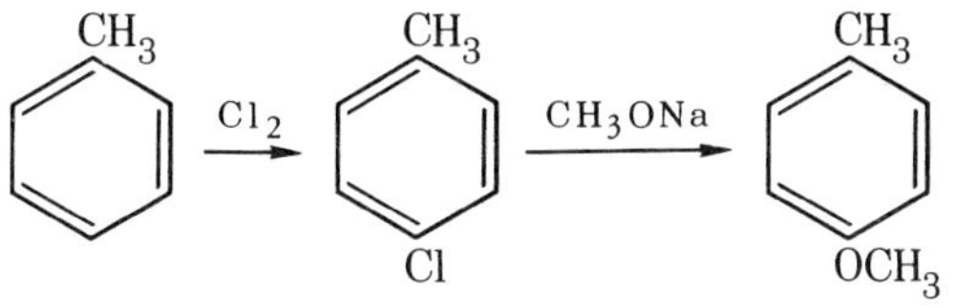

$$CH_3 \quad \xrightarrow{Cl_2} \quad CH_3-C_6H_4-Cl \quad \xrightarrow{CH_3ONa} \quad CH_3-C_6H_4-OCH_3$$

Its main use is in the perfume industry, which consumes some 20 million pounds per year.

Chlorination of the side-chain to p-chlorobenzotrichloride and subsequent fluorination gives p-chlorobenzotrifluoride.

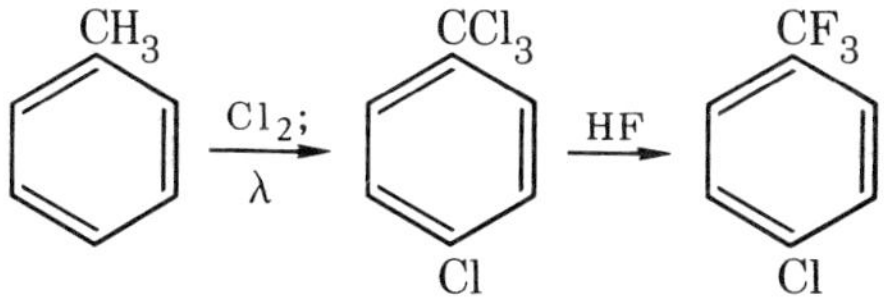

$$CH_3 \quad \xrightarrow[\lambda]{Cl_2;} \quad CCl_3-C_6H_4-Cl \quad \xrightarrow{HF} \quad CF_3-C_6H_4-Cl$$

It is an intermediate for several important orange azo dyes, and for certain pesticides. Herbicides made from this intermediate are being used against crab grass in gardens and in commercial crops such as cotton and soybeans. "Trifluralin" (Lilly) is an example.

Toluene can be alkylated with isobutylene and then oxidized. The resulting *p-t*-butylbenzoic acid is used instead of benzoic as an alkyd resin modifier, especially to control the rate of reaction. The main producer is Shell, and the demand is believed to be around 3.0 million pounds per year.

There are several important herbicides related to benzoic acid on the market. They are made by first carrying out the necessary benzene-ring substitutions, and then oxidizing. One example is TBA (trichlorobenzoic acid).

The manufacture of vinyl toluene is similar to that of styrene; the main producer is Dow Chemical. It is used instead of styrene in modified alkyd resins; in addition to being cheaper, it produces faster-drying coatings. Use in 1965 was 30.0 million pounds.

16 MIXED XYLENES

In addition to a number of chlorinated plasticizers and herbicide intermediates, mixed xylenes are used to make two large classes of products. Nitration and subsequent reduction give xylidenes; during World War II, as much as 350.0 million pounds per year of these compounds were used as anti-knock additives for aviation gasoline, but this has diminished considerably with the advent of jets.

The sulfonic acid salts of xylene and toluene are an important group of compounds known as hydrotropes. Since the builders that are mixed with active material in liquid household detergents are not very soluble in water, it is necessary to promote the solubility of the mixture by addition of what are in effect cloud-point depressants. Heavy-duty liquids require 5 to 10 percent of these hydrotropes; solid detergents require lower concentrations, but since they are sold in much larger quantities most of the hydrotropes end up in these formulations. Solid hydrotropes have several liquid competitors, of which ethyl alcohol is the most important; around 11.0 million gallons per year are still used for this purpose. Glycols are also employed. There are two opposite trends at work affecting the demand for hydrotropes. On the one hand, liquid formulations are gaining in popularity much faster than solids, which are actually expected to decline in volume during this decade. However, on the other hand, the so-called "third-generation" raw materials ("hard alkylate" being the first, "soft alkylate" the second), especially linear alcohol ether sulfates, are much more soluble than alkyl aryl sulfonates, and can be used to make heavy-duty liquid formulations that dispense with solubilizing agents. Furthermore, the public is beginning to accept opaque liquid detergents, freeing manufacturers from the problem of producing clear formulations. These opaque liquids already account for 50 percent of the 950.0 million pounds per year light-duty liquid detergent market.

Of the 80.0 million pounds of hydrotropes consumed in 1965, 22.0 million pounds were in the form of the sodium salt of xylene sulfonic acid; sodium toluene sulfonate was next in importance, potassium and ammonium salts making up the remainder. Total xylene requirements for the manufacture of hydrotropes was about 20.0 million pounds per year.

The production of hydrotropes requires several steps: sulfonation with oleum, removal of moisture and unconverted raw material (which would, if present in the product, diminish solubility considerably) and, finally, neutralization with the required base.

17 PHTHALIC ANHYDRIDE

It may seem unconventional to discuss phthalic anhydride under *o*-xylene, since, at least in the U.S., about 80 percent of the capacity is still based on naphthalene. However, there is a definite trend in favor of vapor-phase oxidation of *o*-xylene. Practically all the new plants built since 1962 were based on fixed-bed oxidation of *o*-xylene instead of on the inherently cheaper fluid-bed process, mainly because the latter operates on more expensive petroleum-derived naphthalene. This trend has been revised by the announcement of a catalyst for making phthalic anhydride from *o*-xylene in a fluid-bed reactor, thereby simultaneously placing within reach of present fluid-bed operators, as well as of future plants, the advantages of fluid-bed operation and that of using a cheaper raw material.

The oxidation reactions of *o*-xylene and naphthalene, respectively, are as follows.

$$\text{(naphthalene)} + \tfrac{9}{2}O_2 \longrightarrow \text{(phthalic anhydride)} + 2CO_2 + 2H_2O$$

Initially, when the only conceivable raw material for phthalic was coal-tar naphthalene, the process was carried out in fixed-bed reactors, using vanadium pentoxide as a catalyst. The earlier plants condensed the product contained in the gaseous stream in a number of large condensation chambers, known as "hay barns." This was considered necessary because phthalic anhydride condenses as a solid, making recovery in conventional heat-exchange equipment impractical. Later, "switch-condensers," which at first were fin-tube condensers hooked up in parallel with water flowing inside the tubes, were introduced. At regular intervals, steam replaced the water in the tubes in order to melt the product, which flowed on to the recovery section. This early type of switch-condenser had the disadvantage of developing leaks as a result of repeated thermal shocks. Better designs were developed, and since have become the standard method of recovering phthalic. Several old plants, in fact, have converted their recovery systems from hay barns to switch-condensers; at present, the preferred method is indirect heat-exchange via a circulating oil system with external heat removal instead of the earlier steam-water alternation.

There are also two approaches to reactor design. In the U.S., where raw materials have always been abundant and cheap, there has been a tendency towards oxidation at high temperatures, at the expense of yields but at the same time requiring a smaller reactor for a given output as well as a reduced catalyst charge. This amounts to a difference of 15 percent in investment costs. In Europe, on the other hand, preference has been given to low-temperature, high-residence-time oxidation, which makes for a safer plant (reactor effluent remains outside the explosive limits of the phthlic anhydride-air mixture) and better yields, but a higher initial investment. The reaction being highly exothermic, reactors must be cooled; evolution has been from mercury-cooled reactor tubes to circulating fused-salts baths, which not only require a lower initial investment (especially after mercury prices began rising in 1963) but also allow for safer waste-heat recovery than the old-fashioned mercury boilers.

The most widely used fixed-bed process, which can be designed for either naphthalene or o-xylene, is that of von Heyden. A flow-sheet is shown in Fig. 8-32. After reaction and condensation,

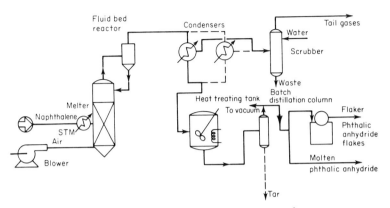

Fig. 8-31 Phthalic anhydride (fluid-bed process).

molten product goes to a distillation system. All plants built up to 1965 used vacuum batch distillation, but the possibility of continuous purification was being considered. At first, distillation is carried out at total reflux, while light impurities condense or polymerize to products heavier than phthalic anhydride. At this point, product begins to be withdrawn overhead. Flaking completes the process, although in the U.S. most merchant phthalic is delivered molten, in tank-cars.

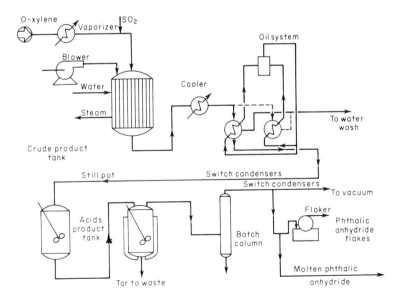

Fig. 8-32 Phthalic anhydride (von Heyden process).

The most important fluid-bed process is that developed by Sherwin-Williams and Badger (see Fig. 8-31). The catalyst is also vanadium pentoxide, but has the form of a finely divided powder. There are several advantages to this process. Fluid-bed reactors can be operated at fairly uniform temperatures, since the bed is well mixed and catalyst can be continuously withdrawn to remove the exothermic heat of reaction. This uniformity of temperature-distribution in turn allows operation at a lower air-to-feed ratio, about 12:1 instead of the 30:1 ratio required by the von Heyden process. This would seem to indicate lower utility requirements, but proponents of the fixed-bed version claim that higher pressure drops in the fluid reactor make up in cost for the lower throughput. One definite advantage of the fluid-bed technique is to be found in the recovery section, since the partial pressure of phthalic anhydride in the effluent stream is more than twice as high and condensation that much easier. Yields are better, less byproducts are formed and much of the equipment common to both processes is smaller, although the reactor itself is almost as large as an American-type, fixed-bed unit, because of the longer residence times. A fluid-bed plant requires equipment missing from the fixed-bed process flowsheet, particularly the provisions for catalyst recovery.

Although the fixed-bed process can run on unrefined coal-tar naphthalene, the fluid-bed catalyst has a very low sulfur tolerance and must, since a catalyst for fluid-bed o-xylene processing had not been proven until 1965, be run on refined or petroleum naphthalene. There are two sources of raw material meeting these specifications: naphthalene from coal-tar can be purified with metallic sodium, at an additional cost of 1.5 cent per pound, or hydrogen treated, which requires considerable capital investment and involves losses of yield due to ring hydrogenation; or petroleum-derived naphthalene. In either case, the raw material would cost more than would be warranted by the low prices at which o-xylene was available between 1963 and 1966, when the fluid-bed process found no new clients. Fluid-bed processes may once again become the choice for large new units if a catalyst for o-xylene is proven successful.

The reason for the difficulty in processing o-xylene is that much more heat is released per lb of product than with naphthalene. Yields by weight for the von Heyden process are around 82 percent using naphthalene and 92 percent using o-xylene. Fluid-bed yields are 95-100 percent. Since the theoretical yield by weight is over 20 percent higher from o-xylene than from naphthalene, this means that the stoichiometric yield from the former is actually much lower, and that consequently a greater portion is oxidized directly to CO_2 and water. The heat that must be removed per unit of product is thus higher when oxidizing o-xylene.

The only situation in which processing o-xylene may be uneconomical, even in view of the favorable price-differential, would be that of an already existing fixed-bed unit designed for naphthalene oxidation. Since its heat-removal capacity is fixed, its capacity would be reduced by 20 percent when running on o-xylene. For a new unit, the incremental investment involved in designing additional heat-removal facilities would be insignificant compared with the savings afforded by o-xylene. Another factor, which particularly concerns less-developed areas, is that freight costs for o-xylene, which can be transported in bulk, are about 0.5 cent per pound lower than for naphthalene, so that the price differential at the point of utilization is actually enhanced. Also, sublimation losses of naphthalene are more difficult to keep down than evaporation losses of o-xylene. In the U.S., the most important factor which keeps o-xylene from becoming the major raw material for PA is probably that many of the companies not using o-xylene have captive naphthalene sources.

All things considered, o-xylene seems to be the raw material of the future for phthalic anhydride production; coal-tar aromatics are declining in importance, petroleum naphthalene is expensive, and the sustained increase in catalytic reforming capacity will in time make more o-xylene available. Since o-xylene is a liquid, it is also cheaper to ship, store and handle because no heating is needed. Furthermore, potential improvements in yields on o-xylene are much larger than these on naphthalene due to the difference in molecular weights. The economics of fluid and fixed-bed phthalic anhydride production are compared below. Apart from fluid-bed oxidation of o-xylene, a possible future development is liquid-phase oxidation. This would undoubtedly lead to higher yields, but so far these have been more than offset by the higher capital costs. Cowles Chemical and Catalytic Construction, and Progil in Europe, claim to have developed such processes with yields of 1.1 pound per pound feed. In 1966, Monsanto announced it would build a 75.0 million pounds per year plant using fluid-bed o-xylene oxidation.

Investment for a 30.0 million pounds per year von Heyden plant is \$4.5 million, and for a fluid-bed plant of the same capacity, \$3.5 million. o-Xylene and petroleum-naphthalene are taken at 4.0 and 5.0 cents per pound, respectively.

	Cost, ¢/lb phthalic anhydride	
	von Heyden	Fluid-bed
o-Xylene	4.4	–
Naphthalene	–	5.3
Utilities, catalyst and chemicals	(-0.6)	(-0.2)
Labor and overhead	1.1	0.6
Capital charges	4.9	3.9
Total	9.8	9.6

A phthalic anhydride plant is a net producer of steam, which is why net utility charges are negative.   Catalyst consumption is higher for the fluid-bed plant.

The demand for phthalic anhydride in the U.S. was 570.0 million pounds in 1965, distributed as follows:

End-use	MM lbs, 1965
Alkyd resins	200.0
Polyesters	63.0
Plasticizers	277.0
Other	29.0
Export	25.0
Total	594.0

Source: *Chemical Week*,
August 20, 1966.

The breakdown between raw materials was as follows:

Coal-tar naphthalene	45%
Petroleum naphthalene	30%
o-Xylene	25%

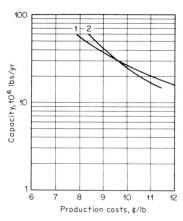

Fig. 8-33 Phthalic anhydride. 1-von Heyden, fixed-bed, o-xylene; 2-Badger-SW, fluid-bed, naphthalene.

The main producers of phthalic anhydride are listed below. There are several plants in mothballs in the U.S., the operation of which even for captive use has become unprofitable.

Producer	Location	Capacity, MM lbs/yr	Raw material
Allied Chem.	Frankford, Pa.	80.0	Coal tar naphthalene
	Ironton, Ohio	36.0	Coal tar naphthalene
	El Segundo, Calif.	26.0	o-Xylene
Monsanto	St. Louis, Mo.	40.0	Coal tar naphthalene
	Bridgeport, N.J.	60.0	Petro naphthalene
Reichhold	Elizabeth, N.J.	30.0	Petro naphthalene
Koppers	Bridgeville, Pa.	50.0	Low sulfur naphthalene
Sherwin Williams	Chicago, Ill.	20.0	Petro naphthalene
Tenneco	Fords, N.J.	12.0	o-Xylene
Chevron	Perth Amboy, N.J.	30.0	o-Xylene
	Richmond, Calif.	18.0	o-Xylene
Union Carbide	Institute, W.Va.	60.0	Petro naphthalene
Thompson	Hebronville, Mass.	10.0	Coal tar naphthalene
U.S. Steel	Neville Island, Pa.	43.0	Coal tar naphthalene
W.R. Grace	Fords, N.J.	53.0	Petro naphthalene
Witco	Chicago, Ill.	20.0	Coal tar naphthalene
	Perth Amboy, Ill.	30.0	Coal tar naphthalene
Stepan	Millsdale, Ill.	42.0	o-Xylene
Total		660.0	

Source: *Chemical Week*, August 20, 1966.

In addition, a 48.0 million pounds-per-year plant was built in Puerto Rico, designed for o-xylene.

17.1   Plasticizers

Plasticizers are compounds added to polymeric materials to make them more workable by molding, extrusion, etc. The addition of plasticizers affects yield-strength; PVC, for example, is only suitable for pipe, corrugated roofing and similar applications if it remains rigid, that is, if no (or very little) plasticizer is added.

There are numerous types of plasticizers. Phosphate esters are added to PVC in order to restore the fire-retardancy of the unplasticized polymer when this has been diminished by addition of other plasticizers not containing a fire-retardant element (P, Cl, Br). Chlorinated waxes, chlorinated diphenyls and others also serve these purposes. Polymeric materials are more resistant to extraction and migration than monomerics. Polymeric esters made from aliphatic dicarboxylic acids such as adipic are used in PVC to impart low-temperatures flexibility; for this reason they are often referred to as "flexibilizers." Epoxidized plasticizers have a synergistic effect on the commonly used PVC stabilizers. Internal plasticizers are compounds that not only affect the physical properties of the host compound, but can also participate in the polymerization reaction via an extra reactive group such as -OH, -COOH, or a double-bond. Maleates, citrates, allyl compounds and crotonates are examples of this type of agent. Finally, the most important of all are the esters of phthalic anhydride, the standard general-purpose PVC plasticizers.

The relative positions of the various types of compounds are illustrated below. Note that not all phosphate esters, or esters of fatty acids, are used as plasticizers; the estimate given refers only to plasticizer products actually used as such.

Plasticizers	MM lbs/yr, 1965	%
Phthalates	680.0	63.0
Other monomerics	90.0	8.0
Polymerics	40.0	4.0
Epoxy plasticizers	76.0	7.0
Phosphate esters	69.0	6.0
Chlorinated waxes	9.0	1.0
Internal plasticizers, etc.	21.0	2.0
Others	98.0	9.0
Total	1,083.0	100.0

Source: Based on U.S. Tariff Commission.

In the main field of application, over 85 percent of these compounds are used in vinyl chloride polymers and copolymers. The remainder is in polyvinyl acetate, cellulosics, and other plastics.

Within the vinyl chloride market, the plasticizer end-use distribution is as follows:

End-use	MM lbs/yr, 1965
Flooring	114.0
Film and sheeting	150.0
Wire and cable coating	136.0
Other extruded and molded products	152.0
Textile and paper	126.0
Others (e.g., surface coatings, etc.)	162.0
Total	840.0

Phthalic esters are made by the reaction between phthalic anhydride and an alcohol or, as in the case of butylbenzyl phthalates, an alcohol and an alkyl chloride.

Over 90 percent of all phthalic esters are used as plasticizers. The most important, accounting for 52 percent of all phthalates, are the diesters of isooctyl and 2-ethylhexyl alcohols. They are almost entirely incorporated into the general-purpose grades of PVC. Film, sheeting and extrusion products use butyl-octyl and butyl-decyl phthalates. The less volatile phthalates are used principally in wire and cable coatings, which are subject to higher temperatures. There is a trend towards the use of less volatile (and more expensive) phthalates in other applications, which means that the ratio between phthalic anhydride consumption in esters and total phthalates produced (now around 0.42) will decline slightly in time. The principal phthalic plasticizer for cellulosics is diethyl phthalate; for nitrocellulose it is dibutyl phthalate, but the dicyclo-hexyl ester is also important. In vinyl acetate a variety of esters are used, depending on the application; dimethyl and dimethoxyethyl phthalate are the most frequently employed.

The estimated production of each type of plasticizer for 1965 is shown below. The production of phthalates as a group tends not to grow as fast as the demand for PVC, since less plasticized applications such as rigid products and flooring are increasing faster than the overall average. Nevertheless, a growth-rate of 12 percent per year until 1970 is not an unreasonable prediction.

Ester	Production, MM lbs, 1965
DOP	212.0
DIOP and mixed C_8's	141.0
DBP	20.0
DEP	18.0
DIDP	90.0
Butyl-octyl	15.0
Octyl-decyl	25.0
Dicyclohexyl	8.0
Dimethoxyethyl	11.0
Dimethyl	4.0
Ditridecyl	13.0
Others	119.0
Total	676.0

The main producers of phthalate plasticizers are either manufacturers of phthalic anhydride, or of C_8-C_{13} alcohols, or of both. The margins between raw material costs and the prevailing sales prices are very small and integration has been almost mandatory. An example of the economics of making DOP is shown below. A 40.0 million pounds per year continuous phthalic ester unit costs $1.2 million, and captive phthalic anhydride is assumed charged in at 8.5 cents per pound.

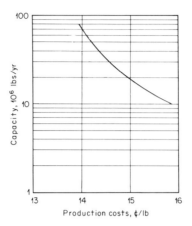

Fig. 8-34   Dioctyl phthalate.

	Cost, ¢/lb DOP
Phthalic anhydride	3.4
2-Ethylhexanol (0.7 lb/lb)	8.6
Labor and overhead	0.4
Utilities	0.9
Capital charges	1.0
Total	14.3

At the time of this calculation, DOP was selling for 14.5 cents per pound.

The main producers of phthalate plasticizers in the U.S. are given in the table below.

Producer	Capacity, MM lbs/yr	Sources	
		Captive PA	Captive alcohols
Monsanto	140.0	x	–
Union Carbide	125.0	x	x
Allied Chem.	75.0	x	–
Texas Eastman	75.0	–	x
Pittsburgh Chem.	50.0	x	x
Reichhold	50.0	x	–
Total	515.0		

The capacities are, of course, very flexible, since many different esters can be made in the same installation. However, the longer the alcohol, the longer it takes for reaction to go to completion; the figures indicate only the relative positions of the main manufacturers.

Some phthalates have other uses. For example, dimethyl phthalate was used in large amounts during World War II as an insect repellent, and is the preferred vehicle for shipping MEK peroxide; diethyl phthalate is an alcohol denaturant and a perfume fixative. However, these cases represent only a small fraction of the total.

17.2   Alkyd Resins

"Alkyd resins" is a generic name for condensation products of dicarboxylic acids and polyfunctional alcohols, which are usually modified by the addition of a monofunctional fatty acid. Reacting phthalic anhydride with glycerine, for example, would yield an infusible, crosslinked resin unsuitable for the manufacture of surface coatings; a portion of the hydroxyl groups, therefore, is made to react with a monofunctional acid. A "short-oil" resin, for example, contains about five acid groups from phthalic anhydride and one from a fatty acid for every six hydroxyl groups present. Since the total phthalic anhydride content of the 640.0 million pounds per year of alkyd resins made in the U.S. is around 200.0 million pounds, one can conclude that the average composition (on the simplifying assumption that only phthalic anhydride, glycerine and fatty acid are used) is close to an equimolar mixture of the three constituents. The resulting resin can be represented as follows:

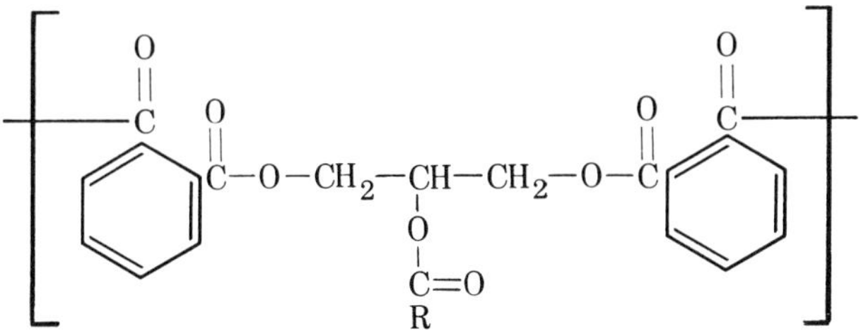

Acids other than phthalic and alcohols other than glycerine can be used. Maleic anhydride and isophthalic acid, aliphatic diacids such as sebacic, or fermentation products such as citric acid can be substituted for part of the acid requirements, and benzoic acids can be substituted for fatty acids to modify properties of the final product. Similarly, pentaerythritol is used widely in alkyds instead of glycerine. This allows a larger proportion of fatty acid to be included (since there are four groups present per mol) and produces alkyds of higher melting point. "Modified" alkyds, on the other hand, usually denote resins that have been made tougher and more durable by the addition of styrene, phenolic resins, amino resins or epoxy resins.

The demand for alkyd coatings has remained static since around 1960, because the incremental overall demand for coatings has been satisfied by more sophisticated materials: latex paints for "do-it-yourself" applications; thermosetting acrylic finishes for the automobile industry; and epoxy coatings in maintenance paints. The total consumption of alkyd resins is around 640.0 million pounds per year, of which 65 percent are unmodified phthalic-type resins, and the rest (25 percent) modified or based on acids other than phthalic anhydride (10 percent).

17.3   Phthalocyanine Pigments

Phthalic acid can be converted to phthalonitrile by reaction with urea:

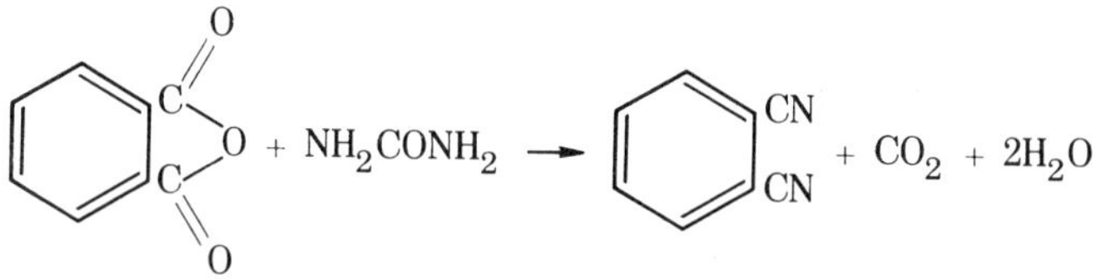

This reaction can also be accomplished by ammoxidation of o-xylene, a process first used commercially in Japan.

Phthalonitrile forms a complex with copper that is the basis for phthalocyanine blue pigment. Chlorination gives various shades of green, depending on the degree of chlorination. In 1964, about 6.0 million pounds of blue and 4.0 million pounds of green phthalocyanine pigments were consumed in the U.S.; they are extremely light-fast and are the only organic pigments, aside from the red and brown quinacridones, that can be utilized in the 350-500°F temperature range. They are used mainly in surface coatings as tints for titanium dioxide pigment; in fact, the main problem in making phthalocyanine pigments is presenting the product in a physical form that avoids settling out of the tint from the TiO_2 matrix.

17.4   Anthraquinone Dyes

Benzene and phthalic anhydride react in the presence of a Friedel-Craft catalyst to form anthraquinone:

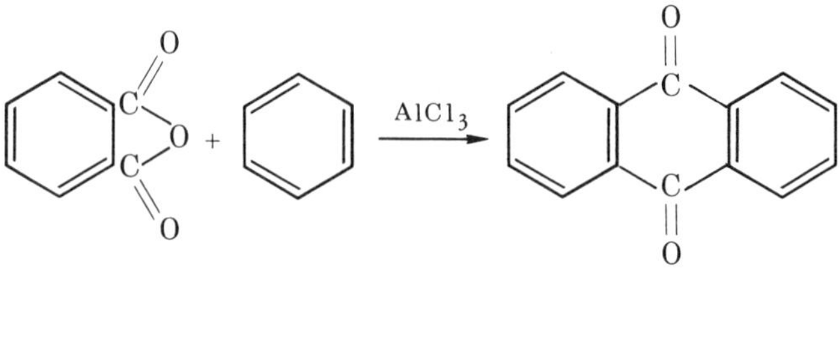

Similar reaction with p-chlorophenol gives quinizarin, and with monochlorobenzene 2-chloroanthraquinone is obtained.

These three products are very important in the dye industry as intermediates for the production of vat dyes. The group of colorants for use on cotton and rayon fabrics have the best light and wash fastness of all. Their use has been fairly static, however, since cotton and rayon are not very dynamic fibers compared with the synthetics. In 1965, about 48.0 million pounds of anthraquinone vat dyes were used; anthraquinone derivatives are also used to some extent in wool and nylon dyes.

17.5   Anthranilic Acid

Phthalic anhydride reacts with ammonia to form phthalimide, which can be hydrolyzed to anthranilic acid.

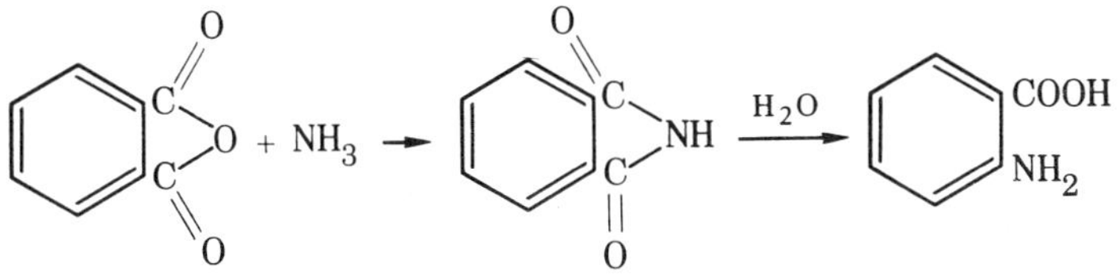

Anthranilic acid is an intermediate obtained during saccharin production via the Maumee Chemical process. It can be converted to anthranilamide, the starting point for "Guthion," one of the more important phosphate insecticides, of which about 6.0 million pounds per year are used as a broad-spectrum poison against cotton, tobacco and vegetable pests.

Anthranilic acid is the starting point for making acrylic fiber dyes of the indazole family.

Another derivative of phthalimide is "Folpet," a fungicide chemically similar to "Captan" but not as widely used.

18   DIMETHYL TEREPHTHALATE AND TEREPHTHALIC ACID

Terephthalic acid is obtained by liquid-phase oxidation of p-xylene, which can be esterified to give dimethylterephthalate (DMT):

$$
\underset{\substack{CH_3}}{\overset{\substack{CH_3}}{\bigcirc}} \xrightarrow{(O)} \underset{\substack{COOH}}{\overset{\substack{COOH}}{\bigcirc}} \xrightarrow{CH_3OH} \underset{\substack{COOCH_3}}{\overset{\substack{COOCH_3}}{\bigcirc}}
$$

Liquid-phase oxidation is always chosen when carboxylic acids instead of anhydride are to be produced. The higher capital and operating costs are due to several factors. First, oxidation usually cannot be carried out without the use of a promoter or of an oxidizing agent other than air; second, reactors cannot be built for the same single-train throughput as in vapor-phase processes; third, because of the lower temperatures employed, corrosion is more of a problem and stainless steel equipment must be selected; finally, the step of separating the finely divided catalyst and oxidizing media is one of the most expensive operations of a liquid-phase process.

In the U.S. Hercules uses the Witten air oxidation process, which occurs in four steps. First, p-xylene is oxidized to p-toluic acid, which is then esterified to its methyl ester. A second oxidation step converts the remaining methyl group, and a second esterification finally gives DMT. The reason for this intermediate esterification step is that the second methyl group is much more difficult to oxidize than the first. Thus it is necessary to protect the molecule from degradation under the more severe conditions necessary for the second oxidation step. In actual practice, oxidation and esterification both take place in a single reactor, all intermediates from the esterification step other than DMT itself being recycled to the oxidation reactor. The process is suited only for DMT (that is, it cannot produce DMT).

Dupont makes DMT by a one-step nitric acid oxidation process which has the disadvantages of requiring 2 pounds nitric acid per pound of product. Tennessee Eastman uses air oxidation which takes place in acetic acid solution in the presence of a cobalt catalyst. The process can be used to make any aromatic carboxylic acid, or a mixture of several of them. Terephthalic acid, being insoluble in acetic acid as in most other solvents, precipitates and can be purified by esterification with methanol. Yields are around 90 percent.

Since p-xylene is an expensive raw material, there is an obvious incentive to start from a different hydrocarbon. Two such processes have been developed. Neither is used in the U.S., but the Henkel process has been adopted in several Japanese plants. The Bergbau Forschung GmbH process starts with toluene, which is chloromethylated with formaldehyde and HCl. The reaction is carried out in two stages, with about 60 percent conversion in each. Next, p-chloromethyl toluene is oxidized with nitric acid, also in two stages, one group at a time. Since o-chloromethyl toluene is also formed, some 0.7 pound phthalic acid are obtained as a byproduct per 1 pound of DMT. Esterification and purification complete the process.

The only route in commercial use that circumvents p-xylene as a raw material is the Henkel process. The older version involved oxidizing a mixed stream of xylenes to the respective dicarboxylic acids, and neutralizing the mixture with caustic potash. The next step is isomerization of the o- and m-isomers to potassium terephthalate; neutralization with sulfuric acid and esterification complete the process. The present version of the Henkel process begins with toluene, which is oxidized to benzoic acid and neutralized with high-purity potassium hydroxide. Next, continuous solid-phase disproportionation of the alkali benzoate to benzene and potassium terephthalate takes place in the presence of carbon dioxide.

Thus the process is a means of upgrading toluene to DMT and benzene, both more valuable than the starting material. As before, neutralization and esterification finally yields DMT. This second version is especially promising for economies that do not possess a source of p-xylene, which is why it has been adopted in Japan. This process is reported to produce better fiber-grade acid than xylene oxidation since it contains no p-carboxybenzaldehyde (formed from p-xylene by incomplete oxidation of the second methyl group) or polycarboxylic acids. A schematic flow-diagram is shown in Fig. 8-36.

All these processes make crude TPA in the first stage. This must be purified of catalyst traces and byproducts in order to obtain fiber-grade TPA or IMT. In the latter case, most processes involve esterification with methanol in concentrated sulfuric acid.

Below, the economics of DMT production from p-xylene using the Witten process are compared with those of making TPA. Investment for a 50.0 million pounds-per-year DMT plant is $8.5 million. It is assumed that the company can charge in captive methanol at 3.0 cents per pound; this is the case for several major producers. Actually, since most of the methanol is recovered in the polymerization stage, the cost of methanol has no effect on the economics of making polyester fibers or film. The investment for a 50.0 million pounds-per-year plant making fiber-grade TPA via the Amoco process is $7.5 million.

	Cost, ¢/lb DMT	Cost, ¢/lb TPA (Amoco process)
p-Xylene (at 9.5 ¢/lb)	6.5	6.9
Methanol	1.3	—
Chemicals and catalyst	0.7	1.9
Utilities	1.7	1.7
Labor and overhead	1.6	1.6
Capital charges	5.8	5.0
Total	17.6	17.1

However, the cheapest route to TPA appears to be the Henkel process. Investment for a 50.0 million pounds-per-year plant is $10.0 million. Toluene is charged in at 2.8 cents per pound, and

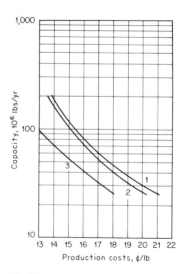

Fig. 8-35 Dimethyl terephthalate and TPA.
1-DMT (Witten process); 2-TPA (Amoco process); 3-TPA (Henkel II process).

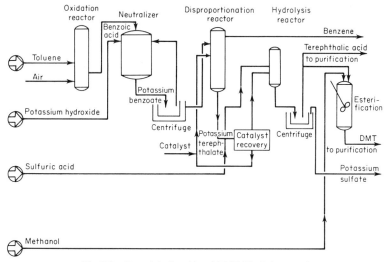

Fig. 8-36   Terephthalic acid and DMT (Henkel process).

benzene credited at 3.4 cents per pound; the catalyst and chemicals costs are the calculated net of the credit for recovered potassium sulfate.

	Cost, ¢/lb TPA	
Toluene	4.0	
Catalyst and chemicals	3.0	
		7.0
Benzene credit	-1.7	
Net raw materials		5.3
Utilities	1.9	
Labor and overhead	1.6	
Capital charges	6.6	
Total		15.4

Producers of DMT and terephthalic acid in the U.S. are given in the table below.

Producer	Location	1966 Capacity, MM lbs/yr	Products
Mobil Chem.	Beaumont, Tex.	100.0	TPA
DuPont	Gibbstown, N.J.	120.0	DMT
	Old Hickory, Tenn.	200.0	DMT
Amoco	Joliet, Ill.	40.0	TPA
	Decatur, Ala.	200.0	TPA + DMT
Eastman	Kingsport, Tenn.	80.0	DMT
Hercules	Burlington, N.J.	60.0	DMT
	Spartanburg, S.C.	75.0	DMT
Total		875.0	

Source: *Oil, Paint & Drug Reporter*, November 21, 1966; May 30, 1966.

In 1965, consumption of polyester fibers and film in the U.S. was 325.0 and 50.0 million pounds respectively. This required a total of 400.0 million pounds of DMT.

18.1   Polyester Fibers

In addition to high strength (inferior only to nylon) and wrinkle resistance, the distinctive property of polyester fibers is that, compared to nylon and acrylonitrile, they are compatible with natural fibers. In fact, almost 80 percent of polyester fibers are sold in the form of staple for blending with cotton and wool.

Polyester fibers are made from dimethyl terephthalate and ethylene glycol:

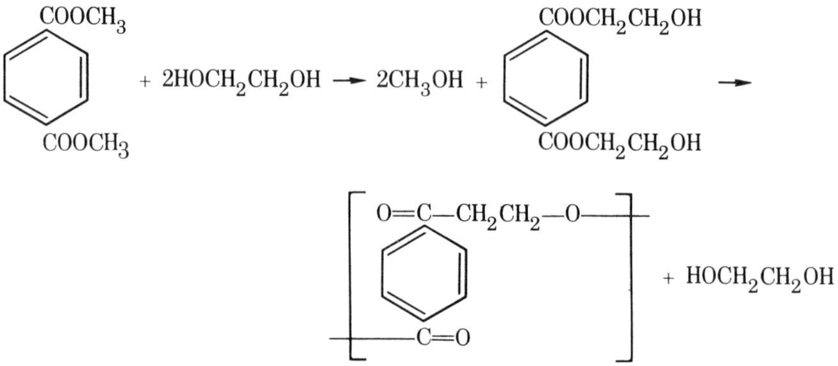

With respect to higher glycols, which lower the melting point of the polymer, ethylene glycol must be especially pure when making polyesters.

Eastman used 1,4 cyclohexane dimethanol as the alcohol for some its "Kodel" fiber and "Terafilm" polyester film. It is made by high-pressure catalytic reduction of DMT:

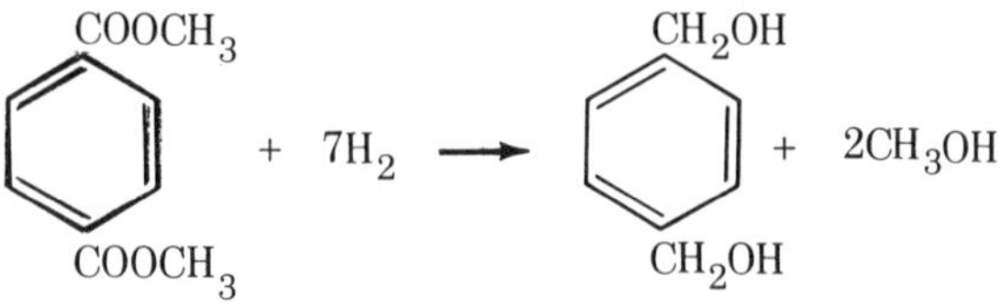

Polyesters made from 1,4 cyclohexane dimethanol are more resistant to hydrolysis than those from aliphatic alcohols such as ethylene glycol.

It would seem that using terephthalic acid directly, without going through the esterification step, could offer a number of advantages. Actually, there are severe difficulties to this approach, which have only recently begun to be overcome.

Since the polymer is somewhat unstable at the spinning temperature (around 540° F), when polymerization is carried out batchwise it must be followed by converting the resin to chips, which are then remelted in the spinning machinery. For this reason, great emphasis has been placed on the development of continuous polymerization-spinning processes. Almost all major polyester fiber producers and technology licensors now use or offer continuous processes, which, by eliminating the chip stage, reduce capital requirements and operating costs appreciably and increase yields by diminishing the formation of low-molecular-weight polymers. The Zimmer process, for example, consists of a transesterification reactor followed by three polymerization reactors in series, the last two of which are designed for plug-flow to ensure a narrow distribution of molecular weights. Although chip-making facilities are needed for better operating flexibility, the polymer can be fed directly to the spinnerets. In proceeding from the raw materials to the spinning melt, the continuous process has an overall advantage of 2.0 cents per pound melt over the batch method.

The growth in demand for polyester fibers has been extremely rapid, some 22 percent per year during the 1960's. Garments make up over 50 percent of all polyester fiber consumption; they are limited to lightweight goods, and therefore are the most suitable synthetic for temperate climate economies. Polymer and spinning production costs are lower than for polyamides filaments; they are the synthetic most likely to be used soon in popular-price clothing, especially since the successful introduction of permanent-press treatments for polyester-cotton mixtures. An important application is "fiber fill," used for making insulated winter apparel, pillow stuffing, and brassiere padding where it has largely replaced rubber foam. Yarn, which accounts for about 12 percent of polyester fiber output, is used both in apparel and in washable, light-fast household goods such as curtains. Industrial uses include fire-hose, power belting and filter cloth. Heavy gauges are beginning to be used in tire cord; polyester tire cord is apparently tied to the introduction of radial-ply tires, and it is also a candidate for replacing rayon in original-equipment tires since, unlike nylon 6/6, it does not flat-spot. The major problem so far has been to find a suitable adhesive. It also has a promising future in farm-equipment and truck tires. Spinning directly from a mixed melt of nylon and polyester is also being tested; this would combine the optimum properties of both. Finally, terephthalic polyester fibers are one of the constituents of DuPont's poromoric material, "Corfam." The main manufacturers of polyester fibers in the U.S. are listed below.

Producer	Location	Est. capacity 1965, MM lbs	Est. capacity 1968, MM lbs
American Enka	Enka, N.C.	–	60.0
Bates	Rocky Mount, N.C.	10.0	25.0
Beaunit	Elizabethton, Tenn.	30.0	54.0
Celanese	Salisbury, N.C.	–	20.0
	Shelby, N.C.	75.0	250.0
DuPont	Kinston, N.C.	220.0	220.0
	Old Hickory, Ky.	–	230.0
	Cape Fear, N.C.	–	150.0
FMC	Lewistown, Pa.	25.0	25.0
Hercules	Spartanburg, S.C.	–	30.0
IRC Fibers	Cleveland, Ohio	6.0	10.0
Monsanto	Decatur, Ala.	15.0	45.0
Tennessee Eastman	Kingsport, Tenn.	50.0	150.0
Vectra	Odenton, Md.	1.0	1.0
Total		432.0	1,270.0

Source: Based on *Chemical Week*, November 12, 1966.

18.2   Polyester Film

There are several polyester films on the market, for example, "Mylar" and "Kronar" (DuPont), "Terafilm" (Eastman) and "Videne" (Goodyear). The most important end-uses for polyester film are in all-purpose electrical insulation, magnetic tape for data-processing equipment, pressure-sensitive tape, and audio and engineering reproduction materials. It can be laminated to other materials, such as polyethylene and polyvinylidene chloride. Miscellaneous uses include plastic book jackets and vacuum packaging of meats and cheeses.

The demand for polyester film was 52.0 million pounds in 1965, "Mylar" being the major factor in the field.

19   OTHER *p*-XYLENE DERIVATIVES

There are few uses for *p*-xylene aside from making DMT. "Dacthal" is a well-established crab-grass killer; it is obtained by ring chlorinating followed by oxidation:

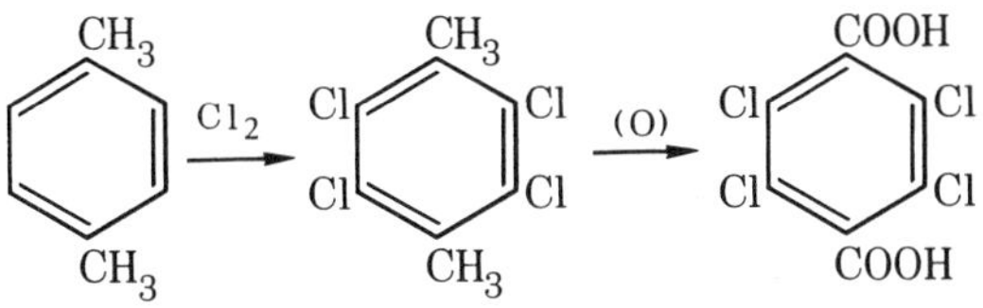

p-Xylene or certain of its ring-substituted derivatives can be polymerized to a family of polymers whose outstanding properties are that they can be formed into very thin films, and that they have excellent dielectric qualities.

They are prepared in three steps: pyrolysis of p-xylene to a cyclic dimer, conversion under vacuum to the polymerizable monomer and polymerization by cooling to ambient temperature.

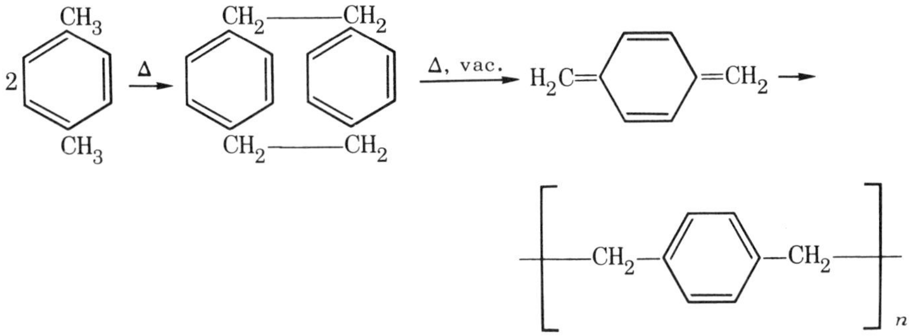

These polymers are still in the experimental stage, but their main potential application seems to be in the electrical industry where they will compete, among others, with polyester film for insulating miniature capacitors. They are made by Union Carbide under the name "Parylene."

20   ISOPHTHALIC ACID

There are two processes for making isophthalic acid in commercial use. California Chemical uses ammonium sulfate to oxidize m-xylene, the reduction product being H_2S:

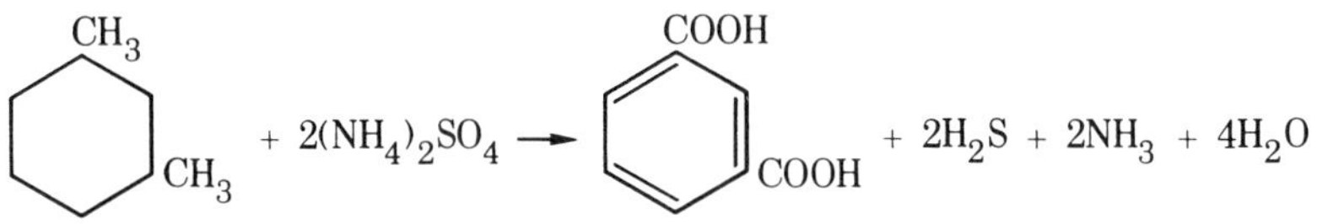

The reaction is carried out in two steps; as in the case of terephthalic acid, the second methyl group is much harder to oxidize.

Amoco produces isophthalic acid by oxidation of mixed xylenes in acetic acid. Phthalic acid is the easiest to separate of the three, after which isophthalic and terephthalic acids are purified by fractional crystallization from a selective solvent. The main uses for isophthalic acid parallel those for phthalic anhydride. At times a considerable demand for isophthalic arises as a result of naphthalene (and therefore phthalic anhydride) shortages caused by steel strikes. The use of isophthalic instead of phthalic in alkyd resins or to make phthalate plasticizers presents no particular advantage and involves considerably higher production costs; even so, some isophthalic acid is used for these applications. In making polyester, on the other hand, isophthalic gives products with much better toughness, resistance to abrasion, resiliency and freedom from cracking and crazing. It has therefore come to be used, almost to the exclusion of phthalic, for pressure-molding polyester applications. Alkyd resins using isophthalic acid have superior resistance to abrasion and corrosion, but processing difficulties offset these advantages.

The U.S. manufacturers of isophthalic acid have the following capacities:

Producer	Location	Capacity, MM lbs/yr
California Chem.	Richmond, Calif.	50.0
Amoco	Joliet, Ill.	50.0
Sinclair	Channelview, Tex.*	35.0

*Scheduled to operate in 1967.

The demand in 1965 was around 43.0 million pounds, distributed as follows:

End-use	MM lbs, 1965
Alkyd resins	15.0
Polyesters	25.0
Other	3.0
Total	43.0

A certain amount of dimethyl isophthalate is used by some manufacturers as a modifier for DMT in making polyester fibers and film. As in the case of TPA, direct esterification of isophthalic acid is difficult and poor solubility of the acid presents a technical problem.

Among the more recent developments in the use of isophthalic acid, there are the polyamide fibers made from acids and amines, one or both of which are aromatic. One example is "Nomex," a polyamide fiber made by DuPont from isophthalic acid and *m*-phenylene diamine.

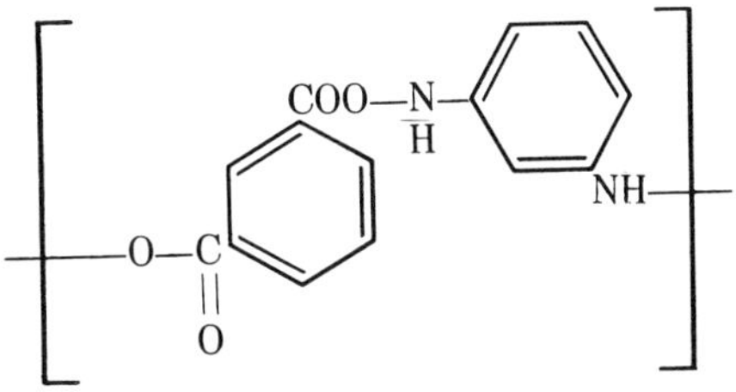

Polybenzimidazoles, a family of high-temperature polymers, are made by condensation of phenyl isophthalate and a tetra-amine such as diamino-3,3 benzidine.

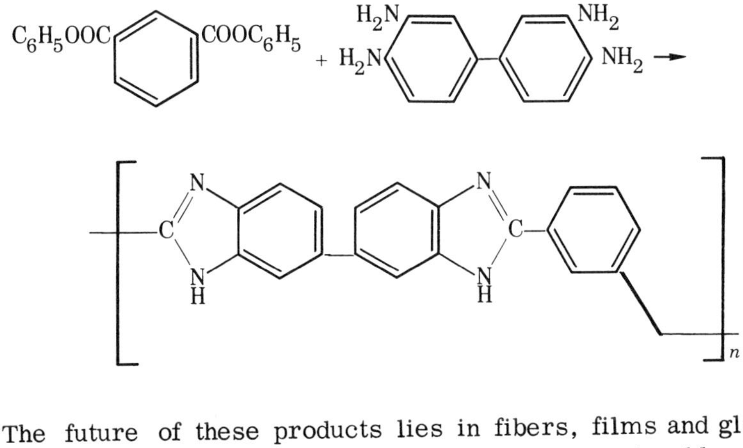

The future of these products lies in fibers, films and glass-reinforced plastics, which are reported to be serviceable up to 440°F for film and fiber, and even higher for reinforced resins.

21   *m*-XYLYLENEDIAMINE

m-Xylylenediamine is an intermediate for making special poly-amides. It is made directly from mixed xylenes by ammoxidation of the contained *m*-xylene and subsequent hydrogenation of the resulting nitrile, according to a process developed in Japan by Showa Denko, which operates a 5.0 million pounds-per-year plant. It can be combined with adipic acid, or an aromatic diacid. Aside from being a raw material for polyamides, *m*-xylylenediamine is an epoxy-resin curing agent.

22   β-NAPHTHOL

High-temperature sulfonation of naphthalene gives naphthalene-2-sulfonic acid almost to the exclusion of the 1-sulfonated derivative. This can be hydrolyzed to β-naphthol.

The 1-substituted derivative is easier to hydrolyze back to naphthalene than the desired 2-sulfonic acid. This property is used to eliminate the unwanted isomer, the resulting naphthalene being removed by the hydrolysis steam and sent back to the sulfonation stage.

About 46.0 million pounds of β-naphthol were used in the U.S. during 1965. The approximate end-use pattern is the following:

End-use	MM lbs, 1965
Phenyl-β-naphthylamine	23.0
Dyes and dye intermediates	19.0
Others	4.0
Total	46.0

In the dye field, β-naphthol is one of the most important azo-dye coupling agents, either as such, or in the form of numerous aminated and sulfonated derivatives.

β-Naphthol is the starting point for a number of other dye intermediates. The Kolbe reaction gives β-oxynaphthoic acid (BON), which forms amides with aromatic amines to give intermediates of the naphthol AS series.

BON is also a coupling agent for a number of red and brown azo pigments.

Many azo-dye coupling agents, of which H-acid is the most important, are made from naphthalene without going through naphthol production.

Total consumption of the various acids derived from naphthalene (H-acid, Tobias acid, etc.) is about 22.0 million pounds per year. The demand distribution among the various intermediates varies from year to year with color fashions, but it can be estimated that 100.0 million pounds of dyes require 12.5 million pounds of naphthalene. Thus the 180.0 million pounds per year produced in 1965 required 22.5 million pounds of naphthalene.

β-Naphthol is one of the raw materials for making phenyl-β-naphthylamine (PBN), a widely used rubber antioxidant.

The effects of vertical integration are noticeable even in the field of dye intermediates and rubber chemicals, where profit margins were once much wider than in basic raw materials. The cost of making β-naphthol and PBN are calculated below. A 10.0 million pounds-per-year β-naphthol plant requires an investment of $1.0 million.

	Cost, ¢/lb β-naphthol
Naphthalene (refined, at 6.0 ¢/lb)	7.2
Sulfuric acid	1.5
Caustic soda, soda ash	4.7
Labor and overhead	2.4
Utilities	0.8
Capital charges	3.3
Total	19.9

Assuming the producer has a captive source of aniline, phenyl-β-naphthylamine can be produced for the following costs:

	Cost, ¢/lb phenyl-β-naphthylamine
β-naphthol	15.2
Aniline	4.4
Production costs (estimated)	6.0
Total	25.6

When these figures were compiled, this compound was obtainable for around 30.0 cents per pound.

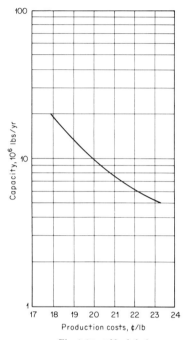

Fig. 8-37   β-Naphthol.

23   α-NAPHTHYLAMINE

Nitration of naphthalene takes place primarily in the 1-position; reduction gives α-naphthylamine. About 8.5 million pounds per year are used, mainly in the dye industry. Other important derivatives are phenyl-α-naphthylamine, made by reacting with aniline in the presence of an iodine catalyst.

"Alanap," a herbicide, and "ANTU," an arodenticide, are other minor outlets for α-naphthylamine.

23.1   α-Naphthol

Acid hydrolysis of α-naphthylamine yields α-naphthol.

The main use of this compound is in the manufacture of "Sevin" (1-naphthyl methylcarbamate), an insecticide. About 80.0 million pounds were made in 1965 for use against beetles, caterpillars, grasshoppers and other pests, in field crops as well as for fruit, vegetable and public health applications. Union Carbide, the sole producer, has a capacity of 100.0 million pounds per year. α-Naphthol is also a dye intermediate.

24   OTHER USES OF NAPHTHALENE

Naphthalene can be treated with sodium metal or hydrogen to produce a refined product suitable for chemical synthesis such as manufacture of β-naphthol, or for mothballs. Use of naphthalene mothballs has been declining, and is being gradually replaced by p-dichlorobenzene.

Alkylated and sulfonated naphthalene derivatives find some use as detergents.   About 10.0 million pounds per year of these products are consumed in the U.S., for example, as dispersants in the latex paint and rubber processing fields.

The most important class of synthetic tanning agents or "syntans" are the sulfonated derivatives of methylene-bis-naphthalene, obtained by sulfonating the condensation product of naphthalene and formaldehyde:

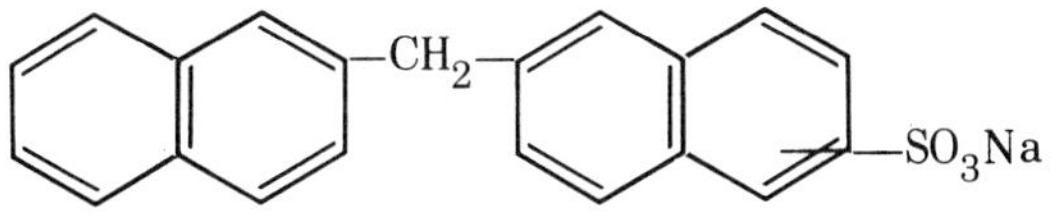

Total "Syntan" consumption is around 27.0 million pounds per year; it is used mainly to tan sole leather, which has been losing ground to SBR for many years.   Since SBR now accounts for 60 percent of all shoe-soles production in the U.S., the demand for "syntans" is unlikely to grow any further.

25   TRIMELLITIC ANHYDRIDE

Trimellitic anhydride is manufactured by Amoco by liquid-phase air oxidation of pseudocumene, a process analogous to the production of terephthalic acid:

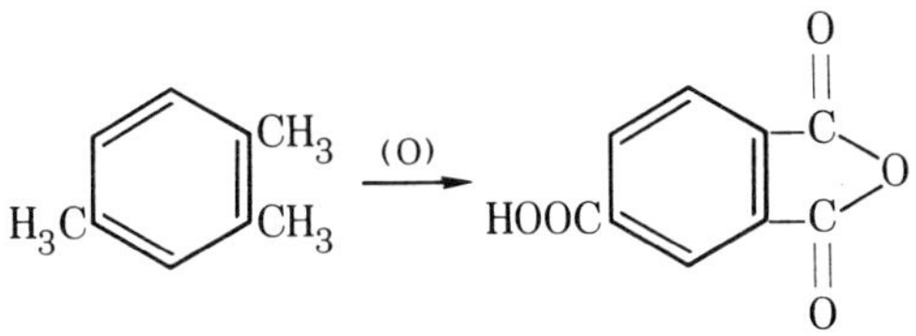

It is used in certain coating systems as a curing agent for epoxy resins and, being trifunctional acid, as a crosslinking agent for polyesters.   Trimellitates are also used as PVC plasticizers, such as "Morflex" (Pfizer), which compete with polymerics. In 1965, about 4.0 million pounds were used. They are as migration resistant as polymerics, but as processable as monomerics; also, they do not thicken in winter as do polymerics. Trimellitate use promises to increase rapidly.

Amoco has developed a family of polyamide-polyimide resins made from the chloride of trimellitic anhydride and an aromatic diamine:

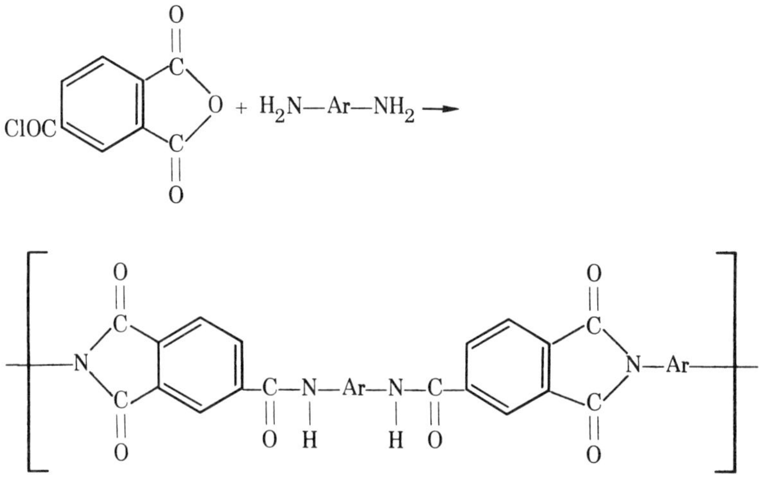

The reaction product of trimellitic anhydride, first with hydrazine, then with phosphorus pentachloride, is an intermediate for certain reactive dyes.

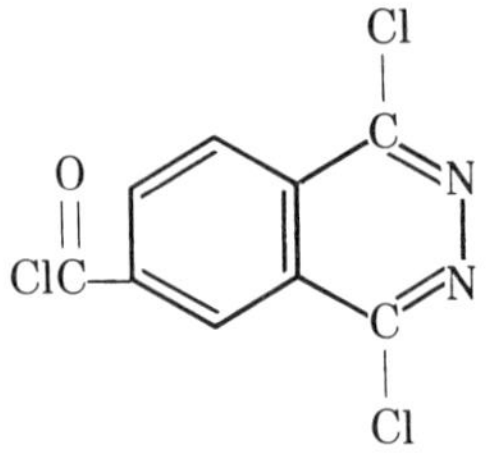

26 PYROMELLITIC DIANHYDRIDE

Durene can be oxidized to pyromellitic dianhydride (PMDA):

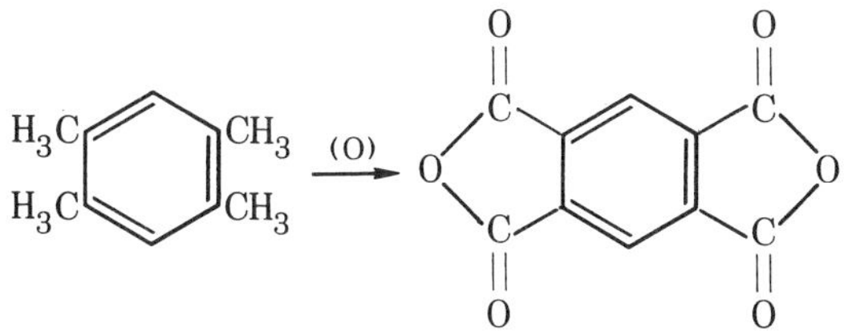

In the Bergbau Forschung process, *p*-xylene is converted to PMDA by alkylation with propylene in the 2 and 5 positions, followed by oxidation in the vapor phase.   Yields from crude alkylate are 50 percent by weight.

Yields are low and purification consequently difficult, as each succeeding methyl group is more difficult to convert to a carboxyl function.

Three U.S. companies make PMDA.   Princeton Chemical has a 0.4 million pounds-per-year plant using a new vapor-phase air-oxidation technique particularly suited to small plants.   DuPont, the first manufacturer of polyimides on a commercial scale, uses liquid-phase nitric acid oxidation.   Hexagon Laboratories oxidizes durene with chromic acid in the liquid-phase.

By far the major use for PMDA is as a raw material for making polyimides.

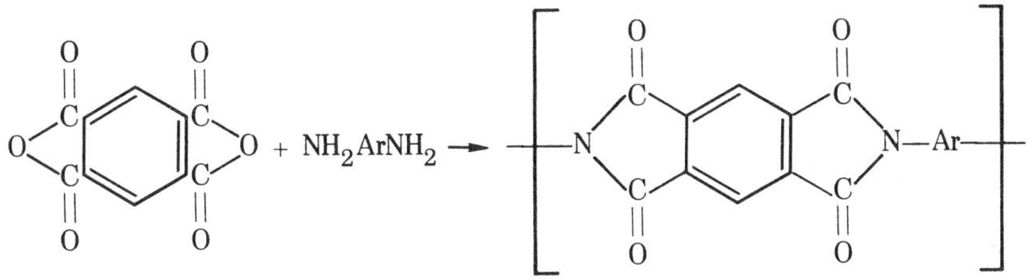

The amine can be diamino-4,4'-diphenylether or *m*-phenylene-diamine.   The earliest commercial applications of polyimides were film, which can be used at 480° F for indefinite periods; wire coating, which enables electric motors to run at 460-465° F with a consequent reduction in copper requirements for a given rating; and special molding applications.

Polyimides are the first successful members of the "second generation" of engineering plastics.   Although less than 1.0 million pounds per year were made in 1965, a prediction of 25.0 million pounds per year has been made for 1972, of which 10.0 million pounds will be for wire coating.

9
OTHER PETROCHEMICALS

1 PETROLEUM WAXES

Aside from small amounts of vegetable, animal and mineral waxes which make up around 3-4 percent of the total U.S. requirements, all wax consumed today is derived from petroleum.

The production of petroleum waxes is tied to that of lubricating oils. Over the years, the demand for lube oils has been growing slowly since the development of better refining methods and additives has made it possible to produce more durable lubricants. This in turn has lengthened considerably the period between oil changes in automobiles and machines in general. Hence, there has been an incentive to substitute synthetic waxes or plastics for petroleum products, which has in fact largely taken place. Milk and orange juice containers, the largest market for wax in 1960, were being made almost entirely of polyethylene-coated board by 1965. Recently, however, the development of ethylene copolymers (with ethyl acrylate or vinyl acetate), to be used as wax additives which improve adhesion and especially resistance to flexure, has brought about a renewed interest in petroleum wax.

The block-flow diagram in Fig. 9-1 shows how wax production is related to other lube-oil operations. Vacuum tower bottoms are first deasphalted to remove materials that would otherwise oxidize to sludge and must therefore be eliminated from lubricants. The most widely used method is selective extraction with propane, which dissolves all hydrocarbons in the feed with the exception of asphaltenes. Next, waxes must be removed. These are straight-chain or slightly branched saturated hydrocarbons which melt above 90° F, and would therefore crystallize out at low temperatures

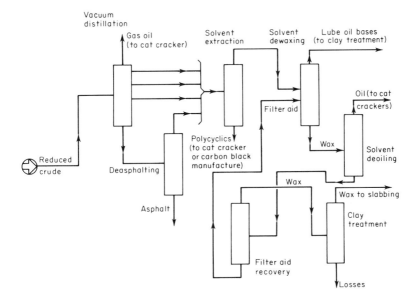

Fig. 9-1. Petroleum wax.

if they were allowed to remain in the lubricant. Dewaxing can be accomplished by means of a solvent, for example, a mixture of benzene and methyl ethyl ketone (MEK). The presence of the ketone causes the wax, upon chilling, to settle out in a form easy to separate by filtration, while the aromatic compound reduces the oil-content of the wax crystals. Low-cold-test oils are made by urea-dewaxing of stocks that have been previously dewaxed with MEK or another solvent.

Since the wax thus produced still contains about 10 percent non-paraffinic material, it must be deoiled. At first this was done by batchwise chilling, known as "sweating," but at present is accomplished by a selective solvent in much the same way that wax and oil were separated earlier. Various treating steps, such as clay percolation, hydrogenation or acid treatment of the deoiled wax then yield the finished product.

In 1965, the demand for wax in packaging applications was around 600.0 million pounds, having, however, reached well over 1000.0 million pounds five years earlier. The total consumption can be broken down as shown below.

End-use	MM lbs, 1965
Packaging uses:	
Milk and juice containers	60.0

End Use	MM lbs, 1965
Bread inner wrap	65.0
Household wax paper	100.0
Other wax paper	90.0
Nested cups and containers	120.0
Butter and ice cream	30.0
Laminated boxes	20.0
Corrugated boxes	40.0
Frozen food packages	15.0
Miscellaneous	60.0
Sub total	600.0
Nonpackaging uses:	
Candles	80.0
Carbon paper and ink	30.0
Polishes	25.0
Explosives	25.0
Electrical	15.0
Rubber antioxidant	15.0
Textiles and cordage	10.0
Miscellaneous	30.0
Sub total	230.0
Chlorinated waxes	40.0
All other	90.0
Total	960.0

Petroleum waxes can be divided into two categories — paraffin and microcrystalline. Paraffin waxes have lower melting points and a crystalline structure; they are derived from low-viscosity lube-stock dewaxing. Heavy stocks yield amorphous, or microcrystalline, wax having a higher melting point (150-210°F). About 80 percent of the wax produced by lube-oil operations is of the paraffin type. Microcrystalline wax is used mainly in the manufacture of flexible packaging materials, for example, wax paper and laminated products such as paper-plastic or paper-foil. Finally, a certain amount of petroleum wax is obtained by deoiling crude-oil tank-bottom deposits.

In all their major applications, wax products suffer severe competition from synthetics. Polyethylene has taken over much of the coated foodboard business; cellophane, aluminum foil and other types of household wraps have cut into the wax-paper business; synthetic emulsions have invaded the floor-polish field. Waxed corrugated cartons, on the other hand, are rapidly replacing wooden crates for transporting farm produce. It is estimated that the market for petroleum wax reached its lowest point in 1965.

The main producers of petroleum wax in the U.S. are Enjay, Mobil, Atlantic, Shell, Cit-Con and Standard of California.

1.1 Chlorinated Waxes

Liquid or solid paraffins in the C_{10} to C_{30} range can be chlorinated in the presence of light to a chlorine content of 40-70 percent giving either viscous liquids or solids, which have a variety of applications.   The most important of these is still as extreme-pressure additives in cutting oils, although use of water-soluble cutting oils is gaining at the expense of straight oils. Ceramic and carbide cutting tools, which have enabled higher cutting speeds to be attained, require cooling media having a higher specific heat than the hydrocarbon media used to formulate straight cutting oils; these emulsifiable oils contain no EP additives. Cutting oils consist of a base, usually a mixture of vegetable and mineral oils, containing sulfur and about 6 percent of an EP additive, mostly chlorinated waxes, although P_2S_5 derivatives can also be used. Chlorinated waxes are also used as EP additives in industrial gear oils, but more effective compounds must be used for the more severe conditions to which automotive gear oils are subject. In vinyl plastics, chlorinated waxes are used as flame-retardant plasticizers in combination with antimony oxide. While this is still a common type of formulation in Europe, in the U.S. phosphate esters are more commonly used for this purpose.   Finally, chlorinated waxes are used as plasticizers for rubber-based surface coatings and in fire-retardant coating formulations for heavy textiles such as awnings and tarpaulins. The consumption of chlorinated waxes in the U.S. is fairly static at around 40.0 million pounds per year, broken down as follows:

End-use	MM lbs, 1965
Cutting oils	18.0
Lube uses	8.0
PVC plasticizers	8.0
Miscellaneous solvent and plasticizer	6.0
Total	40.0

Source: *Oil, Paint & Drug Reporter*, January 18, 1965.

Producers of chlorinated paraffins in the U.S. are the following:

Producer	Location	1965 Capacity, MM lbs/yr
Diamond Alkali	Painesville, Ohio	12.0
Dover Chem.	Dover, Ohio	8.0
Hercules	Parlin, N.J.	3.0
Hooker	Niagara Falls, N.Y.	15.0

Continued

Producer	Location	1965 Capacity MM lbs/yr
Koppers	Wyandotte, Mich.	4.0
Neville	Santa Fe Springs, Calif.	3.0
Pearsall	Phillipsburg, N.J.	5.0
Total		50.0

Source: *Oil, Paint & Drug Reporter*, January 18, 1965.

2   *n*-PARAFFINS

Interest in petroleum fractions containing only linear paraffins first developed when it was found that these were the best media in which to carry out solution polymerization of olefins. Later, however, the principal fields of application for *n*-paraffins became the detergent industry and the production of synthetic fatty acids.

Solvent-range *n*-paraffins (below C_8) are obtained from gasoline, and since they have the lowest octane numbers of any type of hydrocarbon, separating them from naphtha increases the value as motor fuel of the remainder. Detergent-range paraffins (C_{10}-C_{12}), however, are present only in the kerosene fraction, and their removal produces a raffinate that is useless as jet fuel and thus suitable only as cat-cracker feedstock. There are four processes in use in the U.S. for effecting the separation between linear and branched hydrocarbons, three of which employ synthetic zeolites, better known as molecular sieves. The operating principle of these sieves is that they have pore openings capable of rejecting molecules having branched and cyclic structures, while adsorbing straight-chain compounds whose molecular cross-sectional diameters are just below the pore diameter of the synthetic zeolite.

The "IsoSiv" process, developed by the Linde Division of Union Carbide, consists of vapor phase adsorption in one of two fixed-bed reactors while in the other desorption takes place at a lower pressure aided by a purge-stream. Olefin content at the feed must be below 2.5 percent so hydrogenation of the feedstock before adsorption may be required. Production costs vary considerably with the nature and paraffin concentration of the feed; naphtha-range paraffins are cheaper to produce than the more common detergent-range fractions, and costs diminish with increasing *n*-paraffin content in the feed. This process has been more successful than its main competitor, "Molex," developed by Universal Oil Products, which employs liquid-phase, fixed-bed adsorption and, furthermore, differs from "IsoSiv" in that it is entirely continuous. Desorption is no longer carried out thermally, but by means of a recycle stream of n-paraffins. Feed and desorbent enter the adsorbent chamber, while raffinate and an n-paraffin stream leave the chamber continuously. Both exit streams contain light n-paraffins, which are recovered for recycle in a two-tower fractionation system, one of which yields raffinate, the other the desired detergent-range n-paraffins. "Molex" places greater purity demands upon the feedstock, which must be free of olefins and contain less sulfur than the 300 ppm that can be handled by "IsoSiv." A flowsheet of the "IsoSiv" process is shown in Fig. 9-2.

The Esso process also employs vapor-phase adsorption, but desorption is thermal instead of by pressure reduction, as is the case with "IsoSiv." Still a fourth molecular sieve process, developed by British Petroleum, has achieved commercial status in Germany; it resembles "Molex" in that desorption takes place by displacement, but adsorption is carried out in the vapor phase.

The only n-paraffin extraction process not based on molecular sieves is the Shell-Edeleanu urea dewaxing technique. Feedstock, urea and an activator such as methanol or methylene chloride are mixed in a reactor, where urea forms adducts with the linear molecules which can be separated by chilling and filtration. The filtrate is fractionated into solvent and dewaxed oil, while the solid complex is sent to the thermal decomposition step. Effluent from the decomposition reactor is decanted into an aqueous phase, containing urea in solution, and an organic phase consisting of

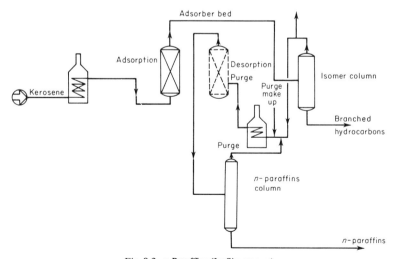

Fig. 9-2. n-Paraffins (IsoSiv process).

n-paraffins and solvent.   Urea and solvent are recycled to the adduction reactor, and water to the decomposition step.

The investment for a 50.0 million pounds-per-year n-paraffins plant, assuming 20 percent n-paraffin content in the feed, is \$1.5 million, and production costs are calculated on the assumption that kerosene is worth 1.3 cent per pound and the cracking feedstock obtained as byproduct only 0.8 cent per pound. Yields are close to 100 percent, so the value of feedstock is 3.3 cents per pound of detergent-range n-paraffins.

	Cost, ¢/lb n-paraffins
Feedstock	3.3
Utilities and chemicals	2.5
Labor and overhead	0.3
Capital charges	1.0
Total	7.1

This cost excludes an eventual aromatics extraction step following product recovery, which may be required in some cases in order to meet biodegradability requirements; it also excludes feed preparation (desulfurization, hydrogen treating) and royalties.

The manufacturers of n-paraffins in the U.S. are listed below. The biodegradability issue has already caused n-paraffin plants to be built in Europe, but in less developed economies detergent alkylate will probably continue to be made from propylene tetramer for several years.

Producer	Location	1965 Capacity, MM lbs/yr
Allied Chem.	Winnie, Tex.	80.0
Atlantic-Richfield	Wilmington, Calif.	100.0
Chevron	Richmond, Calif.	150.0
Continental	Lake Charles, La.	150.0
Enjay	Baytown, Tex.	150.0
South Hampton	Silsbee, Tex.	20.0
Union Carbide	Texas City, Tex.	250.0
Texaco	Pointe-à-Pierre, Trinidad	150.0
Total		1,050.0

Source: Based on *European Chemical News*, December 2, 1966.

2.1 Detergent Raw Materials

There are two routes for making linear alkylate from benzene and in the C_{10}-C_{13} range that begin with chlorination:

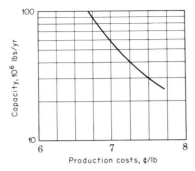

Fig. 9-3. *n*-Paraffins.

$$n\text{-}C_nH_{n+2} + Cl_2 \longrightarrow n\text{-}C_nClH_{n+1} + HCl$$

Since exclusive monochlorination is impossible, yields per pass must be kept low with recycle of the unconverted paraffins, to avoid excessive raw-material consumption.

From there on, there are two alternatives: units that previously made hard detergent alkylate via aluminum chloride (Friedel Craft) alkylation can use the same method for alkylating benzene with *n*-paraffin hydrochloride:

$$RCl + \bigcirc \xrightarrow{AlCl_3} \bigcirc + HCl$$

Producers using alkylation with HF or sulfuric acid which do not catalyze the reaction of an alkyl chloride with benzene, must crack the monochloride to an olefin and use this as the alkylation raw material:

$$R\text{---}CHCl\text{---}CH_2\text{---}R^1 \xrightarrow[-HCl]{\Delta} RCH\text{=}CH\text{---}R^1$$

$$RCH\text{=}CHR^1 + \bigcirc \xrightarrow{HF} \overset{RCH\text{--}CH_2R^1}{\bigcirc}$$

The position of the double bond on the *n*-paraffin molecule is distributed randomly, and therefore so is the point along the chain where the alkyl chain is attached to the benzene ring.

Better properties of the final detergent are reported to be obtained by the olefin route, since this reduces the percentage of molecules attached to the benzene ring at the 2-position, which in turn improves foaming and detergency properties of the resulting sodium alkyl benzene sulfonate. Most linear alkylate, however, is produced by reacting the n-paraffin monochloride directly.

UOP has announced a process whereby detergent-range olefins can be obtained by dehydrogenation under mild conditions of an n-paraffin to the n-monoolefin of the same carbon number. A long catalyst life and yields above 90 percent are claimed. α-Olefins are also made by cracking higher paraffins (e.g., waxes) to give a range of products of various chain lengths.

A 125.0 million pounds-per-year detergent alkylate plant using the UOP n-paraffin dehydrogenation process is estimated to cost $5.2 million, and production costs (based on n-paraffins at 7.0 cents per pound) are as follows:

	¢/lb detergent alkylate
Raw materials, utilities, catalyst and chemicals	8.2
Labor and overhead	0.3
Capital charges	1.4
Total	9.9

These economics are a good deal more attractive than via conversion to an olefin by chlorination which requires Cl_2 to the extent of 1.3 cent per pound of product. α-Olefins are also made by a variant of the Ziegler fatty alcohol process, first used by Continental Oil. Instead of oxidizing the trialkyl aluminum compounds after the telomerization reaction to fatty alcohols, the long-chain radicals are displaced thermally by n-butane to form tri-n-butyl aluminum and long-chain olefins. Ethylene then displaces the n-butyl radicals, as well as the C_{16}-C_{18} alkyl chains which are not displaced by butylene, regenerating triethyl aluminum and butylene in addition to the remaining olefins.

Gulf has built a 100.0 million pounds-per-year α-olefins plant at Port Arthur, Texas, reportedly using a similar process.

The advantages of the Ziegler-type process are, first, that hardly any byproducts are formed; furthermore, the products are pure α-olefins whereas the products from a cracking process necessarily contain impurities in the same boiling range that it would be impractical to separate by distillation. On the other hand, the Ziegler process is more expensive due to the high investment costs involved.

Apart from their possible use as detergent raw materials, the future of α-olefins seems to lie in the manufacture of long-chain oxo-alcohols and polymers. The latter are being used as additives in certain high-performance lubricants. They can also undergo many typical reactions of olefins, giving olefin oxides, glycols, mercaptans, versatic acids, halides and other intermediates. However, few of these potential uses had reached commercial importance by 1967.

By 1965, conversion to soft alkylate on the part of detergent raw material manufacturers had been completed, some hard alkylate still being produced in order to obtain the branched heavy alkylate required mainly as raw material for lubricant additives. At that time, however, it was still uncertain whether linear alkylate, the second generation of synthetic detergents, would prove the final answer to the problem of poor biodegradability and pollution. Long-chain alcohol derivatives, which show better biodegradability in the absence of air, are beginning to appear on the market as the possible third generation of detergents.

n-Paraffins can be oxidized to alcohols by a dilute oxygen stream (3-4 percent) in the presence of a mineral acid which converts them to esters as they are formed, thus avoiding their further oxidation to acids. The product consists mainly of secondary alcohols, which are cheaper to produce by this method than the Ziegler primary alcohols, but are less reactive and therefore more difficult to ethoxylate.

$$RCH_2CH_3 \xrightarrow{O} RCHOHCH_3$$

Paraffins can also be converted to alcohols by first dehydrogenating to olefins. Steam-cracking of n-paraffins or wax can be carried out in such a way that 90 percent of the feed is converted to olefins having the double bond at the α-position. This makes them suitable raw materials for the oxo process. This method gives a mixture of primary and secondary alcohols. The economics of cracking wax to α-olefins depend considerably on the degree of utilization of the various chain-length ranges. Furthermore, the large plants required to make the process attractive are necessarily linked to oil refineries processing at least 100,000 pounds per day of crude with a reasonably high wax content. A 300.0 million pounds-per-year olefins plant from wax requires an investment of $7.0 million. Assuming wax to cost 3.0 cents per pound, production costs of α-olefins at 100 percent utilization of the entire range are as follows:

	$\text{¢/lb } \alpha\text{-olefins}$
Wax (60% olefins yield)	5.0
Less: fuel credit for byproducts	(1.0)
Net raw materials	4.0
Utilities	0.4
Labor and overhead	0.3
Capital charges	0.8
Total	5.5

At the more realistic level of 60 percent utilization, these costs rise to 9.3 cents per pound. Thus the costs of α-olefins from wax cracking are very sensitive to demand-supply equilibrium in the various product ranges. The approximate distribution between the various cuts is 25 percent in the plasticizer alcohol range (C_5-C_9), 25 percent in the detergent range (C_{10}-C_{13}) and 50 percent in the C_{14}-C_{20} range, for which new uses are gradually being found.

Commercial processes for the production of fatty acids by the oxidation of paraffins operate in various locations in Eastern Europe. Plants are still operating where wax is oxidized to fatty acids by potassium permanganate, much as it was during World War II (e.g., in Rodleben, East Germany). Also, the trialkylaluminum (Ziegler intermediate formed during manufacture of long-chain alcohols) can be converted to fatty acids. Other detergent raw materials made from n-paraffins are sodium alkane sulfonates [R–CH(SO$_2$ONa)–R'], obtained by reaction with SO$_2$ and oxygen in the presence of x-radiation followed by neutralization, and sodium alkenyl sulfonates (R-CH$_2$-CH= CH SO$_2$ONa), which are being made commercially by Stepan Chemical from the corresponding α-olefin by SO$_3$ sulfonation. Reaction of paraffins with sulfonyl dichloride gives sulfonyl chlorides $\underset{\underset{\text{SO}_2\text{Cl}}{|}}{(\text{R—CH—R'})}$ which yield sulfonic esters when reacted with substituted phenols. U.S. producers of linear alkylate (LAS) from n-paraffins are listed below:

Producer	Location	Capacity, MM lbs/yr
Allied Chem.	Claymont, Del.	100.0
Atlantic-Richfield	Port Arthur, Tex.	100.0
Chevron	Richmond, Calif.	250.0
Continental	Baltimore, Md.	150.0
Monsanto	Chocolate Bayou, Tex.	150.0
Phillips	Pasadena, Tex.	30.0
Union Carbide	S. Charleston, W.Va.	150.0
Witco	Watson, Calif.	30.0
Total		960.0

Source: Based on *European Chemical News*, December 2, 1966.

Monsanto, the only company which has no captive production of
n-paraffins, acquires its supplies from Enjay (the only producer of
n-paraffins not in the detergent raw materials business). The
total demand for detergent alkylate in the U.S. is expected to remain constant at 550.0 million pounds per year, and may even
decline. Production of synthetic fatty alcohols was around 120.0
million pounds in 1965 compared to the 350.0 million pounds per
year installed capacity. Although there is a certain demand for
n-paraffin derived alcohols outside the detergent raw materials
range, it appears that much of this capacity will only be utilized
when the public compels detergent producers to abandon soft
alkylate (LAS) as it had previously forced them to abandon alkylbenzene sulfonate.

3 CARBON BLACK

Carbon black can be defined simply as elemental carbon made
under conditions permitting control of its particle size, degree of
particle agglomeration and surface chemistry. Although the
product is almost 100 years old, considerable changes in the
methods of its manufacture and in the market distribution among
the various grades have occurred over the years.

Carbon black was first produced by the channel process. This
consisted of burning natural gas in a deficiency of air and having
the resulting stream of carbon and combustion gases impinge on
slowly moving mild steel channels, from which the soot is continuously removed. The resulting product was best suited for
reinforcing natural rubber, and the gradual abandonment of this
process has accompanied the increase in the proportion of synthetic
elastomers of total rubber demand. Also, the channel process requires about 500 cubic feet of gas per pound carbon black; whereas
natural gas was once almost free in the oil-producing regions of
the U.S., it soon began to be carried by pipeline to the space-heat
consuming population centers in all portions of the U.S. and thereby
acquired a cost which brought about the decline of the channel
process.

In the furnace process, which accounts for about 82 percent of
all carbon black produced in the U.S., the raw material is injected
into a stream of combustion gases inside a refractory-lined furnace
chamber. The hot exit gases are cooled in two spray-towers in
series and enter an electrostatic precipitator, where a small
part of the product is recovered and the rest caused to agglomerate

into particle clusters which can be removed in cyclone collectors immediately following the precipitator. The carbon black-water slurry is returned to the top of the precooler. The exit gases are sprayed with water once again, for reasons of public hygiene, before entering the exhaust stack; in the U.S., where regulations are more stringent, bag filters are used instead. The product is pulverized and pelleted in so-called "bead machines," which produce spherical particles that permit bulk shipment without excessive dust; in the U.S., only a minor part of the carbon black produced is still shipped in bags. This operation can be dry or wet, the latter procedure being employed when the product is to be shipped over long distances. The product is finally screened and sent through an electromagnetic separator, since the presence of iron particles in the product cannot be tolerated in the rubber-processing industry. The finished product is stored in special containers known as "Berquist bins." A flowsheet appears in Fig. 9-4.

In recently built plants the trend has been to smaller furnace diameters, which involves a higher pressure drop through the system; in these plants, air compressors have replaced the conventional blowers and exhaust fans. The electrostatic precipitator, a very troublesome piece of equipment, is being replaced in contemporary plants by a system of multicyclones. Raw material can be natural gas or oil. Due to rising price of natural gas, about 75 percent of the carbon black produced in the U.S. is made from liquid feedstocks. These can be, for example, the polycyclic materials recovered from refinery operations which have a low viscosity index. If left in oils these polycyclic materials tend to form carbon deposits and are therefore removed from the lubricant stocks by solvent extraction. These solvent extraction processes have largely replaced sulfuric acid treatment, which has in turn increased the availability of raw material for carbon-black manufacture. The most common feedstock, however, is 0° API catcracker cycle oil, which also contains a high proportion of polycyclic aromatics.

A third process is used to make a product known as thermal black. A brick checkerwork is first heated by complete combustion of natural gas, and, in the second phase of the cycle, surrenders stored heat to more natural gas which is thereby cracked to carbon black and hydrogen. The product has a much higher average particle diameter than channel or furnace blacks, and is used particularly in the manufacture of rubber products requiring high elongation, or in applications where high abrasion resistance is not required. Thermal blacks are relatively cheap;

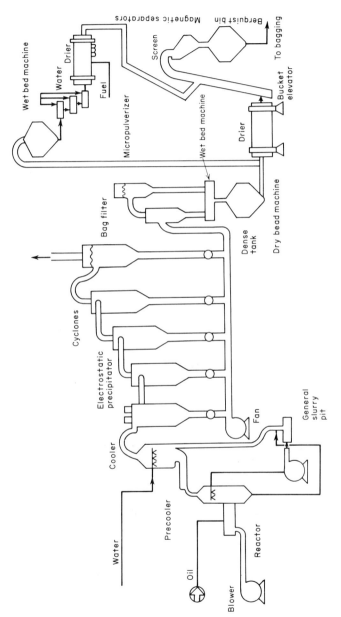

Fig. 9-4. Carbon black.

one reason is that the off-gas is almost pure hydrogen and can be valued at close to the cost of synthesis gas. Their use is growing roughly as fast as that of furnace blacks, while that of channel blacks is declining.

Acetylene black is made by pyrolysis of acetylene, and is used mainly in dry cells. Its chain-like structure imparts exceptionally good electrical conductivity; it is this property that accounts for its use in aircraft tires and flooring tile, since it prevents build-up of static electricity and thereby reduces fire hazards. Union Carbide operates the only U.S. plant, in Ashtabula, Ohio, which has a capacity of 8.0 million pounds per year.

The three most important properties of carbon blacks are pH, specific surface and particle structure. Whereas channel blacks have an acidic surface pH, furnace blacks, which are cooled with water, exhibit an alkaline surface due to the presence of evaporation deposits. This has had a profound effect on the type of vulcanization accelerators used in the rubber industry, since an acidic pH acts as a cure retarder while an alkaline pH activates the vulcanization process, increasing the danger of scorching the product. This has brought about a radical change in the type of vulcanization accelerators used by the rubber industry.

The dominant characteristic with respect to the reinforcement properties of carbon black is specific surface. Grades used in the rubber industry range from 9 to 153 m^2 gram. A higher specific surface imparts better abrasion resistance, but both processing times and mixing power requirements increase rapidly as average particle diameter diminishes. The third property is structure, a concept referring to the degree of agglomeration. Thermal blacks, with their large, individual particles are reported to have low structure, whereas the chain-like disposition of the particles in oil-based furnace blacks is referred to as a high structure. High structure contributes to modulus, extrudability and hardness, but scorch time (processing safety) is reduced. In synthetic rubber stocks, a high-structure black gives a mixture having high Mooney viscosity, which permits incorporation of a higher percentage of extending oils and reclaim rubber, which in turn lowers the overall raw material costs. Thus the trend is towards blacks combining the reinforcing action of high surface with the greater ease of dispersion and processing, faster extrusion and durability of the resulting product conferred by high structure.

Another reason for the trend to higher structures has been the development of stereospecific rubbers; only after fine-particle, high-structure types were introduced was it possible to produce

100 percent polybutadiene tires. On the other hand, heat build-up increases with structure, which is why channel blacks remained the preferred reinforcing material for heavy tires for many years. One of the more significant developments in carbon-black manufacture are the low-structure furnace grades obtained from oil instead of the more expensive natural gas, which have been widely used since 1960 as channel-black replacements; one producer achieves this result by adding metallic sodium to the flame. Thus, the range over which furnace blacks are able to substitute for channel blacks is increasing and will indeed soon reach 100 percent.

A fourth property whose importance has been recognized is surface activity. This is related to the amount of surface oxygen, and to the nature of the surface itself. Its main influence is considered to be on adhesion to the elastomer, especially in butyl rubber; surface-active carbons also reduce heat build-up. By 1965, however, work on surface activity was still mostly in the experimental phase.

The production of carbon black in 1965 is broken down by grades in the table which follows. HAF and FEF were the first grades to be used with oil-extended cold SBR; still the highest-volume grade, although no longer growing in demand, HAF is a compromise product providing good abrasion resistance while maintaining heat build-up at a reasonable level. The fastest growing grades are ISAF and GPF, introduced around 1952 for use in treads and carcasses, respectively, of tires suitable for high-speed driving on turnpikes. These two grades are rapidly overtaking HAF and FEF. An even finer grade, SAF, was also introduced, but its use has not grown as fast as expected; SRF is still widely used for tire carcasses, but most of the growth for this application is being taken up by GPF, which has a higher structure for the same fineness.

	MM lbs/yr, 1965
SAF (Super Abrasion Furnace)	18.0
ISAF (Intermediate Super Abrasion Furnace)	504.0
HAF (High Abrasion Furnace)	595.0
FEF (Fast Extrusion Furnace)	271.0
GPF (General Purpose Furnace)	198.0
SRF (Semi-Reinforcing Furnace)	314.0
HMF (High Modulus Furnace)	32.0
Channel blacks	148.0
Thermal blacks	273.0
Total	2,353.0

Source: U.S. Bureau of Mines "1965 Minerals Yearbook."

Some fineness grades (SAF, ISAF, HAF) are made in several degrees of structure. Domestic sales in 1965 were 2,073 million pounds, distributed as follows:

End-use	MM lbs, 1965
Rubber	1,945.0
Printing ink	54.0
Paint	11.0
Chemical and food	9.0
Plastics	20.0
Carbon paper	8.0
Others	25.0
Total	2,073.0

Source: Bureau of Mines "1965 Minerals Yearbook."

Exports account for the difference between production and domestic demand. Producers of carbon black and their capacities were:

Producer	Location	Furnace	Thermal	Channel
Cabot	Big Spring, Tex.	120	–	–
Cabot	Franklin, La.	170	126	–
Cabot	Pampa, Tex.	48	–	–
Cabot	Skellytown, Tex.	–	–	36
Cabot	Ville Platte, La.	170	–	–
Columbian	Conroe, Tex.	111	–	–
Columbian	El Dorado, Ark.	87	–	–
Columbian	Eola, La.	70	–	–
Columbian	Hickok, Kans.	52	–	–
Columbian	North Bend, La.	110	30	–
Columbian	Seagraves, Tex.	91	–	–
Columbian	Seminole, Tex.	–	–	40
Commercial Solvents (Thermatomic)	Sterlington, La.	–	120	–
Continental	Bakersfield, Calif.	50	–	–
Continental	Eunice, N.M.	–	–	25
Continental	Ponca City, Okla.	75	–	–
Continental	Sunray, Tex.	60	–	–
Continental	Westlake, La.	70	–	–
Huber	Borger, Tex.	115	–	10
Huber	Eldon, Tex.	115	–	–
Phillips	Borger, Tex.	290	–	–
Phillips	Orange, Tex.	60	–	–
Richardson	Big Spring, Tex.	50	–	–
Richardson	Odessa, Tex.	–	–	60
Shell	Pittsburg, Calif.	–	15	–
United	Aransas Pass, Tex.	130	–	30
United	Ivanhoe, La.	163	–	–
United	Johnson, N.M.	–	–	30
United	Mojave, Calif.	64	–	–
United	Shamrock, Tex.	100	–	–
Total		2,371	291	231

Source: Based on *Oil, Paint & Drug Reporter,* June 29, 1964.

The investment for a 70.0 million pounds per year furnace black plant is \$6.0 million. Such a plant operates on an average of 10 furnaces at a time. Production costs are calculated below for a weighted average of all grades, distributed in proportion to the market for the various grades. Yields of black per pound feed decrease with increasing fineness, which is why finer grades are more expensive; a representative figure is 0.55 pound of carbon black produced per pound of oil. Note that not one of the major tire companies produces carbon black; one reason appears to be that prices are close enough to production costs to make integration unnecessary.

	Cost, ¢/lb carbon black
Feedstock	1.5
Utilities	0.4
Labor and overhead	0.9
Capital charges	2.8
	5.6

4 CRESOLS

Slightly over 50 percent of the natural cresols and xylenols, or cresylic acids, obtained in the U.S. derive from the caustic extraction of cracked gasoline; the remainder comes from coal-tar processing.

As has been the case with other coal-tar derivatives, supply of coke-oven byproduct cresylics has failed to keep up with demand. This brought about shortages of various types. First, the p-isomer is relatively more abundant in coal-tar than in petroleum, hence shortage of this isomer was the first to develop; furthermore, purity requirements for certain applications are such that the p-cresol obtained from mixed cresylics, which is difficult to isolate from the m-isomer, became inadequate. Thus, p-cresol was the first isomer to be synthesized commercially. Second, refiners began to remove mercaptans by hydrogen treating instead of caustic extraction, with the result that the supply of petroleum cresylics has also remained static; eventually a shortage of o-cresol, which can be separated from mixed cresylics by simple fractionation, also developed, which resulted in the first commercial unit for its synthesis being built in 1964. Third, xylenols are also more abundant in coal-tar than in petroleum, and since they are required for making phosphate esters employed as hydraulic fluids, supply also began to fall short of necessities, and users had to resort first to imports, then to synthesis.

A flowsheet of a typical cresylics plant is shown in Fig. 9-5. Feed to the unit is a mixed stream of sodium phenates, cresylates and higher homologs, as well as about 10 percent sodium thiophenate and its homologs. Most plants handle both petroleum and coal-tar feedstocks. The feedstream is first devolatilized in order to remove any remaining hydrocarbons, and then treated by solvent extraction which can precede or follow the acidification step that converts sodium phenate and cresylates to phenol and cresols. In the process shown, solvent extraction in a mixture of naphtha and methanol follows springing with CO_2 and the feed is processed into two streams: crude cresylics, and a waste stream from which thiophenol and thiocresols can be recovered. Apart from their disagreeable odor, aryl mercaptans inhibit condensation with formaldehyde and reduce adhesion to metals. The crude product stream is decanted to remove water, and fractionated into ''natural'' phenol, o-cresol (mainly for use in phenolic resins), a mixture of m- and p-cresols used to make phosphate esters, and finally a heavy fraction containing mostly xylenols. Aryl mercaptans make up about 10 percent of the feed, the rest being phenolic compounds. The uses of mixed cresols in 1964 are given below.

End-use	MM lbs, 1964
Phenolic resins	78.0
Phosphate esters	42.0
Wire enamel solvent	11.0

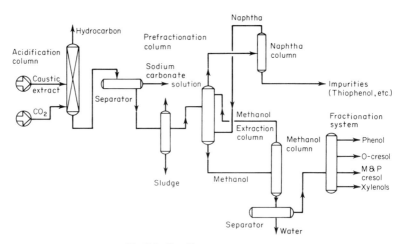

Fig. 9-5. Cresylics recovery.

Flotation agents	5.0
Other	6.0
Metal cleaning, etc.	3.0
Exports	11.0
Total	156.0

Source: *Oil, Paint & Drug Reporter*,
August 10, 1964.

Producers of mixed cresols, including natural phenol, and their respective capacities are:

Producer	Location	Capacity, MM lbs/yr
Petroleum		
Jefferson Lake	Houston, Tex.	28.0
Northwest	Anacortes, Wash.	15.0
Pitt-Consol	Newark, N.J.	35.0
Productol	Santa Fe Springs, Calif.	25.0
Shell	Wilmington, Calif.	5.0
Coal tar		
Allied	Frankford, Pa.	30.0
Koppers	Follansbee, W.Va.	45.0
U.S. Steel	Clairton, Pa.	20.0
Total		203.0

Aryl mercaptans can be recovered from the residue of a cresylics plant, or made synthetically by reduction of sulfonic acids:

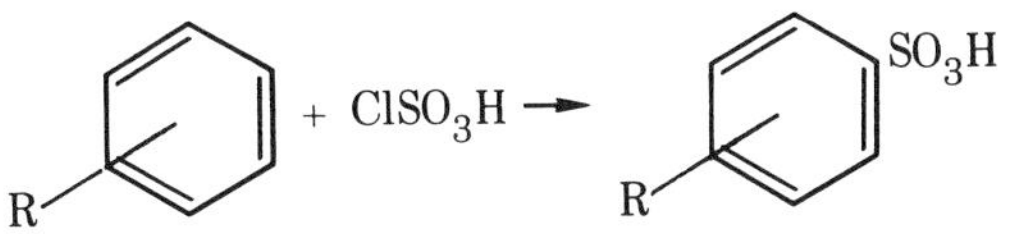

Thiocresylics and their chlorination and alkylation derivatives (for example, pentachlorothiophenol and *t*-butyl-*o*-thiocresol) are used as rubber peptizers. These compounds, of which about 3.0 million pounds per year are used in the U.S., greatly reduce milling time and power requirements. They are also used as rubber reclaiming agents.

Thiophenol can be converted to thiosalicylic acid, an intermediate for making "Merthiolate," a well-known wound disinfectant.

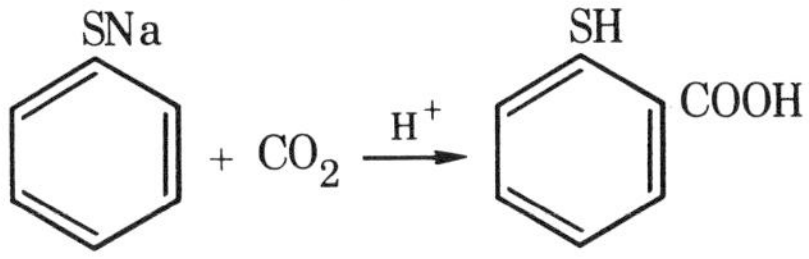

Thiosalicylic acid is also an intermediate for thioindigoid and thioxanthene dyes.

4.1   Synthetic p-Cresol

Two processes are used to make synthetic p-cresol. The one employed by Sherwin Williams is analogous to the sulfonation phenol route, starting from toluene instead of benzene

$$\text{C}_6\text{H}_5\text{CH}_3 + \text{H}_2\text{SO}_4 \longrightarrow \text{CH}_3\text{-C}_6\text{H}_4\text{-SO}_3\text{H} \xrightarrow[\text{H}^+]{\text{NaOH}_2} \text{CH}_3\text{-C}_6\text{H}_4\text{-OH}$$

Sodium sulfate and sulfite are obtained as byproducts.

Hercules makes p-cresol from terpenes, obtained from the company's naval stores operations in the South. Dehydrogenation of monocyclic terpenes gives p-cymene, which can be converted to p-cresol and acetone by a process that resembles the production of phenol from cumene

$$\text{CH}_3\text{-C}_6\text{H}_4\text{-HC(CH}_3)_2 \xrightarrow[\text{H}^+]{[\text{o}];} \text{CH}_3\text{-C}_6\text{H}_4\text{-OH} + \text{CH}_3\text{COCH}_3$$

In addition to acetone, the process yields other alkylated aromatics containing more than one hydroxyl group, which are employed in the mining industry as flotation agents.

The principal outlet for p-cresol is the production of 2-6-di-t-butyl-p-cresol, better known as BHT. It is a widely used antioxidant in food, in rubber products where staining would preclude the use of amines, and in gasoline. It is preferred to p-phenylenediamine derivatives in motor fuel containing a high proportion of catalytic reformate, and is consequently increasing its share of this market. In food, the main outlet, BHT is usually consumed in the form of proprietary mixtures with BHA (butylated hydroxyanisole), n-propyl gallate and others. Producers of BHT in the U.S. follow.

Producer	Location	Capacity, MM lbs/yr
Catalin	Fords, N.J.	5.0
Eastman	Kingsport, Tenn.	3.0
Hercules	Gibbstown, N.J.	3.0
Koppers	Oil City, Pa.	5.0
National Polychemicals	Wilmington, Mass.	3.0
Shell	Martinez, Calif.	6.0

The demand reached 23.0 million pounds in 1965, distributed as follows:

End-use	MM lbs, 1965
Food	8.0
Rubber	2.3
Gasoline	4.6
Plastics	3.5
Industrial lubricants	4.6

p-Cresol is also an intermediate for other antioxidants, such as 2,2'-methylenebis (4-methyl-6-t-butyl)-phenol, produced by American Cyanamid. Finally, p-cresol is a coupling agent for a number of acetate and polyester azo dyes.

4.2 Synthetic o-Cresol

Synthetic o-cresol is produced by methylating phenol with methanol at around 600 psig over an Al_2O_3 catalyst

o-Cresol is the preferred isomer for the manufacture of the phenolic resins used in can linings, and is also used in place of phenol for making certain types of polycarbonate resins. It is also a dye intermediate. The main outlets outside the U.S. are in the pesticide field. For example MCPA, the o-cresol equivalent of 2,4-D, is widely used in Europe as a post-emergence herbicide in corn, flax and small grains; 4,6-dinitro o-cresol is employed in Europe both as a contact herbicide and as an insecticide.

U.S. producers of synthetic *o*-cresol are Koppers, with a 5.0 million pounds-per-year unit at Follansbee, W.Va.; and Pitt-Consol with a 10.0 million pounds-per-year plant at Newark, N.J.

The main byproducts from *o*-cresol manufacturers are 2,4 and 2,6-xylenol. They represent 15 to 25 percent of the *o*-cresol produced. Mixed xylenols are used in the production of phosphate esters that go into hydraulic fluids, and a major part of U.S. demand is satisfied by imports. General Electric has developed a family of engineering thermoplastics, polyphenylene oxide, made by oxidative coupling polymerization of 2,6-xylenol.

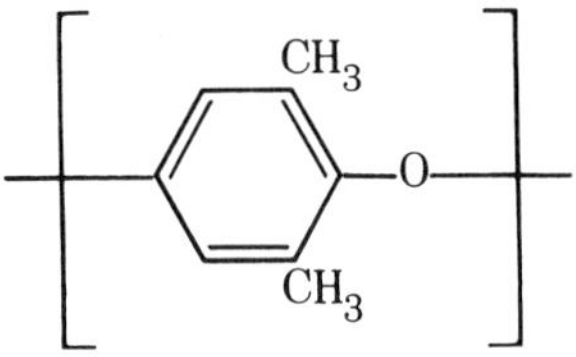

They are resistant to hydrolysis, have excellent mechanical and electrical properties and can be used over the -275 to 375° F temperature range. They are expected to be used in clothes and dishwasher parts, in surgical instruments (replacing stainless steel) and in numerous electronic applications (waveguides, circuit boards, etc.). General Electric has a 10.0 million pounds per year plant at Albany, N.Y.; the ultimate potential for this and other second-generation engineering thermoplastics (the first having been ABS, polyamides, polyacetals, and polycarbonates) has been estimated at 100.0 million pounds per year.

4.3 Tricresyl Phosphate

The general method for producing phosphate esters, of which tricresyl phosphate is the most important example, is the reaction of a hydroxyl-group with $POCl_3$.

$$3ArOH + POCl_3 \longrightarrow OP(OAr)_3 + 3HCl$$

The raw materials react at 500° F and essentially atmospheric pressure in an agitated kettle, HCl being given off and recovered. The effluent is flashed to remove unreacted raw material which is returned to the reactor; the product is neutralized, dried by falling-film evaporation, and purified by filtration.

The overall U.S. production of phosphate esters in 1965 was around 96.0 million pounds broken down as follows:

Phosphate ester	MM lbs, 1965
Cyclic	
Triphenyl	10.0
Tricresyl	35.0
Cresyldiphenyl	20.0
Tri 2-ethylhexyl	1.0
Acyclic	
Tributyl	5.0
All other, primarily acyclic	31.0
Total	102.0

Source: U.S. Tariff Commission, "Synthetic Organic Chemicals, 1965."

Of the cresol derivatives, tricresyl and cresyldiphenyl phosphate are used principally to reduce combustion-chamber deposit formation in gasoline engines. They are being gradually replaced by methyldiphenyl phosphate, which contains more phosphorus per unit weight. The preferred isomers for making fire-retardant plasticizers are m- and p-cresol, which give esters of lower density and therefore lower cost per unit volume. Their main fields of application are in wire and cable coatings; although pure PVC is flame-resistant, the addition of phthalates gives a plasticized product with reduced flame-retardancy, and this is corrected by addition of phosphate esters. Other uses, for which xylenols are often used in place of cresols, are extreme-pressure lubricant additives and hydraulic fluids. The overall demand for all phosphate esters can be broken down as follows:

End-use	MM lbs, 1965
Flame-retardants and plasticizers	43.0
Gasoline additives	23.0
Hydraulic fluids and lubricant additives	25.0
Surfactant	6.0
Others	5.0
Total	102.0

Source: Based on U.S. Tariff Commission, "Synthetic Organic Chemicals, 1965."

Although $POCl_3$ is not the major cost-component in the manufacture of phosphate esters, almost 60 percent of production capacity is by producers who manufacture phosphorus as well as $POCl_3$.

Trialkylaryl phosphites are stabilizers for rigid PVC cellulosic molding powders and ABS resins. Since the demand for rigid PVC in the U.S. began to expand rapidly around 1963, the prospects for these stabilizers are excellent.   Other uses include fire-retardants for textiles, intermediates for other phosphoric acid derivatives, and as starting point via trimethyl phosphate for pesticides such as "Phosdrin," "Phosphamidon" and "DDVP." The principal producers of phosphites are Socony and Stauffer.

The investment for a 10.0 million pounds-per-year phosphate esters plant is $700,000.

5 CYCLOPENTADIENE

Dicyclopentadiene is obtained from the aromatic oils produced when cracking hydrocarbons to olefins.   Since cracking naphtha produces almost ten times as much aromatic oils as ethane, the U.S. produces relatively less cyclopentadiene than Europe and Japan; supposing an average content of 2 percent in olefin plant aromatic oils, the total U.S. potential output can be estimated at 40.0-50.0 million pounds per year in 1965, assuming all were recovered.

In producing dicyclopentadiene the C_5-C_8 fraction from olefin plant aromatic distillate is first sent to a depentanizer, giving a C_5 overhead and a bottoms stream that can, for example, be sent on to aromatics extraction.   Dicyclopentadiene is removed as bottoms from a second column and the overhead is suitable as feedstock for making olefinic petroleum resins. Cyclopentadiene is produced from its dimer by thermal cracking.

The most important use for cyclopentadiene is the production of numerous chlorinated insecticides, such as aldrin, dieldrin, endrin, chlordane, heptachlor and methoxychlor.   The total consumption of these insecticides in 1965 was around 60.0 million pounds.   The two main constituents of these various pesticides are cyclopentadiene itself and hexachlorocyclopentadiene, obtained by chlorination

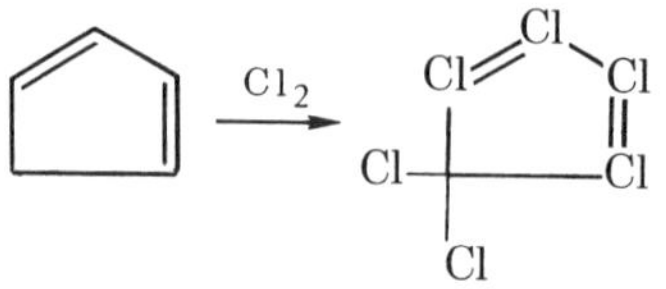

The reaction for making aldrin and dieldrin, the two most important members of the group, are shown below; the epoxidized form (dieldrin) is the more powerful of the two.

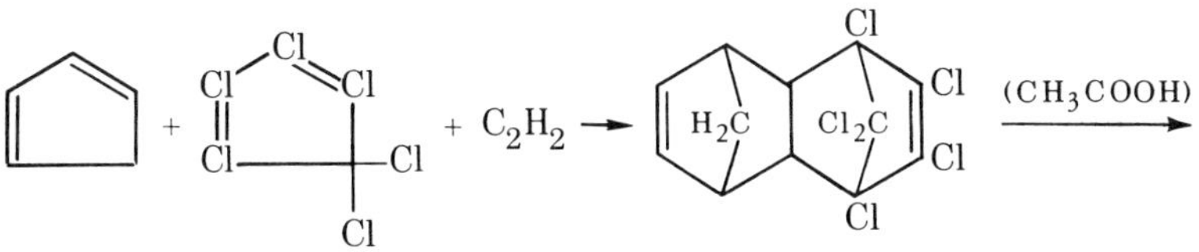

Aldrin

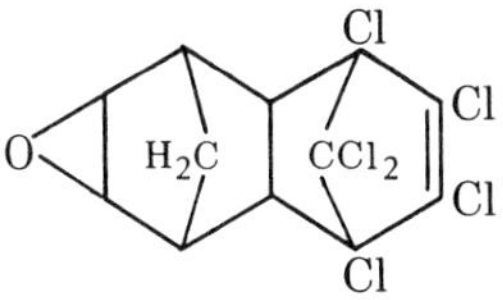

Dieldrin

Chlorendic anhydride, or "HET" acid, is made by the Diels-Alder reaction between hexachlorocyclopentadiene and maleic anhydride:

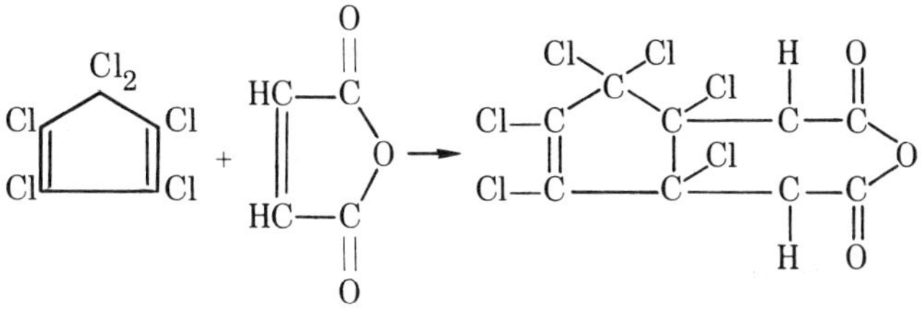

A related compound, known as "Chloran," is made from tetra-hydrophthalic anhydride and hexachlorocyclopentadiene. "HET" acid and "Chloran" are used mainly to make fire-resistant polyesters. These can be made by reaction with polyhydric alcohols, or by esterification to allyl chlorendate followed by vinyl-type polymerization.

In 1965, 14.0 million pounds of flame-retardants were used to make polyesters, most of which were based on hexachloro-cyclopentadiene derivatives. Rigid urethanes required 15.0 million pounds, but the better part of this consisted of phosphorus-containing polyols such as tri- (chloroethyl) phosphate and others. Nevertheless, some polyols are polyesters made from chlorendic anhydride. Chlorocyclopentadiene reacts with sodium to give cyclopentadienyl sodium, which combines with ferric chloride to give "ferrocene," a compound having an unusual "sandwich" structure, which is used to promote smokeless combustion of residual fuel oil:

$$\left(\begin{array}{c} CH{=}CH \\ | \qquad \rangle CH \\ CH{=}CH \end{array}\right)_2\!\!\!-\!Fe$$

Cyclopentanetetracarboxylic acid dianhydride (CPDA), a derivative of the cyclopentadiene-maleic anhydride Diels-Alder adduct,
finds some use as an epoxy resin curing agent, in competition with
other dianhydrides such as pyromellitic dianhydride. Epoxy resins
cured with CPDA have exceptionally high heat-distortion temperatures and stability, being used principally in the manufacture of
high-strength tools and dies and epoxy adhesives for the armament
industry.

Finally, the largest outlet for dicyclopentadiene may in the
future be as a third comonomer in the manufacture of EPT. If this
elastomer becomes established as a general-purpose rubber for
tire manufacture, requirements for third monomers may reach 20-
40 million pounds per year at a utilization level of 4 percent; in
this case, domestic cyclopentadiene sources would become insufficient, which justifies the interest being shown in synthetic unconjugated dienes such as 1,5-cyclooctadiene and 1,4 hexadiene.

6 PETROLEUM RESINS

Petroleum resins, often improperly referred to as coumarone-
indene resins, are normally made by polymerizing aromatics,
conjugated olefins or branched olefins contained in the C_6-C_8 cut
from olefin plant pyrolysis naphtha. True coumarone-indene
resins are similar in their properties and applications, but are
derived from coal-tar; only 10 percent of all hydrocarbon resins
produced at present are of the C-I type, the remaining 90 percent
being based on petroleum. Polymerization is carried out in a
solvent, generally, the unpolymerizable material in the feed itself.
Aromatic resins have higher melting points, are used especially
to make asphalt floor tile, and are insoluble in food products.
They are also used in printing ink formulations. Olefinic resins go
into coating, rubber compounding products, waterproofing adhesives
for plywood, laminating resins for cellulosic materials, and many
other uses.

Petroleum resins come in numerous grades (one producer alone
makes around 700 different kinds) which can be divided into three

classes according to price. About 25 percent of the total consumption is accounted for by materials costing as little as 5 cents per pound, used for pipe coating, rubber reclaiming and for making tempered hardboard. Resins costing 5 to 9.5 cents per pound account for 35 percent of the total market, and are employed as components of foundry core oils, vegetable and soup-can coatings and rubber softeners. Finally, the main application of the more expensive petroleum resins is as binders for asphalt floor tile, the largest single outlet for hydrocarbon resins, despite the fact that 65 percent of all floor tile is now based on PVC. They also end up in adhesives and for foil- or plastic-to-paper lamination (where the use of petroleum resins is gaining at the expense of asphalt), and in the rubber industry, as softeners or tackifiers.

Miscellaneous uses are in printing inks, rug backing and chewing gum; a growing application is as binder for colored pavements, used on an increasing scale instead of traffic paint, which costs less but involves much higher maintenance expenses. The various types of resins differ in molecular weight, reactivity as measured by their iodine number, specific gravity, color and volatile matter content.

The domestic demand for hydrocarbon resins — petroleum resins plus C-I resins — was 311.0 million pounds in 1965, distributed as follows:

End-use	MM lbs, 1965
Floor tile	74.0
Rubber compounding	63.0
Exports	26.0
All others	148.0
Total	311.0

Source: U.S. Tariff Commission, "Synthetic Organic Chemicals, 1965."

The major producers of hydrocarbon resins are Pennsylvania Industrial Chemicals, Allied Chemicals, Neville Chemical, Amoco, Chemfax, Velsicol and Alabama Binder & Chemical.

7 NAPHTHENIC ACIDS

Naphthenic acids are carboxylic acids contained mainly in the gas-oil fraction of crude oil, having a structure consisting of a straight chain to which a carboxyl and cycloaliphatic group are attached:

$$R—CH—(CH_2)_n—COOH$$

$$\begin{array}{c} CH \\ H_2C \quad CH_2 \\ H_2C——CH_2 \end{array}$$

They are recovered in the form of sodium naphthenates by caustic extraction of naphthenic crude Diesel oil fractions.

Sodium naphthenates can be converted to salts of other metals that are employed mainly as drying accelerators in oil, and high-pressure lube catalysts. Since the gas-oil fraction also contains extractable aromatic mercaptans, naphthenates have poor odor and color characteristics, and in some applications the more expensive synthetic naphthenates, usually metal soaps of 2-ethylhexanoic or of other acids in the C_8-C_{10} range, are preferred. This trend has become more apparent in Europe than in the U.S. The most common paint driers are lead and cobalt naphthenates; lead and zinc have the best color characteristics, cobalt and manganese are darker, more effective. These compounds are also used in lubricant and gear oils, as emulsifiers or emulsion breakers, catalysts for various reactions, especially for the "oxo" process, surface detackifiers, carrying agents for polyester resins, and wherever else an oil-soluble metal is required in cleaning compounds. Copper naphthenate is a fungicide for cotton textiles. In the U.S.S.R., sodium naphthenates are reacted with ethylene dichloride to give PVC plasticizers reported to have excellent properties. Naphthenic acids have military uses, such as that as an ingredient in "Napalm" manufacture.

Peacetime consumption of naphthenates in the U.S. is fairly static at 26.0 million pounds per year; much of the growth in demand for surface coatings is being taken up by latex paints, which require no drier since the film is formed by water evaporation. This demand can be broken down as follows:

Type of salt	MM lbs, 1965
Calcium	2.0
Zinc	1.0
Lead	13.0
Cobalt	3.5
Manganese	1.5
Copper	3.0
Others	2.0
Total	26.0

Source: U.S. Tariff Commission, "Synthetic Organic Chemicals, 1965."

End-use	MM *lbs, 1965*
Paint driers	19.0
Fungicides	3.0
Corrosion inhibitors	3.0
Export	3.0
Others	2.0
Total	30.0*

*The difference between end-use requirements and production is accounted for by an unknown quantity of imports, believed to be 8 to 10 million pounds.

Source: Based on U.S. Tariff Commission and *Oil, Paint & Drug Reporter*, September 13, 1965.

The producers of naphthenic acids in the U.S. are:

Producer	Location	1965 Capacity, MM lbs/yr
Chevron	Richmond, Calif.	10.0
Mobil Chem.	Torrance, Calif.	10.0
Atlantic-Richfield	Wilmington, Calif.	5.0
Sun	Marcus Hook, Pa.	3.0
Total		28.0

Source: *Oil, Paint & Drug Reporter*, September 13, 1965.

Note that these producers are all located on the West Coast, the only source of naphthenic crudes in the U.S. Venezuela, however, is the world's main source of naphthenic crudes.

The general trend towards hydrocracking as a result of anti-smog legislation has diminished the potential output of naphthenic acids, which will undoubtedly encourage their replacement by "neo-acids," octoic acid and other synthetic substitutes.

8 HYDROGEN SULFIDE

Hydrogen sulfide is available from two different petroleum sources. First, sour natural gas contains varying percentages of H_2S and CO_2, and must be sweetened in order to be saleable. Second, a portion of the sulfur contained in the crude oil fed to a refinery appears in the off gases as H_2S, which must be separated from the remaining constituents in order to meet LPG specifications.

The methods for removing CO_2 from crude synthesis gas also apply to H_2S. Hydrogen sulfide can be recovered by absorption in ethanolamines, hot potassium carbonate and several other solvents, the choice depending on the combination of investment requirements and cost of utilities and chemical implied by the various absorption systems.

Hydrogen sulfide is occasionally needed at a location where it is not available from either of these sources. Several H_2S-generating plants have been installed using the Girdler process, whereby hydrogen and sulfur react in the vapor-phase:

$$H_2 + S \longrightarrow H_2S$$

The reactor effluent is a mixture of H_2S and unreacted sulfur from which the latter is removed by condensing against water and re-cycled.

8.1 Sulfur

Hydrogen sulfide can be converted to sulfur by the Claus process

$$H_2S + \tfrac{3}{2}O_2 \longrightarrow SO_2 + H_2O$$

$$2H_2S + SO_2 \longrightarrow 3S + 2H_2O$$

The net reaction is

$$H_2S + \tfrac{1}{2}O_2 \longrightarrow H_2O + S$$

When CO_2 and H_2S are recovered in the same acid-gas removal operation and the H_2S content of the feed is low, burning only one-third of the H_2S would result in too low a flame-temperature for combustion to be sustained. In these cases, a by-pass design can be employed, in which one-third of the H_2S feed is completely burned to SO_2 in a separate reactor and recombined with the other two-thirds in order to oxidize the remaining H_2S to elemental sulfur.

In general, streams containing sufficient H_2S to make sulfur recovery worthwhile are concentrated enough so that the entire feed may enter the plant via the reactor furnace. A flowsheet of a sulfur plant is shown in Fig. 9-6.

Initially, the feed is burned under conditions of partial oxidation in a boiler where steam is generated. Already in the furnace, however, part of the unconverted H_2S is oxidized to sulfur, and this is recovered in a condenser where more steam is generated. Approximately 60 percent of the contained H_2S is recovered in this step.

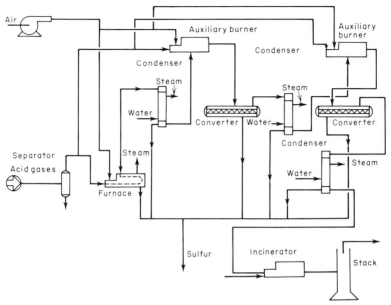

Fig. 9-6. Flowsheet of a sulfur plant.

The remaining H_2S reacts with SO_2 in a series of catalytic converters, each preceded by a reheating step and followed by condensation of the sulfur produced. More steam can be generated in each condensation step. The number of stages is determined both by economics and the necessity to comply with regulations regarding pollution. In general, where H_2S is a free item, two converters are sufficient, whereas three are employed, for example, in the H_2S-recovery section of a carbon disulfide plant. Reheating may be accomplished by extracting part of the hot gases from the furnace and mixing them with the cold gases from the condensers, or by burning part of the feed in auxiliary burners whose function it is to reheat the main stream. This latter approach involves a higher investment for the added burners, but recovery improves by 2.5 percent. Sulfur recovery is around 84, 95.0 and 97.5 percent with one, two or three converters, respectively, with some variations around these averages depending on the type of reheat scheme and amounts of catalyst in the converters. The flowsheet in Fig. 9-6 shows a plant using auxiliary burners and two converters, which corresponds to the solution adopted at Lacq.

The investment for a 50,000 tons-per-year sulfur plant feeding a high-assay H_2S stream is $800,000, and production costs are estimated as below. Note that a sulfur plant is a net producer of steam.

	Cost, $/ton sulfur
Utilities	(-2.70)
Labor and overhead	6.40
Capital charges	5.30
Total	9.00

With sulfur prices in 1967 at $42.50 long ton, fob Gulf Coast, sulfur recovery can be seen to be a very profitable undertaking. The cost of removing the H_2S from the hydrocarbon stream has been assumed charged against the latter, since this operation is usually carried out for reasons other than the desire to produce sulfur. The production cost for a plant of the same capacity receiving a low-assay H_2S feedstock would be around $11.50 per ton. The cost of H_2S removal depends very much on the H_2S concentration in the incoming gas, but for a sulfur plant of the size considered above this cost would be around $6.50 per ton.

The largest sulfur-recovery plants in the world are the one at Lacq (1 million tons per year), followed by several in Alberta, Can., in the 100,000 to 500,000 tons per year range. The U.S. produces a total of around 1,000,000 metric tons per year, at some 60 plants, half of which process H_2S from natural gas and the rest refinery off-gas. About two-thirds of this output, however, is accounted for by the refineries. Recovered sulfur represents 37 percent of the elemental sulfur produced in the world, and 20 percent of all sulfur including that from pyrites, sulfide ore smelters and other sources. These figures do not include those referring to the Socialist countries. The U.S. consumed the equivalent of 7100 M metric tons of sulfur in 1964, distributed by source and end-use approximately as follows:

End-use	1000 Metric tons, 1964	
Sulfuric acid		6,050.0
Fertilizers	2,800.0	
Chemicals	1,400.0	
Pigments	600.0	
Steel	300.0	
Cellulosics	300.0	
Petroleum refining	150.0	
Others	500.0	
Non-sulfuric acid		1,050.0
Pulp and paper	450.0	
Carbon disulfide	200.0	
Pesticides	200.0	
Others	200.0	
Total		7,100.0

Sources	1000 Metric tons, 1964
Frasch mining	4,550.0
Sulfur recovery	1,550.0
Pyrites	500.0
Smelter off-gases	400.0
Other	100.0
Total	7,100.0

Source: Based on U.S. Bureau of Mines and *Chemische Industrie*, September 1965.

Phosphorus Pentasulfide

Phosphorus pentasulfide is made by the reaction between elemental molten phosphorus and sulfur.

Reaction takes place in an inert gas atmosphere, at around 650° F. Purity requirements are severe, and are attained by distillation, which has the inconvenience of bringing about considerable losses and corrosion problems, or by eliminating the impurities contained in furnace phosphorus by concentrated sulfuric acid treatment. A correct stoichiometric ratio between the reactants is important, as the yield of the reaction P_2S_5 and the appropriate alcohol falls rapidly as the content in other phosphorus sulfur combinations increases. The exothermic heat of reaction is removed by circulating a suitable coolant, although at these temperatures it is difficult to find one that will not decompose; the Knapsack continuous process has as its main feature an air-cooled reactor. Following reaction, the product is flaked and ground fine in order to increase its reactivity.

All P_2S_5 producers in the U.S. have their own phosphorus furnaces; capacities and locations are:

Producer	Location	1965 Capacity, MM lbs/yr
American Agricultural	Carteret, N.J.	10.0
Hooker	Columbus, Miss.	14.0
	Niagara Falls, N.Y.	30.0
Monsanto	E. St. Louis, Ill.	40.0
Stauffer	Morrisville, Pa.	14.0
	Mount Pleasant, Tenn.	4.0
	Nashville, Tenn.	6.0
Total		118.0

Source: *Oil, Paint & Drug Reporter*, August 23, 1965.

Half of all P_2S_5 produced in the U.S. goes into numerous types of lubricant additives of which by far the most important are the zinc dithiophosphates, used as high lead-carrying oxidation and bearing corrosion inhibitors. They are made by the following sequence:

$$P_2S_5 + 4ROH \rightarrow 2(RO)_2\overset{\displaystyle S}{\overset{\|}{P}}-SH + H_2S$$

$$\downarrow ZnO$$

$$(RO)_2\overset{\displaystyle S}{\overset{\|}{P}}-S-Zn-S-\overset{\displaystyle S}{\overset{\|}{P}}(OR)_2 + H_2O$$

Over 100.0 million pounds of these additives were used in 1965.

The shorter alcohols, such as isopropyl and isobutyl, are preferred since the resulting product contains higher percentages of the active elements, phosphorus and sulfur.

Other lube additives derived from P_2S_5 are the reaction product with cyclic terpenes and polybutenes. P_2S_5 is an intermediate in the manufacture of phosphate insecticides. "Malathion" (American Cyanamid) and "Guthion" (Bayer), for example, are made from the reaction product between P_2S_5 and methanol.

$$P_2S_5 + 4CH_3OH \rightarrow 2(CH_3O)_2\overset{\displaystyle S}{\overset{\|}{P}}-SH + H_2S$$

This compound, or its ethyl homolog, can be chlorinated giving raw materials for such products as parathion, "Diazinon" (Geigy), "Disyston" (Bayer) and "Ronnel" (Dow).

$$2(RO)_2\overset{\displaystyle S}{\overset{\|}{P}}-SH + 3Cl_2 \rightarrow (RO)_2\overset{\displaystyle S}{\overset{\|}{P}}-Cl + S_2Cl_2 + 2HCl$$

The breakdown of P_2S_5 consumption in the U.S., which reached 105.0 million pounds in 1965, was as follows:

End-use	MM lbs, 1965
Dithiophosphates	45.0
Other lubricant additives	5.0
Pesticides	40.0
Flotation agents	7.0
Miscellaneous	3.0
Total	100.0

Source: *Oil, Paint & Drug Reporter*, August 23, 1965.

8.2   Mercaptans

Normal aliphatic mercaptans are synthesized by reacting the respective alcohol or alkyl chloride with H_2S

$$ROH + H_2S \longrightarrow RSH + H_2O$$

$$RCl + H_2S \longrightarrow RSH + HCl$$

For example, the reaction between methanol and hydrogen sulfide at pressures around 25 atmos. produces methyl mercaptan.

The yield on methanol is around 70 percent. American Oil operates a 5.0 million pounds-per-year methyl mercaptan unit at Texas City, Texas; hydrogen sulfide comes from refinery off-gas. The major use for methyl mercaptan is as an intermediate for methionine. It is also a raw material for "Ametryne," one of the triazine herbicides, made by reacting Atrazine with methyl mercaptan. The higher members of the series are intermediates for numerous pesticides.

Raw material	Product	Producer	Function
Ethyl mercaptan	"Thimet"	American Cyanamid	Systemic insecticide
	"Disyston"	Chemagro	Systemic insecticide
	"Eptam"	Stauffer	Pre-emergence herbicide
n-Propyl mercaptan	"Tillam"	Stauffer	Pre-emergence herbicide
n-Butyl mercaptan	"DEF"	Chemagro	Cotton defoliant
	"Merphos"	Virginia Carolina	Cotton defoliant
Propargyl mercaptan	"Avadex"	Monsanto	Pre-emergence herbicide

Nonionic detergents can be made by ethoxylating higher straight-chain mercaptans, made from the alkyl chloride of the respective fatty alcohol.

$$ROH \xrightarrow{HCl} RCl \xrightarrow[NaOH]{H_2S} RSH \xrightarrow{C_2H_4O} RS(CH_2CH_2O)_nH$$

Secondary and tertiary mercaptans are made by the Friedel-Craft reaction between H_2S and an olefin.

Phillips and Pennsalt are the largest U.S. producers of mercaptans. t-Dodecylmercaptan, made from propylene tetramer, is the standard chain-length distribution regulator in styrene-butadiene rubber recipes.   About 12.5 million pounds per year were used in 1965 for this purpose.

Mercaptans are used in concentrations of 0.1 ppm or less as odorants in refinery streams; the nature of the mercaptan depends on the molecular weight of the stream to be odorized. Mercaptans can be oxidized to disulfides, which have a variety of uses as lubricant additives and rubber plasticizers.

$$2RSH + \tfrac{1}{2}O_2 \rightarrow RSSR + H_2O$$

8.3 Thioglycolic Acid

Monochloracetic acid reacts with H_2S in basic medium to give sodium thioglycolate, which can be neutralized to the respective acid.

$$H_2S + ClCH_2COONa \xrightarrow{\text{OH}^-; \text{H}^+} HSCH_2COOH$$

This product is used chiefly in the form of its ammonium salt, in shampoos and hairwaving lotions.   Demand for the acid and its salts in the U.S. is around 4.0 million pounds per year, although it may in future be partly replaced by less odorous compounds. Recently, thioglycolic acid has begun to be used in crease-proofing treatments for wool.

Esters of thioglycolic acid are used as softeners for nitrile rubber.

8.4 Thiourea

Two commercial methods are used to make thiourea.   The first is by reacting ammonia and carbon disulfide

$$CS_2 + 2NH_3 \rightarrow (NH_2)_2C{=}S + H_2S$$

The more important of the two, used in France by Société Nationale des Pétroles d'Aquitaine (SNPA) and in the U.S. by Elco Chemical, begins with calcium cyanamide and H_2S.

$$CN_2Ca + CO_2 + H_2S + H_2O \rightarrow CO_3Ca + (NH_4)_2C{=}S$$

The use of this process depends on the availability of practically costless H_2S. SNPA uses that coming from the Lacq natural gas plants, while Elco Chemical reportedly obtains H_2S as a byproduct from a phosphorodithioate plant at Cincinnati, Ohio. The U.S. consumed around 8.0 million pounds per year of thiourea in 1966, distributed among the following applications:

End-use	MM lbs, 1964
Antiyellowing agent in diazo copying paper	4.0
Water treatment (removing copper scale)	0.8
Textile flame-proofing	1.6
Other uses	1.6
Total	8.0

The miscellaneous uses of thiourea include that as plasticizer for amino-resin adhesives, in dry-cleaning fluids, hair preparations and as a chelating agent. Among its uses as an intermediate, thiourea is a raw material for sulfathiazole, thiobarbiturates, thioindigoid dyes and "Omadine," an important fungicide derived from pyridine.

The demand for thiourea is growing at 7 percent per year; the only U.S. plant has a capacity of 6.0 million pounds per year.

8.5 Dimethyl Sulfoxide

In the U.S. this solvent of growing importance has up to now been made from pulp-mill wastes. However, in Europe it is made by Société Nationale des Pétroles d'Aquitaine (SNPA) from hydrogen sulfide and methanol. Two mols of methanol react with one of H_2S to give dimethyl sulfide, which is oxidized to the sulfoxide:

$$2CH_3OH + H_2S \rightarrow CH_3SCH_3 + 2H_2O$$

$$CH_3SCH_3 + \tfrac{1}{2}O_2 \rightarrow CH_3SOCH_3$$

Dimethyl sulfide is a sulfite pulp-mill byproduct; in the U.S. it is oxidized to the sulfoxide by Crown Zellerbach in an 8.0 million pounds-per-year plant at Bogalusa, La.

It has been shown that DMSO enhances the activity of numerous drugs, insecticides, bactericides such as quaternary ammonium salts and hexachlorophene, antibiotics, etc.

The established industrial uses for DMSO are a result of its excellent properties as a solvent and reaction medium. It is a spinning solvent for acrylic fibers in Japan; it dissolves polyurethanes and thus can be used to clean foaming equipment as well as for spinning urethane fibers or coating textiles.

INDEX

INDEX